International Federation of Automatic Control

ANALYSIS, DESIGN AND EVALUATION OF MAN-MACHINE SYSTEMS 1992

IFAC Symposia Series, 1993. Number 5

IFAC SYMPOSIA SERIES

Janos Gertler, *Editor-in-Chief*, George Mason University, School of Information Technology and Engineering, Fairfax, VA 22030-4444, USA

DHURJATI & STEPHANOPOULOS: On-line Fault Detection and Supervision in the Chemical Process Industries *(1993, No.1)*
BALCHEN *et al*: Dynamics and Control of Chemical Reactors, Distillation Columns and Batch Processes *(1993, No.2)*
OLLERO & CAMACHO: Intelligent Components and Instruments for Control Applications *(1993, No.3)*
ZAREMBA: Information Control Problems in Manufacturing Technology *(1993, No.4)*
STASSEN: Analysis, Design and Evaluation of Man-Machine Systems *(1993, No.5)*
VERBRUGGEN & RODD: Artificial Intelligence in Real-Time Control *(1993, No.6)*
FLIESS: Nonlinear Control Systems Design *(1993, No.7)*
DUGARD, M'SAAD & LANDAU: Adaptive Systems in Control and Signal Processing *(1993, No.8)*
TU XUYAN: Modelling and Control of National Economies *(1993, No.9)*
LIU, CHEN & ZHENG: Large Scale Systems: Theory and Applications *(1993, No.10)*
GU YAN & CHEN ZHEN-YU: Automation in Mining, Mineral and Metal Processing *(1993, No.11)*
DEBRA & GOTTZEIN: Automatic Control in Aerospace *(1993, No.12)*
KOPACEK & ALBERTOS: Low Cost Automation *(1993, No.13)*
HARVEY & EMSPAK: Automated Systems Based on Human Skill (and Intelligence) *(1993, No.14)*

BARKER: Computer Aided Design in Control Systems *(1992, No.1)*
KHEIR *et al*: Advances in Control Education *(1992, No.2)*
BANYASZ & KEVICZKY: Identification and System Parameter Estimation *(1992, No.3)*
LEVIS & STEPHANOU: Distributed Intelligence Systems *(1992, No.4)*
FRANKE & KRAUS: Design Methods of Control Systems *(1992, No.5)*
ISERMANN & FREYERMUTH: Fault Detection, Supervision and Safety for Technical Processes *(1992, No.6)*
TROCH *et al*: Robot Control *(1992, No.7)*
NAJIM & DUFOUR: Advanced Control of Chemical Processes *(1992, No.8)*
WELFONDER, LAUSTERER & WEBER: Control of Power Plants and Power Systems *(1992, No.9)*
KARIM & STEPHANOPOULOS: Modeling and Control of Biotechnical Processes *(1992, No.10)*
FREY: Safety of Computer Control Systems 1992

NOTICE TO READERS

If your library is not already a standing/continuation order customer or subscriber to this series, may we recommend that you place a standing/continuation or subscription order to receive immediately upon publication all new volumes. Should you find that these volumes no longer serve your needs your order can be cancelled at any time without notice.

Copies of all previously published volumes are available. A fully descriptive catalogue will be gladly sent on request.

AUTOMATICA and *CONTROL ENGINEERING PRACTICE*

The editors of the IFAC journals *Automatica* and *Control Engineering Practice* always welcome papers for publication. Manuscript requirements will be found in the journals. Manuscripts should be sent to:

Automatica	*Control Engineering Practice*
Professor H A Kwakernaak	Professor M G Rodd
Deputy Editor-in-Chief	Editor-in-Chief, CEP
AUTOMATICA	Institute for Industrial
Department of Applied	Information Technology Ltd
Mathematics	Innovation Centre
University of Twente	Singleton Park
P O Box 217, 7500 AE Enschede	Swansea SA2 8PP
The Netherlands	UK

For a free sample copy of either journal please write to:

Pergamon Press Ltd	Pergamon Press Inc
Headington Hill Hall	660 White Plains Road
Oxford OX3 0BW, UK	Tarrytown, NY 10591-5153, USA

Full list of IFAC publications appears at the end of this volume

ANALYSIS, DESIGN AND EVALUATION OF MAN-MACHINE SYSTEMS 1992

Selected Papers from the Fifth IFAC/IFIP/IFORS/IEA Symposium,
The Hague, The Netherlands, 9 - 11 June 1992

Edited by

H.G. STASSEN
Delft University of Technology,
Delft, The Netherlands

Published for the

INTERNATIONAL FEDERATION OF AUTOMATIC CONTROL

by

PERGAMON PRESS

OXFORD • NEW YORK • SEOUL • TOKYO

UK	Pergamon Press Ltd, Headington Hill Hall, Oxford OX3 0BW, England
USA	Pergamon Press, Inc., 660 White Plains Road, Tarrytown, New York 10591-5153, USA
KOREA	Pergamon Press Korea, KPO Box 315, Seoul 110-603, Korea
JAPAN	Pergamon Press Japan, Tsunashima Building Annex, 3-20-12 Yushima, Bunkyo-ku, Tokyo 113, Japan

This compilation copyright © 1993 IFAC

First edition 1993

Library of Congress Cataloging in Publication Data

A catalogue record for this book is available from the Library of Congress

British Library Cataloguing in Publication Data

A catalogue record for this book is available from the British Library

ISBN: 9780080419008

These proceedings were reproduced by means of the photo-offset process using the manuscripts supplied by the authors of the different papers. The manuscripts have been typed using different typewriters and typefaces. The lay-out, figures and tables of some papers did not agree completely with the standard requirements: consequently the reproduction does not display complete uniformity. To ensure rapid publication this discrepancy could not be changed: nor could the English be checked completely. Therefore, the readers are asked to excuse any deficiencies of this publication which may be due to the above mentioned reasons.

The Editor

Transferred to digital print 2009
Printed and bound in Great Britain by CPI Antony Rowe, Chippenham and Eastbourne

5th IFAC/IFIP/IFORS/IEA SYMPOSIUM ON ANALYSIS, DESIGN AND EVALUATION OF MAN-MACHINE SYSTEMS 1992

Organized by
Koninklijk Instituut van Ingenieurs (KIvI)
Delft University of Technology (DUT)

Sponsored by
International Federation of Automatic Control (IFAC)
Technical Committees on
- Systems Engineering (SECOM)
- Social Effects of Automation (SOC.EFF)
- Computers (COMPUT)
- Economic and Management Systems (EMSCOM)

Co-sponsored by
International Federation for Information Processing (IFIP)
International Federation of Operational Research Societies (IFORS)
International Ergonomics Association (IEA)

International Programme Committee

H.G. Stassen (NL)   (Chairman and Editor)
J.L. Alty (UK)
A.B. Aune (N)
J.R. Blau (D)
P. Brodner (D)
P.C. Cacciabue (I)
F. Conrad (DK)
W.J. Edwards (AUS)
J.L. Encarnacao (D)
S. Franzen (S)
R. Genser (A)
G. Guida (I)
B.S. Hu (PRC)
G. Johannsen (D)
K. Kawai (J)
A.H. Levis (USA)
M. Lind (DK)
B. Liu (PRC)
N. Malvache (F)

L. Martensson (S)
P. Milgram (CAN)
N. Moray (USA)
L. Motus (ESTONIA)
L. Nemes (AUS)
G. Olsson (S)
W.K. Oxenaar (NL)
J. Ranta (SF)
J. Rasmussen (DK)
J.E. Rijnsdorp (NL)
A.P. Sage (USA)
G. Salvendy (USA)
T.B. Sheridan (USA)
Shi-Quan Su (PRC)
B.G. Tamm (ESTONIA)
H. Tamura (J)
T. Vamos (H)
J. Wirstad (S)
M. Wozny (USA)

National Organizing Committee

P.L. Brinkman   (Chairman)
N.W.S. Bruens
L. Hammendorp
H.W. Schneider
H.G. Stassen
P.A. Wieringa

PREFACE

The field of man-machine systems or human-machine interaction in complex technological systems is very strongly changing, and asks for a multidisciplinary contribution of research, development and industrial activity. The shift from manual control towards supervisory control has a strong influence on human operator tasks. It is forcing human personnel to compete with machines equipped with an ever increasing artificial intelligence.

The various aspects of human-machine interaction was elucidated in the different sessions, such as: Manual and supervisory control; human performance and mental load modelling; human-machine interfaces; knowledge engineering and decision support; and finally task allocation, work design and training. Application areas of interest were teleoperation of manipulators, process control, power stations, intensive care units, vehicles control, etc.

The aim of the symposium was to present and to discuss the state of the art of human-machine interactions and interfaces; it included the Analysis, the Design and the Evaluation. The symposium emphasized recent advances in theory, in analysis and experiments. The topics discussed ranged from university research to advanced industrial applications.

The symposium was the fifth in a generally three-yearly planned IFAC/IFIP/IFORS/IEA symposia on Analysis, Design and Evaluation of Man-Machine Systems, i.e.

> 1982: 1st Conference, Baden-Baden, FR Germany
> 1985: 2nd Conference, Varese, Italy
> 1988: 3rd Conference, Oulu, Finland
> 1989: 4th Conference, Xi'an, China
> 1992: 5th Symposium, The Hague, The Netherlands

The International Programme Committee hopes the results presented, the discussions held and the ideas gathered were beneficial to all attending engineers, scientists, designers, and all others who are active or interested in the field of man-machine systems.

Henk G. Stassen
Chairman IPC

CONTENTS

HUMAN-MACHINE INTERFACES I

DECISION SUPPORT SYSTEMS I

MANUAL CONTROL II

HUMAN RELIABILITY

HUMAN-MACHINE INTERFACES II

DECISION SUPPORT SYSTEMS II

ROUND TABLE DISCUSSIONS

TELEOPERATION, TELEROBOTICS, AND TELEPRESENCE: A PROGRESS REPORT

T.B. Sheridan

Man-Machine Systems Laboratory, 3-346, Massachusetts Institute of Technology,
Cambridge, MA 02139, USA

Abstract. This paper briefly surveys and reports progress in the field of *teleoperation*, meaning human control of remote sensors and actuators. Included is the subclass of teleoperation called *telerobotics*, which means human *supervisory* control of remote semiautomatic systems, and the phenomenon of *telepresence*, in which special sensing and display technology enables the human to feel present at the remote location even though not really there. Current and new applications are reviewed. Techniques for human-computer cooperation in planning, commanding, and sensing are described. The telerobot is considered as a paradigm for any complex vehicle or process having many separate automatic control loops all of which are supervised by a human; some current examples are presented. Finally, opinions are given as to the current status of the field.

Keywords: man-machine systems, telecontrol, robots, manipulation, human factors, computer interfaces, artificial intelligence.

Introduction: Definitions and Taxonomy

A review of telerobotics by the author appeared in *Automatica* three years ago (Sheridan, 1989). The present paper is a short report of progress in the field since then, with the emphasis on man-machine interaction rather than hardware or software. The reader interested in a much fuller account is referred to (Sheridan, 1992a).

A *teleoperator* is a machine enabling a human operator to move about, sense and mechanically manipulate objects at a distance. It usually has artificial sensors and effectors for manipulation and/or mobility, plus a means for the human to communicate with both. Most generally, any tool which extends a person's mechanical action beyond his reach is a teleoperator.

A *robot* is a machine which senses and acts upon its environment autonomously (which means it also has a computer), and, according to most dictionaries, behaves with what appears to be human intelligence. A *telerobot* is a subclass of teleoperator in which the machine acts as a robot for short periods, but is monitored by a human supervisor and reprogrammed

from time to time. (We call the latter *supervisory control*.) Non-telerobot teleoperators are fully manual, like a master-slave manipulator, or a joystick (rate-controlled) manipulator.

Teleoperators may look like "robots", i.e., be anthropomorphic, with serial-link arms mounted on a mobile platform, and video-camera eyes mounted above on a moveable "head". Or they may be non-anthropomorphic, i.e., not have human-like form. Thus any semiautomatic machine which has artificial sensors, actuators, and a computer, and is controlled in supervisory fashion, may be called a telerobot. Later in the paper modern aircraft, automobiles and power plants will be discussed as non-anthropomorphic telerobots.

Fig. 1 diagrams the author's current model of the supervisor's task in controlling a telerobot. The generic telerobot may have multiple degrees of freedom (DOF) and act upon multiple task components, as shown in the bottom half of the diagram. A microcomputer may be locally attached to most task elements, such as task 3 in Fig. 1; some may still have to be controlled manually to effect transient control or regulation relative to a set point. At the top of the diagram are shown the steps the supervisor must go through, in many cases aided by a computer. The blocks can represent mental activities or computer activities or a combination. The supervisor functions are:

    (1) plan, which includes the sub-activities of (a) modeling the physical system, (b) trading off objectives to decide what is satisfactory ("satisficing"), and (c) formulating a strategy.

    (2) teach, including the distinctly different activities of (a) deciding what to have the telerobot do, and (b) deciding how to tell the telerobot to do it.

    (3) monitor, which includes (a) deciding how to allocate attention among all the various signals that can be observed, (b) estimating current system state or "situation", and (c) detecting /diagnosing any abnormality in what is currently happening.

    (4) intervene, which in the case of abnormality means (a) deciding on and effecting minor adjustments if they will suffice, or (b) complete manual takeover, or (c) system shutdown; or, (d) if the programmed action has come to a normal conclusion, it means reverting back to step (2).

(5) <u>learn</u> from experience to improve future planning.

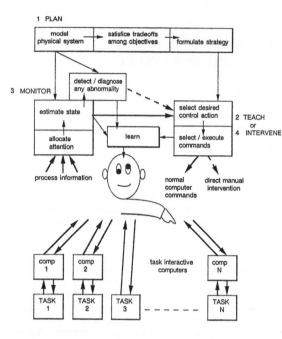

Fig. 1. Supervisory functions in control of a telerobot.

Applications and Configurations

Space. This is the application that most frequently comes to mind when teleoperation is mentioned. In fact most of the deep space probes have been telerobots, relatively simple ones with respect to their control, having low-bandwidth capability to receive pictures and other sensory data, and ability to be reprogrammed in space. The Viking's tour of the outer solar system is the most striking example, from which signals have been received and updates sent from over a billion miles distance.

Currently Japan is the leader in space teleoperator developments (Whittaker et al, 1991). The Japanese Experimental Module (JEM) is an integral part of Space Station Freedom. JEM includes a long teleoperator operating from a "porch" (Fig. 2). In its initial phases this teleoperator will be controlled manually from ground and later be telerobotic. A 6-DOF "small fine arm" is to be added later to the end of the long teleoperator. The Japanese also plan a "free flyer" vehicle with a telerobot arm

The German space agency DLR is scheduled to demonstrate the first space telerobot: the "Rotex" experiment. While being relatively small and operating wholly inside a container, it can demonstrate the ability of a computer to perform control in space.

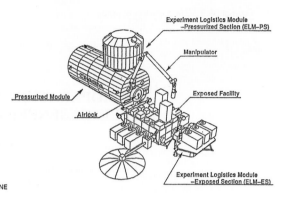

Fig. 2. Teleoperator on Japanese Experimental Module for Space Station Freedom.

The Canadian firm SPAR built the 60 foot Remote Manipulator System on the space shuttles, which have been fully manually controlled. They are now developing similar manipulator arms for the Space Station.

Regrettably, the large US NASA effort in space teleoperation, the three-limbed and rather ambitious Flight Telerobotic Servicer, which was well along in its development, was cancelled in 1991, allegedly to save money. Currently there is no significant American space teleoperation project.

Undersea. Current undersea applications include offshore oil exploration, inspection and maintenance on drillheads, oil platforms and pipelines, marine biology experiments, geological surveys, archeological search and recovery, and classified navy tasks. The French, British, Norwegians, Japanese and Americans have led in this research.

The Argo-Jason project of Woods Hole Oceanographic Institution, named after Jason and his Argonauts of Greek mythology, has been the premier test-bed for deep ocean teleoperation in recent years. Argo is a heavy passive assembly of high-energy sonar and photographic equipment suspended by up to 6000 m of cable from its support ship, while the telerobot Jason maneuvers on a flexible cable within easy return range from Argo, all controlled from the surface. Jason is pictured in Fig. 3. It was Jason's

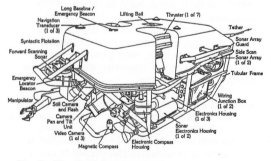

Fig. 3. Woods Hole Oceanographic Institution's remotely operated undersea vehicle Jason.

prototype, Jason Junior, which swam inside the Titanic, but was recently lost at sea. Jason is programmed with a variety of supervisory control modes, and also makes use of some sophisticated techniques such as sliding-control to compensate for unmodeled dynamics (e.g., for the mass of water being pushed along in front of Jason as it maneuvers).

Jacobson et al (1989) have developed for the Naval Ocean Systems Center what is surely the highest fidelity (bandwidth) master-slave teleoperator with force feedback. Master and slave are isomorphic, having seven DOF in the arm. The hand and fingers of both master and slave have three DOF. The system, shown in Fig. 4, is electrohydraulic.

Fig. 4. University of Utah – Naval Ocean Systems Center electrohydraulic master-slave teleoperator.

Toxic waste cleanup. The US Department of Energy (DOE), the US manufacturer of nuclear weapons for several decades, is not subject to the same regulations as are the commercial nuclear power plants for storage and disposal of radioactive and chemically toxic wastes. Over the years DOE stored high-level wastes in large tanks in desert areas, and low level wastes in ordinary 55 gallon steel drums. After several decades these tanks and drums have corroded and leaked their toxic contents into surrounding groundwater, and the government now feels an obligation to recover the wastes and transfer them to more secure containers. The hazards to human health are such that the job must be done by teleoperation. The task is mammoth, the number of tanks and containers being quite large, and no one quite knows exactly how it will be done or how it will be financed.

Other applications. In recent years developmental telerobots and manual teleoperators have moved into mines, are being commercially sold as night sentries for factory security, and are delivering mail and sweeping floors in factories — many of the applications promised but not realized only three years ago. Simple teleoperators are being used in ever greater numbers to perform laparoscopic and

other endoscopic surgical procedures at a fraction of the cost and suffering of three years ago. Applications such as fire fighting and rescue are still not realized.

As teleoperators find their way into more and different applications, they become accepted as the normal and mundane way of doing tasks, not particularly "robotic" or otherwise glamorous.

Human-Computer Cooperation in Planning, Command and Control

So-called modern control theory is based on the use of a model of the controlled process to provide a best state estimation. Measurements of state are combined, using Bayesian updating, with the model's prediction of response to past control inputs to produce a best current "belief state". An optimal gain matrix then operates upon this best estimate of current state to produce the new control input. The discrepancy between measured state and estimated state (after modifying the latter by a model of the measurement delay/distortion) is used to modify the process model. Fig. 5 diagrams this well documented idea as it can apply to control aids for man-machine systems. This idea is a powerful one, and is being used in various ways to aid the human operator of dynamic systems, as described below.

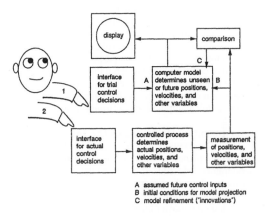

Fig. 5. Use of computer-model of process as aid to human operator for system control.

Predictor display. One of the more straight-forward applications of the computer model is where there are significant time delays in human control loops and the unaided operator therefore cannot provide stable control. A remedy is model-based prediction, in the form of what has come to be called a predictor display. Sheridan (1989) discussed development of such a predictor display (see Noyes and Sheridan, 1984, for an earlier report) for controlling a teleoperator over a several-second transmission delay. The operator's control commands are sent simultaneously to the controlled process and to a model. Since the model has no delay the operator sees immediately what the commands will do to the process. This lead provides compensation which enables him to stabilize control.

One interesting task for the Argo-Jason project described above is to maneuver the passive Argo along ocean canyons by control of the thrusters on the support ship. However the (up to 6000 m) umbilical poses a time delay of up to 10 minutes between ship movement and Argo response, and thus it has been virtually impossible to control Argo in other than very large radius curves. Thus it seemed that a predictor display might help. However the cable dynamics are quite nonlinear and complex, depending on ocean currents and many other factors. Cheng and Sheridan, in a separate paper at this Symposium, describe a predictor display technique which, even though representing the cable by an extremely simple adaptive model, nevertheless enables manual control to be quite stable and smooth.

Conway et al (1987) developed a novel way to use the predictor display in which they "disengage control synchrony" in time and space by use of a "time clutch" and a "space clutch". The time clutch allows the operator to disengage from real time by making control inputs and getting back simulator responses more rapidly than the actual process would allow for easy maneuvers, and conversely slowing down the pace of human interaction where more samples are needed in higher bandwidth maneuvers. A computer buffers the commands and later feeds them to the actual process at the real-time pace, interpolating between samples as necessary. Disengaging the position clutch allows the user to move the simulator in space without later commitment of the commands to the actual process -- for the purpose of trying alternative commands to see what they will do -- in essence going into an "off-line" mode for short periods. An experimental evaluation of using this scheme in telemanipulation confirmed its efficacy.

Other recent applications of the internal model. The computer has been used in other ways to aid in the planning, command and control of teleoperators. Machida et al (1988) demonstrated a system on which the supervisor could teach a computer-modeled telerobot and receive force feedback on the master hand if mechanical interference occurs between the modeled slave hand and its environment. The operator can edit the trajectory on any selected time-scale, with the sequence of actions selected to run in either forward or reverse. The latter proved useful, for example, if after some trial instructions the telerobot became hung up. In this case it could easily be returned to the last acceptable point and reprogrammed from there.

Park (1991) demonstrated a computer-aided technique for commanding a telerobot to move to a goal point while avoiding objects. The technique assumes a model of the geometry of the teleoperator, and all other objects (obstacles to motion) in the near environment. Additional obstacles can be inserted into the model when observed by television, after fixes are taken on their positions (from their edges).

In real telemanipulation television cameras can be panned and tilted but not otherwise moved, and hence there are unseen spaces (penumbra) behind the newly seen obstacles. In Park's technique these are regarded as *virtual obstacles* (Fig.6). At any time the computer can be called upon to display the updated field of obstacles (by using a trackball any viewpoint can be set in) and the human operator can suggest a trajectory of the vehicle and/or manipulator through the obstacles which is drawn on the computer screen (Fig. 7). There are also some simple AI trajectory search heuristics which may be used. Then the computer immediately displays whether the trajectory is feasible, and if so how close it comes to collision (how tight is the path). The human may iterate with other trial paths. Once a satisfactory path is selected, the computer can automatically guide the teleoperator for part or all of the trajectory, stopping if it finds itself in trouble. In simulated tasks Park found this technique to prevent errors and speed up teleoperation considerably.

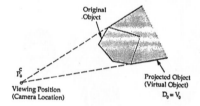

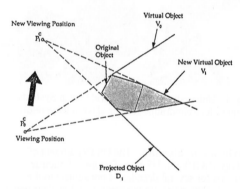

Fig. 6. Park's "virtual obstacles" in computer-graphic model used for planning teleoperator moves.

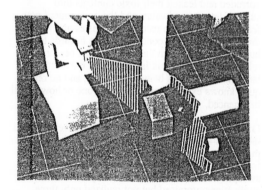

Fig. 7. Planned trajectory displayed in Park's computer model.

4

Funda et al (1992) extended the Machida et al work and the earlier supervisory programming ideas. Again the operator programs by kinesthetic as well as visual interactions with a (virtual) computer simulation. A key feature of their work is that the instructions to be communicated to the telerobot are generated automatically in a more compact form than record and playback of analog signals. Several free-space motions and several contact, sliding and pivoting motions, which constitute the terms of the language, are generated by automatic parsing and interpreting of kinesthetic command strings relative to the model. These are then sent on as instruction packets to the remote slave.

The Funda et al technique also provides for error handling. When errors in execution are detected at the slave site (e.g., because of operator error, discrepancies between the model and real situation and/or the coarseness of command reticulation), information is sent back to help update the simulation. This is to represent the error condition to the operator and allow him to more easily see and feel what to do to correct the situation. With a time delay, of course, some additional actions may have been taken in the meantime. In this situation a predictor display could be useful to help the operator explore on the simulation different alternatives for how best to recover from the error. The authors' current research is considering a system which can automatically take back control when a safe situation is recognized.

Van de Vegt et al (1990) were concerned with modeling the supervisory operator of teleoperator in a time-delay situation. They extended the Kleinman, Baron and Levison (1970) optimal control model (OCM) of the human operator (which assumes a perfect representation of process dynamics) to allow for an imperfect internal model of the system being controlled. They also relaxed the requirement that the human internalize a prediction to compensate for his own time delay (delays significantly less than one second being significant in the aviation applications for which the OCM was developed). Instead they assume the human to take predictor display information (typically used for loop delays of several seconds) as though it truly represents current process state. They tested their revised model on a simulated undersea teleoperator and found it to work well.

Other Aspects of Sensory Feedback: More Roles for the Computer

Visual aids to sense depth. Several graphical techniques have become common to enhance visual sense of depth, which can be very difficult to obtain from a video picture. Stereoptical video (two cameras) accompanied by stereo display is the most promising. The latter may be accomplished in several ways. One is to feed two separate images to the two eyes, say by miniature video monitors A second is to alternate video images from the two cameras onto a single monitor (e.g., each successive interlace from a different camera) and have the viewer wear liquid crystal spectacles which alternately block one eye or the other, synchronized to the interlace. A non-stereopsis technique is to superpose onto the video

picture a computer-graphic image with a perspective rectangular grid comprising a "floor"; each of many objects in the visual field has a vertical line "attached" at one end, the other end of the line being projected to the artificial floor at the corresponding (known) distance.

Force feedback. Force feedback in a master-slave teleoperator means the net force vector imposed by the slave on the environment is reflected back and imposed by the master on the operator's hand. Many of the earliest teleoperator systems were force-reflecting.

Force feedback in teleoperation has been studied in laboratory experiments for many years (Sheridan, 1992). A recent study by Hannaford et al (1989) compared both task completion time and level of force used for a variety of teleoperation tasks as well as a variety of control modes. The latter included (1) position control with visual but no force feedback, (2) regular visual feedback plus force feedback by means of visual display, (3) visual plus kinesthetic (conventional bilateral force) feedback, (4) what the authors call "shared control", and (5) bare-handed manual control. Shared control in this case means force feedback is imposed or suppressed as a computer-based function of object contact, recent past forces applied in all degrees of freedom, and other stability considerations (Hayati and Venkataraman, 1989). Some sample results from Hannaford et al are shown in Fig. 8 for a peg-in-hole task. Massimino and Sheridan (1989) showed how mean completion time in such tasks is significantly reduced by force feedback independently of visual parameters such as frame rate and spatial resolution of the image.

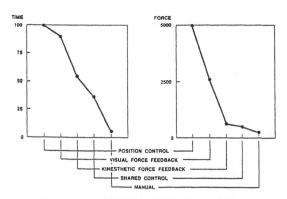

Fig. 8. Hannaford et al results for task completion time and level of force used under a variety of teleoperator control modes.

Force reflection has been accepted in remote handling of nuclear and toxic wastes, where master-slave position teleoperation is common. However in space and undersea applications force feedback is not generally accepted. This is because in space forces are to be avoided, so rate-control joysticks are used to control motions to very slow rates. It is also because of the added complexity and cost imposed, and because when there is a significant time delay in the control loop force feedback to the same hand as is operating the control produces a (thus far unavoidable) instability. Massimino and Sheridan (in another paper at this Symposium) show that the latter problems can be overcome when vibrotactile or auditory displays are substituted for display of actual force.

When force feedback is used, it is important to avoid biasing forces caused by friction, damping and inertial properties of the intervening teleoperator mechanism. Therefore force is measured near to the slave hand and near to the operator's hand (e.g. using a multi-DOF force sensor at the wrist), and control gains are adjusted carefully to avoid instability while minimizing the discrepancy between the two (Raju, Verghese, and Sheridan, 1989).

Teletouch. In contrast to feedback of the force resultant from the slave hand to the master hand, teletouch refers to feedback of the spatial distribution of forces imposed on the slave hand by contact with the environment. Electrical, optical and other types of touch sensors have been developed to measure patterns of forces and mechanical distortion on the gripping surface of the slave hand, and some of these are now available commercially at low cost. Unfortunately there is no available counterpart display of touch information to the skin, though it is relatively simple to devise computer-graphic visual display of the sensor's spatial force pattern.

Telepresence, virtual presence and the combination. *Telepresence* is the sense of being at a real location other than where one actually is. *Virtual presence* is the sense of being at a location which does not actually exist, which instead is a compelling graphic or auditory illusion. New computer and display technology is driving much current activity in this field, which also goes by the terms *virtual reality* and *artificial reality* (neither term being acceptable to the author because the dictionary definitions of the terms are self-contradicting).

Currently visual telepresence is said to be experienced when a human operator wears a head-mounted video display, which, upon panning or tilting of the head, drives a slaved remote video camera in corresponding motions. Thus the observer sees the same moving images as would be seen were the eye placed at the remote camera's viewpoint. Stereo, high resolution, high frame-rate, good contrast and color rendition all enhance visual telepresence. Having tried a number of such telepresence systems, the writer believes the best to be that of Prof. S. Tachi of the University of Tokyo (see Tachi et al, 1989).

Virtual presence may be achieved the same way, but with a computer-generated graphic images replacing the video. The software must be able to modify the images quickly to keep up with even modest head movements, which pushes today's graphic workstations to their limits. A most interesting demonstration of virtual presence can be seen at Matsushita Electric's kitchen show- room in Tokyo. A prospective customer can don a head-mounted display, also equipped with an auditory device to generate compelling sound "presence", and have the experience of walking into different kitchens, looking

around, peering into closets, even knocking over glassware and hearing it smash in just the correct location – none of it real, all of it produced by software.

Some informal experiments in the writer's laboratory have revealed that even relatively poor visual telepresence and poor force feedback, when used together, are mutually enhancing. Though there are glove and exoskeletal devices to sense the relative poses of fingers and communicate these to a computer-graphic hand image, no one has yet produced satisfactory virtual teletouch. Patrick (1990) came close, by showing that vibrotactile sensors mounted on the fingertips, when activated by the fingers performing a grasping motion in space, do give the impression of grasping a tuning fork (Fig.9).

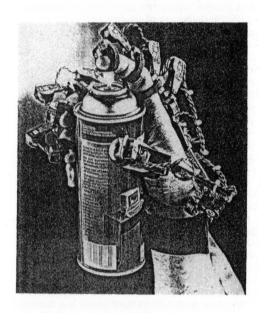

Fig. 9. Patrick's experimental setup: vibrotactile sensors mounted on the fingertips are activated when grasping position reaches a criterion corresponding to a computer model (actual object pictured is not really there). This gives a virtual sensation of grasped object.

Fig. 10 suggests that virtual presence is a function of three independent components (Sheridan, 1992b). The first is resolution of the sensory channel, including pixels per visual, auditory or tactile "frame", frames per second, and bits per pixel of grayscale or magnitude resolution – the product of which is bits per second. The second component is the ability of the sensor to move about in space, much as a visual sensor takes and compares images from different viewpoints (thus using parallax and other cues) or engages in haptic touch, etc. The third component is the capability to actually modify the relative positions of objects in the environment. An ultimate sense of "being there" requires the full extent of all three.

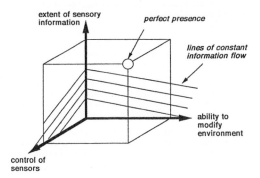

Fig. 10. Proposed independent components of virtual presence.

Non-anthropomorphic Telerobots: Airplanes, Automobiles and Power Plants

A *non-anthropomorphic teleoperator* is a teleoperator which does not have human form – arms or legs or scanning eyes or creature-like motions. It may still be a vehicle which moves through its environment such as an aircraft or an automobile, or it may be a process control or other manufacturing plant in which the manipulated environment (in this case the product) moves through the teleoperator, while the teleoperator itself is stationary in space. The latter still involves human control of sensors and actuators which are remote, functionally if not physically (i.e., the human operator is located in a centralized control room, far from the machines that process the product). Just as with the anthropomorphic kind, as such teleoperators become semiautomated and the human becomes a supervisor, we call them *non-anthropomorphic telerobots*. Here is a description of what is happening in three such applications. The parallels with space, undersea, and toxic environment teleoperation are striking.

Aircraft. The newer aircraft, both commercial and military, are largely controlled in supervisory fashion through computers. The human pilot is called *flight manager*, and he more and more controls the aircraft through a multi-purpose interactive computer called a *flight management system (FMS)*. By means of the FMS he can program the autopilot in one of several modes, or examine electronic check lists which remind him to take certain actions and call his attention if he misses some. He can also do routine maintenance or fault diagnosis by calling up different synoptic displays of electrical or hydraulic diagrams, indicating the quantitative values of voltages, currents, temperatures, pressures, etc. at the key component locations, as well as showing by color, or flash-rate whether a variable is in an abnormal state (Billings, 1992).

A variety of commands and other information, which in the past would have been sent by verbal communication with ground controllers, is now being sent directly to the FMS by digital datalinks, and then interpreted by computer for display to the pilot. The FMS also has the capability to provide navigation maps or approach plate information on different

scales or with some detail suppressed, or provide detailed advisories or instructions on operation or maintenance of various systems.

Automobiles. In the past the automobile driver controlled the vehicle entirely manually. That is now changing, with the European Community Prometheus and Drive projects, the Japanese Amtics and Racs projects, and the American IVHS (see IVHS America, 1992). In ongoing field trials, in-vehicle computers now receive navigational and accident-alert data from infrared, radar, or ultrasonic sensors mounted either in the vehicle or by the roadside, or get messages via satellite from an advanced traffic management center in urban areas. Tiny antennas mounted on the test vehicles get latitude and longitude fixes from GPS (global positioning system) satellites to within 30 feet, and this together with an on-board gyrocompass provides continuous route information relative to programmed starting point and destination.

Display of route information is presented visually by computer-graphic map, graphic symbols indicating lane to be in or upcoming left or right turn, or name of the street to turn onto. Simultaneously information is given by computer-generated speech, relieving the operator from having to look down at displays except for confirmation. Such messages also contain alerts about accidents and congestion. Such systems can also be queried to determine nearest fuel stations, or medical/police services, or restaurants. The new technology also allows for automatic braking or steering in the event the driver is not responding to obstacles that are measured to be too close, and there are interesting questions here about whether control should be seized from the driver suddenly or applied gradually so that for a time both driver and computer are working in parallel, and indeed when and under what circumstances to give control back to the driver.

As with many of the other applications of telerobotics and supervisory control, the "hard" technology is mostly in hand. What is lacking is a base of human machine research (Sheridan, 1991) and an understanding of the legal and socioeconomic implications.

Process Control Plants. Many chemical plants, oil refineries, and nuclear and fossil power generating stations have been re-engineered in recent years to move the human operator away from being a direct controller manipulating valves and switches out in the plant to being more of a supervisor of computer-based semi-automation, operating from a centralized control room which may be physically remote from many parts of the plant. This configuration appears least like a telerobot (is most non-anthropomorphic), but still, from the human operator's viewpoint, poses the very same requirements as diagrammed in Fig. 1.

There may be conventional anthropomorphic teleoperators within it, for example, for the remote handling of nuclear fuel elements during refueling operations or repairing of radioactive steam generators in nuclear plants, or remote handling operations on highly toxic substances in chemical plants. However, "teleoperation" in process control plants is also characterized by giving commands to a central computer in the control room which in turn sends set points to hundreds or thousands of simple computer-regulators out in the plant, and getting back information from sensors, which the central computer translates into integrated graphical information.

Conclusions

Telerobotics is alive and well, especially in Japan. The field could certainly be healthier in Europe and the US, but distrust of technology and demand for quick profits have discouraged investment in telerobotics research and development. In the aviation and automotive fields, non-anthropomorphic telerobots are being developed actively, most commonly under the rubric of *human-centered automation.*

Telerobots of various forms have been touted as answers to problems of safety, productivity, transportation and military security. Thus far little fear has been manifest that they will destroy us or take our jobs away. But, as with the seemingly benign automobile and refrigerator, which by quietly generating greenhouse gasses or releasing chlorofluorocarbons that destroy the ozone, the telerobot can easily be put to tasks which are unnecessary or in some cases harmful, all the while wasting valuable resources. The socioeconomic impacts have yet to be taken seriously by more than a small number of people.

In systems of the type discussed above, the human is called upon to perform those "higher-level" functions of pattern recognition, association, evaluation of objectives, and creativity which we do not yet know how to program the computer to do. At the same time, computers are used to give advice or to control systems cooperatively with the human. In these circumstances there remain serious questions about how to study the human operator, what intelligence can be attributed to him or her vis-a-vis the computer, and how far one can go in modeling such complex human-machine systems.

For the same reason we do not know how to program the computer to do these functions, we also do not know how very well how to model them. Further, in order to evaluate a computer aid or a cooperative control system, a relatively complete process model and objective function must be used as a norm. However, almost by definition of the system being complex enough to require the human supervisor, such a model and explicit objective function are not available. And if the latter were available, why would the human be necessary? We have only begun to face these compelling challenges of human-machine interaction.

References

Billings, C. E. (1992). *Human-Centered Aircraft Automation: A Concept and Guidelines.* Book manuscript in preparation. Moffet Field, CA: NASA Ames Research Center.

Conway, L., Volz. R. and Walker, M. (1987). Teleautonomous systems: methods and architectures for intermingling autonomous and telerobotic technology. In Proceedings 1987 IEEE International Conference on Robotics and Automation, March 31–April 3, Raleigh, NC: 1121–1130.

Funda, J., Lindsay, T.S. and Paul, R.P. (1992). Teleprogramming: towards delay-invariant remote manipulation, *Presence: Teleoperators and Virtual Environments*, Vol.1, No.1.

Hannaford, B., Wood, L., Guggisberg, B., McAffee, D. and Zak, H. (1989). Performance Evaluation of a Six-Axis Generalized Force-Reflecting Teleoperator. JPL Publication 89-18, Pasadena, CA: California Inst. of Technology JPL, June 15.

Hayati, S. and Venkataraman, S. (1989). Design and implementation of a robot control system with traded and shared control capability. In Proceedings of 1989 IEEE International Conference on Robotics and Automation, Scottsdale, AZ, May 14–19: 1310–1315.

IVHS America (1992). Strategic Plan for Intelligent Highway-Vehicle Systems in the United States, Washington, DC: .

Jacobson, S. C., Iversen, E. K., Davis, C. C., Potter, D. M. and McLain, T. W. (1989). Design of a multiple degree-of-freedom, force-reflective hand master/slave with a high mobility wrist. In Proceedings of ANS/IEEE/SMC 3rd Topical Meeting on Robotics and Remote Systems, March 13–16, Charleston, SC.

Kleinman, D. L., Baron, S. and Levison, W. H. (1970). An optimal control model of human response, Part I. *Automatica*, Vol. 6, no. 3: 357–369.

Machida, K., Toda, Y., Iwata, T., Kawachi, M., and Nakamura, T. (1988). Development of a graphic simulator augmented teleoperator system for space applications. In Proceedings of 1988 AIAA Conference on Guidance, Navigation, and Control, Part I: 358–364.

Massimino, M. and Sheridan, T. B. (1989). Variable force and visual feedback effects on teleoperator man–machine performance. In Proceedings of the NASA Conference on Space Telerobotics, Pasadena, CA, January 31–February 2.

Noyes, M. V. and Sheridan, T. B. (1984). A novel predictor for telemanipulation through a time delay. In Proceedings of Annual Conference on Manual Control, Moffett Field, CA, NASA Ames Research Center.

Park, J. H. (1991). Supervisory Control of Robot Manipulators for Gross Motions. PhD Thesis, MIT, August.

Raju, G. J., Verghese, G. and Sheridan, T.B. (1989). Design issues in 2–port network models of bilateral remote manipulation. In Proceedings of IEEE International Conference on Robotics and Automation, Scottsdale, AZ, May 14–19: 1316–1321.

Sheridan, T. B. (1989). Telerobotics. *Automatica* 25, no. 4: 487–507

Sheridan, T. B. (1991). Human Factors of Driver-Vehicle Interaction in the IVHS Environment, US Dept. of Transportation Report.

Sheridan, T.B. (1992a). *Telerobotics, automation, and Human Supervisory Control.* MIT Press.

Sheridan, T.B. (1992b). Musings on telepresence and virtual presence, *Presence: Teleoperators and Virtual Environments*, Vol.1, No.1.

Tachi, S., Arai, H. and Maeda, T. (1989). Development of anthropomorphic tele–existence slave robot. In Proceedings of International Conference on Advanced Mechatronics, May 21–24, Tokyo: 385–390.

Van de Vegte, J.M.E., Milgram, P. and Kwong, R.H. (1990). Teleoperator control models: effects of time delay and imperfect system knowledge, *IEEE Trans. Systems, Man and Cybernetics*, Vol. 20, No. 6, pp. 1258-1272.

Whittaker, W., Kanade, T. et al. (1991). Space Robotics in Japan, Japanese Technology Evaluation Center, Loyola College, MD.

TASK ALLOCATION PROBLEMS AND DISCRETE EVENT SYSTEMS

A.H. Levis*, N. Moray** and Baosheng Hu***

*George Mason University, Fairfax, Virginia, USA
**University of Illinois, Urbana, Illinois, USA
***Xi'an Jiaotong University, Xi'an, PRC

ABSTRACT : The allocation of tasks to humans and machines in complex systems is discussed from three perspectives. Systems engineering as manifested in the design of flexible manufacturing systems, organization design as applied to decision making teams that monitor and control engineering systems, and cognitive ergonomics are the three perspectives used. The limitations of fixed task allocations are articulated; while the case is made for integrated experimental and theoretical investigations that address the dynamic aspects of task allocation.

KEYWORDS: Task Allocation, Discrete Event Systems, Ergonomics, Human-Machine Systems, Petri Nets, Flexible manufacturing Systems.

INTRODUCTION

The question addressed in this paper is the decomposition of a task or function into component parts so that it can be carried out by different agents; these agents can be humans, machines, or both. This is a very old problem that appears under many guises: as the classical problem of the division of labor, as the systems engineering principle of dividing a problem into its component parts, or even as the design of the interface between a human and a machine. The latter assumes that the division has been done and the focus has shifted to how to make this division less disruptive. On the theoretical side, decomposition algorithms, parallel processing, holographic decompositions, etc. represent efforts to break a problem into smaller units that can be solved more effectively individually. The difficulty, and the core of the systems engineering problem, is not how to decompose, but how to integrate the results of the solutions of the component problems into an overall solution.

The complexity of the problem is articulated clearly by Descartes' dictum "Break down each problem you are trying to solve into as many parts as you can and as many as you need to solve them easily," to which Leibnitz responded "The rule of Descartes is not effective, since the art of separation does not yield to interpretation." These two dicta capture the challenge of the task allocation problem: how to allocate sub-tasks without compromising the wholeness of the task. When the problem is viewed from this perspective, it becomes clear that theories and analytical tools are needed for representing the problem both as a whole and as a set of interacting or interdependent sub-problems. This is common practice in the modeling of engineering systems, where all the agents carrying tasks or solving sub-problems are engineering systems themselves, such as processors. It is also well understood, even though not well practiced, when all the agents are human - it is called organization design. It becomes a real challenge when some of the agents are humans and some are machines and some of the human agents must interact with the machines, whether these machines are computer workstations, or the control panel of a machine tool, or the cockpit of an aircraft. The evolving mathematical theories of discrete event systems, some coming from computer science and some from system theory, are providing now the means for re-examining this classic problem of task allocation in a new light. (Johannsen et al., 1992). Unfortunately, the resulting mathematical models have not been integrated yet with human factors and cognitive ergonomics.

In the past, most of the effort was focused on task decomposition and then on resource allocation. In recent years, as concern for total factor productivity in economics, total quality management in manufacturing, fault tolerance in large data communications networks, or distributed decision making, to name a few, increased, attention has been focused on the system integration problem. These efforts have been giving rise to the recognition that there are many phenomena previously ignored or not understood that affect the integration. For example, coordination, migration of control, meta-decision making, and domain shifts are indicative of the issues that need to be addressed. To understand these issues, both empirical and analytical studies need to be pursued. Indeed, there is most hope for results when experimental investigations and theoretical developments are closely linked, when experiments drive the development of theory, and when theory generates hypotheses for model-driven experimentation.

TASK ALLOCATION PROBLEMS

The notion that a function is decomposed and its components (or sub-functions) assigned to different agents is an old one. However, when the agents are humans or intelligent machines, then each agent contributes to the execution of *several* different functions, sometimes in sequence and sometimes concurrently. Thus, the problem is not solved by doing a simple allocation of a decomposed function to the available resources - human and machine ones. One must allocate several decomposed functions in such a manner that the resulting *workload* does not exceed the capacity of each agent. The modeling of (cognitive) workload must take into account not only the workload associated with each sub-function carried out by a human, but also the workload associated with coordinating the execution of the sub-functions and with coordinating the interactions between the agents - between humans and between humans and machines.

Still, the first step in solving the task allocation problem is carrying out a functional decomposition. There are many taxonomies and related terminology used to carry out the decomposition; one can talk about goals, objectives, and tasks when referring to humans and organizations; one can also describe the decomposition in terms of a mission which is decomposed into functions and functions which are decomposed into tasks. This hierarchical decomposition, Mission -> Function -> Task, is shown in Figure 1.

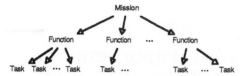

Figure 1. Functional decomposition.

This functional decomposition can be carried to any level of detail. In practical terms, it is carried out to the point that the lowest level tasks (the leaves of the tree) must be executed by a single resource. Another property of this hierarchical decomposition is that it is nested; any node in the tree can be called a mission if that becomes the starting point for the problem at hand. This is shown graphically in Figure 2.

The allocation problem could then be formulated simply (and simplistically) as an optimization problem where the lowest level tasks are assigned to agents - humans or machines - so that a specific objective is optimized subject to a large number of constraints. The result of such allocation leads to two other problems: the determination of the material flow from agent to agent so that the physical processing is accomplished (as in a manufacturing plant) and the determination of the information flow so that the monitoring, control, and supervisory functions can be accomplished. The last problem to be addressed in that process is the human-machine interaction where material flows or information flows go from machines to humans or from humans to machines. While cognitive limitations may have been taken into consideration in the optimization problem -a rare occurrence - they are usually dealt with in the design of the human-machine interface. The result of such a process is a set of fixed allocation strategies. Given a mission, a fixed decomposition leads to a fixed allocation. For different missions, different fixed allocations can be determined. A simple example is the determination of the process used in a job shop to machine different parts. For each part, there is a fixed sequence of processes carried out by predetermined machining stations.

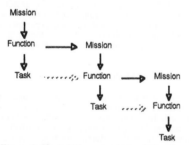

Figure 2. Nested Decomposition of Functions

The above approach leads not only to fixed allocation strategies but also to suboptimal solutions because the partition of the overall design problem into four sequential problems is obviously not optimal. But optimization is not the issue when complex man-machine systems are being designed. A satisficing solution that deals with the bounded rationality of humans (Simon, 1972) is a more appropriate goal. The more interesting and more realistic problem is the one in which the task allocation is dynamic and depends not only on the nature of the task to be performed by the system as a whole, but also on the state of the individual agents/components that constitute the system.

The traditional system theoretic approaches, based on differential and difference equations, are not appropriate for addressing this class of problems. Discrete mathematics and its application to the modeling of discrete event processes provides a more appropriate conceptual and analytical framework for the

representation. Petri Nets and, more specifically, Colored Petri Nets (Jensen, 1991) are one mathematical representation of discrete event systems that is particularly well suited to this class of problems and for which a wealth of results has appeared in the last ten years (Jensen and Rozenberg, 1991). Petri Nets are bipartite directed multigraphs appropriate for representing conditions and events – logical operations – that are otherwise difficult to represent in the conventional dynamical models. In addition, when tokens and the concepts of marking and firing are added to the Petri Net formalism, then executable models result that can be used to analyze discrete event dynamical systems.

In Colored Petri Nets, a form of High Level Nets, tokens are distinguishable. A set of attributes is associated with each token, where each attribute can take a number of values. The color of the token denotes a particular choice of attribute values. All the possible combinations of attribute values (all the colors) constitute the universal color set for a particular problem. There is a Color Set associated with each place in the Petri Net: the Color Set specifies the colors of the tokens that may reside in that place. This Color Set is a subset of the universal color set. Similarly, an Occurrence Color Set is associated with each transition. Arcs are inscribed with expressions of the form:

[Boolean]%Expression

where Boolean is a Boolean expression. When this expression evaluates as true, then the arc inscription evaluates to a set of colors according to the normal expression Expression. A transition is enabled, if there exists at least one binding for the variables in the arc inscriptions from the input places such that each input place contains at least as many color tokens as specified by the arc inscription. In addition, when the transition contains a guard function, the condition indicated by the guard function must also be satisfied. The following example, Fig. 3, demonstrates these definitions.

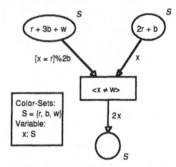

Figure 3. Colored Petri Net example

The universal Color Set contains three colors, red, white, and blue. All three places, denoted by circle nodes, have as their Color Set the universal set; they can hold all types of tokens. A variable x has been defined that can take values in (be bound to) the same universal set S. The transition, denoted by a box node, has a guard function that restricts the variable x from taking the value w if the transition is to be enabled. The inscriptions translate as follows: If the variable x takes the value r and there is at least one r token in the input place on the right, and if there are two blue tokens in the place on the left, then the transition will fire. Two blue tokens will be removed from the left input place, one red token from the right input place, and two red tokens will be generated in the output place. If x takes the value b, then the right place must have at least one blue token. However, the arc inscription on the left will not evaluate as True and, therefore, there is no enablement condition for that binding of the variable. Finally, because of the guard function, there is no point checking what will happen to the arc inscriptions when x takes the value w.

Consider now the task allocation problem. There are n distinct tasks, x_1 to x_n, represented as inputs, in the set X and four Color Sets, each set corresponding to one of the partitions X_i of X. We would like to route an arriving task according to which subset X_i it belongs. The arcs from the input place to each one of the receiving agents are inscribed as follows:

$$[x_j \in X_i] \% x_j \quad \text{for } i = 1 \text{ to } 4 \text{ and } j = 1 \text{ to } n.$$

Note that instead of an arc inscription, a guard function of the form $<x_j \in X_i>$ could have been placed in a transition preceding the i-th agent The places on the right in Figure 4 are the input places of the individual agents; a particular agent will receive a task and will be activated only if a token or tokens belonging to the correct Color Set appear in its input place.

Note that the Colored Petri Net (CPN) formalism makes it very simple to model concurrent execution of several tasks; if tokens x2, x6, and x17 are present in the source place, then the transitions corresponding to agents 1, 2, and 3 will be concurrently enabled and can fire. Furthermore, the formalism allows the modeling of variable-structure systems which can serve as the paradigm for dynamic resource allocation.

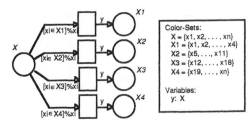

Figure 4. Colored Petri Net Model of Task Allocation

In the following sections, three classes of problems are discussed. The first one draws from systems engineering as applied to Flexible Manufacturing Systems, and illustrates the issues and problems from the perspective of large scale engineering systems. The second draws from the class of problems where small organizations of humans monitor and control engineering systems through information and decision systems. The third addresses man-machine systems per se and the inherent difficulties, experimental and theoretical, of dynamic task allocation.

TASK ALLOCATION PROBLEMS FOR FLEXIBLE MANUFACTURING SYSTEMS

Since the 1980s, control scientists, system engineers, ergonomists, and cognitive psychologists are paying a good deal of attention to the man-machine interactions and the human-computer interfaces for advanced manufacturing systems. Hundreds of papers have been published at dedicated symposium or conference proceedings and in archival journals.

At the 11th IFAC Congress at Tallinn (August 1990), many papers were presented and discussed; T.J. William's (1990) paper, "A Reference Model for CIM from the Viewpoint of Industrial Automation" discussed explicitly man-machine interactions in the Computer Integrated Manufacturing (CIM) environment. In T. Martin's (1990) paper, "Human Software Requirements Engineering for Computer-Controlled Manufacturing Systems", the human requirements for operators of a FMS are considered, and a function/criteria matrix analysis method was suggested as a useful tool for analysis.

At the second International Conference on Human Aspects of Advanced Manufacturing and Hybrid Automation (Honolulu, Hawaii, August 1990), 127 papers were presented. Of these, 16 papers discussed the design of advanced manufacturing systems from the socio-technical point of view in which sociological factors are considered in order to improve the total effectiveness of modern manufacturing systems.

A Flexible Manufacturing System (FMS) may be defined as "an automated system of machine tools provided with transferring, handling, and controlling capabilities via digital computers." FMS can be considered as the latest stage in a continuing process of improving the efficiency of manufacturing through automation. Although the early FMS have been in operation for over ten years, the use of such systems is still a new experience for many enterprises and for the new generation of engineers. While the term "manufacturing" embraces a wide range of engineering activity, FMS were originally developed for metal cutting applications, and this is still by far the major area of their application. For this reason, the discussion of task allocation problems will be based on metal-cutting FMS.

An FMS is not just another off-the-shelf piece of capital equipment. Flexible manufacturing systems are complex combinations of various types of capital equipment which have to be brought together and made to work in coordination. The hardware components in the FMS include equipment from every aspect of manufacturing such as machine tools, robots, inspection stations, storage facilities, material handling and transportation, cutting tools and fixtures stations, or loading and unloading stations, as shown in Figure 5.

The complexity of such systems and the variety of components that they include means that they must be designed by a team comprising various specialists. To match an FMS to a particular enterprise requires that the customer and vendor collaborate closely in many aspects of system design and specification. The design of an FMS has similarities to other design procedures, but to solve the task allocation problems between machines and between humans and machines is the most important aspect of FMS design.

Until recently, the hardware components within an FMS had not been especially designed for FMS. They were designed for independent installation and could include their own computer control for operating as flexible cells. As a result, in designing an FMS, the task allocation to hardware components and the integration problems have received much needed attention. Proper task allocation preserves the control characteristics of each component as if it were operating as a single cell or island, while integration allows the components to act as one system. It becomes the function of the central

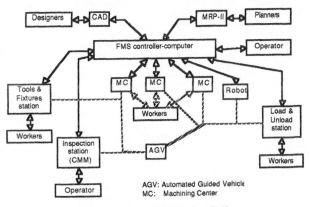

Figure 5. Block diagram of an FMS

computer (the FMS controller computer) to coordinate all hardware and the embedded computer controllers to work together because no component computer is capable to carry out the integration decisions.

For example, the controller for a machining center can perform sequences of tool changes and cuts, enabling the completion of a machining operation on a part automatically. However, in the FMS, the part must be properly referenced or positioned on the machine. Instead of expanding the controller's functions to incorporate reference decisions, these decisions are made in a less direct manner. First, the task for assuring accuracy is allocated to the hardware. This means that all machining centers must be as identical as possible and that all fixtures must be within design tolerances. But design tolerances can never be completely maintained. Alternatively, the task is easily accomplished by machine control: the predetermined offset calculations are made by the computer to account for the lack of repeatability between fixtures, then the controller uses these offsets to compensate and bring the fixture reference to within tolerances. From this simple example, the significance and importance of task allocation between machines and of the integration of hardware during the design of an FMS become obvious.

For task allocation to machines (and good hardware integration), the capabilities and limitations of machines with current technology (as supplied by vendors) should be specified and analyzed carefully and in detail. The main principle for proper task allocation among machines is that the machines (and computers) must be able to work together efficiently and to tolerate and compensate for each other's deficiencies or limitations. A matrix table, in which the functions, capabilities, and limitations of machines and their interactions are listed, becomes a very useful tool for solving the task allocation problem in an FMS.

Because the flexible manufacturing systems of today have not yet reached 100% automation, human operators are needed to carry out complex operations where automation is infeasible, too costly, or unreliable. And in some area of FMS, manual labor is still needed. At this time, the FMS is a manufacturing system which intersperses manual operations amongst the most advanced factory equipment. Such operations as deburring, swarf removal, cutting process monitoring, tool alignment, part fix and fixture alignment on a pallet, loading and unloading, a large part of machine diagnostics, and even computer keyboard manipulation still involve humans as operators. In order to carry out the task allocation between humans and machines in an FMS, a detailed analysis or man-machine interactions and information exchanges is very necessary, and the table suggested by Hu (1990) is useful for reference.

The principle for proper task allocation between humans and machines is: in the FMS, men and machines must help each other to achieve a result for which each one is separately incapable. They must act together, tolerate and compensate for each other's deficiencies. The designer should strive for the highest man-machine synergy possible. Training and education of operators of the FMS is an important additional measure for that purpose. During this design process, the ergonomic principles suggested by Teo (1989) can be used as references.

And a descriptive model which was suggested by Tunälv (1990) is an interesting method that could help describe the task allocation and coordination of tasks between operators within an FMS and between an FMS and the environment in which the FMS functions. The model suggested is based on the theoretical framework developed by Henry Mitzberg (1979). In this context, task allocation, which is called division of labor by Tunälv, determines what specific tasks that are supposed to be performed by each of the operators in the FMS. Task allocation can be horizontal as well as vertical. Horizontal task allocation refers to whether the operator

performs few or many tasks along the production line. Vertical task allocation refers to whether the operator performs managerial and controlling tasks as well. A list of horizontal and vertical tasks considered to be important for the operation of an FMS is shown on Table 1.

Table 1. Division of labor in the FMS at Huge Machines

HORIZONTAL AND VERTICAL TASKS	Function/ Level
Horizontal tasks	
Pre-setting of tools	1/2/3
Changing/Loading tools	1/2
Fixing of workpiece to fixture	1
Loading of fixture to palette	2/3
Unloading of workpiece etc.	1
Manual machining and/ or sub-assembly	1
Transportation, materials handling	3
Preparing maintenance	3
Minor electrical repairs	3
Minor mechanical repairs	3
Major repairs, maintenance	3
Inspection of raw materials/ castings	1/3
Final inspection of components	3
Assist at hiring	2
Supervising of newly hired	1/2
Vertical tasks	
Programming	3
Testing programs	2/3
Changes in programs	1/2/3
Ordering of material	3
Planning, sequencing	2
Reporting performance to PICS	2

1 = Operator in FMS
2 = Foreman/Line hierarchy
3 = Operator outside FMS

The five coordinating mechanisms defined by Mintzberg (1979) have been used for the coordination of tasks. These are: mutual adjustment, direct supervision, standardization of work processes, standardization of work output, and standardization of operator skills. The coordination of tasks performed by the operators is based on a judgment of how important each coordinating mechanism is for the performance of the tasks. If the mechanism is not used for the coordination of tasks, it has been given the value of "1". If the mechanism is of some importance or used seldomly for coordination, it is given the value of "2". Finally, if the mechanism is of crucial importance, it is given the value of "3". Then the coordination matrix can be constructed. Table 2 shows the coordination matrix for a specific case in which the study is limited to two levels: (1) coordination within the FMS, and (2) coordination between the FMS and tasks done by operators outside the FMS.

Such coordination matrix may be useful for structuring the ways in which an organization properly coordinates the work of the FMS at the operational level according to its available human and machine (hardware and software) resources and limitations. The explanation of the entries in Table 2 follows Tunälv's paper (1990).

(1) Coordination of tasks within the FMS: When the production supervisor has made the planning and sequencing in the FMS computer, the flow of components is to a large degree determined. Therefore, the work is to a large extent coordinated through standardization of work processes. However, which operator does a specific task is decided by the operators. This means that the coordination is heavily dependent on mutual adjustment. The production supervisor does not interfere in how, when, and by whom the tasks are performed, i.e., the coordination through direct supervision is not utilized. Since the operators have no real possibility to control the process so that production goals are achieved, coordination of work by standardization of output is limited. Finally, coordination through standardization

Table 2 The use of coordinating mechanisms at Huge Machines

	Direct Supervision	Standard. of work processes	Standard. of output	Standard. of skills	Mutual Management
Tasks within the FMS	1	3	1	1	3
Tasks performed outside the FMS	2	2	3	1	1

of skills is of little or no importance since the operators have only general knowledge and training.

(2) Coordination between tasks in the FMS and external tasks: The coordination of the work performed in the FMS and other functional departments is standardized to a large degree. These are standards for quality and quantity that regulate the contacts between the FMS and the warehouse or the assembly department. hence coordination is achieved through standar-dization of output. When problems arise, for example with delivery time, other mechanisms become more important. Since the supervisor does all the planning and machine loading, it is most reasonable for the assembly or the warehouse department to inform him about any delays, quality defects etc. This is also true of coordination of tasks in the vertical dimension, such as changes in programs and the production of new articles. For coordination between the FMS and functions in the work flow, coordination by direct supervision is of limited importance. For coordination of tasks in the vertical dimension and when problems arise, coordination through direct supervision is of great importance.

To some extent, the work is also coordinated through the standardization of work processes, since the production outcome is automatically communicated through the FMS computer. When the supervisor or the instructor are not in charge (16:00 to 07:00) the operators have greater responsibility for the system. In case of a machine breakdown, they have authority to call maintenance personnel on duty or, if possible, make minor repairs by themselves. However, when the data collection was made,

coordinating the work at another level.

If the connection between long term production plans and the FMS computer could be complemented by a simultaneous change in the task allocation (operators take care about the production planning and sequencing) the tasks would have to be coordinated in completely different ways. This will create a new coordination matrix, as shown in Table 3.

While a number of concepts and techniques have been described for addressing the task allocation problem in FMS, it is clear that one of the main difficulties is the lack of a well developed theory.

TASK ALLOCATION IN ORGANIZATIONS

A concept that has been introduced as the basis for modeling task allocation in information processing and decision making organizations is that of a *role*. The role represents the lowest level of functional decomposition for a particular application; a role must be executed in its entirety by a single agent. Therefore, a role is used to model the execution of such a task by a single resource. The basic model of the role can be represented in block diagram form as shown in Figure 6. It consists of three processing stages and two interaction stages for a total of five stages.

Two types of interactions among roles have been defined: (a) those that are among roles of the same agent, called

Table 3. Coordinating mechanisms when the division of labor has changed

	Direct Supervision	Standard. of work processes	Standard. of output	Standard. of skills	Mutual Management
Tasks within the FMS	1	2	3	1	3
Tasks performed outside the FMS	1	2	3	1	3

the supervisor expressed his dissatisfaction with the way the operators handled these things. The coordination of tasks through mutual adjustment between the FMS and tasks performed outside is limited.

Finally, the coordination through standardization of output between the assembly department and the FMS is, among other things, manifested through the entry of delivery time and batch quantities into the FMS computer. For the work within the FMS, coordination is achieved through standardization of work processes, since the machines are controlled by these objectives. In another perspective, the coordination can be considered to be achieved by direct supervision since it is the production supervisor that decides and enters the goals into the FMS computer. From this it is obvious that (1) the coordinating mechanisms are closely linked and mutually dependent, and (2) a coordinating mechanism at one level may, by its physical manifestation, be transformed to another mechanism for

internal interactions, and (b) those that are between roles carried out by different agents, called external interactions. The latter are the ones that determine the organizational structure or architecture of a system.

A role receives inputs or data x from the external environment (sensors) or from other agents of a system. The incoming data are processed in the first block marked situation assessment (SA) to obtain the assessed situation z. This variable may be sent to other agents, as shown by the outgoing arrow. If the role receives data about the

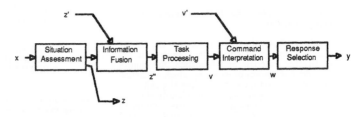

Fig. 6 The five-stage model of a role.

assessed situation form other agents, these data z' are fused together with its own assessment z in the information fusion (IF) stage to obtain the revised assessed situation z''. The assessed situation is processed further in the task processing (TP) stage to determine the strategy v to be used to select a response. However, if hierarchies exist in the organization, the particular role may receive command inputs v' from supervisors (or higher echelon decision makers) that may restrict the strategies available for selecting a response. This is depicted by the use of the command interpretation (CI) stage. The output of that stage is the variable w which contains both the revised situation assessment data and the response selection strategy. Finally, the output or response of the role y is generated by the response selection (RS) stage. Note that if there are no situation assessment and command inputs from other agents, the five stage model reduces to the common two-stage situation assessment and response selection model of the single, non-interacting decision maker (Boettcher, 1982). If there is no information fusion, then the IF stage disappears and the SA and TP stages can be merged into a single SA stage. Finally, if there are no command inputs, then the CI stage disappears and the TP and RS stages can be merged into a single RS stage.

This model of the role is particularly amenable to representation as discrete event system. Indeed, the Petri Net formalism can be used to produce a formal mathematical representation of the role model. If ordinary Petri Net notation is used, the model of Figure 6 takes the form shown in Figure 7. The rounded box enclosing the five stages has no formal meaning; it is used to indicate the internal structure of a role and show explicitly the inputs and outputs of the role.

In accordance with Petri Net conventions, transitions are denoted by solid bars and places are shown as circles. A detailed description of Petri Net modeling of decision making organizations is given in Levis (1988).

In order to carry out a task allocation, whether a fixed one or a variable one that changes either quasi-statically or dynamically, first the functional decomposition shown in Figure 1 must be converted to an functional diagram that describes the operational concept for executing the task. There are numerous ad-hoc methods for accomplishing that based on structured analysis. The method chosen at this time is the Structured Analysis and Design Technique (SADT) as described by Marca and McGowan (1988) and the IDEF$_0$ implementation of it (Meta Software, 1989). The reason for the choice is that it

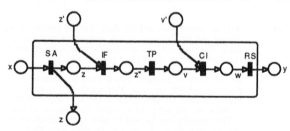

Figure 7. Petri Net representation of the five-stage role model.

is possible to convert the IDEF$_0$ diagram into a rudimentary Petri Net.

The resulting flow diagram indicates the inter-relationships among functions: the inputs and the outputs, whether materials or information, the control inputs and the mechanisms or means by which a function is carried out. This constitutes the functional architecture of the system. The next step is the identification of functions as roles and the assignment of roles to resources - whether humans or machines. Andreadakis (1987) developed an algorithm for generating the allocation.

The functional allocation is based on two concepts - that of redundancy and that of complexity.

The need for redundancy within the organizational structure arises from fault tolerance (or survivability) needs. The degree of redundancy of a Situation Assessment transition is defined as the number of Information Fusion stages that receive the output of the SA transition. The term redundancy is justified by the fact that the same information is communicated to more than one fusion node. A similar definition applies to the Task Processing transition.

The degree of complexity of an Information Fusion transition (and of a Command Interpretation transition) is defined as the number of transitions that send data to the IF transition. The term "complexity" is justified by the observation that the more types of data to be fused at a fusion transition, the more complex the required processing is.

By parametrizing complexity and redundancy, a set of data flow structures are obtained that identify the tasks to be done. The task allocation procedure is based on the Petri Net representation. If tasks are assigned to humans, distinction is made between internal communications and external ones. If the external ones are between humans, a communications medium must be selected and modeled. If the external ones are between humans and machines, the man-machine interface problem must be addressed. The last two problems address the system integration part of the design process.

Note that the presence of redundancy in this design procedure does not lead to variable structures; it allows different resources to undertake the same role. A true variable structure is obtained by using Colored Petri Nets and the corresponding version of the Lattice Algorithm developed by Demaël (1990). In this case, the number of physical resources, or nodes, is fixed, but each node can instantiate a number of different roles depending on the task to be performed and on the values of the parameters that characterize the environment. An illustrative example of such a variable structure is the operational architecture for controlling airport surface traffic. Two ground controllers and a local controller are responsible for the control of surface traffic - the directing of planes from the gates of airport terminals to the runways for take-off and from the runways to the gates after landing. The traffic patterns depend on the wind (which affects the choice of runway) on the location of the terminals and on the schedules of the airlines. Consequently, if a fixed task allocation is used - by assigning specific terminals and gates to each ground controller - the cognitive workload and the resulting stress will vary greatly during a shift leading to possible degradation of performance. On the other hand, a variable structure that adapts the allocation to the two controllers so that the workloads are balanced is preferred. The Lattice algorithm generates such structures; an outcome of the design procedure is a set of requirements for the information system and for the decision support system that the controllers use. The current limitation of the theory is that it does not address the transition from one allocation to the next - each change of allocations is a discrete event. The difficulties associated with truly dynamic situations are addressed in the next section.

THE COGNITIVE ERGONOMICS APPROACH TO TASK ALLOCATION BETWEEN HUMAN AND MACHINE

Originally ergonomics approached the problem of allocating functions between human and machine by using the "Fitts List" approach. This required a listing of what humans do better than machines and vice versa, with a view of allocating roles appropriately once and for all during system design. There is some doubt whether that approach was ever a viable one, and it is certainly inappropriate these days, since there is almost no limit to the technical abilities of advanced computers and machines, if the client is willing to pay the price. Certainly, any list which claims to specify relative superiorities of machines and humans is likely to become rapidly out of date. But more modern approaches see the problem of allocation in fundamentally different terms. There is no such thing as a fully automated system. There is a always a role for the human operator at least as a supervisor and maintenance person. In fact, in the vast majority of cases, humans play a far more active role in the operation of discrete manufacturing systems, including fault management, reprogramming, rescheduling, and the determining of priorities.

It is not really satisfactory to allocate function in advance once and for all, even when powerful mathematical and graphical techniques are available, such as those described in earlier sections of this paper. The reason is that it is never clear in advanced complex systems exactly what role will be required of the operators during the lifetime of the system. If the discussion is confined to advanced discrete manufacturing systems with a considerable degree of automation, there are some roles which are not expected to fall to the human. For example, the problem of determining scheduling over a period of many hours to optimize some objective function is more appropriately allocated to off-line computer programs, because of the extreme difficulty of solving scheduling problems, and the frequent need for brute force numerical solutions. But even here there is room for human intervention. In a recent review Sanderson discussed the literature on appropriate roles, past and present, for humans in scheduling problems of hybrid automated systems (Sanderson, 1989, 1991; Sanderson and Moray, 1990). She pointed out that even when a schedule has been optimized over a long time horizon (of the order of many hours or days), there may remain "windows of opportunity" where intervention by the operator may result in increased use of local optimization which is unavailable to the algorithm or heuristic with a longer time horizon. Moreover, several industrial engineers have recently pointed to the need for a better understanding of the role of the operator and supervisor, and the problems which arise when automation is implemented without adequate consideration of human factors problems (Kellso, 1989; Kuo an Hsu, 1990; Vierling, 1990; Zuboff, 1988).

There are basically two reasons why allocation of function according to some algorithmic method such as Petri Net design will not prove optimal. The first is that the design will be optimized for normal operation, but there will always arise abnormal and fault conditions for which a fixed design is not optimal. A major reason for retaining human operators is to have them handle plant faults. The second is that even during normal plant operation, the operator will find reasons to intervene from time to time, and perform manual intervention in the operation of the plant.

In discrete manufacturing, there is usually relatively little danger from abnormal plant operation, so that one solution to the detection of faults in the operation of the system is simply to shut down the plant. But with so much emphasis on productivity and the meeting of JIT criteria, it would clearly be far better if operators could intervene and perform fault management, not merely fault detection, so that the plant can be kept running. The problem here is that even if expert systems are implemented, they are designed in the context of a well defined description of the properties of the plant. But when faults develop, the nature of the plant changes, and it is impractical, if not impossible, at the time that decision aids and expert systems are designed to allow for unforeseen changes in the coupling of plant variables. Complex systems have self-organizing properties, in the sense used by Ashby (1953), and a fixed design has insufficient requisite variety to cope with changes in plant organization. It is now also well known that the

progressive upgrading of expert systems can leave them extremely vulnerable to "brittle" failures as the number of rules increases, since it is almost impossible to work out the implication of changing rules in a large data base. Hence, the artificial aids themselves come to need human guidance and interpretation.

There will, then, remain important roles for humans in the foreseeable future. Moreover, if the very notion of getting rid of humans is fundamentally mistaken, then what we really need to do is to design symbiotic hybrid systems where the intelligence of the human and the intelligence of the automatic systems can be combined to solve problems during systems operation.*

In order to do this, two problems must be solved. The first is to understand what is the best way to allocate responsibility for decisions between the human and the automated system. This problem was discussed in the previous sections of this paper. Some work has been carried out on the first problem in the US under the rubric "Distributed Tactical Decision Making" with support from the Office of Naval Research. This work brings together system theorists, systems engineers, and mathematically oriented cognitive scientists and psychologists to develop theories and experimental paradigms for designing human organizations that monitor and control complex engineering systems. The results of this work by many researchers from several organizations have appeared regularly in the IEEE Transactions on Systems Man and Cybernetics over a period of ten years.

The second problem is to understand the psychological factors which govern the interaction between humans and their machines. Very little work has been carried out on this problem, despite the long interest in supervisory control which stems from the work of Sheridan in the 1970's (Sheridan and Johannsen, 1976; Moray, Ferell, and Rouse, 1991). Two pieces of research suggest promising directions for further research.

Muir (1987, 1989); Moray and Lee (1990); Lee and Moray (1992); and Lee (1991) have recently completed a series of experiments and modeling efforts to define the psychodynamics of the relations between operators and machines in continuous process control. They investigated the interaction of operators with pumps and heaters in a simulated pasteurization plant in which the components could be run automatically or under manual control. Once the operators were fully trained in both modes of operation, faults were introduced into the system, and operators were required to maximize productivity in the face of the faults. It was found that subjective ratings of trust in the ability of the automatic controls to handle the fault, and also the self confidence of operators in their ability to handle the problem under manual control, could be predicted using a first order time series model. In addition, it was possible to predict when the operators would choose to use manual control and when they would choose to use automatic control as a function of the difference between self confidence and trust. Lee developed the following equations to predict the trust mode of operation at time t as a function of recent past events:

$$\text{Trust}(t) = \Phi_1 \text{Trust}(t-1) + A_1 \text{Productivity}(t)$$
$$+ A_1\Phi_2 \text{Productivity}(t-1) + A_2 \text{Fault}(t)$$
$$+ A_2\Phi_2 \text{Fault}(t-1) + a(t).$$

$$\%\text{Automatic}(t) = \Phi_1\%\text{Automatic}(t-1) + A_1*((T\text{-}SC)(t))$$
$$+ A_2*(\text{Individual Bias}) + a(t).$$

Where A_1, A_2 are weights on productivity and the occurrence of a fault, a is a noise term, and Φ_1, Φ_2 are the

---

* For a discussion of such systems, see Levis (1988b) and the Proceedings of the two IFAC Symposia on Distributed Intelligence Systems (1988 and 1991) published by Pergamon Press.

backshift operators of a time series model. t and t-1 are arguments denoting the current and the immediately prior trial. T is the value of Trust in the automatic controls and SC the value of the human operators' self confidence in their ability to control the system manually. %Automatic is the percent of the trial which is spent in automatic control, and Individual Bias is a term which represents the fact that some human operators seem inherently to prefer to use automatic and others to use manual mode. The two equations account for up to 80% of the variance found in the experiments, providing a very strong model of operator intervention.

Note that such intervention occurred under instructions simply to maximize the productivity, subject to certain constraints such as preventing overflow or depleting the feedstock supply, and subject to the physical limitation on power available, pump rates, and the properties of the automatic controllers. It seems likely that any operators who are asked to supervise a system and to maximize productivity will naturally develop a series of private strategies which they will implement in accordance with their interpretations of the management's meaning of "maximize productivity." In so far as they will be blamed if they do not intervene when they could have managed to keep the plant running, and presumable praised for managing to improve productivity, it is highly unlikely that operators will possibly implement a fixed allocation of function policy. Reason (1990) has discussed the psychology of "violations," or deliberate breaking of rules by operators.

Research along these lines could provide a scientific basis of system designers' decisions on allocation of function. The problem at present is that this set of studies is almost the only one of its kind. We simply lack the knowledge to help designers to implement dynamic allocation policies, even though these are known to be better than fixed allocation policies.

Moray and Sanderson are currently extending this approach into the area of discrete manufacturing systems, where even less information is available. For the time being, mathematical and graphical algorithms will have to be used to allocate function. But it should be clearly understood that such approaches, which have tended to emphasize pure engineering at the expense of cognitive ergonomics and engineering psychology, are inherently inferior to methods which incorporate cognitive engineering to define the capabilities and limitations of the human operator.

The third major problem is to understand how to distribute *responsibility* between humans and machines. The clearest statement of the alternatives was made by Sheridan and Verplank (1978), and is summarized in Table 4.

The pure engineering solution, level 10. is known to be in general infeasible with current systems, and there is no reason to believe that it will become feasible in the foreseeable future. The classical human role in the 19th century was level 1 -- without even the computer being present. This is known to be almost impossible, given the demands and constraints of JIT and FMS. The question which the designer of human-machine systems must answer is, therefore, at what level of organization will we find the best symbiosis between human and machine? As we saw earlier, Sanderson (1989) quotes work which suggests that either a solely human or a solely automated scheduling system can often be outperformed by a suitable hybrid combination. But our understanding of how to choose a hybrid relationship from the list in Table 4, and how to implement it in the light of data from experiments such as those of Muir, Moray, and Lee, is far from adequate at present to guarantee improved human-machine system design. Pure engineering and pure mathematical solutions must be regarded as stop-gap, pending deeper understanding of how to create hybrid human-machine systems, and cognitive ergonomics offers a rapidly maturing technology for a scientific basis to aid system designers: but much remains to be done.

Table 4. A taxonomy of the distribution of responsibility between human and computer.

1.	Human does all planning, scheduling, optimizing, etc., and turns task over the computer merely for deterministic execution.
2.	Computer provides options, but human chooses between them, plans the operations, and then turns task over to computer for execution.
3.	Computer helps to determine options, and suggests one for use, which human may or may not accept before tuning task over to computer for execution .
4.	Computer selects option and plans action, which human may or may not approve, computer can reuse options suggested by human.
5.	Computer selects action and carries it out if human approves.
6.	Computer selects options, plans, and actions and displays them min time for human to intervene, and then carries them out in default if there is no human input.
7.	Computer does entire task and informs human of what it has done.
8.	Computer does entire task and informs human only if requested.
9.	Computer does entire task and informs human if it believes the latter needs to know.
10.	Computer performs entire task autonomously, ignoring the human supervisor who must completely trust the computer in all aspects of decision making.

THEORY AND REALITY

The last section already argued the point that theories that combine cognitive ergonomics for describing the human operator or decision maker together with compatible theories for describing the physical systems with which the humans interact in a variety of ways are needed. The so called "hybrid systems" or "distributed intelligence systems" are beginning to address these issues. Mathematical models of the human decision maker suitable for use in modeling and designing such systems have appeared and have been tested experimentally; see Bushnell et al., Louvet et al., (1988). This work addresses much more the human as decision maker in the context of an information system rather than the human as an operator. Furthermore, only under some circumstances can the role of the human be considered to be supervisory. It is interesting to note that this primarily mathematical approach (which drives experiments) led to the independent derivation of psychological phenomena identified earlier in the cognitive science literature, such as meta-decision making and its effect on cognitive workload and performance (Weingaertner and Levis, 1988). Another consequence of these models is the clarification of the concept of migration of control (Levis and Skulsky, 1990) which highlights the issues raised by Table 4, namely, in real systems consisting of humans and machines, the responsibilities of humans and machines keep shifting depending on the plans and decisions each system component makes. Indeed, domains of dominance can be defined that indicate the conditions when the human is dominant and when the machine is dominant with the human in a supervisory mode.

Until recently, while hundreds of papers were presented discussing human-machine interactions and human aspects of advanced manufacturing systems (including

16

FMS and CIMS), relatively few of them addressed the task allocation problems for FMS. Task allocation and coordination for FMS are complex because they cover all horizontal dimension tasks, which need quite different skill level of the operators, and all vertical dimension tasks which usually need higher knowledge-based skills. The fact is that integrated operation of FMS requires operators with a broad range of skills. Many of these skills may be obtained through experience. The dynamic operation of an FMS provides a very poor learning environment. In fact, the learning curve is so flat in an FMS, that several years are needed to learn appropriate responses to certain situations. To overcome this difficulty, one method is to use task rotation, i.e., to allocate a group of operators to all the tasks of FMS operation. Each operator in the FMS will then be capable to perform any of the required tasks. Of course, not all operators are required to perform the same task simultaneously, so task priorities need to be assigned to each operator. This priority changes over time so that operators rotate through the various tasks. The advantage of rotating task allocation is that no specific job or skill level will become a limiting factor to the FMS operation. This approach of using flexibility in the workforce is consistent with the use of flexibility in the workstations. It is this characteristic of flexibility that has caused the rotating task allocation to become a common form of FMS operation. In this way, task allocation becomes a less important consideration, especially for the horizontal dimension of an FMS.

On the other hand, Petri Nets, which have been claimed to be an ideal modeling tool for FMS operation (among various existing DES modeling methods), have not been used for modeling the human-machine interaction and task allocation in the FMS. Some research efforts in this direction may be meaningful. For example, the monitoring of FMS could be studied by combining Petri Nets with rule based approaches in AI. Perhaps using the similarities between rules and transitions and the similarities between the token player and the inference engine, the complexity of controlling and monitoring FMS can be overcome

CONCLUSIONS

The previous paragraphs illustrate clearly the unfortunate, but not unexpected, phenomenon, that while the problems addressed in this paper are fundamentally interdisciplinary, in reality, they are being addressed almost independently by various disciplines. One would like to see much more interaction and cooperation between the various disciplines. Engineers should integrate cognitive ergonomics in the design process from the very beginning; cognitive scientists should look carefully at the opportunities that new theoretical developments are providing for addressing some of the more vexing problems in carrying out useful experimental investigations.

REFERENCES

Andreadakis, S. K. and Levis, A. H., "Design Methodology for Command and Control Organizations," *Proc. 1987 Symposium on C^2 Research*, National Defense University, Fort McNair, Washington DC, June 1987.

Boettcher, K. L. and Levis, A. H. (1982) "On Modeling the Interacting Decision maker with Bounded Rationality," *IEEE Trans. on Systems, Man, and Cybernetics*, SMC-12, May/June.

Bushnell, L. G., D. Serfaty and D. L. Kleinman (1988). Team information processing: a normative-descriptive approach. In *The Science of Command and Control*, S. E. Johnson and A. H. Levis, Eds., AFCEA International Press, Fairfax, VA.

Demaël, J. J. and A. H. Levis (1990). On Generating Variable Structure Architectures for Distributed Intelligence Systems,. *Proc. 11th IFAC World Congress*, Pergamon Press, Oxford.

Hu, Baosheng, Qu Liangsheng, He Yue, Shao Fuqing. (1988) Challenge to manufacturing Automation. *Proc. Int'l Symp. on Automation and Robotics in Production Engineering*, Xi'an, China.

Hu, Baosheng and Weiliang Le (1990). Man-Machine Systems Consideration for FMS Development. *Proc. Second Int'l Conference on Human Aspects of Advanced manufacturing and Hybrid Automation*. Honolulu, Hawaii, pp. 479 - 484.

Jensen, K. (1990). *Coloured Petri Nets*. Book Report. Aarhus University, Denmark.

Jensen, K. and G. Rozenberg, Eds. (1991). *High-level Petri Nets*. Springer-Verlag, Berlin.

Johannsen, G,, A. H. Levis, and H. G. Stassen (1992). Theoretical problems in Man-Machine Systems and their experimental validation. *Proc. 1992 IFAC Symposium on Man-Machine Systems, The Hague, The Netherlands*.

Kellso, J. R. (1989). CIM in action: microelectronics manufacturer charts course toward systems integration. *Industrial Engineering*, Vol. 21, pp. 18-22.

Kuo, W. and J. P. Hsu (1990). Update: simultaneous engineering design in Japan. *Industrial Engineering*, Vol. 22, pp. 23-28.

Lee, J. D. (1991). *Trust, self-confidence and operator's adaptation to automation*. Unpublished Ph.D. thesis. Engineering Psychology Research Lab. Technical Report EPRL-92-01, University of Illinois, Urbana-Champaign, IL.

Lee, J. D. and N. Moray (1992). Trust, control strategies and allocation of functions in human-machine systems. *Ergonomics* (in press).

Levis, A. H. (1988). Quantitative Models of Organizational Information Structures. In *Concise Encyclopedia of Information Processing in Systems and Organizations*, A. P. Sage, Ed., Pergamon Books Ltd., Oxford.

Levis, A. H. (1988b). Human Organizations as Distributed Intelligence Systems,. *Proc. 1988 IFAC Symposium on Distributed Intelligence Systems*, Varna, Bulgaria,.

Levis, A. H. and S. L. Skulsky, (1990). Migration of control in decision making organizations. *Information and Decision Technologies*, vol. 15.

Louvet, A. C., J. T. Casey and A. H. Levis, (1988). Experimental investigation of the bounded rationality constraint. In *The Science of Command and Control*, S. E. Johnson and A. H. Levis, Eds., AFCEA International Press, Fairfax, VA.

Marca, D. A., and C. L. McGowan (1988). *Structured Analysis and Design Technique*. McGraw-Hill, New York..

Martin, T. (1990). "Human Software Requirements Engineering for Computer-Controlled Manufacturing Systems" in *Proc. 1990 IFAC World Congress*, Pergamon Press, Oxford.

Meta Software, Inc. (1989). *Design/IDEF User's Manual*. Cambridge, MA.

Mintzberg, H., (1979) *The structuring of organizations,* Prentice-Hall, NJ.

Moray, N. and J. D. Lee (1990). Trust and the allocation of function in the control of automatic systems. Engineering Psychology Research Lab. Technical Report EPRL-90-05, University of Illinois, Urbana-Champaign, IL.

Muir, B. M. (1987). trust between humans and machines , and the design of decision aids. *International Journal of Man-machine Studies,* vol. 27, pp. 527-539.

Muir, B. M. (1989). Operator's trust in and use of automatic controllers in a supervisory process control task. Ph.D. Thesis, University of Toronto, Toronto,.

Reason, J. (1990) *Human error.* Cambridge University Press, Cambridge.

Sanderson, P. M. (1989). the human planning and scheduling role in advanced manufacturing systems: an emerging human factors role. *Human Factors,* vol. 31, no. 6, pp. 635-666.

Sanderson, P. M. (1991). Towards the model human scheduler. *International Journal of Human Factors in Manufacturing,* vol. 1, pp. 195-219.

Sanderson, P. M. and N. Moray (1990). The human factors of scheduling behavior. In *Ergonomics of hybrid automated systems II,* W. Karwowski and M. Rahimi, Eds. Elsevier, Amsterdam.

Sheridan, T. B. and G. Johannsen (Eds.) (1976). *Monitoring behavior and supervisory control.* Plenum, New York.

Sheridan, T. B. and W. L. Verplanck (1978). Human and computer control of undersea teleoperators. Man-machine Systems Lab. Dept. of mech. Eng., MIT, Cambridge, MA.

Simon, H. (1972). Theories of bounded rationality. In *Decision and Organization,* C. B. Radner and R. Radner, Eds. North-Holland, Amsterdam, pp. 161-176.

Tunälv, C., (1990). FMS, Division of Labor and Coordinating mechanisms. In *Ergonomics of Hybrid Automated Systems II,* W. Karwowski and M. Rahimi, Eds., Elsevier, Amsterdam, pp. 241-248.

Teo, Moh Gin (1989). Human Resource Issues in the Reinvented factory - Computer Integrated Manufacturing (CIM). *Proc. of Advanced Manufacturing Technology Conference, AMT'89.* Hong Kong, pp. 223-234.

Vierling, A. E. (1990). Machines can only produce as efficiently as the people who operate them. *Industrial Engineering,* vol. 22, pp. 24-26.

Weingaertner, S. T., and A. H. Levis (1988). Evaluation of decision aiding in submarine emergency decision making. *Automatica,* vol. 27.

Williams , T. J. (1990). "A Reference Model for CIM from the Viewpoint of Industrial Automation" in *Proc. 1990 IFAC World Congress,* Pergamon Press, Oxford.

Zuboff, S. (1988). *In the age of the smart machine.* Basic Books, New York.

THEORETICAL PROBLEMS IN MAN-MACHINE SYSTEMS AND THEIR EXPERIMENTAL VALIDATION

G. Johannsen*, A.H. Levis** and H.G. Stassen***

*Laboratory for Man-Machine Systems (IMAT-MMS), University of Kassel (GhK),
D-3500 Kassel, Germany
**Department of Electrical and Computer Engineering, George Mason University,
Fairfax, VA 22030, USA
***Man-Machine Systems Group, Delft University of Technology, 2628 CD Delft, The Netherlands

Abstract. This survey paper focusses on the main theoretical issues in today´s man-machine systems research and applications. The following problem areas are discussed: (1) modelling human performance and mental workload, with identifying the state of the art as well as major methodological difficulties; (2) task allocation and decision support, with a human-centred perspective on cooperative problem solving, integrated automation, and distributed decision making in teams; (3) man-machine interfaces, with outlining some presentation and dialogue issues; (4) design problems, with stressing the need of early active participation of man-machine-systems specialists and the usefulness of guidelines; and, finally, (5) evaluation and experimental validation, with covering laboratory and field evaluations, with covering laboratory and field evaluations, experimental design and validation, as well as model-driven experimentation. The importance of man-machine-systems contributions to the design of better technical systems and their user acceptability is emphasised.

Keywords. Man-machine systems; human performance modelling; mental workload; task allocation; decision support; integrated automation; distributed decision making; Man-machine interfaces; human-centred design; experimental validation.

1. INTRODUCTION

The complexity of industrial processes has enormously been increased during the last decades. This tendency originates from a number of reasons, such as

- the scale enlargement of modern plants,
- the required specifications dealing with the product quality, the energy conservation, the environmental pollution control, and the safety of the plant, and finally
- the progress in process control and informatics creating totally new possibilities.

This essential change in process operation has led to the definition of new human operator tasks. In just two decades, human manual control became much less important and human supervisory control developed as the main concept for man-machine interactions. The tasks of the human supervisor are predominantly cognitive ones, and contain at least the following six subtasks [Sheridan, 1980, 1992]:

- the monitoring of all data presented to the human supervisor,
- the learning and interpretation of the data presented,
- the process tuning or set-point control or teaching of the process in normal circumstances,
- the intervention into the process for instance during abnormal process conditions,
- the fault management during malfunctioning of the plant, and finally,

- the planning, such as for starting-up and shutting down the plant.

All the above changes in complexity and requirements of industrial process plants, in automation concepts and technologies related to new systems, computer and software engineering approaches, as well as in the tasks of the human operator(s) showed the evidence of a number of theoretical problems in man-machine systems more clearly. Concepts of cognitive engineering and human-centred design approaches evolved as possible answers to these problems. They suggest that human information processing behaviour as well as knowledge and goal structures of the human operator(s) have to be investigated. The results need to be applied in advanced automation and decision support systems as well as in advanced man-machine interfaces, in order to guarantee enough flexibility and job satisfaction for the human operator(s) which are prerequisites for safe systems operation.

A discussion session with the same title as this survey paper was held during the 11th IFAC World Congress at Tallinn, Estonia, in August 1990. The three authors of this paper and S. Franzén from Sweden were the panelists in that discussion session. This survey paper includes some of the material presented at Tallinn, but further elaborates it and tries to focus on the main theoretical issues in today´s man-machine systems research and application. Particularly, the following problem areas were identified:

- modelling human performance and mental

workload,
- task allocation and decision support,
- man-machine interfaces,
- design problems, and
- evaluation and experimental validation.

These problem areas will be discussed in the subsequent sections of this paper. Evaluations and/or experimental validations of suggested improvements in man-machine systems are indispensible because, at last, we are dealing with an applied and experimental field of research and development when trying to contribute to better man-machine systems in industrial and public domains.

Methodologies to evaluate and to validate human behaviour have been developed in great detail in manual control. However, the state of the art is totally different in supervisory control [Stassen, Johannsen, Moray, 1990]. The high complexity of the plant and the vague definition of the subtasks of the human supervisor as mentioned above may be of crucial significance. It is therefore that some special attention is focussed on the phenomenon of complexity [Stassen, 1992], before one is able to touch again the evaluation and validation of supervisory control behaviour.

In a recent article, it was stated that complexity is directly related to the combination of four factors [Tolsma, 1991]: great numbers, diversity, coupling, and interaction. This statement is difficult to defend; with a system theoretic apoproach, one might come to the following analysis. The factor of great numbers deals with the number of functions a system satisfies, whereas the factor of diversity may be interpreted in two different ways: either it means the different functions, or it means flexibility. Finally, the factors of coupling and interaction are synonymous with static and dynamic interaction. Hence, one can argue that complexity is mainly dependent on two factors: the number of functions and the interaction, and the design of a system is the optimisation of a criterion where complexity and flexibility are weighted.

By analogy with the well-known Richter-scale for earthquakes, one can define a scale from 0 until 7, describing the degree or intensity of complexity, whereby complexity is assumed to be dependent on at least two factors. Therefore, one defines lines where the complexity is constant, in the surface described by the coordinate system of degree of interaction over number of functions. These lines are called the Iso Complexity Curves, the ICCs. In Fig. 1, seven ICCs are proposed; in the left lower corner, the 0-ICC is drawn, in the right upper corner, the 7-ICC. It will be suggested that this classification of complex systems can be taken as a

basis for developing a methodology to standardise evaluation and validation studies.

2. MODELLING HUMAN PERFORMANCE AND MENTAL WORKLOAD

The basic problem in modelling human performance and mental workload is to conceptualise a theory that allows one to smoothly and consistently move from the characteristics of the human to those of the system (the machine). While human and machine have essentially different characteristics (and there is no implication that we need machine-like models of humans) we should be able to describe or model those in a consistent analytical framework. Many of the mathematical formalisms that we have are well-suited to the description of machines and machine behaviour, but are not well-suited to the description of human behaviour.

Nevertheless, it can be stated that, with regard to human performance models, the goal of conceptualising such a theory has been achieved in manual control to a certain extent [McRuer, Jex, 1967; Kleinman, Baron, Levison, 1971; Stassen et al., 1990]. However, the mathematical formalisms for describing human supervisory control behaviour are still far away from what is desired; not any human performance model, mathematical or verbal, is able to fit human supervisory control behaviour, even not for just one of the six sub-tasks as mentioned in section 1. One of the major reasons may be the fact that all human performance modelling is based on the exact knowledge of the system dynamics, on well-defined tasks, and on knowledge of the statistics of the disturbances [Kalman, Bucy, 1961; Conant, Ashby, 1970; Francis, Wonham, 1975]. However, to what extent are these presumptions or hypotheses correct? In fact, one may even question whether it is at all possible to describe human supervisory control behaviour by general and accurate models [Stassen et al., 1990]. The large amount of models which has been reported for the six subtasks, and which are based on control and identification theory, detection and queueing theory, fuzzy set theory, expert and knowledge based systems, and artificial intelligence concepts, obviously show that no unique solution will be found.

Another view which supports what just has been argued, is based on the three-level concept of human cognitive behaviour [Rasmussen, 1983, 1986], where a distinction was made between a target-oriented Skill-Based Behaviour, SBB, a procedure oriented Rule-Based Behaviour, RBB, and a goal-controlled Knowledge-Based Behaviour, KBB. This qualitative model can be used in order to classify human operator tasks. It is widely accepted that manual control tasks and intervention tasks in stationary process conditions mainly lead to SBB, whereas monitoring, interpreting, and teaching a plant in stationary as well as in non-stationary conditions are most often RBB. Fault management and planning are not easily to be classified [Johannsen, 1988]. They require not only knowledge of the tasks to be performed, but also appeal to the creativity and intelligence of the human operator, hence, they lead to KBB. The Table 1 shows that different subtasks are performed at different cognitive levels and, as a consequence, simple models as validated in manual control certainly cannot be expected to be developed for supervisory control. Another aspect that complicates human performance modelling is the often extremely vague definition of tasks, a situation where the human´s creativity is explicitly required. This yields some very contradictory and intriguing statements, i. e.:

- If human creativity can be modelled, one can no longer speak about creativity; in fact one might say

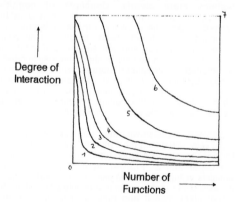

Fig. 1. Complexity defined as depending on the numbers of functions and the interaction. In the figure, the Iso Complexity Curves are indicated from 0 to 7. (after Stassen, 1992).

Table 1. The relation between human behaviour and human operator tasks. The number of * indicates the significance of the relation (Stassen, Johannsen, Moray, (1990).

Human Operator Tasks / Human Behaviour	Manual Control	Supervisory Control			
		Intervention	Interpreting Monitoring	Teaching	Fault Manag. Planning
SBB	* *	* *	*	*	*
RBB	*	*	* *	* *	*
KBB			*		* *

that the cognitive level of KBB is shifted to that of RBB.
— A good and practical system design should be robust for errors, however, one hopes for errors in order to stimulate human creativity.

A design is always the result of some kind of optimisation according to a more or less well-defined criterion, in this case a criterion where performance of human and system are weighted with the costs, i. e. mental workload. Unfortunately, the state of the art in mental workload modelling is even worse than that in performance modelling. Since Moray organised a workshop on mental workload, theory and applications in 1978 [Moray, 1979], a lot of work has been performed, but a profound theory is still not developed [Hart, 1987; Stassen et al., 1990]. The only goal achieved at this moment is a number of applicable and rather consistent measuring methodologies [Wickens, 1984; Hart, Wickens, 1990]. New critical reviews of workload research may be helpful but will not solve the problem. It is necessary to derive a conceptual model of mental workload on the basis of the integration of all existing material. The performance-workload-reliability-satisfaction relationships need to be better clarified within this conceptual model context. In addition, the question to what extent the internal representation, i. e. the knowledge available to the operator, plays a role in mental workload and also in performance is quite important. The whole issue is of high practical use for evaluating alternative man-machine system designs, particularly also those with computer-aided decision support.

From the literature, Table 2 can be reconstructed [Hart, 1987]. The table gives a global view on the state of the art in available human performance models and mental workload measurement techniques.

Table 2. Models for human performance and mental workload, as a function of the process control mode.
SBB = Skill-Based Behaviour; RBB = Rule-Based Behaviour; KBB = Knowledge-Based Behaviour.
+ = available; o = to a certain extent available; - = not available.

Process Control Mode	Perform. Level	Models Available	
		Perform.	Mental Workload
Manual Control	SBB	+	+
Stationary Supervisory Control	SBB	+	o
	RBB	o	-
	KBB	-	-
Non-Stationary Supervisory Control	RBB	o	-
	KBB	-	-

3. TASK ALLOCATION AND DECISION SUPPORT

3.1 Human-Centred Perspective of Cooperative Problem Solving and Integrated Automation with Embedded Decision Support

Considering the six subtasks of the human supervisor, one may argue that the most important tasks are fault management and planning. In particular for these tasks, creativity of the human supervisor is required or, in terms of Rasmussens´s three level model, KBB occurs. This yields that there is a need to support the human supervisor not only at the SBB- and RBB-levels but also at the KBB-level; see Fig. 2.

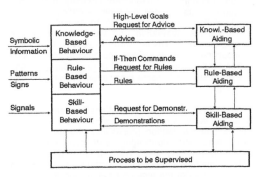

Fig. 2. Decision support systems classified according to the three-level model of Rasmussen (after Sheridan, 1987).

All kinds of decision support systems, including expert systems and knowledge-based systems, have been proposed and developed. It is amasing to see how much effort has been put into the design of those systems, and how little time was spent for their evaluation and validation. Several factors can be raised as possible causes. To start with, scientists and in particular computer scientists, are designers of tools and systems; it is their joy forever to create new systems. Probably, and hopefully, this attitude will change at the moment one realises that their products are not used the way it was expected. Often, these decision support systems degrade the human to someone who has to supply missing data or knowledge, and has finally to accept one problem solution or may be allowed to select among a few alternatives.

A much more prominent problem is the user-machine problem. As said before, the human supervisor needs to possess a correct internal representation of the process to be supervised, of the tasks to be performed, and of the disturbance statistics. In analogy, the designer of the decision support system needs to know how the user will use such a system, hence, the designer of the decision support system needs to build up an internal

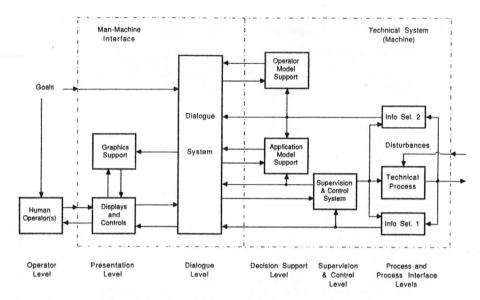

Fig. 3. Extended Operator (User) Interface Management System structure for dynamic technical systems (after Johannsen, 1992).

representation of the behaviour of the user. This aspect is often forgotten or neglected; honestly, it should be noted that this problem is extremely difficult to handle, and solutions are nowadays still not available. Another factor is the evaluation and validation of decision support systems. The fact that this activity is very time-consuming, where a general approach is not available, often leads to not doing it at all [Van Daalen, 1992]. Later, in section 6, some more attention will be paid to this important aspect.

The function and task allocations in complex man-machine systems need to consider different behavioural levels and information processing phases of human and automatic controlling and problem solving [Johannsen, 1991, 1992]. Within an extended Operator (User) Interface Management System architecture, the functionalities of the technical system include those of the technical process, the traditional automation (computer supervision & control), and the computer

DECISION SUPPORT SYSTEM
(Computer Support)

	Operator Model Support	Application Model Support	
Knowledge-Based Behaviour	Procedural Support	Plant Management	
	Plan Recognition	Fault Diagnosis	
Rule-Based Behaviour	Error Evaluation	Heuristic Control	
Skill-Based Behaviour			

	Supervisor Level	
	Group Level	
	Component Level	

HUMAN OPERATOR   SUPERVISION & CONTROL SYSTEM (Computer Control)   TECHNICAL PROCESS

Fig. 4. Relationship between human operator, decision support system, supervision & control system, and technical process (from Johannsen, 1992).

support (decision support based on application and operator modelling). This architecture ist shown in Fig. 3 and, with more detailed hierarchical levels, in Fig. 4.

The term integrated automation is also used for this concept. The man-machine interface is left out in Fig. 4 for reasons of simplicity; it is outlined with its presentation and dialogue levels in Fig. 3 and will be discussed in section 4 of this paper.

The function and task allocations have to be achieved under consideration of all the possibilities available with the capabilities of the human operator(s), the different decision support systems, and the supervision & control system. Starting from a human-centred perspective, a cooperative problem solving approach is very much needed. The cooperation is meant between human and machine problem solvers. In any case, the main responsibility lies with the human problem solver. Thus, several types of cooperation with different degrees of machine subordination can be thought of. Examples as shown in Fig. 4 are the application-oriented decision support systems for fault diagnosis, particularly those which combine human and computer reasoning processes based on test procedures, value histories, and transition networks [Borndorff-Eccarius, 1990]. Other examples which are more operator-oriented decision support systems were developed as operator´s associates for procedural support and intent recognition [Rubin at al., 1988; Sundström, 1991; Johannsen, 1992]. In order to improve such cooperative systems, the knowledge structures and the problem solving strategies of humans as well as their cognitive biases and deficits need to be better understood for a number of different application domains. The difficult and time-consuming techniques of knowledge elicitation in cognitive task analyses have to be further elaborated [Johannsen, Alty, 1991] and have to be applied in order to serve the requirements of cognitive modelling of the human problem solver. The conceptual models, such as that of Rasmussen [1983, 1986], have to be further elaborated, extended and/or modified, and at least partially validated. The final objective can be a form of dynamic task allocation for cooperative problem solving where the machine serves the human by knowledge enhancement and by interactive procedural support.

New approaches exist for evaluating human errors on the key-stroke level, for checking the consistency of input sequences, and for intent or plan recognition [Heßler, 1989; Johannsen, 1992]. These approaches can be based on models of correct task execution. For further development, it will be important to introduce the natural fuzziness of task execution into such models. Furthermore, the question exists whether the recognition of human errors on higher cognitive levels will be at all possible, or whether we reach limits of our understanding. Understanding human error as well as the limits of our understanding is of major practical use with respect to our responsibility for the control of complex man-machine systems. It contributes also to the cooperative problem solving approach mentioned above.

The introduction of systems that learn or, at least, have their knowledge base change with time creates further problems. Consider the case that several identical decision support systems — each with the same knowledge base — are installed in different departments of an organisation. In the beginning, there is indirect coordination among the decision support systems. But as time passes and the knowledge base evolves differently in the different systems, the indirect coordination will disappear with the possibility now present that conflicts and instabilities may be created within the organisation. We need a theoretical framework that allows us to allocate resources — human and machine resources — dynamically to tasks while maintaining information consistency and indirect coordination. To develop such a theory we need to understand both the human aspects and the machine aspects and merge them together — not the one being an afterthought of the other. Such a theory should apply equally well to systems consisting of a single human with a single machine and to teams of humans with distributed machine systems.

3.2 Distributed Decision Making in Teams of Humans with Computer Support

A more complex problem arises when computer support, in the form of a decision support system (DSS), is introduced in an organisation. Let the organisation represent a team and let the task allocation be such that the team engages in distributed decision making. This means that the decision problem has been decomposed into individual subproblems, each assigned to a different organisation member. However, these subproblems must be solved in a coordinated manner to produce the organisational response. The decision support system may contain a number of features: it can contain a centralised or distributed data base that the individual decision makers can access; it probably includes a communication system that allows the different team members to interact — to communicate directly and to exchange data and results; and it may also include decision aids that can support the tasks of individual members. The presence of the decision support system raises two design problems, both at the theoretical and the practical level.

The first problem, and the most obvious one, is the interaction of each individual team member with the decision support system itself, seen as a decision aid. The original task, allocated to that particular team member, will now be shared with the machine. This is a classic allocation problem: what does the human do and what does the machine do? The result of this allocation leads to the definition of the Human-Computer Interaction (HCI) problem. Note that designing the HCI can follow general theoretical principles based on theories of cognitive processing and on human factors research, but that the actual design is very dependent on the answer to the allocation problem between the human and the machine. Embedded in this well defined problem is a more subtle one: the manner in which the decision aid

is used. While there is a whole spectrum of interactions, it is convenient to describe three distinct modes. Clearly, the decision maker may ignore the decision aid — he or she may not query the data bases or not use the information it provides. While this is an extreme case, it is a definite possibility, especially if the support system has been designed in such a manner that the task can be performed without the use of the decision aiding capabilities of the support system. The other extreme occurs when the interaction between the human and the machine follows prescribed steps — the operational concept requires that the human interacts with the machine in order for the next step in the process to be enabled. The machine is not just supporting the decision maker, it carries out some parts of the decision making task itself. Some forms of this interaction may be labelled supervisory control — the decision aid acts as a controller and the human decision maker intervenes by issuing commands to the aid or resetting parameters, as appropriate. The third mode is, of course, the more interesting one. The decision aid plays a consulting role: the decision maker consults the aid and then he may or may not use the information or recommendations provided as to the preferred course of action.

The three modes are shown schematically in Fig. 5. The bold lines indicate the flow of task related activities. In Fig. 5 (a) the decision maker carries out his task without interacting with the aid or without availing himself of the services of the decision support system (as indicated by the dashed lines). In Fig. 5 (b) one of the possible variants between an integrated operation between a human and a machine is shown. The task is received by the decision maker, he carries part of the task, then control is transferred to the machine by the human. The machine does its part of the task and transfers information and control to the human for completion of the task. The dashed line between the two-part model of the decision maker indicates that the human cannot bypass the machine and carry out the task without it.

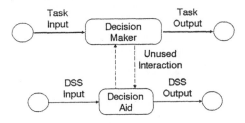

Fig. 5 (a). Unaided decision maker.

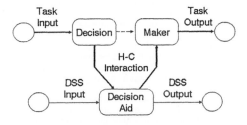

Fig. 5 (b). Integrated human-computer interaction.

The third Fig. 5 (c) shows the case that the human-computer interaction is optional: the decision maker chooses to query the decision aid (or consult it) and then uses or ignores the output of the aid. This gives more flexibility to the decision maker but it also increases his mental workload: in addition to processing the task itself, he now has to make additional decisions regarding the proper use of the decision aid and of the results it produces. This phenomenon is called meta-decision making. The three types of interactions have been studied by Weingaertner (1989), Grevet (1988) and Perdu (1991).

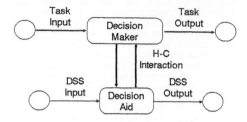

Fig. 5 (c). Optional human-computer interaction.

The second problem has to do with the interactions among the decision makers. There are three types of interactions that can be identified in distributed decision making. Each one of them imposes different information exchange and communications requirements that the support system must meet. The first type is the exchange of information regarding the state of the system, as seen by each decision maker. This can be called information sharing and helps indirectly in coordinating the distributed decision making by broadening the situation assessment of each decision maker: it reduces the differences in their estimates of the system state. Note that in a distributed system, each decision maker may be seeing only a few of the state variables and some aggregates of the rest of the variables. The second type is the exchange of results or results sharing. In this case, a decision maker communicates to others what he is going to do (or what decision he has made) rather than describing them what he perceives the situation to be. Note that the advantage of this approach is that it reduces the communication requirements and corresponds to communicating the individual control actions to the members rather than their part of the state. This approach provides a more direct means of coordinating, provided the various decision makers share common mental models of the system's operation so that they can interpret the control actions correctly. The reduction in communication is counterbalanced by an increase in the cognitive processing — the interpretation of the controls. In very broad terms, it can be interpreted as inferring from the control what the state was that led to that decision. The third type of exchange is a direct one in which one decision maker issues orders or commands to others. In most distributed decision making environments, there is an embedded hierarchical structure — a shift supervisor, a controller, a foreman, or a manager. These are special communications that restrict the options of the subordinates; they reduce the cognitive load of the subordinates by reducing the number of alternatives (sometimes down to a single alternative as is the case of a direct command) and by reducing the mental workload associated with situation assessment. It may increase, however, the workload of the supervisor, mitigating the advantages of distributed decision making. Computer support in an organisation makes possible all these types of interactions that support coordination. This is a design problem that cannot be left to chance given a task allocation solution. This type of problem is being investigated both theoretically and experimentally under the rubric of coordination in decision making organisations [Lu and Levis, 1991; Wang et al. 1991].

4. MAN-MACHINE INTERFACES

A key problem in man-machine systems is the determination of the boundary between the human and the machine and the manipulation of the boundary — the interface — to "match impedance". The questions of where that boundary should be and whether the boundary should be fixed or flexible with respect to a wide variety of tasks or even changing with time, contribute to the issues of task allocations and decision support systems (as described above). Additionally, the issues of the organisation of the information flow across the boundary and the form of the information presentation are very important. The investigations of the man-machine interfaces in the narrower sense have to deal with these problems. Following the UIMS (User Interface Management System) concept, dialogue issues can be separated from presentation issues [Alty, Johannsen, 1989; Johannsen, 1992]. The presentation issues are concerned with displays and controls as well as with knowledge-based graphics support. The dialogue system may also contain knowledge-based modules (e. g., so-called dialogue assistants) which relate to the different subsystems of the machine — including a number of different decision support systems — and deal with local knowledge about these subsystems for observing and controlling the dialogue information flow, and for picking up loose ends of dialogue in case of priority interrupts.

A lot of basic research results exists in the literature [e. g., Gilmore et al., 1989; Diaper et al., 1990; Fejes et al., 1992]. Nevertheless, there is still a need of practical guidelines for the design of information presentation in control rooms, particularly when hundreds of pictures are needed by the operators. This aspect will be taken up again in section 5. One question is: what is the appropriate mixture between parallel and serial information presentation? This question is mainly of concern for the organisation of the graphical output of the technical system. However, it may also be of interest for the organisation of inputs into the technical system, particularly when the inputs are also graphically supported, e. g., as with touch-screens [Hartz, Borys, 1990].

5. DESIGN PROBLEMS

A thorough systems approach in design is necessary which suggests a goal-oriented top-down procedure supported by bottom-up means [Gilmore et al., 1989; Rouse, 1991]. The design of man-machine interfaces and of decision support systems requires at least a well-defined task. Then, on the basis of such a task description, one may design interfaces and support systems by taking into account, on the one hand, the dynamics of the system to be supervised and, on the other, the capabilities and limitations of the human supervisor. Hence, the underlying problem is again that of a responsible allocation of tasks between man and machine, as has been elaborated already in section 3. This task allocation is dependent on the amount of automation intelligence applied [Sheridan, 1980] and can be of help at all three cognitive levels which are shown in Fig. 2. A taxonomy of the introduction of automation intelligence in the design of a controlled process is shown in Fig. 6; it elucidates that the choice of the degree of automation intelligence is to be based on the cognitive processes the human supervisor is able to achieve.

For a long time, it was believed that the automation level could be chosen a-priori by the designer. However, in the practice of the real world, it is experienced that this approach fails; it is too simple. The needs for help which have to be supplied by the designer may differ among the human operators and the expected process control modes, and they may vary in time. Therefore, the designer cannot decide and just make a choice. At least, the system to be developed should be flexible, and probably it is even necessary to make the system adaptive to the circumstances.

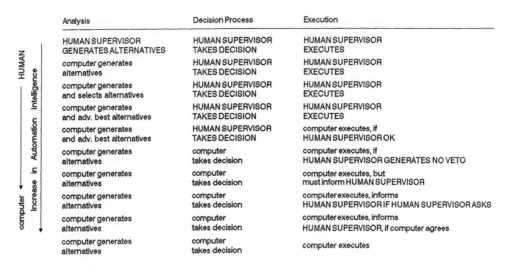

Analysis	Decision Process	Execution
HUMAN SUPERVISOR GENERATES ALTERNATIVES	HUMAN SUPERVISOR TAKES DECISION	HUMAN SUPERVISOR EXECUTES
computer generates alternatives	HUMAN SUPERVISOR TAKES DECISION	HUMAN SUPERVISOR EXECUTES
computer generates and selects alternatives	HUMAN SUPERVISOR TAKES DECISION	HUMAN SUPERVISOR EXECUTES
computer generates and adv. best alternatives	HUMAN SUPERVISOR TAKES DECISION	HUMAN SUPERVISOR EXECUTES
computer generates and adv. best alternatives	HUMAN SUPERVISOR TAKES DECISION	computer executes, if HUMAN SUPERVISOR OK
computer generates alternatives	computer takes decision	computer executes, if HUMAN SUPERVISOR GENERATES NO VETO
computer generates alternatives	computer takes decision	computer executes, but must inform HUMAN SUPERVISOR
computer generates alternatives	computer takes decision	computer executes, informs HUMAN SUPERVISOR IF HUMAN SUPERVISOR ASKS
computer generates alternatives	computer takes decision	computer executes, informs HUMAN SUPERVISOR, if computer agrees
computer generates alternatives	computer takes decision	computer executes

On the left margin: HUMAN — Increase in Automation Intelligence — computer

Fig. 6. Task allocation between human (denoted by capital letters) and computer (after Sheridan, 1980).

The present way of designing man-machine interfaces, supervision and control systems, decision support systems, and/or control rooms are as follows. The industrial management defines the product specifications, on the basis of which the process engineer develops the process. Then, control engineers design the control and safety systems, hardware as well as software, and they often also determine which information will be displayed by the interfaces in the control room, and which variables can be controlled by the human supervisor. Moreover, they often indicate the lay-out of the control room and its man-machine interfaces. Sometimes, the human factors, ergonomics, and man-machine system disciplines can contribute to the design at this phase. However in many cases, no influence of these experts is requested at all. Thus, in the overall design process, ergonomists or man-machine-system specialists usually have to cope with a man-machine interface already fully determined by the control engineer, in this way leaving open only questions concerning the definition of supervisory tasks, the recruitment, training and selection of potential operators, and the classical human factors aspects of displays and controls. This procedure will lead to conflicts in the allocation of tasks between operator and machine, among others because the man-machine systems discipline is called in at a too late phase of the design process of the control room. In fact, the design process should be achieved just the other way around, in starting with a correct allocation of the tasks to be performed. This fact, combined with the consequences of the ongoing automation, the vital role of the human in supervision of the plant and the necessary multi-disciplinary approach, have asked for the development of man-machine guidelines.

The purpose of these guidelines is not to replace the specialised knowledge of the designer, nor to come to a standardisation which often delays progress in new developments of hardware and software. However, it is just to point out the need for a particular multi-disciplinary expertise at all levels of the decision making in the design process [Stassen, 1984; Gilmore et al., 1989]. In this way, for example, the EWICS-guidelines were developed by the members of the European Workshop on Industrial Computer Systems during the period from 1981 until 1986 [Scanlon, 1981; Wirstad, 1982; Stassen et al., 1986]. Of course, these and other guidelines may be helpful, but they are certainly not an adequate answer to the task allocation problem. What is needed is a methodology for building up models to describe the allocation of tasks.

A methodology, full of expectations, is probably the one which is based on coloured Petri nets [Boettcher and Levis, 1982; Levis et al., 1992]. With this method, tasks, task interactions, and organisational structures can quantitatively be described [Levis, 1988]. In addition, the effectiveness and the safety of systems supervised by a human supervisor can be estimated. The strong point of the method is that system and human behaviour are modelled with just one method in the same way.

6. EVALUATION AND EXPERIMENTAL VALIDATION

6.1 Laboratory and Field Evaluations

There is a very limited experience in the evaluation and validation of man-machine interfaces and decision support systems in an operational environment. Extensive laboratory evaluations will have to precede evaluations of systems in the real world. In particular, it may be helpful to survey the experiences in the field of medical diagnosis with reference to the use of expert and knowledge-based systems [Van Daalen, 1992]. The evaluation of man-machine interfaces and all kinds of decision support systems is cursed with problems, many of which cannot easily be solved. Direct practical problems are the large amount of variables, the wide range of these variables, the typical human cognitive properties such as memory and adaptivity, the discrete and abrupt changes in behaviour, the very seldomly occuring disturbances, and the time-consuming processes of carrying out evaluations or of even running more formal experiments. All these problems yield that rather seldomly evaluations and validations of newly designed man-machine interfaces and decision support systems are performed at all — even more seldomly they are performed in a rigorous way.

However, evaluation studies should be a continuous process which has to be carried out in parallel with the development of any man-machine system and, particularly, of its man-machine interface and all technical subsystems. In the beginning, evaluation will be directed towards obtaining information for the improvement of the system; later on, an evaluation will be aimed at investigating whether the system satisfies the design objectives. It also implies that the evaluation will move from informal studies to formal investigations. The informal studies include expert evaluations in cognitive task analyses, knowledge elicitation, and participative system prototyping [Johannsen and Alty,

1991]. However, some of these techniques have been developed to a fairly high degree of formality during the last years. The formal evaluation of knowledge-based systems in field studies will involve the validation, whereby it should be shown to a satisfactory degree that the behaviour of the system is correct with respect to the specifications prescribed for the system [Shwe et al., 1989]. Furthermore, the user interaction should be investigated, and the performance of a system in a laboratory environment should be evaluated, where safety, accuracy, reliability and transferability are important. Here, it should be noted that formal field evaluations should only be carried out after demonstrating superior performance in a laboratory experiment. Only a few systems have reached the level of maturity which is required to justify a field evaluation. Examples, in the medical field, which have undergone field evaluations may be found in the literature [Adams, 1986; Sutton, 1989; Murray, 1990].

Field evaluations should involve investigations of the efficacy of the system in the target environment by measuring the impact of system use on the quality of the user's decisions and the impact of the system on the results of the user's task performance. Furthermore, the human-machine interaction is of utmost importance and should be evaluated also. It will involve assessments of the acceptibility and the usability of the system and of the quality of the human-computer interface. Cost-benefit analyses should be carried out, and the impact on the organisation and the social environment as well as legal and ethical aspects should be assessed.

Just to give an example, in the evaluation of the efficacy of a decision support system, the following classification can be used as a framework of the evaluation: Selection of the goals for evaluation; the experimental setup, containing the choice of the experimenal unit, the specification of the control group, the selection of the test input, the selection of the way to enter the test data, the specification of a standard of performance, and the specification of the variables to be measured; the analysis of the results; and finally, the bias and the confounding variables. Altogether, it is clear that an evaluation study is a large and very time-consuming activity. It is, therefore, obvious that a standard methodology has to be developed in order

— to decrease the evaluation time and effort;
— to compare results obtained from different processes, process control modes, and process circumstances, hence from processes with different degrees of complexity; and
— to compare results obtained at different locations and laboratories.

In order to come to such a standard experimental set-up, one may consider the concept of the Iso Complexity Curves. At the moment that different processes — different in terms of the number of functions to be supervised and of the degree of interaction — can be standardised by the Iso Complexity Curves mentioned earlier, one could develop a methodology to measure performance as a function of the Iso Complexity Curve value. Hence, standard processes, such as the Generic Power Plant or the William's plant for the chemical industry, can be categorised by the Iso Complexity Curve number and, thus, by a certain performance index. To what extent this philosophy can be extended to mental workload studies is difficult to be estimated but, at least, it should be given a trial.

In those cases where the degree of complexity is very high, measures can be achieved in order to decrease the degree of complexity experienced, such as the reduction of the number of components [Tolsma, 1991], cancelling out the interactions by decoupling [Van der Veldt and Van den Boomgaard, 1985], presenting information about the interaction by means of predictive displays [Veldhuijzen and Stassen, 1977; Johannsen und Govindaraj, 1980], or by using artificial intelligence concepts as done in integrated automation or plant-wide control [Johannsen, 1991, 1992].

It is believed that by simulation for each Iso Complexity Curve, performance and mental workload indices can be found and, thus, a more or less standardised procedure can be achieved. However, there are many very practical problems in executing this type of research; to mention the most important ones:

— The facilities to simulate a control room, a process to be supervised, a series of realistic disturbances, and the tasks to be achieved are rather expensive, in terms of hardware and software. In addition, it requires a very good feedback from industry.
— The training of potential subjects is very time-consuming; often, one cannot replace the actual human operators by just students.
— In particular, human performance on very seldomly occuring disturbances, as it may happen in nuclear power plants, is extremely difficult to be estimated. Note that time scaling is mostly not allowed at all.
— The practical definitions of performance and of mental workload indices are difficult to be achieved.

6.2 Experimental Design and Validation

Experiments involving several human decision makers and computer simulations are generally complex and difficult to design and control. One of the difficulties is the large number of parameters involved. A second difficulty is determining which parameters should be held fixed, and which ones should be varied and over what range. A third problem is that the participation of human decision makers in the experiment precludes the execution of a large number of trials. While there is a strong tradition of experiments with single decision makers, no useful guidelines for experiments with decision making teams are available. In most cases, if the task is complex, the organisation is a simple one, while if the organisational structure is complex, the task is very simple. This is not a satisfactory situation, especially when the real problems today involve small decision making teams employing complex procedures to monitor and control complex engineering systems. Examples of problems include air traffic control, the control of energy systems (nuclear power plant control), or the control of a highly automated manufacturing plants.

In order to design a controllable experiment in a complicated environment, a model is necessary for determining appropriate variables which ought to be controlled or measured. In the physical sciences and in engineering, procedures have been developed over the years for using models to design experiments. For example, to address the problem of many parameters and the problem of physical scale, dimensional analysis has been developed and is routinely used in mechanical and aeronautical engineering (Hunsacker and Rightmire, 1947). This well established technique from the physical sciences has been extended to include the cognitive aspects of the distributed decision making environment (Jin and Levis, 1992).

A dimension is the measure which expresses a physical variable qualitatively. Fundamental dimensions are the primary dimensions which characterises all variables in a physical system. For example, length, mass, and time are fundamental dimensions in mechanical systems. A dimension such as length per time is a secondary or derived dimension. If the dimension of a physical variable cannot be expressed by the dimensions of others in the same equation, then this variable is dimensionally independent.

The foundation of dimensional analysis is the Principle of Dimensional Homogeneity, which states that if an equation truly describes a physical phenomenon, it must be dimensionally homogeneous, i.e., each of its additive terms should have the same dimension. The basic theorem of dimensional analysis is the π theorem, also called Buckingham's theorem which states that if a physical process is described by a dimensionally homogeneous relation involving n dimensional variables, such as

$$x_1 = f(x_2, x_3,..., x_n) \qquad (1)$$

then there exists an equivalent relation involving (n-k) dimensionless variables, such as

$$\pi_1 = F(\pi_2, \pi_3, ..., \pi_{n-k}) \qquad (2)$$

where k is usually equal to, but never greater than, the number of fundamental dimensions involved in the x's. It is clear from comparing Eqs. (1) and (2) that the number of independent variables is reduced by k, where k is the maximum number of dimensionally independent variables in the relation. The π theorem provides a more efficient way to organise and manage the variables in a specific problem and guarantees a reduction of the number of independent variables in a relation.

To apply dimensional analysis to decisionmaking organisations, the fundamental dimensions of the variables that describe their behaviour must be determined. A system of three dimensions is shown in Table 3 that is considered adequate for modeling cognitive workload and bounded rationality. The approach was applied first to a 1988 experiment [Louvet et al., 1988] to demonstrate the application of dimensional analysis to the experimental investigation of bounded rationality. The purpose of the single-person experiment was to investigate the bounded rationality constraint. The experimental task was to select the smallest ratio from a sequence of comparisons of ratios consisting of two two-digit integers. Two ratios were presented to a subject at each time. The subject needed to decide the smaller one and to compare it with the next incoming ratio until all ratios were compared and the smallest one was found. The controlled variable (or manipulated variable) was the amount of time allowed to perform the task. The measured variable was the accuracy of the response, i.e., whether the correct ratio was selected.

Table 3. Dimensions for Cognitive Problems

Dimension	Symbol	Unit
Time	T	sec
Information (uncertainty)	I	bit
Task	S	symbol

The controlled variables were the number of comparisons in a sequence, denoted by N, and the allotted time to do the task, denoted by T_W. For each value of N, T_W took m values with constant increment. The performance was considered to be accurate or correct if the sequence of comparisons was completed and if the smallest ratio selected was correct.

The hypothesis to be proved in that experiment was that there exists a maximum processing rate for human decision makers. When the allotted time is decreased, there will be a time beyond which the time spent doing the task will have to be reduced, if the execution of the task is to be completed. This will result in an increase in the information processing rate F, if the workload is kept constant. However, the bounded rationality constraint limits the increase of F to a maximum value F_{max}. When the allotted time for a particular task becomes so small that the processing rate reaches F_{max}, further decrease of the allotted time will cause performance to degrade. The performance drops either because all comparisons were not made or because errors were made. It was hypothesised that the bounded rationality constraint F_{max} is constant for each individual decision maker, but varies from individual to individual. The bounded rationality constraint can be expressed as

$$F \leq F_{max} = G / T_W^* \qquad (3)$$

where T_W^* is the minimum allotted time before performance degrades significantly. G and T_W^* vary for different tasks, but F_{max} should remain constant for a decision maker, no matter what kind of tasks he does. Therefore, significant degradation of performance indicates that the allotted time approaches T_W^*. Observation of this degradation during the experiment allows the determination of the time threshold and, therefore, the maximum processing rate, provided the workload associated with a specific task can be estimated or calculated.

The application of dimensional analysis reduces the complexity of the equations and facilitates experimental design and analysis. Properly designed experiments using dimensional analysis provide similitude of experimental condition for different combinations of dimensional variables which result in the same value of p's. Similitude reduces the number of trials needed. This is a major advantage when the physical (dimensional) experimental variables cannot be set at arbitrary values. This technique addresses all the problems stated earlier but one: the problem of selecting the ranges of variables to be changed during the experiment so that the behaviours of interest can be captures by a limited number of trials. This is a practical consideration given the difficulties associated with using many teams of trained decision makers over many trials. Model-driven experimental design is one useful approach to this problem.

6.3 Model-Driven Experimentation

The experimental design starts with the development of a model that represents the experiment. The model contains Models of the individual decision makers, the decision strategies that each team member has available, the protocols of interactions between team members and the task the team is to execute. The model also contains in an explicit form the experimental variables that are to be manipulated and the measured variables that are to be collected in the actual experiment. Finally, the model contains the data processing and analysis techniques that will be used in determining the validity of the hypotheses being investigated. In simple terms, the model simulates the experiment and generates pseudo-data. One key consideration in the above approach is the need for the assumptions embedded in the model to be consistent with those underlying the actual experiment. This is difficult to achieve in general, but it is essential. For example, human dicision makers may exhibit behaviour that is not included in the model (they may select strategies outside the range assumed in the model or they may use past experience — memory — while the model assumes memoryless operation).

Running a small-scale pilot experiment may be necessary at this stage to determine the ranges for certain time variables for which no theoretically derived quantitative estimate is available. For example, it may not be possible to predict for a particular cognitive task the minimum amount of time necessary to carry it out correctly. A separate single person experiment may be run for this type of task to obtain an estimate of the minimum time; this estimate is then used in the team model where members carry out similar cognitive tasks.

Simulations and analysis are then used to predict team performance for various values of the controlled variables. Specifically, the simulations may indicate for what combination of parameter values a discrete change in behaviour can occur. This information is used in the actual experiment by assuring that these sets of values are included in the experimental program. In the actual experiment, it is the dimensionless groups that are manipulated. This is very useful because the parameters that reflect human characteristics cannot be manipulated at will; however, the environmental parameters can be manipulated by the experimenter so that the dimensionless group takes the required set of values. The predictions from analysis and simulation results lead then to the formulation of quantitative hypotheses that can be tested experimentally.

This engineering-based methodology is feasible and practical and was shown to be applicable to the design of model-driven experiments in team decision making; it has actually led to the formulation of hypotheses and to the design of an experiment, and has guided the collection and analysis of data for proving or disproving the hypotheses [Jin and Levis, 1992]. A special class of organisations was considered — a team of well-trained decision makers repetitively executing a set of well-defined cognitive tasks under severe time pressure. The cognitive limitations of decision makers imposed a constraint on organisational performance. Performance, in this case, was assumed to depend mainly on the time available to perform a task and on the cognitive workload associated with the task. When the time available to perform a task is very short (time pressure is very high), decision makers are likely to make mistakes (human error) so that performance will be degraded.

The experimental results showed, as predicted, that the accuracy of the response decreases as the available time to do a task is reduced. The variation in performance is less between different teams than between different individual decision makers within a team, which means *that organisational performance is more predictable than individual performance*. It has also been found that degradation of accuracy as a function of available time is less abrupt for organisations than for individuals. *Interaction among decision makers in an organisation compensates for differences in individual performance characteristics.* These results are consistent with the predictions from the theoretical model. Furthermore, *the critical value of the ratio of response time to available time for doing a task is an observable measure of the bounded rationality constraint.* Therefore, this ratio, which is observable from simple experiments, can be used in development of future organisation designs as a key design parameter.

In this section, it has been shown that the use of computer simulation and analytical models, when combined with engineering techniques for large scale experimentation, provide a feasible approach for addressing the need for reliable and repeatable experimental data on the behaviour of team members executing complex tasks in the context of an engineering system, an environment in which team members interact with the system and with each other through man-machine interfaces.

7. CONCLUDING REMARK

This paper clearly shows that we face several unsolved problems in our multidisciplinary field of man-machine systems. Basically, the most severe ones are determined in some way by our limited understanding of the higher cognitive behaviour of humans. Of course, we need a lot of future research and good ideas to improve this situation.

However, we also outlined here the many facets of available knowledge. Maybe, the main statement which we can make is that all experts of our discipline should be courageous and strong enough for applying our know-how in real systems and for training designers and managers about it. If we hesitate because of our limited knowledge, other people will continue to build new technical systems with much less or even no knowledge about the man-machine relationship.

REFERENCES

Alty, J. L., and G. Johannsen (1989). Knowledge based dialogue for dynamic systems. Automatica. 25. pp. 829 - 840.

Adams, I. D., et al. (1986). Computer aided diagnosis of acute abdominal pain: A multicentre study. British Medical Journal. Vol. 293. pp. 800 - 804.

Boettcher, K. L. and A. H. Levis (1982). On modelling the interacting decision maker with bounded rationality. IEEE Trans. on Systems. Man. and Cybernetics. SMC-12, Nr. 3, pp. 334 - 343.

Borndorff-Eccarius, S. (1990). CAUSES - State-Based Diagnosis Support Expert System. ESPRIT-GRADIENT P857, Report No. IMAT-MMS-11, Labor. Man-Machine Systems, University of Kassel (GhK).

Conant, R. and W. R. Ashby (1970). Every good regulator of a system must be a model of that system. Int. J. Systems Science. Vol. 1. pp. 89 - 97.

Daalen, C. van (1992). Field evaluation of knowledge based systems. The medical system PLEXUS. Proc. 5th IFAC/IFIP/IFORS/IEA Symposium on Man-Machine Systems. The Hague, The Netherlands.

Diaper, D., D. Gilmore, G. Cockton, and B. Shackel (Eds.), (1990). Human-Computer Interaction INTERACT'90. North-Holland, Amsterdam.

Fejes, L., G. Johannsen, and G. Strätz (1992). A graphical editor and process visualisation system for man-machine interfaces of dynamic systems. The Visual Computer (will appear).

Francis, B. A. and W. M. Wonham, (1975). The internal model principle of linear control theory. Proc. IFAC 6th World Congress. Boston, MA, Paper 43.5.

Gilmore, W. E., D. I. Getmann, and H. S. Blackman (1989). The User-Computer Interface in Process Control. Academic Press, Boston.

Grevet, J.-L. and A. H. Levis (1988). Coordination in organizations with decision support systems. Proc. 1988 Symposium on C2 Research. Monterey CA.

Hart, S. G. (1987). Research Paper and Publications 1981 - 1987. NASA Technical Memorandum 10001b.

Hart, S. G. and C. D. Wickens (1990), Workload assessment and prediction. In H. R. Boohrer (Ed.), MANPRINT: An Approach to systems Integration. Van Nostrand Reinhold, New York, pp. 257 - 296.

Hartz, J. O., and B.-B. Borys (1990). Task Analysis. Internal Report, BRITE-EURAM FANSTIC Project, Labor. Man-Machine Systems, University of Kassel (GhK).

Heßler, C. (1989). Use of a task model for detection of operator errors and for flexible task allocation in flight control. Proc. ESA/ESTEC Workshop Human Factors Engineering: A Task-Oriented Approach. Noordwijk.

Hunsacker, J. D. and B. G. Rightmire (1947). Engineering Applications of Fluid Mechanics.

McGraw-Hill, New York.

Jin, V. Y. and A. H. Levis (1992). Impact of organizational structure on team performance: Experimental findings. To appear in Command Control and Communications: Advanced Concepts and Paradigms. Carl R. Jones, Ed., AIAA Press, Washington, DC.

Johannsen, G. and T. Govindaraj (1980). Optimal control model predictions of system performance and attention allocation and their experimental validation in a display design study. IEEE Trans. Systems. Man. Cybernetics. SMC-10. pp. 249 - 261.

Johannsen, G. (1988). Categories of human operator behaviour in fault management situations. In L. P. Goodstein, H. B. Andersen, and S. E. Olsen (Eds.), Tasks. Errors and Mental Models. Taylor & Francis, London, pp. 251 - 258.

Johannsen, G. (1991). Integrated automation in complex man-machine systems. Proc. Eur. Contol Conf. ECC 91, Grenoble, France, pp. 1013 - 1021.

Johannsen, G., and J. L. Alty (1991). Knowledge engineering for industrial expert systems. Automatica. 27. pp. 97 - 114

Johannsen, G. (1992). Towards a new quality of automation in complex man-machine systems. Automatica. 28. March.

Kalman, R. E. and R. S. Bucy (1961). New results in linear filtering and prediction theory. Trans. on ASME. J. of Basic Eng.. 83. pp. 85 - 107.

Kleinman, D. L., S. Baron, and W. H. Levison (1971). A control theoretic approach to manned vehicle systems analysis. IEEE Trans. on Automatic Control. AC - 16. pp. 824 - 832.

Levis, A. H. (1988). Quantitative models of organizational information structures. In: A. P. Sage (Ed.). Concise Encyclopedia of Information Processing in Systems and Organisations. Pergamon Books Ltd., Oxford.

Levis, A. H., N. Moray, and B. Hu (1992). Task allocation problems and discrete event systems. Proc. 5th IFAC/IFIP/IFORS/IEA Symposium on Man-Machine-Systems. The Hague, The Netherlands, 9 p.

Louvet, A.C., J. T. Casey and A. H. Levis (1988). Experimental investigation of the bounded rationality constraint. In Science of Command and Control: Coping with Uncertainty. S. E. Johnson and A. H. Levis, Eds., AFCEA International Press, Fairfax, VA.

Lu, Zhuo and A. H. Levis (1991). A colored Petri net model of distributed tactical decision making. Proc. 1991 IEEE International Conference on Systems. Man. and Cybernetics.

McRuer, D. T. and H. R. Jex (1967). A review of quasi-linear pilot models. IEEE Trans. on Human Factors in Electronics. HFE No. 3. pp. 231 - 249.

Moray, N. (1979). Mental Workload: Its Theory and Measurement. Plenum Press, New York, 500 p.

Murray, G. D. (1990). Assessing the clinical impact of a predictive system in severe head injury. Medical Informatics. Vol. 15. No. 3, pp. 269 - 273.

Perdu, D. M. and A. H. Levis (1991). Analysis and evaluation of decision aids in organizations. Automatica. 27. March,

Rasmussen, J. (1983). Skills, rules and knowledge; signals, signs and symbols; and other distinctions in human performance models. IEEE Trans. on SMC. SMC-13. No. 3, pp. 257 - 266.

Rasmussen, J. (1986). Information Processing and Human-Machine Interaction. New York, North-Holland, 215 p.

Rouse, W. B. (1991). Design for Success. Wiley, New York.

Rubin, K. S., P. M. Jones, and C. M. Mitchell (1988). OFMspert: Inference of operator intentions in supervisory control using a blackboard architecture. IEEE Trans. Systems. Man. Cybernetics. 18. pp. 618 - 637.

Scanlon, J. P. (Ed.), (1981 a). Guidelines for the Design of Man-Machine Interfaces. Level 0. EWICS/TC-6, SINTEF, Trondheim, Norway, 14 p.

Scanlon, J. P. (Ed.), (1981 b). Guidelines for the Design of Man-Machine Interfaces. Level 1 (Early Project Stages). EWICS/TC-6, SINTEF, Norway, 40 p.

Sheridan, T. B. (1980). Computer control and human alienation. Technology Review. MIT, pp. 60 - 76.

Sheridan, T. B. (1987). Supervisory control. In G. Salvendy (Ed.), Handbook of Human Factors. Wiley, New York, pp. 1243 - 1268.

Sheridan, T. B. (1992). Telerobotics. Automation and Human Supervisory Control. MIT Press, Cambridge, Mass.

Shwe, M. A., S. W. Tu, and L. M. Fagan (1989). Validating the knowledge base of a therapy planning system. Methods of Information in Medicine. Vol. 28. pp. 36 - 50.

Stassen, H. G. (1984). Man-machine guidelines: A need in the design of man-machine interfaces. Proc. European Seminar on Industrial Software Engineering and the EWICS. Freiburg, FRG, pp. 181 - 188.

Stassen, H. G. (1989). On the modelling of manual control tasks. In: G. R. McMillan, et al., (Eds.), Applications of Human Performance Models to System Design. Plenum Press, New York and London, pp. 107 - 122.

Stassen, H. G. (1992). To what extent does a process operator experience an industrial process to be complex? (in Dutch) In: M. J. A. Alkemade (Ed.). Take Advantage of the Complexity: Man. Technology. Information and Organisation. SAMSSOM-Publ. Comp. Alpha 200 p., in press.

Stassen, H. G., G. Johannsen, and N. Moray (1990). Internal representation, internal model, human performance model and mental workload. Automatica. Vol. 26. No. 4, pp. 811 - 820.

Stassen, H. G., and E. T. van Ravenzwaay (Eds.), (1986). Guidelines for the Design of Man-Machine Interfaces. Level 3 (Design of Interfaces and Control Room). EWICS/TC-6, DUT, Delft, The Netherlands, 72 p.

Sundström, G. A. (1991). Process tracing of decision making: Am approach for analysis of human-machine interactions in dynamic environments. Internat. J. Man-Machine Studies. pp. 843 - 858.

Sutton, G. C. (1989). How accurate is computer-aided diagnosis? The Lancet. Oct. 1989, pp. 905 - 908.

Tolsma, H. (1991). In what way do large, complex systems remain under control? (in Dutch) Ingenieurskrant. Nr. 15, pp. 8 - 9.

Veldhuyzen, W. and H. G. Stassen (1977). The internal model concept: An application to modelling human control of large ships. Human Factors. Vol. 19, pp. 367 - 380.

Veldt, R. J. van der, and W. A. van den Boomgaard (1985). Predictive information in the control room. Proc. 5th Eur. Conf. Human Decision Making and Manual Control. Berlin, FRG, pp. 249 - 266.

Wang, W. P., D. Serfaty, P. B. Luh, and D. L. Kleinman (1991). Hierarchical team coordination: Effects of teamstructure. Proc. 1991 Symp. on Command and Control Research. SAIC, McLean, VA 22102.

Weingaertner, S. T. and A. H. Levis (1989). Evaluation of decision aiding in submarine emergency decisionmaking. Automatica. 25. May.

Wickens, C. D. (1984). Engineering Psychology and Human Performance. Merrill, Columbus, OH.

Wirstad, J. (Ed.), (1982). Guidelines for the Design of Man-Machine Interfaces. Level 2 (Specification Stages). EWICS/TC-6, SINTEF, Trondheim, Norway, 91 p.

HUMAN RELIABILITY IN PROCESS CONTROL DURING MAL-FUNCTIONING – A SURVEY OF THE NUCLEAR INDUSTRY WITH A CASE STUDY OF MAN-MACHINE SYSTEM DEVELOPMENT

K. Monta, H. Hayakawa and N. Naito

Nuclear Energy Division, Toshiba Corporation, Kawasaki, Japan

Abstract. Improvement in human reliability is of great concern in modern high technology systems. Efforts in the nuclear industry, mostly in Japan were surveyed and classified into improvement in human work situations and that in man-machine systems. In the former, human error prone situations were investigated using training simulators and from actual near-misses and incidents in real operating plants. The results have been used to improve operator training and other work situations.

The objective of man-machine system improvement is to optimize the total performance of nuclear power plants such that automation and supervisory control by human operators will be designed to reduce human errors as well as to effectively utilize human ingenuity.

Computerized operator support systems were surveyed based on the operators' role in plant operational safety. The main support functions considered here are a kind of computerization of operation manuals revised after the Three Mile Island-2 accident, and mainly supports the operators' rule-based behavior.

The development of intelligent man-machine systems was then aimed at to support operators in their knowledge-based behavior to cope with unforeseen abnormal events, considering the recent advance in artificial intelligence and cognitive engineering. A case study is described.

Keywords. Human factor; man-machine systems; error analysis; nuclear plants; cognitive systems; artificial intelligence; supervisory control; intelligent machines.

INTRODUCTION

Improvement in human reliability is of great concern in modern high technology systems such as those in aviation, nuclear power and chemical industries, since potential risks due to human errors are high in such systems and thus should be reduced constantly (Reason, 1990).

This paper reports the result of a survey on efforts to improve human reliability in the nuclear industry, mostly in Japan. These efforts were classified into improvement in human work situations and that in man-machine interfaces.

In the former, human error prone situations were investigated using full-scale training simulators and from investigations of near-misses and incidents in real operating plants. Various performance shaping factors (PSF) were analyzed using collected data, and especially the team behavior of the operating crew were analyzed. The results of these investigations and analyses have been used to improve operator training, operational procedures, man-machine interface design guidelines, and working situations in general.

Developmental efforts to improve man-machine systems for nuclear power plants, starting from a modern CRT-based control room, have been carried out since the Three Mile Island-2 (TMI-2) accident in 1979.

The objective is to optimize the total performance of nuclear power plants as man-machine systems such that automation and supervisory control by human operators will be designed to reduce human errors as well as to effectively utilize human ingenuity.

Regarding the recent advances in information technology and cognitive system engineering, development of intelligent man-machine systems seems promising through principle driven, top-down approaches to improve human reliability.

As a case study, developmental efforts are described in order to support human operators in their knowledge-based behavior to cope with unforeseen abnormal events.

SURVEY ON RESEARCH FOR IMPROVEMENT IN HUMAN WORK SITUATIONS

Investigations of plant operators' behavior during incidents and accidents have been carried out during training courses with full-scale training simulators. A joint study by a group of Japanese BWR owners of six utilities and two vendors (Kawano and others,

1991; Yoshimura and others, 1991) started in 1984. In the first phase, the behavior during standard training scenarios at the BWR Training Center was collected and analyzed. Data from 216 cases were analyzed. Behaviors deviating from a standard procedure table have been classified according to an error taxonomy, and a database has been developed. Some errors were observed in high work load situations such as after an emergency plant trip. Some errors were due to poor man-machine interface designs.

In the second phase, data for studying the operators' internal cognitive actions such as event detection, diagnosis, and task decision were collected in addition to the above from scenarios with multiple failure events. Data from 62 cases involving 18 crews were collected. The operators' cognitive processes were traced by Rasmussen's ladder model on the basis of verbal protocol and data from interviews. Differences in team performance were observed and a relationship with communication difference was found.

A similar study has been performed by a joint study by a Japanese PWR owners' group and vendors(Fujita and others, 1991). The primary objective was to develop countermeasures against human factor deficiencies found in current nuclear power plant control room operating crews. The secondary objective was and still consists of obtaining a better understanding of the PSFs to ensure a sound technical basis. The study incorporated a full-scale replica training simulator in which job performance data were collected. Psychological instruments were used to collect PSF data on control room operators. The relationships between scores on the PSF instruments and job performance measures were statistically analyzed and influential PSFs were identified. Previous job experience and operator background factors from six PSF areas were found most contributing. Cognitive abilities and stress factors also played a role. It has been concluded that in a cross- cultural study (Western vs. Orient, engineering vs. psychology, etc.), researchers must be aware of potential problems, and therefore the conduct of a pilot study is highly recommended.

Team performance of an operating crew is practically very important, but studies of this sort seem few due to the complexity in analyzing the behavior of plural subjects and multiple performance indexes regarding team members. A full scope training simulator-based study(Sasou, Nagasaka and Yukimachi, 1991) was performed to determine the characteristics of a team activity coping with simulated abnormal operating conditions. The behavior of the team was described by their communication flow, action flow, and position flow during the simulated events. Analysis was performed from these data to derive a "goal" as the team's final purpose, "subgoal" as the team's current purpose and "strategy" as the team's particular plan implemented to achieve the subgoal. Then, strategy decision patterns were derived by an analysis using the communication flow, and four types were identified: two were the leader initiative type and two were the follower initiative type. An investigation of the communication contents further revealed that these patterns were related to the characteristic of the subgoals for which the strategy was implemented. That is, the subgoals were classified into an immediate problem solving type and a goal-oriented problem solving type. Usually during plant transients, the former is first set up to settle a disturbed plant state into a safer state, and then the latter takes over to determine more specific strategies. The follower initiative strategy type is used for the former type subgoal due to its rule-based, familiar behavior pattern. For the latter type subgoal, the leader takes the initiative and communications between the crew are more frequent and contribute to a higher cognitive level.

Improvement in simulator training (Miyazawa and others, 1991) has been pursued based on these research results as well as on its own experience. As an example, training for breaking away from a stereotyped monitoring patterns in daily operation is introduced. Parameters indicating the plant condition during daily operation represent more of a static behavior than dynamic ones. It is necessary to strengthen the watch on plural parameters in order to break away from habitual monitoring patterns in daily operation, such as single point observation. Scenarios for this purpose have been developed from lessons in domestic and overseas plants.

Team training by actual crews for themselves is important. In the training, even auxiliary equipment operators are given training regarding operation in the main control room under the guidance of a senior personnel. This enables them to understand the relation between operations being executed in local areas and the responses in the main control room. They can grasp the meaning of their work and items in this way when communicating with the control room.

A device referred to as THURMOS(Total Human Performance Monitoring System) has been developed by the Central Research Institute of the Electric Power Industry (CRIEPI) in order to evaluate the effects of automation on operator performance (Inoue, Yoshino and Takano, 1991). It monitors external signs of human performance such as physiological functions, vision behavior, and human movements of up to three persons simultaneously to evaluate objectively (while maintaining freedom from restrictions) the relationship between human performance and errors during operations. The results of such tests have affirmed that continuous and simple measurement of changes in the heart rate and respiration rate was possible and had a definite correlation with the type of task, the burden of information assigned to the subjects, and variation in the operators' mental workload, and the wakefulness level under that load.

CRIEPI has developed, since 1986, a Japanese version of the Human Performance Enhancement System, J-HPES (Takano, Sawayanagi and Kabetani, 1991) which was originally developed by INPO (Institute of Nuclear Power Operation) in the USA. It is a system which first appropriately analyzes and evaluates the entire relationship of events and facts, and then the causes of human errors in the course of a trouble caused and/or affected by human acts in nuclear power plant operations and maintenance, and secondly proposes effective and concrete preventive countermeasures for them. It is offered in an easily usable procedure form. Appropriate considerations were paid in the conversion to native operation management structures and organizational management styles while maintaining database compatibility with INPO-HPES. The system was distributed to Japanese utility companies.

SURVEY ON MAN-MACHINE SYSTEM DEVELOPMENT

Current Status of Man-Machine System Technology

At present, new control rooms based on computer driven CRT displays are in operation in recent operational nuclear power plants in Japan (Tomizawa and others, 1984; Nitta and others, 1991), Germany (Aleite, and Gremm, 1991), Canada (Olmstead, and others, 1991), and so on. For instance, ten units are already in operation in Japan and another nine units are under construction. Their operational experiences have already been reported (Makino, and others, 1987).

In the meanwhile, various computerized operator support systems (COSS) have been developed based on the lessons learned from the TMI-2 accident in various countries (Cain, 1986). In Japan, COSSs were developed supported by the Ministry of International Trade and Industry (MITI) during 1980-1984, which widely support the operators' rule based behavior including plant state identification and synthesis of operational procedures during plant incidents and accidents (Monta and others, 1985; Masui and others, 1985; Higashikawa and others, 1985).

The design of the COSS support functions (Monta and others, 1985) was based on the role of the operators proposed by Corcoran (1981).
The first role of the operators is to keep the plant set up so that it will respond to disturbances correctly. It is clear that the functionability of standby safety systems is mandatory, and computer monitoring of such functionability is expected to be a valuable support for this role coupled with a computerized support for surveillance tests.

The second role is to minimize the occurrence frequencies and severity of adverse events. In order to minimize the frequencies of adverse events, manual operations which need complex procedures and a long execution time should be supported by computers to reduce operational errors. A control rod operational guidance system has been developed for this purpose.
Another possibility is to maximize the plant operational margin during normal operation. An operational margin monitoring system was developed for this purpose.
Disturbance analysis systems (Bastl, 1988) are intended to detect disturbances occurring in nuclear power plants at their early stages and inform the plant operators regarding disturbances so that operators can take corrective actions based on pre-planned cause-consequence trees for predictable disturbances. This technique has been applied to BWR power plants to support operators in their second role (Naito and others, 1987).

The third role is to assist the operation of the installed plant equipments in implementing safety functions during an event.
The post trip operational guidance system is to support this role. That is, the functions of this system are to monitor the performance of the engineered safety feature, to identify the plant status, and to guide appropriate corrective actions for plant operators to achieve safe plant shutdown (Takizawa and others, 1989). The extension and revision of emergency operating procedures after the TMI-2 accident were useful in developing these operational guides due to their symptom based nature.
Similar systems have been developed such as the emergency operating tracking system (Sun and others, 1991), SAS-II (Owre and others, 1991), and a computerized procedure, COMPRO (Lipner and Orendi, 1991).

In addition, information integration was aimed at where monitoring, diagnosis, and action information was organized to match the operators' cognitive process such as an overview display using a large, projection type CRT display. Experimental validation, using a full scale plant simulator, was implemented to evaluate the effectiveness of the supports.

Prototypes of an advanced new control room were developed for the advanced boiling water reactor, ABWR, by integrating these techniques developed for COSSs into CRT-based new control rooms utilizing practical operational experiences (Iwaki, 1991). The first ABWR unit will be in commercial operation in 1996, and the system is now under detailed design.
Similar advanced control rooms are being developed in France (Bozec and others, 1991), Canada (Olmstead and others, 1991), and APWR in Japan (Saito and Tani, 1991).

Survey of Intelligent Man-Machine System Development

Considering the objective of the man-machine system mentioned in the introduction of this paper and the recent advance in cognitive engineering (Woods and Roth, 1988) and artificial intelligence along with the ever increasing importance of human factor issues in nuclear power plant operation and maintenance, it seems both necessary and promising to develop intelligent man-machine systems which support human operators in their knowledge-based behavior and allow them to cope with unanticipated abnormal events, including recovery from erroneous human actions.

Several authors have pointed out by reviewing previous automation experiences in various industries that technology centered automation which has thus far prevailed has caused new human error types and new categories of system break-down, and have suggested that human centered automation should be realized (Woods, 1988).
The main features of the latter are;
-A human is the leader of plant control instead of being a machine counterpart.
-A human is the supervisor of an automated plant.
-A human is the error discoverer and recoverer, supported by appropriate tools.
This means that a knowledge based behavior should be established, when necessary, by appropriate cognitive tools for the operators.

The IAEA advisory group on "Balancing Automation and Human Action in Nuclear Power Plants" recommends that a systematic design approach should be adopted in order to improve task allocation between automation and supervisory control (Bastl and others, 1991). This should also be applied to man-machine system design in order to achieve the above.

The characteristics regarding the human cognitive process should be considered to realize this design approach, especially in case of abnormality in the plant.

The cognitive demands which operators would face in such events are due to the complexity of the situations caused by dynamism in the events, integration and interconnections in the plant functional structure, uncertainty in regard to information accuracy, and risk in the situations (Woods, 1987).

Thus, the problems are;

(1) What is the best way for presenting the complexity.

(2) What is the most effective resource the operators should use to deal with the complexity.

J.Rasmussen (1986) proposed a top down design approach for man-machine systems based on cognitive work analysis.

First, the identification of control requirements and available resources should be carried out. Decision making tasks, such as plant state identification, priority setting for operational goals, etc., are identified. Then, the information requirement is analyzed based on the tasks, the operator's preferred mental strategies and cognitive mechanisms brought to use by the operator. Finally, task allocation should be determined in order to realize human/computer cooperation and resource profile matching.

A typical use of computer support derived by the above approach is to provide supports which correspond to the major cognitive decision making tasks, such as information processing for the state of affairs, decision supports, and supports for implementing the intention for action.

Therefore, the following three main functions are proposed for intelligent man-machine systems as the basis of the present survey, namely;

(1) an ecological interface: support of the operators' direct perception and analytical reasoning (Rasmussen and Vicente, 1989; Vicente, 1990),

(2) a machine problem solver: support of the operators' cognitive resource, and

(3) a robust automatic sequence controller.

These functions are expected to realize a joint cognitive system for interoperations between the human and the computer as a cognitive tool for operators to use in order to realize a human centered design.

Ecological Interface

The objective of ecological interface design (EID) is to realize direct perception capability, which is the property of the human's very effective and reliable sensori-motor system in a natural environment, for an artifact such as nuclear power plants, as well as to support analytical reasoning. An artifact has specific goals to achieve. For nuclear power plants, the goal is to produce electricity while maintaining public safety (i.e., prevention of radioactive material release into the environment beyond a safe limit).

The artifact has a definite functional structure utilizing appropriate natural phenomena in order to fulfill the goal. This structure is hierarchical and involves many to many interconnections, since one component serves many functions and one function needs several subfunctions or components to achieve its task. This "means-ends" relation and the first principles for the natural phenomena utilized, such as the mass and energy conservation laws for the thermodynamic process, are invariants for the artifact. Accordingly, if these invariants are provided to the operators through the interface, a direct perception capability and an analytical reasoning capability are obtainable for the artifact.

A basis for the ecological interface is an abstraction-aggregation functional hierarchical structure, where the purpose and goal, the abstract function, the generalized function and the physical function are the abstraction hierarchy levels, while the plant, the subsystem, and the equipment are the aggregation hierarchy levels. That is, the operator's focus of attention moves around in this space according to his cognitive process during the complexity. Therefore, CRT pictures should be designed to provide appropriate information to the operator for appropriate points in this two-dimensional space, as shown in Fig. 1 for BWR nuclear power plants.

Special emphasis is paid to the abstract function level, since plant models in this level are most appropriate to represent the invariants of the plant. At the abstract function level, mass and energy flow in the plant are modeled by appropriate flow functions, such as source, transport, storage, etc., as shown in Fig. 2. This figure represents the built-in alternative decay heat removal paths from the decay heat source in the nuclear reactor core to the environment. These flow functions are organized to represent a flow structure and are supported by different flow structures to maintain their performance. This structure constitutes the hierarchical functional structure of the plant. That is, the flow structures correspond to the nodes in the means-ends relationship of the plant as well as to the causal models in the mass and energy conservation principles within themselves.

This model, called a multilevel flow model (MFM) by Lind (1982), represents an operators' normative mental model for the plant. Operators can reduce their cognitive work load, such as knowledge, memory, computation, and information retrieval during their cognitive processes by externalizing this mental model through the man-machine interface. Operators can infer failure propagations along the means-ends relationship or causal relations within the flow structures, i.e., which flow function is the most probable cause of failure, etc. They can plan operational strategy based on due considerations regarding multiple, competitive goals or multiple causes, if any.

Another basis for the ecological interface is the concept of the level of cognitive control, i.e., skill, rule, and knowledge based behavior of the operators depending on their familiarity with the situations. That is, operators have different mechanisms for dealing with complexity so that their information usage is also different, i.e., signals, signs, and symbols, respectively for skill, rule, and knowledge, which provides a basis for determining the form of information on the interface (Rasmussen, 1986).

The ecological interface goal derived from the level of cognitive control mentioned above is to design interfaces that do not force cognitive control to a higher level than the required demands of the task but which also provide the necessary support for all three levels. Three prescriptive design principles are proposed to support the three levels of cognitive control.

Machine Problem Solver

The objective of this function is to support the operators' cognitive resources during a complexity brought about by abnormal events. However, it is not intended to be an independent, automated diagnostic system but is a partner of a joint cognitive system with operators.

The GRADIENT project (Hollnagel, 1991) developed a set of interacting intelligent systems where the assistance support functions seemed to correspond to this category. They are the procedural support, the consequence prediction and the state-based diagnosis.

ISACS (Haugset, and others, 1991) was an integration of the various COSS systems with respect to the man-machine interface and process coupling. An information coordinator called "intelligent coordinator" supervised the COSSs which were grouped according to their function such as state identification, action planning and action implementation. It also created new high level information such as an overview of plant events and a safety state for each event and interpreted and acted upon operator actions and requests.

A cognitive model-based advisor is introduced later as the case study.

Robust Automatic Sequence Controller

The maneuvering of process plants, such as startup, power level change, and shut down, is a process to attain an appropriate plant operational goal from a current operational state. As such, designers can carry through all the decision steps and store the resulting design in a decision table in a computer, achieving automation. However, during an automatic execution of such long sequences, operators are requested to supervise all the functions, and if some anomalies should appear, they are requested to diagnose the situation and override automation, if necessary. In order to enable such supervisory control, the operators must be supplied with information about the system states and about the designers' intention in order to be able to verify the decisions taken by the computer through use of his own preferred decision strategy.

A systematic design using a plant functional hierarchy model similar to the work space model in the previous sections can be implemented to fulfill the above requirement, using a knowledge engineering approach (Monta, 1991a). That is, the intention of the plant design can be assembled in a knowledge base together with plant operation knowledge, and they can be used to derive the operational sequences based on the current status of the plant and its various subsystems and components even when some anomalies exist in them. This approach is expected to provide automatic control transparency for the operators and robustness of automation, which is expected to ease cooperation with the operators by "failing gradually" (Bastl, and others, 1991).

A CASE STUDY OF MAN-MACHINE SYSTEM DEVELOPMENT

A Japanese national project, "Advanced Man-Machine System Development for Nuclear Power Plants (MMS)" supported by MITI is introduced here as a case study (Monta and others, 1991a). This eight year project consisted of Phase I (conceptual design, 1984- 1986) and Phase II (detailed design, implementation, and validation, 1987-1991).

This is a follow-on project of COSSs described earlier, and with regard to the system architecture, it corresponds to the second of a successive realization of two kinds of support functions corresponding to the two human cognitive control levels defined by J.Rasmussen (1986); rule and knowledge, as shown in Fig. 3. This architecture will eventually constitute an integrated man-machine system for nuclear power plants.

In the following, TOSHIBA's development of an intelligent man-machine system in the MMS project (Monta and others, 1991b; Hattori and others, 1991) is explained. It mainly consists of an ecological interface, a cognitive model based advisor, and a robust automatic sequence controller.

Ecological Interface

First of all, it is appropriate to describe the multilevel flow model (MFM) for BWR plants, since it represents the invariants of the plant.
Three MFMs were constructed, corresponding to electricity production, safety, and decay heat removal after a reactor scram. The first two correspond to the two ultimate goals of nuclear power plants, while the third represents an essential goal after reactor scrams by the safety protection system. The safety goal is relevant at this phase, too.

The safety MFM represents many barriers and storages between the fission products in the reactor core and the environment. Figure 4 shows, as an example of the abstract function level display, the topmost flow structure of the safety MFM. The support relations in this MFM, shown in the lower part of the figure, also establish the hierarchical means-ends relationship for the plant safety goal.

The electricity production MFM represents the energy flow from the nuclear fission energy source to the electric power grid by flow functions, as shown in Fig. 5, where the topmost flow structure is represented. The support shown in each flow function (FF) represents appropriate functions to maintain FF performance and is implemented by other lower level flow structures. This process is repeated iteratively to represent the plant functional structure hierarchically and in sufficient detail.

In the middle part, the causal structure in the energy conversion process is indicated, using energy balance and the Rankine cycle display. The main energy storage in BWR plants is the reactor pressure vessel. The energy balance there determines the pressure in the reactor coolant system. Hence, the energy inventory is indicated along with its time derivative, i.e. the energy balance, by a tilted line.

The Rankine cycle display proposed by Beltracchi (1990) is appropriate to indicate BWR plant performance, since they constitute the Rankine cycle. The reactor core, the nuclear boiler, is a recycle boiler, and its performance is indicated by the core recirculation flow instead of the turbine steam and feedwater flow. The leftmost figure illustrates the conversion from feedwater enthalpy to core inlet subcooling by simple geometry calculations (Vicente, 1990). This geometry is also used for calculating the reactor heat to the recirculation flow ratio, since the heat cycle is illustrated for a unit coolant flow. Rather complex reactor heat transients due to disturbances in the recirculation flow, the feedwater enthalpy, and the reactor pressure, etc., are well understood by this figure.

The rightmost figure shows the turbine heat cycle. It is a regenerative heat cycle, and the cycle efficiency is indicated in the figure by the ratio of the area surrounded by the cycle to the area below the line segments a-b-c in the figure.

The lower part represents the mass balance in the plant, which supports the above energy balance. The main mass storages in the plant involve the reactor pressure vessel and the hot well in the condenser. Their inventories, as well as their time derivatives, are indicated by appropriate icons with appropriate reference values.

In this way, the most important internal functional relationship for energy production and conversion in the plant is displayed using a standard format for engineering analysis of the heat engine cycles and easy-to-understand icons. This display is expected to support operators in plant status perception and in thought experiments for hypothesis and test in regard to their knowledge based behavior.

According to the EID goal, the information on the display should be used by operators in their favorite level of cognitive control or should support all three levels. Thus, the content of information on the display should be rich, and the operators should be easily able to pick up necessary information from among these contents. Therefore, appropriate pictures in the generalized function and the physical function level are produced as well as in the abstract function level display based on MFMs as shown in Fig. 1.

Another important EID function is a dialog function between the operator and the cognitive model based advisor. However, before discussing the dialog, it is appropriate to describe the cognitive model-based advisor to clarify the partner in the dialog.

Cognitive Model-Based Advisor

This function is based on two cognitive models as a machine problem solver; the work space model, MFM, previously mentioned, and a cognitive decision function model. In addition, a physical plant structure model is used for the operational procedure synthesizer (Hattori, 1991).

The cognitive decision function model describes the operators' cognitive process during a complexity in abnormal events and consists of the following:
-Plant anomaly detection
-Plant status identification
-Operational goal interpretation and planning
-Planning operational strategy
-Operational procedure synthesis

The advisor consists of five modules corresponding to the above, shown in the left-hand part of Fig. 6. The first four of them are based on MFMs as a work space model.

The topographic search strategy (Rasmussen, 1986) using MFM has been adopted for detecting plant anomalies. That is, every flow function is searched to determine its performance in the order of its position in the hierarchy. A systematic search based on the functional hierarchy from the plant performance (productivity, safety, and decay heat removal) viewpoint is accomplished by this approach with a result of an overall plant good/bad map which shows disturbance propagation, if any.

The results are used to identify the plant status, that is, to identify any failed flow functions. The performance index for the flow function and the flow function failure certainty factor are used for this purpose.

Alternate operational goals are predefined for each failed flow function to compensate for its failure. Since failures may propagate following hierarchical links from the lower levels, the lowest flow function in the same hierarchical link should be assigned a high priority if no irreversible effects exist at the upper part of the link. The operational goal selection is accomplished based on the priority of the candidate operational goals.

Upon selecting an operational goal, planning an appropriate operational strategy to attain the goal should be established.

Lastly, the operational procedures are synthesized to realize the planned operational strategy if they do not exist as standards or as experienced. The synthesis is implemented based on the plant physical model which describes the knowledge regarding the components concerning their operational capabilities and their operational procedures.

The above advisor architecture has the following advantages;
(1) The results obtained from MFM-based modules are compatible with EID displays, since both are based on the same model MFM.
(2) The results obtained from MFM-based modules are expected to match the operators' actual mental model, since an MFM is onsidered to be a normative mental model of the plant,
(3) When the modular configuration is compatible with the operators' cognitive process, it is easier

for them to understand and hence easier for both operators and the advisor to intervene, since each module is independent and receives input from the preceding stage module which the operators can override using the dialog function of the ecological interface described in the following.

Ecological Interface Dialog Function

As shown in the right-hand part of Fig. 6, a dialog function is needed first for navigation in two dimensional space through abstraction-aggregation in order to obtain adequate information in the ecological interface. A menu and mouse facility is provided for the purpose.

The second usage is for operator intervention in the advisor's inference mentioned previously. The operators can be expected to accomplish independent diagnosis by the ecological interface's direct perception and analytical reasoning capability. They might have a hypothesis in regard to the operational goals or strategies which they want to test by the advisor before carrying out direct plant applications.

The operators can investigate the results furnished by the advisor and can modify results, such as an inferred flow function as a failure, a selected operational goal, and an operational strategy to adapt to the situation they have recognized.

In particular, this may happen in a case of coordination by resource management, in which the operators monitor carrying out a pre-planned strategy and adapt the strategy according to the deviation from the plan. In fact, as mentioned in the previous section, the advisor selects the most suitable pre-planned strategy stored in the MFM knowledge base. Hence, there may be a need for some adaptation if any changes in situations assumed for the pre-planned strategy may occur.

For example, in case of modifying an inferred failed flow function, the operator points out the icon for the flow function using a mouse cursor. The dialog function pops up an appropriate dialog menu. The menu pops up another menu containing a list of actions for his intervention by pointing to "intervention". The operator can choose the one corresponding to his/her independent diagnosis. Then, the advisor can display an appropriate interpretation and plan for the operational goal and so on corresponding to the operator's hypothesis. The results of this new process can be compared with previous ones or the operator's expectations, and thus enables the operator to reevaluate his/her diagnoses.

The operators supported by the ecological interface can easily intervene with the advisor functions through the dialog function, and the advisor function can accept this intervention by its modular structure. Thus, close collaboration between the operators and computers is expected by these three functions.

CONCLUSION

Approaches using training simulators seem promising for improving human work situations by developing effective analysis methodologies coupled with more realistic scenarios obtainable from real plant investigations and some physiological monitoring for human performance measurements. Simulator fidelity should be improved to realize these approaches. The results also are valuable to validate man-machine interfaces.

The development of a computerized operator support system (COSS) was started from the immediate needs triggered by the lessons learned from the TMI-2 accident. In the early stage of its development, fundamental issues on the man-machine interface of nuclear power plants such as;
-support functions consistent with the operators' decision making process, such as monitoring, decision, and action,
-information integration consistent with the characteristic of the operators' cognitive process, such as skill, rule, and knowledge based behavior, and
-verification and validation of the support system in situations as close as possible to actual, abnormal events
were identified and have been challenged since then.

With the advent of knowledge engineering and cognitive systems engineering, these issues can now be more systematically tackled in order to realize a joint cognitive system of a human and a computer. In addition, valuable experience regarding automation in various industries suggests a human-centered design, which seems to aim at the same direction.

Systematic design approaches are needed for developing intelligent man-machine systems for nuclear power plants and an approach based on plant functional knowledge, knowledge on the cognitive processes of the operators, and advanced information technology has been studied. Prototype developments have shown that the feasibility of this approach is positive and that a potential exists for applications to a wider range of man machine interaction.

ACKNOWLEDGMENT

The authors wish to thank prof. T. Inagaki of the University of Tsukuba for his valuable advice in preparing this survey paper.

REFERENCES

Aleite,W., and O.Gremm (1991). Status of NPP automation in the Federal Republic of Germany. In Balancing Automation and Human Action in Nuclear Power Plants, IAEA, Vienna, pp.33-54.

Bastl,W. (1988). Integrated disturbance analysis (IDA), A basis for expert systems in nuclear power plants. In M.J.Majumder (Ed.), Artificial Intelligence and Other Innovative Computer Applications in the Nuclear Industry, Plenum Press, New York. pp.721-730.

Bastl,W., and others (1991). Balance between automation and human actions in NPP operation - Results of international co-operation. In <u>Balancing Automation and Human Action in Nuclear Power Plants.</u> IAEA, Vienna. pp.11-32.

Beltracchi,L. (1990). A direct manipulation system-process interface. In <u>ANS Topical Mtg. on Advances in Human Factors Research on Man/Computer Interactions.</u> American Nuclear Society, Chicago. pp. 286-292.

Bozec,J., and others (1991). Main results of the 1989 tests performed on the S3C computerized control room simulator for 1400 MW N4 PWR plants. In <u>Balancing Automation and Human Action in Nuclear Power Plants,</u> IAEA, Vienna, pp.147-154.

Cain,D.G. (1986). Review of trends in computerized systems for operator support. <u>Nuclear Safety, 27,</u> 488-498.

Corcoran,W.R. (1981). The critical safety functions and plant operation. In <u>Current Nuclear Power Plant Safety Issues, Vol. 3,</u> IAEA. pp.383-406.

Fujita,Y.,J.Toquam, and W.B.Wheeler (1991). Collaborative cross-cultural ergonomics research:problems,promises and possibilities. In Y.Queinnec and F.Daniellou (Ed.), <u>Designing for Everyone,</u> Vol.1, Taylor & Francis,London. pp.875-877.

Hattori,Y., and others (1991). Development of an incident and accident management support system for BWR nuclear power plants. In H.-J.Bullinger (Ed.), <u>Human Aspects in Computing,</u> Vol.2, Elsevier Science Pub.,Amsterdam. pp.1041-1045.

Haugset,K., and others, (1991). Realization oh the integrated control room concept ISACS. In <u>AI91, Frontiers in Innovative Computing for the Nuclear Industry,</u> American Nuclear Society, Chicago. pp.130-139.

Higashikawa,Y.,and others (1985). a computerized operator support system for boiling water reactor abnormal conditions. In <u>ANS Topical Mtg. on Computer Applications for Nuclear Power Plant Operation and Control,</u> American Nuclear Society, Chicago.

Hollnagel,E. (1991). <u>GRADIENT Technical Overview,</u> Computer Resources International A/S, Copenhagen.

Inoue,K., K.Yoshino, and K.Takano (1991). Evaluation of effects of automation on operator's performance. In <u>Balancing Automation and Human Action in Nuclear power Plants,</u> IAEA, Vienna, pp.449- 459.

Iwaki,K. (1991). Control room design and automation in the advanced BWR (ABWR). In <u>Balancing Automation and Human Action in Nuclear Power Plants,</u> IAEA, Vienna, pp.399-412.

Kawano,R., and others (1991). Plant operator's behavior in emergency situations by using training simulators. In Y.Queinnec and

F.Daniellou (Ed.), <u>Designing for Everyone,</u> Vol.1, Taylor & Francis,London. pp.851-853.

Lind,M. (1982). Multilevel flow modeling of process plant for diagnosis and control. In <u>Int'l Mtg. on Thermal Nuclear Reactor Safety,</u> Rep. NUREG/CP-0027.

Lipner,M.H., and R.G.Orendi (1991). Issues involved with computerizing emergency operating procedures. In <u>AI91, Frontiers in Innovative Computing for the Nuclear Industry,</u> American Nuclear Society, Chicago. pp.556-565.

Makino,M., and others (1987). Operational experience of human- friendly control and instrumentation systems for BWR nuclear power plants. In <u>ANS Topical Mtg. on Anticipated and Abnormal Transients in Nuclear Power Plant,</u> American Nuclear Society, Chicago.

Masui,T., M.Tani, and Y.Okamoto (1985). The development and evaluation of pressurized water reactor advanced control room concepts in Japan, Part II: Computerized operator support system. In <u>ANS Topical Mtg. on Computer Applications for Nuclear Power Plant Operation and Control,</u> American Nuclear Society, Chicago.

Miyazawa,K., and others (1991). Recent BWR operator training using full scope simulators. In <u>The 1st JSME/ASME Joint Int'l. Conf. on Nuclear Eng., Vol.2,</u> pp.553-557.

Monta,K., and others (1985). Development of a computerized operator support system for BWR power plants. In <u>ANS Topical Mtg. on Computer Applications for Nuclear Power Plant Operation and Control,</u> American Nuclear Society, Chicago.

Monta,K., and others (1991a). Development of knowledge-based decision support system for nuclear power plants. In <u>Balancing Automation and Human Action in Nuclear Power Plants,</u> IAEA, Vienna, pp.235-246.

Monta,K., and others (1991b). An intelligent man-machine system for BWR nuclear power plants. In <u>AI91, Frontiers in Innovative Computing for the Nuclear Industry,</u> American Nuclear Society, Chicago. pp.383-392.

Naito,N., and others (1987). A real time expert system for nuclear power plant failure diagnosis and operational guide. <u>Nuclear Technology, 79,</u> 284-296.

Nitta,T., and others (1991). Design concept for the human factor in recent Japanese PWR. In <u>Balancing Automation and Human Action in Nuclear Power Plants,</u> IAEA, Vienna.pp.413-424.

Olmstead,R.A., and others (1991). Computer automation in single unit CANDU station. In <u>Balancing Automation and Human Action in Nuclear Power Plants,</u> IAEA, Vienna. pp.103-117.

Owre,F., and others (1991). Experiences gained from developing and integrating an expert system and modern graphic display system for a Swedish nuclear power plant control room. In <u>AI91, Frontiers in Innovative Computing for the</u>

Nuclear Industry. American Nuclear Society, Chicago. pp.393-402.

Rasmussen,J. (1986) Information Processing and Human-Machine Interaction -An Approach to Cognitive Engineering-. Elsevier Publishing, New York.

Rasmussen,J., and K.J.Vicente (1989). Coping with human error through system design: Implications for ecological interface design. Int'l. J. Man-Machine Studies. 31. pp.517-534.

Reason,J. (1990). Human Error. Cambridge University Press, Cambridge.

Saito,M., and M.Tani (1991). Human factors considerations related to design and evaluation of PWR plant main control boards. In Balancing Automation and Human action in Nuclear power Plants. IAEA, Vienna, PP.369-376.

Sasou,K., A.Nagasaka, and T.Yukimachi (1991). The development of a method to evaluate the team activity - A study on the characterization of the team activity - . The Japanese Journal of Ergonomics. 27. 159-168. (In Japanese)

Sun,B., and others (1991). Human factored control room operator evaluation of an automated emergency operating procedure expert system. In Balancing Automation and Human Action in Nuclear Power Plants. IAEA, Vienna. pp.507-515.

Takano,K., K.Sawayanagi and T.Kabetani (1991). Status of human factors research program in Central research institute of electric power industry (Part 2). The Thermal and Nuclear Power. 42. 700-713. (In Japanese)

Takizawa,Y., and others (1989). A post trip operational guidance system for BWR plants. Nucl. Eng. & Design. 110. 385- 393.

Tomizawa,T., and others (1984). Enhanced operational safety of BWRs by advanced computer technology and human engineering. In Operational Safety of Nuclear Power Plants. Vol.1. IAEA, Vienna. pp.349-370.

Vicente,K.J. (1990). Ecological interface design as an analytical evaluation tool. In ANS Topical Mtg. on Advances in Human Factors Research on Man/Computer Interactions. American Nuclear Society, Chicago. pp.259-265.

Woods,D.D. (1987). Commentary: Cognitive engineering in complex and dynamic world. Int'l. J. Man-Machine Studies. 27. p.571-587.

Woods,D.D. (1988). Invited lecture. In Flight Deck Automations: Promises and Realities. NASA Conf. Report No.10036.

Woods,D.D., and E.M.Roth (1988). Cognitive system engineering. In M.Helander (Ed.). Handbook of Human-Computer Interaction. North-Holland, New York. pp.3-43.

Yoshimura,S., T.Ohtsuka and J. Itoh (1991).Reliability aspect of man-machine interface. SMiRT 11th Post-Conf. Seminar on PSA. Kyoto. To appear in Rel. Eng. & Sys. Safety.

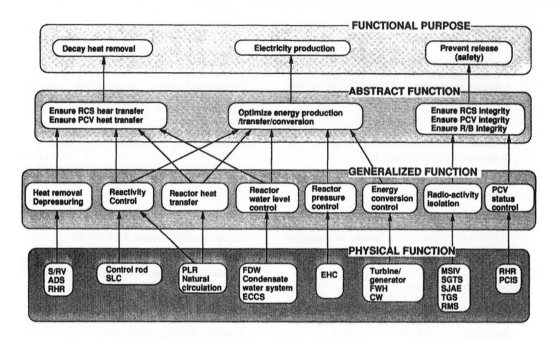

Fig. 1   Abstraction hierarchy levels and CRT picture hierarchy for BWR

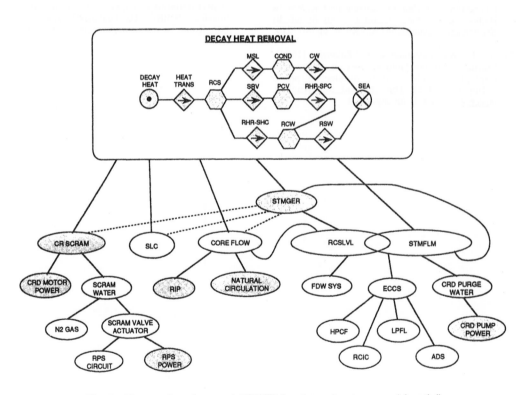

Fig. 2   Abstract function model(BWR) for decay heat removal (partial)

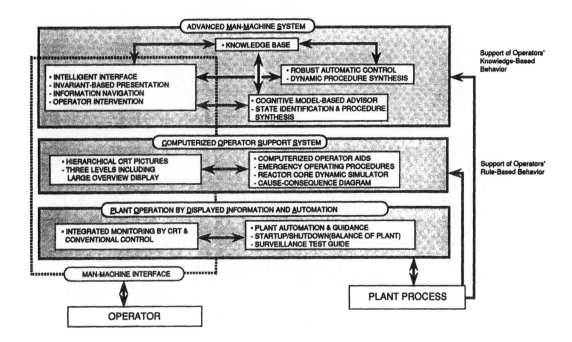

Fig. 3   System architecture for integrated man-machine system

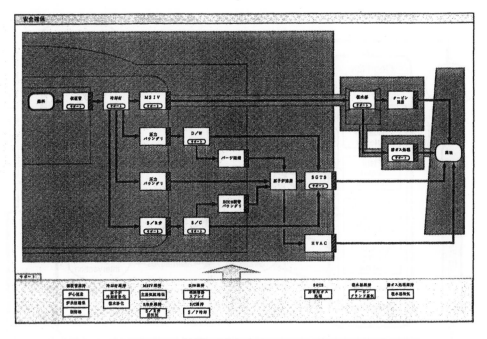

Fig. 4   Example of abstract function level display (safety goal)
(in original Japanese)

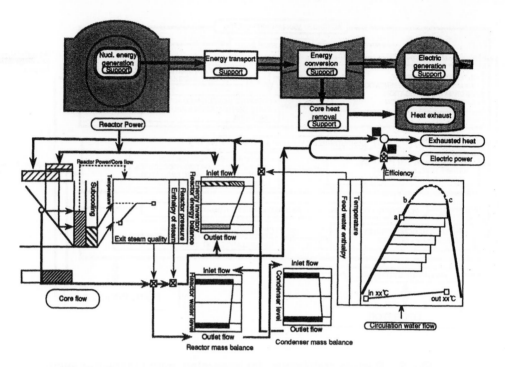

Fig. 5   Example of production goal display based on MFM and EID

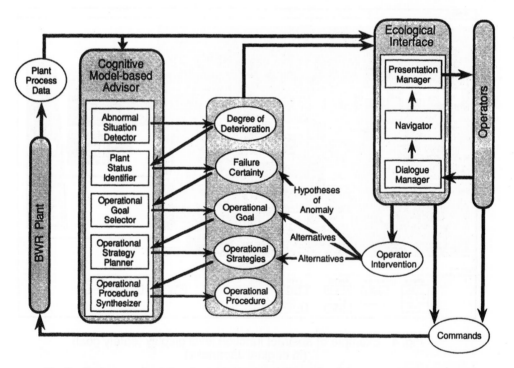

Fig. 6   System configuration for cognitive model-based advisor and ecological interface

HUMAN INTERVENTION IN SUPERVISORY CONTROL

N. Moray and J. Lee

*Department of Mechanical and Industrial Engineering, University of Illinois at Urbana-Champaign,
Urbana, IL 61801, USA*

Abstract. In this paper we describe the results of experiments on supervisory control of a simulated automated pasteurisation plant which could be run under either manual or automatic control. By varying the probability and nature of faults in the system we have been able to discover the "transfer functions" which describe the development of trust in the automated system and self confidence in operators' ability to exercise manual control. We have developed a quantitative model which predicts when an operator will intervene to take control of the plant. Our results suggest that operators may be suboptimal in their self-allocation of control. There is "inertia" in the dynamics of trust and self confidence, and there is some evidence that some operators have a bias towards manual control which may lead to suboptimal strategies of intervention.

INTRODUCTION

The term "supervisory control" refers to a mode of operating human-machine systems in which the moment to moment closed-loop control is allocated to a computer or other automated controller, and where the human, rather than intervening constantly to trim the plant and to make decisions about set-points, etc., monitors the automated system (Sheridan and Johannsen, 1976; Sheridan, 1987; Moray, 1986). Despite the increasing prevalence of automation and the emergence of a supervisory role for operators, there has been remarkably little research done on the best way to integrate operators into such systems.

One aspect of human-machine dynamics is the extent to which an operator trusts the system. We might expect that if operators trust an automated system they will allow it to run undisturbed in automatic mode; but if they distrust it they will tend to intervene and take control. On the other hand, operators may be unwilling to take control if they lack self confidence in their ability to control the system manually. Recent work by Zuboff (1988) on the nature of trust during the introduction of automation in industrial and business systems points to the great importance of these ideas, but there is no systematic body of empirical research on the topic. Muir (1987, 1988, 1989) in our laboratory conducted experimental studies showed that sociological models of trust between people appear adequate to describe trust between operators and a simulated machine. We have therefore undertaken a series of experiments to develop quantitative models of the way in which trust and self confidence determine the dynamics of supervisory intervention.

METHOD

A simulation of a pasteurization plant runs on a Macintosh II computer system. The task of the operator is to achieve the maximum output of successfully pasteurised juice in each run. The juice enters the system through the tank at the top left of the screen, arriving at a random rate and temperature, each within a narrow range. It is fed through the system by the feedstock or "juice" pump at the bottom left of the screen. It then passes through the passive heat exchanger and enters the active heat exchanger. Here it is heated by steam which is pumped from the boiler by the steam pump. On emerging its temperature is sensed at the three-way valve. If it is below 75° it is recycled back to the input vat for reheating. If it is above 85° it is burnt and is sent to the waste vat. Otherwise it passes to the other leg of the passive heat exchanger where it gives up its heat to the incoming cold juice, and is collected in the output vat. The juice pump, the steam pump and the boiler power can be each run either under automatic control or under manual control. Control mode is determined by the operator typing commands on the keyboard, indicating that a particular subsystem is to be set to

automatic or to manual control, and if to manual, to what value the pump rate or heater power should be set. The dynamics of the subsystems are fairly realistic, including lags in the response of pumps to command for rate changes, etc. Operators were all students, and were paid a flat rate per hour for their participation, plus a bonus if they pasteurised more than 90% of the incoming juice. Each trial run lasted approximately 6 minutes. In some experiments both manual and automatic commands were corrupted by the fault: in others the fault only occurred in one or other mode at a time, although this could alter in the course of the experiment.

EXPERIMENT 1: THE DYNAMICS OF TRUST

Each operator ran the system alternately in fully manual and fully automatic mode for the first 10 trials. They were then allowed to use any combination they wished of manual and automatic mode for the remaining trials, and could change one or more of the subsystems between manual and automatic mode at any time they wished. On the 26th trial a fault occurred. The pump control became unreliable, and instead of going to the commanded value it would go to the commanded value ± a random number whether under manual or under automatic control. On the following trial the fault disappeared, and the system ran well until the 40th trial, after which the fault reappeared and remained for the rest of the experimental sessions. On different trials this random number was varied from 15% of the commanded value to 35% of the commanded value.

On each trial measures were taken of the percent feedstock successfully pasteurised, and total feedstock available, and the per cent wasted. At the end of each trial the operators rated their trust in the automatic system using scales based on Muir (1989), and so provided a reliable measure of the subjective dimensions of the relationship of the operators to the plant.

RESULTS

Typical results from the first experiment, in which both manual and automatic modes suffered the effects of faults in the juice pump controller, are shown in Figures 1 and 2.

Even when the fault was large the operators maintained a remarkably high level of output. A fault of 35% in the dynamics of the pump only resulted in about a 10% drop in productivity after the initial shock of its occurrence, although at the cost of a great increase in workload. The number of interventions, measured as the frequency with which commands were typed by the operators rose substantially. A considerable variety of individual strategies of control appeared. In general people developed a characteristic personal strategy quite early in their experience of the task, and then stuck to it, showing little tendency to explore alternatives. The behavior was quite reminiscent of the classical work of Crossman and Cooke (1974), and the later work of Moray, Lootsteen and Pajak (1986), who found that as skill developed the number of control actions decreased. The most efficient controllers made fewer control interventions, and what interventions they made tended to be in the early part of each run. After the fault occurred, the number of interventions increased and were more spread out in time. It seems that when skilled operators tend to go to an open loop feedforward control mode, and that when the fault occurs they revert to their earlier and less efficient closed loop feedback control mode.

Multiple regression analysis was used to determine the most important variables determining the level of trust, and time series analysis (Pandit and Wu, 1987) to derive a dynamic equation predicting trust as a function of the values of the variables identified by regression. Both methods are important in modelling operator-machine interaction, for the regression model alone does not capture the dynamics of changing trust. The resulting transfer function and equations are shown in Figure 4. An operator's trust is a function of his or her trust on the present and immediately prior trial, on the productivity of the present and prior trial, and on the magnitude of any fault which occurs. While the inertia term is clearly present, there is no evidence that we need to use more than a first order model to account for it. The model accounted for nearly 80% of the variance in the value of trust, a very strong result.

Figure 3 shows that the dynamics of trust are somewhat different from those of performance. In particular, there appears to be inertia in trust. Following the transient fault productivity recovers immediately the fault disappears, but trust may take several trials to return to its pre-fault value.

Furthermore, it is very obvious that following the onset of the chronic fault trust does not reach its lowest value for several trials, and then begins slowly to recover as the operators become adjusted to the new situation. It never returns, however, to its prefault value.

EXPERIMENT 2: THE DYNAMICS OF INTERVENTION

Muir found that operators tend to adopt manual control when faults occur, but our data did not show the same effect. Detailed analysis and modelling suggested that this is due to an inherent bias in some operators in favour of using manual control. Such operators are more likely to be in manual than in automatic control when the fault occurs, and therefore one is not likely to see a transition from automatic to manual mode.

If such is the case, then in order to obtain greater insight into the nature of intervention and mode switching it is necessary to bring about a situation in which the operator will first switch out of manual mode, and then switch back again if a fault occurs. The experiment was therefore repeated with the following changes. Once the initial learning phase was complete, half the operators were faced with a pump which was faulty only if they were in manual mode, and the other half only if they were in automatic mode. Some time later these conditions were reversed. It also seemed likely that the use of manual mode might be affected by the self confidence of operators in their ability to run the plant manually. We therefore obtained subjective ratings of self confidence at the end of each trial in addition to the ratings of trust.

RESULTS

We made the assumption that if trust were greater than self confidence, then operators would use automatic mode, and if self-confidence were greater than trust, then they would use manual control. Regression and time series analysis were again used to develop the model shown in Figure 5, which accounted for approximately 80% of the variance again. Note that we found it necessary to include a bias factor, which represented the tendency of some operators to be more biassed towards manual rather than automatic control.

Figure 6 shows a single operator where as the trust and self confidence wax and wane with respect to one another the operator switches from one mode to the other as predicted by the model.

DISCUSSION AND CONCLUSIONS

It is possible to obtain strong quantitative models which predict at what moments operators who are running an automatic simulated plant under supervisory control will respond to the occurrence of faults by switching into manual control mode. The equations include subjective ratings which can be modelled in terms of objective plant parameter values, and the filtering effect of the operators' subjective feelings does not diminish the accuracy of the prediction. The time of first occurrence of a fault, the mode in which it is first experienced, the magnitude of the fault, the initial allocation of control to the manual or automatic mode, the interactions among the efficient and the faulty subsystems, and the stability of the controllers all play a part. Operators could identify which component of the plant was faulty, and spent far more time monitoring the untrusted component. The differences in the frequency of observations of a subsystem correspond to changes in the extent to which a subsystem is trusted. It is possible to obtain very good models for human intervention in supervisory control. That in turn opens the possibility of developing further understanding which will aid designers in defining the role of operators. We have now begun to extend our investigations to similar phenomena in discrete manufacturing plants, a realm which has been sadly neglected by human factors until recently. Given that such systems are sometimes designed on the assumption of up to 30% down time, a set of predictive equations for human intervention should make a major contribution to the design of better systems. We offer our results as a starting point for a quantitative theory of supervisory control.

ACKNOWLEDGEMENTS

This work was supported in part by a grant from the Research Board of the University of Illinois at Urbana-Champaign, and by the Beckman bequest to the university.

REFERENCES

Crossman, E.R. and Cooke, F.W. 1974, Manual control of slow response systems. In *The human operator in process control*, E. Edwards and F. Lees, Eds. London: Taylor & Francis.

Moray, N, 1986. Monitoring behavior and supervisory control. In K.R.Boff, L.Kaufman, and J.P.Thomas (eds). *Handbook of Perception and Human Performance*, chapter 45. New York:Wiley.

Moray, N., Lootsteen, P. and Pajak, J. 1986, Acquisition of process control skills. *IEEE Transactions on Systems, Man and Cybernetics*, SMC-16, 497-504.

Muir, B.M. 1987, Trust between humans and machines, and the design of decision aides. *International journal of man-machine studies*, 27, 527-539.

Muir, B.M. 1988, Trust between humans and machines, and the design of decision aides. In Holnagel, E., Mancini, G. and Woods, D. D. (Eds.), *Cognitive Engineering in complex dynamic worlds*. London: Academic Press.

Muir, B.M. 1989, *Operators' trust in and use of automatic controllers in a supervisory process control task*. Doctoral thesis. University of Toronto.

Pandit, S. M., and Wu, S. M. 1987, *Time series analysis with applications*. New York: John Wiley and Sons.

Sheridan, T.B. 1987. Supervisory control. In G.Salvendy, (Ed). Handbook of Human Factors. New York: Wiley.

Sheridan, T.B. and Johannsen, G. (Eds.), 1976, *Monitoring Behavior and Supervisory Control*, New York: Plenum.

Zuboff, S. 1988. *In the age of the smart machine*. New York: Basic Books.

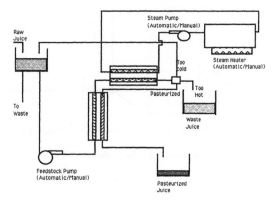

Fig. 1. Simulated Pasteurisation Plant

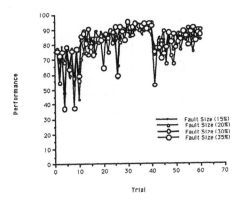

Fig. 2. Productivity as a function of experience and the occurrence of faults. Faults occurred on Trial 26 and from Trial 40 onwards

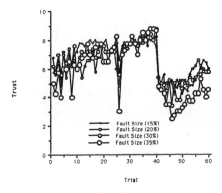

Fig. 3. Trust as a function of experience and the occurrence of faults Faults occurred on Trial 26 and from Trial 40 onwards

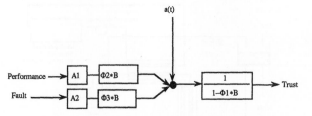

Trust(t)= ϕ_1Trust(t-1) + A_1Performance(t) + $A_1\phi_2$Performance(t-1) + A_2Fault(t) + $A_2\phi_3$Fault(t-1) + a(t)

Fig. 4. Transfer function and equation for the dynamics of trust

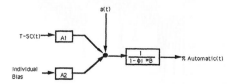

Figure 5. Transfer function and equation for intervention

$$\% \text{Automatic} = \phi_1 \% \text{Automatic(t-1)} + A1*((T\text{-}SC)(t))$$
$$+ A2*(\text{Individual Bias}) + a(t)$$

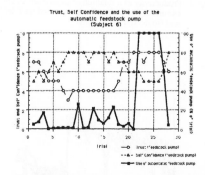

Fig. 6. Switching modes between manual and automatic mode as a function of trust and self confidence. Faults during manual control began on Trial 20, and stopped on Trial 26. Faults in Automatic mode began on Trial 26.

46

INTERFACING BETWEEN FUZZY LINGUISTIC CONTROLLER AND EXPERT'S BEHAVIOURAL SKILLS VIA QUALITATIVE MENTAL MODELS

T. Sawaragi[1], O. Katai and S. Iwai

Department of Precision Mechanics, Faculty of Engineering, Kyoto University, Yoshida Honmachi, Sakyo, Kyoto 606, Japan

Abstract: This paper presents a system for tuning the fuzzy controller in off-line by analyzing the human operator's control actions as well as the evolving system states caused by those actions. That is, a human operator, who is forced to control a plant having complex dynamics and, at the same time, to supervise its behaviour so that it can attain the higher-level control task, provides a training example. The recorded history data of control operations are to be analyzed in terms of the operator's mental models that qualitatively and symbolically reason about his goals to be sought and the dynamics of the plant. This analysis enables the clarification of his goal-and-plan structure for the provided task that is implied in the training example, and specifies tuning parameters needed for designing a fuzzy controller.

Key words: Artificial intelligence, fuzzy control, man-machine system, qualitative reasoning, associate system, skill acquisition.

INTRODUCTION

It has been acknowledged that there exist different styles of cognition levels concerning with the operator's behavioural modelling; SBB (skill-based behaviour), RBB (rule-based behaviour) and KBB (knowledge-based behaviour) [Rasmussen, 1983]. Although the distinctions among these three levels are quite vague, there seems to exist the common consensus as follows. SBB is a behaviour in which the specific features are experienced together frequently and the response is more or less automatic, while RBB is a procedural-oriented task including monitoring and interpreting. KBB includes the full range of problem solving and planning activity with the manipulation of some kinds of "deep" models. Some of the researchers have attempted to model such a skill-based behaviour as a stimuli-response process, where the perception has been directly associated with an action by running artificial neural networks [Miyake, 1987], or by a reinforcement learning [Sutton, 1988].

As for a typical technique for a rule-based control, a fuzzy controller, which was originally proposed by Mamdani [Mamdani, 1974] as one of the applications of Zadeh's fuzzy set theory [Zadeh, 1965], has come into practical use. In contrast to the conventional classical/modern control methods, a fuzzy controller could be designed only if the empirical heuristics on how to control the plant were prepared, even though the precise mathematical modelling of the plant were impossible. Furthermore, the

control rules expressed as linguistic rules are comprehensive enough for the human to understand its controlling process. However, the fuzzy controller could implement only local control tactics. The more global control strategies are supposed to be incorporated implicitly in a stage of designing rules as well as in tuning them, both of which have been performed by the system designer in an *ad hoc* manner.

In a research field of man-machine systems, an importance of the operator's *mental* model or *internal* model has been stressed concerning with the supervisory control of the complex technological plant and of the vehicles having complicated dynamics. As for the actual implementation of this kind of models, many ideas have been proposed so far; a control theoretic mathematical model [Veldhuyzen et al., 1977], a linguistic model that explicitly describes the mental operations declaratively based on fuzzy set theory [Stassen, 1987], a functional model that representing the environment in terms of mass and energy flows [Lind, 1981], and a qualitative reasoning model that predicts the dynamic changes of the system's behaviour by inferring the transitions of qualitatively equivalent states in an elaborated manner [deKleer, 1977].

In order to let the fuzzy controller work more flexibly and purposively like a human skilful expert operator, we have to incorporate some kind of *models* that can plan and supervise the lower rule-base behaviours and can provide it with global control strategies. For this purpose, we attempt to develop a prototype system of a man-machine collaboration system in which a fuzzy controller is adequately tuned by analyzing the domain expert's (an operator's) behaviour in terms of his mental models that determine the global control strategies.

[1] The author is now a visiting scholar at Stanford University. The author's current address is: Dept. of Engineering-Economic Systems, Stanford University, CA 94305-4025, USA.

A DESIGN OF A FUZZY CONTROLLER FOR SHIP-MANEUVERING

A Fuzzy Controller for a Ship-Maneuvering System

As an example of a dynamic system to be controlled, we construct a simulation system for maneuvering a slow responding large ship, whose dynamics are numerically modelled as follows:

$$Td\omega/dt + \omega(t) = u(t), \quad \theta(t) = \int \omega(t)dt + \theta(0), \qquad (1)$$

$$dx/dt = V\sin\theta(t), \quad dy/dt = V\cos\theta(t)$$

T: time constant, V: velocity (const.), u: rudder angle, ω: rate of turn, θ: heading angle, x,y: position.

A human operator, who does not know about the above *precise* model inside the system, executes the manual control task of passing through a number of fixed gates smoothly without overshooting them. Each gate is denoted by a pair of small circles on the display as shown in Fig.1. Monitoring the behaviour of the ship moving with a constant velocity from the bottom to the top in the display, the operator enters the discrete control input of rudder angle *u* by one of five key-commands, HARD APORT (full left turn), MID APORT, STRAIGHT, MID STARBOARD, and HARD STARBOARD (full right turn) at any time with arbitrary durations. The slowly responding character of the above system with a large time constant makes the control task not an easy task, since the task requires the human operator to perform not only a manual control but also a supervisory control such as monitoring and goal-setting tasks at the same time. Note that the primary motivation of our taking the above simulator as an object to be controlled does not exist in simulating an actual "ship" [Veldhuyzen et al., 1977]. Rather, our purpose of introducing this system exists in its characteristics that an operator has to deliberate on the global plans as well as to execute control actions in a real-time fashion.

To control the simulator, the operator is assumed to have verbal knowledge such as *"if the goal is on the forward port side and the ship is turning right, then put the helm hard (moderately, or slightly) aport."* As an implementation of these verbal rules, 35 fuzzy rules in the form of "if dir is A_i and vel is B_j, then put the helm C_k" are prepared. As shown in Fig.2, the antecedent part of each rule contains two fuzzy variables, A_i (i=1,..,5) representing qualitative states of the ship's rotation, and B_j (j=1,..,7) denoting qualitative states of the ship's heading measured as a deviation from the direction towards

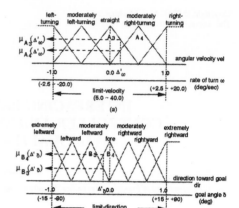

Fig.2 A fuzzy inference process of the antecedent parts of fuzzy control rules

the given goal position. The membership functions of these fuzzy variables, μ_{A_i}(vel) and μ_{B_j}(dir) ($\in [0,1]$), are defined as triangular-types over numerical variables dir and vel, both of which vary between [-1, 1] and represent the normalized values of rate of turn ω and goal angle δ, respectively. Wherein, the scaling or normalizing parameters limit_dir and limit_vel represent the maximum range of the variables of ω and δ, respectively, that are to be determined so that they can reflect the extents of the operator's viewpoints. The fuzzy variable C_k (k=1,..,7) in the consequent part denotes a qualitative amount of inputs to be performed varying between "hard aport" and "hard starboard."

Divergence of a Fuzzy Controller from a Human Operator

To let the above fuzzy controller work well, we have to carefully select the two parameters of limit_dir and limit_vel as well as a number of intermediary goal positions in a coordinated fashion. For instance, if limit_dir and limit_vel are set too large or too small, the fuzzy controller's performance results in "unstable" ones such as an over-reactively fluctuating one and an too-insensitive one to the situational changes.[1] The goal positions are intermediary subgoals to be passed through by the ship.

Fig.3 denotes the comparison of the controlled result by the fuzzy controller, that is tuned ideally in a trial and error fashion, and the one by the human expert operator's manual control. Wherein, small triangles shown beside the ships' trajectories denote time points when the rudder inputs are provided by the controller/operator. From this figure, the differences between two controllers are distinctive, in spite of the apparent similarity of the shapes of the resulted trajectories. That is, since the fuzzy controller repeats the basic cycle of "situation assessment", "judgment" and "action execution" with fixed time intervals regularly, the inputs are provided so frequently in reply to the observed deviations reflexively. Wherein, no judgement is made concerning with whether some actions are *critically* required or not, nor with any

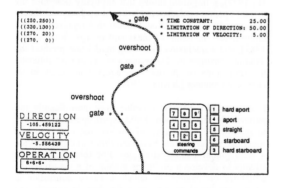

Fig.1 A display of the ship-maneuvering system

---

[1] The "stability" referred to in the above is not identical to the one discussed in the control theory.

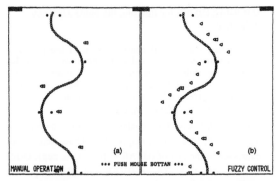

Fig.3 The results of (a) an expert's manual control and (b) of a fuzzy controller

deliberation on whether it is in accordance with any global plan, but its assessment is restricted to the near-sighted one. On the contrary, the performance by the human operator is quite purposive and economical in a sense that the input operations are restricted only to the essential ones and it seems that not only the judgment on the amount of the inputs but also the decisions on the timing of providing inputs are made with the deliberation on his overall plan to perform the task. Fig.4(a) schematically illustrates the internal planning structure that is assumed to be governing his actual actions. The overall control task is divided into a number of subplans, each of which contains a subgoal setting activity and consists of a series of executing actions. These goals are set to the intermediary positions between the neighbouring gates to avoid overshooting (cf. Fig.1), and the attempts to go through those intermediary goals result in the successful passage of all the gates successively with easiness. This planning structure is produced by the manipulation of his mental models that roughly predict the transitions of the ship's behaviour caused by its dynamics and the operator's control actions.

IMPLEMENTATION OF AN OPERATOR'S MENTAL MODELS BY QUALITATIVE REASONING

A Plan-and-Goal Graph

The above observations suggest that in order to design a fuzzy controller that behaves in a purposive way like the human operator, tuning of the fuzzy controller, that is, to determine the parameters of limit_dir and limit_vel and to set subgoals, should be performed in accordance with the operator's mental activity of global planning as shown in Fig.4(b). If we assume that an operator exhibits purposive behaviour, a history of control operations performed by

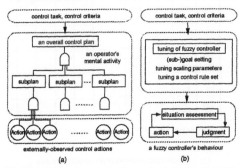

(a)   (b)

Fig.4 Correspondences between expert's manual control and fuzzy controller

the operator should be an instance of the class of behaviours that are practising and implementing the operator's internal global control strategies. So if we were able to analyze the operator's action sequences in terms of these models and to explicate his planning structure implied in it, we could obtain information that is useful and operational enough to determine the tuning parameters for the fuzzy controller from it.

Currently, the researchers on "associate systems" such as a pilot's associate (PA) have been developed being guided by DARPA's (the Defense Advanced Research Projects Agency's) Strategic Computing Program [Banks, 1991]. Wherein, the conventional expert system technologies are adapted to the most critical of any real-time environment so that they could provide information in a timely manner and maintain the pilot's situation awareness. The key to the system architecture of PA is to infer a pilot's latent intents from an explicit observation of his actions or behaviours. This requires knowing the goals of the pilot, reasoning about the decisions he must make, and reasoning about the actions he must perform. Usually, an output of this intent inferencer is a *plan-and-goal graph*, which describes the elements used to link a pilot's actions or plans, with a particular mission goal in a hierarchical fashion, and explains what the pilot is trying to do [Rouse et al., 1990]. In order to construct a plan-and-goal graph from an operator's behaviour and to verify it as a plausible one, we have to incorporate models of human mental activities concerning with situation data, on-going plans, system status, and so on.

In this paper, to construct a plan-and-goal graph from an actual operator's behaviour, we attempt to implement two kinds of qualitative models; one for reasoning about the relationships between the situation (i.e., a provided gate configuration) and the resulted behaviour of the plant (i.e., a ship's trajectory), and another one for reasoning about the relationships between the dynamic characteristics of the plant and its resulted behaviours caused by the control inputs. Outputs from both models are represented as transient sequences of qualitative states and reasoning are performed based on a finite set of qualitative reasoning rules stored within each model. Note that the purpose of introducing these qualitative models is not to generate precise reference trajectories along which the ship should be controlled nor to predict the actual dynamic behaviour of the ship. Rather, our mental model only specifies a general "class" of actual behaviours each of which is performed being guided by an equivalent strategy. It has been well-recognized that the existence of the internal model is a *necessary* but non-sufficient condition for building human performance models [Stassen, 1987]. Since this is also the case of our approach, it is further required to establish a good interface that could relate the abstracted behaviours generated by the models with the actual behaviours produced by the operator or the controller. This will be described in the next chapter.

A Qualitative Model for Reasoning A Ship's Trajectory

As shown in Fig.5(a), the direction θ_i in which the ship should pass through the i-th gate can be determined from the configuration of the gates. Wherein, we do not consider precise values of the direction but qualitative ordinal values between θ_i and θ_{i+1}, that is, either $\theta_i < \theta_{i+1}$ or $\theta_i > \theta_{i+1}$, where the counter-clockwise direction

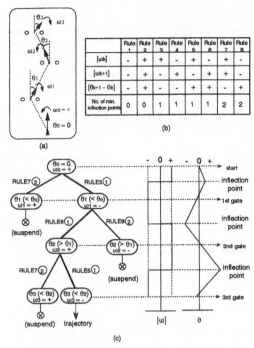

(a)

(b)

(c)

Fig.5 A qualitative model for reasoning a ship's trajectory

Qualitative Physics for Inferring the Ship's Dynamic Behaviour

Considering the general characteristics of the plants to which a fuzzy control method is attempted, we assume that the structure of the plant is only partially known to us as a collection of qualitative constraints. These qualitative constraints corresponding to equations (1) are:

$$T \cdot A + \Omega = U, \quad d\Omega/dt = A, \quad d\Theta/dt = \Omega, \qquad (2)$$

where A, Ω, U and Θ are qualitative variables corresponding to $d\omega/dt$, ω, u and θ, respectively. Those take values consisting of the pair of (QVAR $\in$ {pos, zero, neg}, QDIR $\in$ {inc, std, dec}), where QVAR and QDIR represent qualitative ordinal value and qualitative trend value, respectively. The system behaviours in qualitative physics are represented as sequences of *qualitative states* that are defined by a set of system variables each of which takes a pair of (QVAR, QDIR).

From those constraints, the network shown in Fig.6(a) is obtained as a qualitative physics model according to Kuipers' QSIM [Kuipers, 1986]. The model consists of six nodes including time constant T and multiplied quantity of $T \cdot A$ in addition to those four variables, and of four constraints (a multiplier, an addition, and two derivative constraints). Initiating the model by providing the initial states of T (positive constant), Ω, and U (assumed as positive without change), the model generates a sequence of qualitative states of each variable as a dynamic behaviour of the system, which is summarized in Fig.6(b). The reasoning is performed at first by propagating the values to the unspecified variables within the same qualitative state. The variable whose value cannot be determined uniquely is attached with all the possible values temporally, and when the inconsistency occurs later that value is retracted. After deciding all the variables in that state, the reasoner predicts the qualitative values in the subsequent state based on the local continuation

is positive. In addition to that, we can also define qualitative values for the rate of turn of the ship when passing through the i-th gate [ω_i] $\in$ {+, −}, depending on whether the ship is turning left (+) or turning right (−), respectively. Since the ship is assumed to change its attitude continuously, we can define the eight rules of Fig.5(b) that relate these two kinds of qualitative values for the neighbouring two gates with the minimum number of inflection points of the ship's heading angle between the gates.

Based on these qualitative rules, we can now envision the possible trajectories when provided with an arbitrary gate configuration. The value of [$\theta_{i+1} - \theta_i$] is determined from the gate configuration. However, the qualitative states of its rotating behaviour [ω_i] at the i-th gate is not determined uniquely and two assignments are possible. In this case, the envisioned path is branched according to these possibilities as shown in Fig.5(c). The rules also indicate the number of minimum inflection points for each state transition, that is denoted by an encircled number. If there exists a transition which has more than two minimum inflection points, further envisioning from this transition will not be performed (cf. this is marked with a crossed circle in the figure). In this way, a path (an envisioned trajectory) with the minimum number of inflection points can be determined, each of which is assumed to be representing the operator's global plan consisting of a number of sub-plans. This is denoted by thick lines in Fig.5(c). During each subplan, the qualitative states of the ship's rotating behaviour remain the same as shown in the right of Fig.5(c). Note that the envisioned trajectories produced according to the above qualitative model cannot be always determined uniquely.

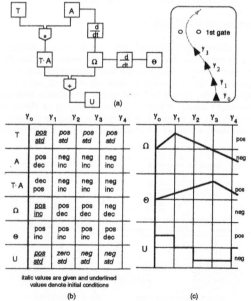

(a)

	γ_0	γ_1	γ_2	γ_3	γ_4
T	*pos* <u>std</u>	pos std	pos std	pos std	
A	pos dec	neg inc	neg inc	neg inc	
T·A	dec pos	neg inc	neg inc	neg inc	
Ω	<u>pos</u> <u>inc</u>	pos dec	pos dec	neg dec	
Θ	pos inc	pos inc	pos inc	pos dec	
U	<u>pos</u> <u>std</u>	zero std	neg std	neg std	

italic values are given and underlined values denote initial conditions

(b)

(c)

Fig. 6 Reasoning by qualitative physics model by Kuipers

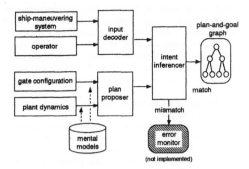

Fig.7 Deductive analysis of an operator's behaviour

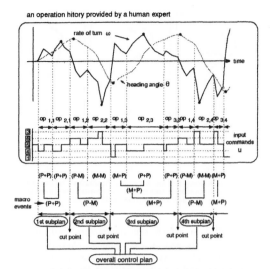

Fig.8 An operation history of Fig.11(a) and its plan-and-goal graph
(• denotes maximum values of ω and θ within each subplan)

principle. Then, from these it resumes the propagation in a subsequent state and repeats those procedures recurrently. Fig.6(c) qualitatively plots the system behaviour according to the reasoning results of Fig.6(b).

DEDUCTIVE ANALYSIS OF AN OPERATOR'S BEHAVIOUR

As shown in Fig.7, the identification of an operator's plan/goal structure can be performed in intent inferencer by integrating the reasoning results that originate from two different sources; one from the operator's history data and the other from the provided task and and the plant characteristics.

Plan Proposer

When a gate configuration and a set of qualitative constraints forming the dynamics of the plant are provided, plan proposer constructs and proposes a plan that attain the entire control task, i.e., to smoothly pass through the all gates with the shorter trajectory. This is performed by running the two mental models described in the previous chapter. Since the former model and the latter model determines the task-behaviour relations and the control-behaviour relations, respectively, the system can derive the task-control relations via behaviours. That is, these relations represent a plan describing when and where the control inputs are required in a qualitative sense.

Input Decoder

This module at first transfers the serial numerical data of an operation history into a sequence of discrete *primitive events*. Each of primitive events consists of the initial qualitative state of the ship's rotation when the input is entered (leftward: P_i, rightward: M_i), the qualitative value of the rudder input (aport: +, starboard: -), and the final state after the rudder is kept straight following the preceding rudder input. All the primitive events can be classified into one of the following six types; $(P_i + P_f)$, $(P_i - P_f)$, $(P_i - M_f)$, $(M_i + P_f)$, $(M_i + M_f)$ and $(M_i - M_f)$. Usually, a history contains not only essential control actions such as the ones shown in Fig.3(a) but also other actions that are to modify the preceding control actions. Input decoder integrates such a sequence of successive primitive events into a *macro* event having a larger duration, where the operations and resultant behaviour of the ship are equivalent to those ones of the original sequence. In contrast, when the final state caused by the modifying operation is qualitatively different from the initial state, this alteration of operations indicates that

the operator has confirmed the attainment of the current goal after the initial operation and has switched his perspective to the next goal. That is, a boundary between the sub-plans should be exist between these operations. After decoding the input stream of situation data of the simulator and the control data of an operator based on the above heuristic rules, input decoder sends an plan structure identified in his behaviour to the intent inferencer.

Intent Inferencer

Being provided the reasoning results from the above-mentioned two separate modules, intent inferencer attempts to check the result from input decoder against the result from plan proposer exhaustively. When the consistency is verified, the module outputs a plan-and-goal graph, in which each control action of the operator is properly located in his global strategy. On the other hand, when a mismatch between those occurs, another module playing a role of error monitor is triggered and the explanations on this error is generated. Fig.8 illustrates the process of input decoder, whose input is an operation history of an expert's controller (Fig.9(a)). The macro events derived from the operation history are shown in Fig.8. The boundaries of macro events can be regarded as boundaries of his subplans. We will call these boundaries as *cutpoints* hereafter. The "|" marks in Fig.9(a) denote the positions of cutpoints determined by the analysis of the operation history of Fig.8.

ADAPTIVE TUNING OF FUZZY CONTROLLER BASED ON THE IDENTIFIED PLANS

Once an operation history is verified as consistent with the global control knowledge, the system sets the locations of these cutpoints, that are readable from the operation history, as the mental goals. The other parameters, i.e., limit_vel and limit_dir, can also be obtained from it. Since these parameters represent the extents of the operator's viewpoints and are thought to be altered adaptively according to the currently activated subplans, we can specify them from the appropriate parts of the analysed operation history. Then, based on these parameters and goals, we construct an adaptive fuzzy

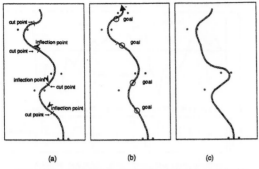

Fig. 9 (a) A training example (b) a result by the adaptive fuzzy controller (c) a result by the non-adaptive fuzzy controller (with constant limit_dir and limit_vel during the trial) (T=5 sec. for all cases)

controller in which the parameters and goals are adaptively changed for each subplan phases,[2] while the conventional fuzzy controller adopts the constant parameters and the goals fixed on the gates during the overall trial.

Figure 9(b) shows the control characteristics of the adaptive fuzzy controller that utilizes the tuned knowledge obtained from the training sample of Fig.9(a), and Fig.9(c) shows the result produced by the non-adaptive fuzzy controller where the parameters of limit_vel and limit_dir are kept constant over the entire plan. These results clearly demonstrate that the trajectory produced by the adaptive fuzzy controller is quite satisfactory when compared with the trajectory produced by the non-adaptive controller. That is, the adaptive fuzzy controller utilizes and integrates both the global control strategy obtained from the human operator and the desirable control characteristics of the fuzzy controller within the locally limited regions. We experimented the above method for the plant with a variety of time constants and the results are quite satisfactory. Especially, although it is actually difficult to tune the fuzzy controller for the plant with a larger time constant in a trial-and-error fashion, our method efficiently tunes the fuzzy controller even for this kind of plants.

CONCLUSIONS AND FUTURE WORKS

In this paper, we attempted to develop a prototype system of a man-machine collaboration system in which a fuzzy controller is adequately tuned by analyzing the domain expert's (an operator's) behaviour in terms of his global mental models. The purpose of our system proposed in this paper is not to "mirror" the human operator's behaviour on the fuzzy controller, but to "extract" the essence of the human behaviour only with respect to some partial aspect of his global control strategies. Then, this is transferred to the fuzzy controller in a way that fits with its design. The current system could perform only tuning in an off-line and a batch-process manner. However, a proposed idea to relate between the operator's behaviour and his mental models can be extended towards more sophisticated on-line, real-time systems as follows.

1. A Design of a Fully-Autonomous System.

---

[2] The parameters of limit_vel and limit_dir are specified from the maximum values of ω and θ, respectively, within each subplan.

It might be possible to design a fully-autonomous control system without a participation of the human operator. That is, such a system may be useful that monitors its control performance with respect to the supervisory models and autonomously tunes its parameters on detecting the deviation from the planned and the occurrences of the unexpected disturbances.

2. A Design of an Associate System: A Human-in-the Loop Controller.

Another promising extension of the current system is to an architecture of an associate system for the human operator dealing with the plant having complex dynamics. In this architecture, the process presented in Fig.7 is performed in concurrent with the operator's control operations, and monitors whether his actions are deviated from what is intended. When this kind of mismatch is detected, an associate system provides useful information varying from a simple alarm to a suggestions for a recovering treatment, i.e., an alternative plan. The key to the design of this kind of associate systems is how to control the presentation of such information to the operator. This control should be done considering the following issues; criticality of the error or consequences of neglecting errors, estimated operator's mental workload, and expected time the operator takes to complete the task. For this purpose, another mental model beside those presented in this paper should be installed and the task by itself should be altered into a much more complicated one from a current toy example.

REFERENCES

Banks, S.B. and Lizza, C.S. (1991). Pilot's Associate: A Cooperative, Knowledge Based System Application, IEEE Expert, 6-3, 18-28.

deKleer, J. (1977). Multiple Representation of Knowledge in a Mechanics Problem Solver, Proc. of 5th IJCAI, 299-304.

Kuipers, B. (1986). Qualitative Simulation, Artificial Intelligence, 29.

Lind, M. (1981). The Use of Flow Models for Automated Plant Diagnosis, In Rasmussen, J. and Rouse, W.B. (Eds.). Human Detection and Diagnosis of System Failures, Plenum Press, New York.

Mamdani, E,H. (1974). Application of Fuzzy Algorithms for Control of Simple Dynamic Plant, Proc. of IEE 121-12, 1585-1588.

Miyata,Y. (1987). Organization of Action Sequences in Motor Learning, Proc. of 9th Annual Conference on the Cognitive Science Society, 496-507.

Rasmussen, J. (1983). Skills, Rules, Knowledge; Signals, Signs, Symbols and other distinctions in Human Performance Models, IEEE Trans. of System. Man and Cybernetics. SMC-13, pp. 257-266.

Rouse, W.B., Geddes, N.D. and Hammer, J.M. (1990). Computer-Aided Fighter Pilots, IEEE Spectrum, March 1990, 38-41.

Stassen, H.G. (1987). Human Supervisor Modelling: Some New Developments, Int. J. Man-Machine Studies, 27, 613-618.

Sutton, R.S. (1988). Learning to Predict by the Methods of Temporal Differences, Machine Learning, 3, 9-44.

Veldhuyzen,W. and Stassen, H.G. (1977). The Internal Model Concept: An Application to Modelling Human Control of Large Ships, Human Factors, 19, 367-380.

Zadeh, L. (1965). Fuzzy Sets, Information and Control, 8, 338-353.

FUZZY SET THEORY FOR MODELLING THE NAVIGATOR'S BEHAVIOUR

R. Papenhuijzen* and H.G. Stassen**

**Ministry of Transport and Public Works, Transportation and Traffic Research Division, P.O. Box 1031,
3000 BA Rotterdam, The Netherlands*
***Delft University of Technology, Faculty of Mechanical Engineering and Maritime Technology,
Laboratory for Measurement and Control, Mekelweg 2, 2628 CD Delft, The Netherlands*

Abstract — It is expected that a computer program by means of which the behaviour of the complete system crew-ship-fairway can be simulated, will make a powerful tool for investigations in the field of navigation and fairway design. Hence, a fuzzy set model of the navigator's behaviour is proposed. Based on a series of preliminary experiments, the applicability of the fuzzy set navigator model is compared to that of an alternative, control theoretic development.

Keywords — Fuzzy control; fuzzy systems; human factors; ship control; simulation.

1. INTRODUCTION

In some fields of shipping, it seems that there has not come a stop yet to the increase in ship dimensions, and to the introduction of new ship types. Examples of such developments are scale enlargement in crude and LNG carrier design, experiments with 270m long push barge units on the river Waal, and the rapid growth of low air draught coaster transport on inland waterways in Europe. In addition, there is a tendency towards plying existing waterways with decreasing safety margins, as analternative for fairway modification or construction. This gives rise to the need for developing more sophisticated and cost-effective methods for evaluating safety aspects of waterway lay-outs and admission regulations.

When performing safety assessment investigations, the following problem has to be dealt with. The question whether a large ship can be steered safely through a confined waterway, depends not only on ship characteristics, but also on the skills of both the navigator and the helmsman. Being the supervisor of the complete navigation process (see Fig. 1), the navigator sets the engine speed, and he dictates course or rudder commands to the helmsman. The helmsman controls the heading of the ship, or simply applies the rudder commands which, under critical circumstances, are given by the navigator. Thus, it is evident that human behaviour aspects play a complicated yet essential role in assessing the relative safety of a combination of a ship, a fairway and a navigation task.

Traditionally, in order to accurately make allowance for human operator aspects, ship bridge simulators have been used. In a ship bridge simulator, ship behaviour is simulated by means of a computer, and the ship is steered by subjects who have a mock-up of a wheelhouse, and a synthetic outside view at their disposal. However, the high cost of operating such a facility may present serious problems, especially when a lot of experiments have to be carried out. This is one of the reasons why the Dutch Ministry of Transportation and Public Works launched a project to build a computer simulation facility, enabling fast time simulation of the complete navigation process (initially only single ship situations). The reproducibility of test results may be

considered an additional advantage of this tool. Eventually, other fields of application for navigation simulation devices are expected to emerge, such as the development of optimal manoeuvring strategies for training purposes.

The basic concepts for this project have mainly been studied by the former Institute of Mechanical Engineering TNO, in cooperation with Delft University of Technology. The most complicated part of the navigation simulation utility, was considered to be the submodel that describes the behaviour of the navigator. Within the frame-work of the project, two alternative approaches for implementing a navigator model have been followed, one of which is a control theoretic one, whereas the other one is based on fuzzy set theory.

This paper focuses on the fuzzy set branch of the project. For details on the control theoretic navigator model we refer to the literature (Papenhuijzen and Stassen, 1987—1989). To start with, in Section 2 the decision to employ fuzzy set theory for implementing a navigator model is motivated. In Section 3, we give an overview of the evolution of the fuzzy set navigator model, and of its working principles. A number of experiments have been carried out in order to investigate the feasibility of a fuzzy set based navigator model, and a first compari-

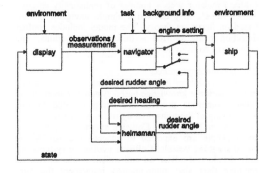

Fig. 1. Overview of the navigation process.

son between the performance of this model and that of its control theoretic counterpart has been attempted.

2. NAVIGATION AND KNOWLEDGE-BASED SYSTEMS

The number of maritime man-machine studies in which researchers resort to knowledge-based systems seems to be growing rapidly. In principle, we may distinguish between two main fields of interest, namely off-line applications and on-line applications. The first category mainly consists of design methods, to be used under experimental circumstances only. On-line applications are typically used on board the ship, playing an essential role in the navigation process. These categories are elaborated on in the next two paragraphs. Next, the suitability of fuzzy set theory for implementing navigation simulation is discussed, drawing on related projects, and generally recognized qualities of fuzzy set theory.

2.1. *Off-line Applications*

Since the early seventies, a lot of effort has been spent on modelling the helmsman's behaviour. The main objectives have been to better understand and develop training aspects, and to investigate the influence of human operator behaviour on required fairway dimensions. Veldhuyzen and Stassen (1977) proposed a nonlinear helmsman model, explicitly using the internal model concept. A phase plane method (in terms of course error and rate of change) was used for discriminating between different control strategies to be applied, and control settings were derived on the basis of predictions, yielded by a linear internal model. The fuzzy set auto-pilot[1] of Van Amerongen et al (1977), and the helmsman model of Sutton and Towill (1988) utilize fuzzy set theory to set the rudder angle depending on course error and rate of turn.

As stated in the introduction, a simulation model of a ship under human control, including an accurate description of the navigator's behaviour, is meant to be used for fairway design and waterway management in the first place. In this context, apart from our own project, the navigator model by Wewerinke (1988) may be referred to. Wewerinke's model, a straightforward implementation of the human operator optimal control model of Kleinman and Baron (1971), features detailed simulation of a variety of human controller aspects such as perception and attention allocation. The model's suitability for fairway design is limited however, due to the fact that the model requires carefully predefined tracks to be supplied by the operator of the simulation program.

Noguchi and Mizoguchi (1990) have suggested a new and interesting application of navigator models for training purposes. In order to confront trainees in a ship bridge simulator with a number of realistically reacting "other ships", they devised a simple navigator model, based on fuzzy set theory. The model includes conflict detection using blurred - fuzzy - observations, and elementary collision avoidance behaviour.

2.2. *On-line Applications*

Collision avoidance is by far the most important on-line use of maritime knowledge-based systems. Especially in Japan, numerous projects are being devoted to this topic, many of them inspired by traffic congestion problems in Tokyo Bay. Collision avoidance systems by some people are expected to succeed the ARPA[2] on board seagoing vessels. Collision avoidance systems not only decide and advise on anti-collision manoeuvres, if required they also autonomously implement the suggested manoeuvre by directly setting the auto-pilot. Implementation (Koyama and Yan, 1987, Iijima et al,

1991) is generally carried out by means of the production tool OPS5, which establishes rule priority and performs rule selection. The work of Koyama and Yan is very interesting in particular, in that their system not just extrapolates other ships tracks. In fact, it takes into account the way other ships may be expected to select and to apply traffic rules.

Automation projects on navigation are not restricted to collision avoidance systems. At present, a lot of research institutes are engaged in studies concerning the development of a so-called intelligent ship. This concept (Numano et al, 1990) implies full integration of monitoring and executing navigation, maintenance and cargo handling. Most of the projects are in a preliminary stage yet, due to technical difficulties as well as legislative constraints.

2.3. *Applying Fuzzy Sets*

Generally, a good control theoretic design is impossible, if not based on thorough knowledge on the system to be controlled, the environmental influences, and the relevant task elements. In many situations however, the structure of the system is unknown, parameter variation is unpredictable, or goals and constraints are not quantifiable by a single number. Fuzzy set theory, introduced by Zadeh (1965), may be invoked to deal with such problems. In fact, fuzzy set theory enables the conversion of a linguistic control strategy, based on expert knowledge, into an automatic control strategy, the resulting algorithm preserving, or being marked by, the expert's subjective, inaccurate, incomplete, and maybe even contradictory ideas.

Fuzzy sets have a short yet remarkably rich history in process control and decision making (Kosko, 1991). For the control of a number of ill-defined processes in particular, fuzzy logic controllers have proved to be the only satisfying substitute for a skilled human operator. One of the first, and most striking, examples of fuzzy logic control, is the control of a cement kiln by Holmblad and Østergaard (1982). A brief introduction to fuzzy set theory, and to the application of fuzzy set theory to control algorithms, is given in the Appendix.

In some respects, the problem of modelling the navigator's behaviour is related to that of devising collision avoidance systems, but there are significant differences nevertheless. Collision avoidance systems are based on the availability of fairly accurate measurements and the application of well-defined traffic rules for instance, whereas most aspects of the navigator's behaviour are much more inexact by nature. Therefore, fuzzy set theory was considered to be an appropriate technique for our purposes, rather than rule-based expert system approach which features selection of a single rule at a time. However, the concepts of collision avoidance systems are not irrelevant to implementing a navigator model, since eventually the model has to be extended so as to be able to describe multi-ship situations as well.

The available literature on fuzzy set helmsman models has been taken as an indication that fuzzy set theory offers a sound technique for implementing a navigator model. On the other hand, that does not say anything against the helmsman model of Veldhuyzen and Stassen, which, in our opinion, manages to show a remark-

---

[1] Automatic course-keeping system that nowadays under most circumstances is preferred to the traditional helmsman.

[2] ARPA = Automatic Radar Plotting Aid; a radar based navigation instrument that features simple track prediction for own ship and target ships.

ably realistic performance. As we see it, this is a result of the fact that for this model, a description of the goals to be attained, instead of the helmsman's behaviour as such, has been taken as a starting-point. Apart from that, it should be noted that, in principle, with the principle of linguistic phase planes (see eg Lee, 1990), integration of the different helmsman model concepts opens up.

Experiments by Papenhuijzen and Stassen (1989) have established the feasibility of a control theoretic navigator model that produces realistic, non-trivial tracks. Here again, as was the case with the helmsman model by Veldhuyzen and Stassen, good performance is achieved by focusing on underlying objectives. In an early stage already, it was concluded that probably the only way to model the navigator's behaviour more directly, would be by implementing rules, suitable for dealing with inexact information, and with a scope, wide enough to cover all relevant experimental conditions. In view of what has been stated about applying fuzzy set theory in general, this explains for our decision to probe a fuzzy set approach, as an alternative to the control theoretic method.

3. FUZZY SET NAVIGATOR MODEL

In 1986 the overall structure of the fuzzy set navigator model was drawn up (Salski et al, 1988). Since then, the various components underwent fundamental changes, which, however, did not affect the model structure as such. In the following, to start with, an overview of the navigator model is given. Further, the various fuzzy submodels are dealt with, not so much concentrating on the developing path, as on the actual state of the project.

3.1. *Assumptions*
Modelling the fairway geometry is accomplished by describing it as a number of contiguous trapezoids. Fairway segments are defined so as to ensure sufficient keel clearance in every point within the fairway, given the draught of the ship. Environmental influences (wind and current) are defined in every segment vertex, and calculated for other points by interpolation.

It was considered essential for the model to be able to feature realistic discrimination between heading output and rudder control. In practice, the choice between the two output modes is governed by the ship type and the heading change to be realized yet. For sea ships, if deviations of more then about 10 degrees are imposed, the navigator usually gives rudder commands. In the case of smaller deviations rudder control is left to the helmsman or auto-pilot. On an inland vessel, the roles of navigator and helmsman are usually combined into one person, which implies that rudder control is always the output mode.

For a helmsman (or auto-pilot) model, a PID-controller, as described by Veldhuyzen and Stassen (1977) is utilized. In fact, PID-control is exactly the principle on which most auto-pilots are based. Furthermore, though different in control characteristic, the dynamics of the combination of a helmsman and a ship does not significantly differ from that of an auto-pilot/ship combination. In practice, when navigating a given waterway, usually the engine setting is only manipulated in the case of collision avoidance or berthing. As a consequence, for the time being constant engine setting is assumed as a permissible simplification.

Ship models to be incorporated in the simulation program should range from inland vessel models to supertanker models. In principle, any ship model that is capable of generating a next state, given an actual state

and a combination of environmental influences and control settings, fits in with the simulation program.

3.2. *Overview of the Model*
As shown in Fig. 2, the fuzzy set navigator model has been divided into three major components, describing the navigator's state estimating, track planning, and track following behaviour, respectively.

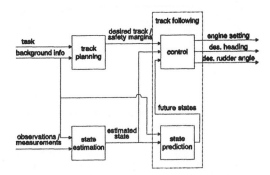

Fig. 2. Structure of the fuzzy set navigator model.

Within the framework of the project, Salski (1988) has proposed a fuzzy state estimator, merging the basic concept of Kalman filtering into a fuzzy logic algorithm. Although developing a fuzzy state estimation submodel was considered feasible, at first an already existing Kalman filter or, as an alternative, exact state information has been adopted for the state estimation submodel. Therefore, it is only in the other submodels that fuzzy sets come in.

The track defining process is aimed at drawing up a plan for carrying out the given navigation task. The plan comprises a planned track and a fuzzy definition of safe manoeuvring zones.

Supervisory and manual control behaviour, the execution of the travel plan, is simulated by the track following submodel. From time to time, future states are predicted, and related to the perception of safety as determined by the track planning submodel. If necessary, a control action is carried out.

3.3. *Track Planning*
In making his voyage preparation, the navigator of a sea ship usually interprets and plans his course on confined waterways as a concatenation of straight lines and arcs of circle (Pfeiffer et al, 1988). Apart from that, it has been observed that in inland navigation, captains tend towards performing their on-line planning task more or less similarly.

Hence, Würzner (1988) proposed a track planning algorithm to yield a series of straight lines and arcs of circle for a planned track. The result was a fairly complicated planning algorithm, because of the close interrelation between the two individual problems of planning tracks and defining safety zones. Obviously, the geometry of a curved track element is partially determined by the safety zones. On the other hand, proper definition of the safety zones is impossible without taking into account the corresponding radii of bend, as safety, in an outside bend in particular, decreases with increasing curvature. In addition, explicit calculation of track curvature requires basic knowledge of ship dynamics to be taken into account, as illustrated in Fig. 3.

A problem like this suggests an iterative solving method. It was tackled by Van der Meulen (1989)

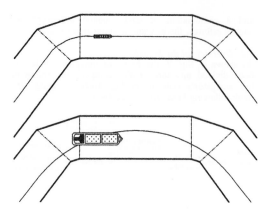

Fig. 3. Illustration of the interrelation between ship dimensions and track planning.

initially, who proposed a two step solution. Firstly, the model devises a so-called principal strategy; for every segment it is decided whether the manoeuvre in that segment is essentially a starboard deviation, a port deviation, or a straight course. This does not imply, however, that for the eventual planned track the transitions from one principal strategy component to another have to coincide with a segment edge. As inputs to the FLC that performs this function, the aspect ratio of the segment and the inclination of the fairway are used. This latter variable is ship dependent, in that the horizon for determining the inclination is a crisp value, produced by another FLC, and related to the ship's dimensions. As the second major step, both the safety zones and the eventual planned track are determined by means of an FLC, made to supply complete output membership function data. A first estimation of the curvature in every segment being available, in every cross-section the linguistic value "safe" of the linguistic variables "distance to bank" can be established. Apart from curvature, current and wind influences are used as an input here. For either side of the fairway the first value of the support of the resulting membership function is taken as a point to define the zero-safety line. The defuzzified output membership function value is used to specify the full-safety line. The result, a bivariate safety function of the position coordinates, is illustrated in Fig. 4.

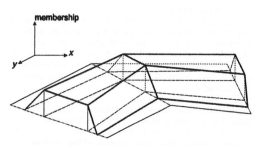

Fig. 4. Illustration of fuzzy definition of safe positions.

Then, the most favourable track to sail is determined as the least articulated concatenation of straight lines and arcs of circles to comply with a given minimum performance index. The principal strategy serves as the starting point for the steepest descent algorithm that solves this geometric problem. For that purpose, both the minimum safety grade and the so-called starboard preference compliance are evaluated by an FLC to a performance index to be minimized.

3.4. *Track Following*

The track following submodel, which was based on preliminary work by Salski et al (1988), Würzner (1988) and Wormgoor (1988), comprises five FLC's. In Fig. 5, which illustrates the working principle of the submodel, the FLC-fulfilled functions are indicated by double lines. The Figure covers the track following decision process as it takes place every time step, not including initialization and termination. Given the overview in subsection 3.2, and the elaboration on output mode in subsection 3.1, Figure 8 is considered self-explanatory as far as the general idea of the algorithm is concerned. Therefore, in the following we confine the discussion of track following to a number of supplementary comments.

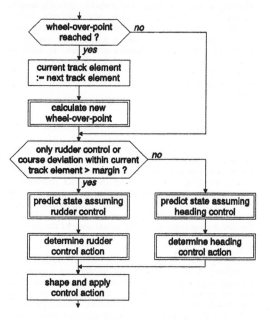

Fig. 5. Flow diagram of the track following decision process. Double outlining indicates FLC utilization.

Changing from one planned track element to another is triggered by reaching the so-called wheel-over-point, which is the position in which the desired change in rate of turn has to be anticipated. In the case of large seagoing ships in particular, the wheel-over-point may lie miles ahead of the beginning of the bend or lane to come. Therefore, the FLC that determines the wheel-over-point, processes perceived ship characteristics as well as the planned track geometry.

State prediction is based on heading, rate of turn, actual desired heading (when applying heading control) and actual rudder setting (when applying rudder control). Note that in the case of heading control the rule base concerns not only the ship dynamics, but also behavioural aspects of the rudder control element (helmsman or auto-pilot). The prediction horizon is readjusted every time step so as to comply with the following criteria:

○ the lower boundary is determined by the wheel-over-point (distance from that point to the next track element, actually);

○ the prediction horizon is increased if no control action has been applied; otherwise it is reset to its lower boundary;

○ the upper boundary is determined by the remaining length of the current track element.

A predicted position is evaluated in terms of safety. The resulting safety, and the predicted speed component perpendicular to the track element, are used for deciding on a control action.

4. EXPERIMENTS

Initially, two seperate prototypes, one for the track planning, and one for the track following algorithm had been devised. The definitive fuzzy set navigator model has been written in the C programming language. All through the program, the implementation utilizes the FLC generating and operating facilities which are detailed in subsection A.4. A first series of experiments has been carried out now, aimed at testing the program and demonstrating the feasibility of the navigator model as such. Consequently, not much effort has been put yet into specifying the contents of the knowledge bases of the FLC's. The knowledge bases, as well as the other elements of the experimental setting, however, were considered sufficiently realistic for the tests to produce the desired information. In the following, we present some results of the tests, after briefly discussing the relevant experimental conditions.

State estimation was by-passed in the experiments, in that perfect knowledge of position, speed, heading and rate of turn was assumed. Not including realistic state estimating in the experiments, in this stage is regarded as a permissible, and even necessary simplification. For the experiments, two ship types were used, namely a relatively agile 210m long container carrier, and, by contrast, a 315m long tanker. Thus, investigations into the influence of manoeuvrability on the overall system performance were enabled. A variety of fairways have been defined, both imaginary and existent ones. They are schematically represented as indicated in subsection 3.1.

In Figs 6 and 7, the results (safety lines and tracks) of two trials with the model in a (schematic) real life environment are presented.

5. DISCUSSION

The results of the experiments have been judged by the following criteria. Naturally, a thorough qualitative inspection of the results should only strengthen our confidence in the technical appropriateness of the implementation. Furthermore, such an inspection should lead to the conviction that the various algorithms function properly, and that they, in principle, allow a validated navigator model to spring from the project.

A survey of the results of all our experiments so far, has shown that, from a technical point of view, the program works satisfactory. Apart from that, the model itself appears to feature fairly realistic simulation of the navigator's behaviour. There are some doubts about the quantitative correctness of the safety lines, however, but we trust that this can be solved by applying a more sophisticated knowledge base for the corresponding FLC. Figures 6 and 7 may serve to illustrate what is stated above. It can be observed, for instance, that the effect, pointed at in Fig. 3, is realistically described by the model.

At the moment, it is hard to make a well-balanced comparison between the fuzzy set navigator model and its control theoretic counterpart. There are some preliminary observations to make, however. It was shown by Papenhuijzen and Stassen (1989) that the control theoretic approach has lead to a model that is especially capable of handling hazardous situations and simulating

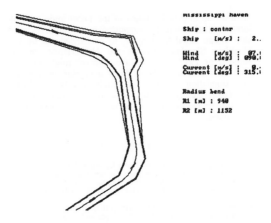

Fig. 6. Trial with a container carrier approaching Mississippi bay (Port of Rotterdam).

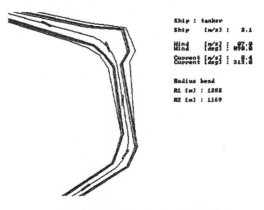

Fig. 7. Trial with a tanker approaching Mississippi bay (Port of Rotterdam).

difficult manoeuvres, which require multi-output commands (both rudder and engine setting) to be given. Under these conditions, a real navigator does not think in terms of straight lines and arcs of circle any more. Performing such a navigation task is not the forte of the fuzzy set navigator model. Nevertheless, in carrying out a more or less ordinary navigation task, like navigating a canal or approaching a harbour, for instance, the fuzzy set navigator model is expected to be superior to the control model. This is concluded from the fact that the fuzzy model performs explicit track planning, in about the same way a navigator will do that. The control model, by contrast, is aimed at finding a planned track that is optimal in terms of risk, control effort and terminal state error. The result may approximate, but never attain an actual planned track that is composed of lines and arcs.

6. CONCLUSION

Some first experiments have indicated that a fuzzy set approach to the problem of modelling the navigator's behaviour is a promising one. In comparison with the control theoretic navigator model, the fuzzy set model appears to be especially suitable for simulating relatively large-scale, and not too complicated passages. The model is still under development, however. The most important issue to be worked on now, is the improvement of the knowledge bases of the fuzzy logic controllers. Validation will be supported by a comprehensive ship bridge simulator project, to be carried out by the Maritime Simulator Centre the Netherlands, in the summer of 1992. Next, future research might concentrate on combining several fuzzy navigator models into one simulation program. A high fidelity multi-ship simulation facility namely, is considered indispensable for further exploration of fairway capacity and safety assessment.

REFERENCES

Amerongen, J. van, H.R. van Nauta Lemke and J.C.T. van Veen (1977). An auto-pilot for Ships Designed with Fuzzy Sets. In H.R. van Nauta Lemke (Ed.), *Digital Computer Applications to Process Control*. Elsevier (North-Holland), Amsterdam. 479—487.

Holmblad, L.P. and J.J. Østergaard (1982). Control of a Cement Kiln by Fuzzy Logic. In M.M. Gupta and E. Sanchez (Eds.), *Fuzzy Information and Decision Processes*. Elsevier (North-Holland), Amsterdam. 389—399.

Iijima, Y., H. Hagiwara and H. Kasai (1991). Results of Collision Avoidance Manoeuvre Experiments Using a Knowledge-Based Autonomous Piloting System. *The Journal of Navigation*, 44, 194—204.

Kleinman, D.L. and S. Baron (1971). *Manned Vehicle Systems Analysis by Means of Modern Control Theory*. NASA CR-1753.

Kosko, B. (1991). *Neural Networks and Fuzzy Systems*. Prentice-Hall, New Jersey.

Koyama, T. and J. Yan (1987). An Expert System Approach to Collision Avoidance. *Proc. 8th Ship Control Systems Symposium*, The Hague, 3.234—3.263.

Lee, C.C. (1990). Fuzzy Logic in Control Systems: Fuzzy Logic Controller — Part I & II. *IEEE Trans. Systems, Man and Cybernetics*, SMC-20, 404—435.

Meulen, B.O. van der (1989). A Fuzzy Set Model of the Navigator's Track Planning Behaviour. M.S. Thesis, Faculty of Mechanical Engineering and Maritime Technology, Delft University of Technology, Delft. (In Dutch.)

Noback, H., A. Salski and H.G. Stassen (1988). A model of the Navigator's behaviour based on fuzzy set theory. In J. Patrick and K.D. Duncan (Eds.), *Training, Human Decision Making and Control*. Elsevier (North-Holland), Amsterdam. 205—222.

Noguchi, N. and S. Mizoguchi (1990). Ship Maneuvering Trainer with Intelligent Objects Using Fuzzy Data. *Proc. MARSIM & ICSM 1990; Joint International Conference on Marine Simulation and Ship Manoeuvrability*, Tokyo, 111—116.

Numano, M., K. Okuzumi and T. Fuwa (1990). Concepts and Safety Evaluation of Intelligent Systems in Future Ship Navigation. *Proc. MARSIM & ICSM 1990; Joint International Conference on Marine Simulation and Ship Manoeuvrability*, Tokyo, 209—215.

Papenhuijzen, R. and H.G. Stassen (1987). On the modelling of the Behaviour of a Navigator. *Proc. 8th Ship Control Systems Symposium*, The Hague, 2.238—2.254.

Papenhuijzen, R. (1988). On the Modelling of the Behaviour of a Navigator; Background and Proposal for a Control Theoretic Approach. In J. Patrick and K.D. Duncan (Eds.), *Training, Human Decision Making and Control*. Elsevier (North-Holland), Amsterdam. 189—203.

Papenhuijzen, R. (1988). Modelling the Planning and Control Behaviour of the Navigator. *Proc. 7th European Annual Conference on Human Decision Making and Manual Control*, Paris, 195—200.

Papenhuijzen, R. and H.G. Stassen (1989). On the Modelling of Planning and Supervisory Behaviour of the Navigator. *Proc. 4th IFAC/IFIP/IFORS/IEA Conference on Man-Machine Systems*, Xi'an, China, 19—24.

Pfeiffer, M.J., R. Papenhuijzen and N. Stassen (1988). Eliciting the navigator's knowledge on ship dynamics. *Proc. 7th European Annual Conference on Human Decision Making and Manual Control*, Paris, 201—206.

Procyk, T.J. and E.H. Mamdani (1979). A Linguistic Self-Organising Process Controller. *Automatica*, 15, 15—30.

Salski, A. (1988). State Estimation using a Fuzzy Model of System. *Proc. 7th European Annual Conference on Human Decision Making and Manual Control*, Paris, 207—212.

Veldhuyzen, W. and H.G. Stassen (1977). The Internal Model Concept: An Application to Modeling Human Control of Large Ships. *Human Factors*, 19, 367—380.

Wewerinke, P.H. and J. Perdok (1990). Navigator-ship Models for the Assessment of Maneuvering Performance and Vessel Traffic Systems. *Proc. MARSIM & ICSM 1990; Joint International Conference on Marine Simulation and Ship Manoeuvrability*, Tokyo, 237—244.

Wormgoor, E. (1988). Modelling the Navigator's Track Following Behaviour. M.S. Thesis, Royal Netherlands Naval College, Den Helder / Faculty of Mechanical Engineering and Maritime Technology, Delft University of Technology, Delft. (In Dutch.)

Würzner, I.M. (1988). Evaluation and Evolution of a Fuzzy Set Navigator Model. M.S. Thesis, Faculty of Mechanical Engineering and Maritime Technology, Delft University of Technol-

ogy, Delft. (In Dutch.)

Zadeh, L.A. (1965). Fuzzy Sets. *Informat. Control*, 8, 338—353.

Zadeh, L.A. (1973). Outline of a New Approach to the Analysis of Complex Systems and Decision Processes. *IEEE Trans. Systems, Man and Cybernetics*, SMC-3, 28—44.

APPENDIX: FUZZY SET LOGIC

In the next Sections, we first give a brief introduction to the basics of fuzzy sets. Then, since our model has been designed as a number of more or less separate fuzzy decision making processes, we concentrate on the principle and the implementation of fuzzy logic control.

A.1. *Fuzzy Sets*

Traditionally, mathematics dictate that an object be only a full member of a given set, or not a member at all. There is nothing in between. What distinguishes fuzzy sets from conventional, "crisp" sets, is the fact that fuzzy set theory allows *degrees* of membership, varying from 0 (non-member) to 1 (full member). Fuzzy sets provide an elegant mechanism to describe and handle all sorts of so-called *linguistic* variables, the variable "speed" for instance. A linguistic variable is characterized by a series of terms, such as "low", "moderate", or "high", each term representing a fuzzy set (eg the fuzzy set "high speeds", of which a speed of 80 km/h may be a member to a degree of only 0.3). The function that associates a grade of membership of a fuzzy set with every object that to some extent could be in that set, is called the membership function. The most commonly used membership function form is that of a trapezium. In some cases, when smoothness of the membership functions is considered essential, exponential relationships are employed, for instance by Van Amerongen et al (1977).

A.2. *Notation, Terminology and Basic Operations*

A fuzzy set A of a universe of discourse U is characterized by a membership function $\mu_A:U\rightarrow[0,1]$, which associates with each element y of U a number $\mu_A(y)$ in the interval $[0,1]$, which represents the grade of membership of y in A. The *support* of A is the set of points in U at which $\mu_A(y)>0$. A *fuzzy singleton* is a fuzzy set whose support is a single point in U. In particular, A is called a *nonfuzzy singleton* if the grade of the only element is 1.

Let A and B be two fuzzy sets in U with membership functions μ_A and μ_B, respectively. The set theoretic operations of union and intersection for fuzzy sets are defined as follows. The membership function $\mu_{A\cup B}$ of the union $A\cup B$ is pointwise defined for all $u\in U$ by

$$\mu_{A\cup B}(u) = \max\{\mu_A(u),\mu_B(u)\}. \qquad (1)$$

The membership function $\mu_{A\cap B}$ of the intersection $A\cap B$ is pointwise defined for all $u\in U$ by

$$\mu_{A\cap B}(u) = \min\{\mu_A(u),\mu_B(u)\}. \qquad (2)$$

For a detailed discussion of fuzzy sets and operations on fuzzy sets we refer to Zadeh (1973).

A.3. *Basic Idea of the Fuzzy Logic Controller*

In the following, the principle of fuzzy logic control is briefly reviewed. We concentrate on the most common type of fuzzy logic controller (FLC). It was employed by Holmblad and Østergaard (1982), and it was chosen for our project as well. A comprehensive survey of FLC design has been made by Lee (1990).

As shown in Fig. A.1, an FLC consists of the following major units. Fuzzy reasoning is performed by the *inference engine*, the kernel of the FLC. The *fuzzification interface* converts real numbers into fuzzy sets, to be supplied as an input to the decision making logic. The *knowledge base* of an FLC comprises two components,

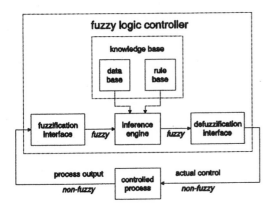

Fig. A.1. Conceptual configuration of fuzzy logic controller.

```
rud_ctrl
2 inputs
    heading
    -3.14      3.14
    3 terms
        low            left_wing      -2.0    0.0
        medium         bucket         -2.0   -1.0    1.02.0
        high           right_wing      0.0    2.0
    speed
    0.0        8.0
    5 terms
        zero           left_wing       0.0    0.1
        dead_slow      bucket         -0.1    0.1    0.51.0
        slow           bucket          0.0    1.0    2.03.0
        moderate       bucket          1.0    2.0    3.55.0
        high           right_wing      4.0    6.0
2 outputs
    rudder
    -35.0      35.0
    5 terms
        starbd_max     left_wing     -30.0  -25.0
        starbd_mod     bucket        -30.0  -20.0  -10.0-6.0
        zero           bucket         -5.0   -1.0    1.05.0
        port_mod       bucket          5.0   10.0   20.030.0
        port_max       right_wing     25.0   30.0
    quality
    0.0        1.0
    2 terms
        low            left_wing       0.2    0.6
        high           right_wing      0.4    0.8
3 rules
    rule_1
    2 conditions
        heading        low
        speed          slow
    2 conclusions
        quality        low
        rudder         starbd_mod
    rule_2
    1 conditions
        heading        medium
    1 conclusions
        quality        high
    rule_3
    2 conditions
        heading        high
        speed          moderate
    2 conclusions
        rudder         port_max
        quality        low
101 discr. intervals for output function
1 order of moment
```

NOTE
For ease of implementation, three membership function shape types have been defined. The membership function of the fuzzy set of low speeds may be an example of *left wing* function shape. Likewise, the membership functions of moderate speeds and high speeds are examples of (inverted) *bucket* shape and *right wing* shape, respectively.

Fig. A.2. Simplified example of linguistic input file.

namely, a data base and a fuzzy control rule base. The data base provides the inference engine with the membership functions of fuzzy sets used in the rule base, whereas the rule base holds the control rules. The *defuzzification interface* transforms fuzzy control actions into nonfuzzy controls.

Fuzzification is performed by interpreting an input x_0 as a nonfuzzy singleton A, that is, $\mu_A(x)$ is equal to zero, except at the point x_0, at which $\mu_A(x_0)$ equals one. The antecedent part of a control rule consists of a number of conditions, each of them relating an input to a term which is represented by a fuzzy set ("*speed* is *high*" for example). Analogously, the antecedent consists of one or more conclusions. Control rules, just as the contents of the data base, are usually formulated by consulting

experts. The so-called self-organising controller (see Procyk and Mamdani, 1979), however, is an exception to this rule. For a self-organising controller, it is not necessary to have an exhaustive set of rules as the starting point. By applying a performance criterion, the algorithm is capable of conceiving additional rules, and of deleting redundant or contra-productive rules.

In order to illustrate the operation of the inference mechanism, let us consider the following dual input single output mechanism of n fuzzy control rules:

<u>input:</u> x is A' and y is B'
<u>rules:</u> if x is A_i and y is B_i then z is C_i
<u>output:</u> z is C'

where x, y and z are linguistic variables representing the process state variables and the control variable, respectively; A_i, B_i and C_i are linguistic values (terms) of the linguistic variables x, y and z in the universes of discourse U, V and W, respectively, with $i=1,2,\ldots,n$. The inference process starts with assessing for every rule the weighting factor α_i, that expresses the significance of that rule. Since nonfuzzy singletons are assumed for a defuzzification strategy (ie the fuzzy inputs A' and B' are equal to the nonfuzzy inputs x_0 and y_0, respectively), this may simply be performed as:

$$\alpha_i = \mu_{A_i}(x_0) \ \wedge \ \mu_{B_i}(y_0) \tag{3}$$

with $i=1,2,\ldots,n$, and where $\mu_A(x_0)$ and $\mu_B(y_0)$ play the role of the degrees of partial match (per condition) between the controller input and the corresponding term in the ith rule. The contribution of the ith rule to the -fuzzy - controller output is then calculated as

$$\mu_{C_i}(w) = \alpha_i \cdot \mu_{C_i}(w). \tag{4}$$

As a result, the membership function μ_C, which is the union of the individual control decisions (4), is

$$\mu_C(w) = \bigcup_{i=1}^{n} \alpha_i \cdot \mu_{C_i}(w). \tag{5}$$

Deducing a crisp control value z_0 from C' is achieved by performing one of two defuzzification strategies. The first method ("centre of area" or "zero-order moment" method) calculates the abscissa value that divides the area under the output membership function in two equal parts. The other method ("centre of gravity" or "first order moment" method) yields the centre of gravity of the output membership function.

A.4. *Implementation of Fuzzy Logic Control*

Since FLC's play an important role in the project, due attention has been given to the implementation of fuzzy logic control. A flexible tool has been designed, as a library of functions to be used in a computer program (C-language). The library comprises functions for creating, interfacing, firing and recording an FLC. Generation and configuration of a multi-input multi-output FLC is performed by including a single command, and providing a plain language ASCII-file. This input file describes the contents of the knowledge base, and specifies the defuzzification algorithm. A listing of a sample input file is shown in Fig. 4. The data structure is designed so as to allow easy on-line expansion or reduction of both the data base and the rule base. Thus, possible use of the tool kit for any type of self-organising control is anticipated.

Once generated, each separate FLC is interfaced to its environment (real input and output variables), including, if necessary, chaining to other FLC's. The function performing this operation, has to be provided with two dispatch tables, that relate linguistic input and output data names to computer program variables. Firing an operational FLC can be performed then by a single command. For maximum flexibility, the FLC can be made to flush the data of its output membership function into a user defined array.

Fig. A.1. Conceptual configuration of fuzzy logic controller.

Fig. A.2. Simplified example of linguistic input fig.

ANALYSIS OF OPERATOR BEHAVIOUR FOR COGNITIVE MODEL IMPLEMENTATION

F. Decortis*, J. Kjaer-Hansen*, J.A Lockett** and K.H. Wheatley**

*Commission of the European Communities, Joint Research Centre, Institute for Systems Engineering and Informatics, Ispra, VA 21020, Italy
**Control and Dynamics Department, AEA Reactor Services, Winfrith, Dorchester DT2 8DH, UK

ABSTRACT

We present results of an analysis of the cognitive processes of an operator controlling the simulator of a complex plant (i.e., a Steam Generator Heavy Water Reactor, SGHWR). This research has been guided towards the adaptation of a computer-based model of an operator (COSIMO), to the specific task of controlling the SGHWR. The objective of the research is to analyse an operator's cognitive processes in a problem solving situation, by focusing on aspects related to information checking and actions to be taken. Five problems to solve on the simulator were selected. Four operators participated in the experiments, three reactor foremen and one fuel foreman. Current techniques of observation and of protocol analysis have been employed for exploring cognitive processes.

KEYWORDS

Cognitive simulation, Cognitive processes, Problem solving, Expertise, Man-machine systems, Process Control.

INTRODUCTION

Several models of an operator confronting a complex environment have been developed in the last decade. Models like CES (Cognitive Environment Simulation, Woods et al., 1987); OMSpert (Rubin et al., 1988); AIDE (Amalberti et al., 1989) address the problem of the operator's cognitive task in different working environments. Their main objectives are (1) to provide a tool for analysing man-machine interactions, i.e. to analyse human error; (2) to assist in the development of tools and techniques for helping an operator in the control of complex installations, in particular, to contribute to the design of Intelligent Decision Support Systems (Mancini et al., 1987). One model of this type COSIMO (Cognitive Simulation Model), has been developed by Cacciabue et al., (1991). Here we explore this model in relation to the behaviour of real operators. This analysis was carried out at the nuclear installation at Winfrith Technology Centre (WTC).

The starting point of the research presented in this paper was to adapt the COSIMO model, originally developed for modelling the operator task when controlling an Auxiliary Feedwater System (AFWS), to an unforeseen situation of man-machine interaction. This raises the following question: is it possible to generalize an operator's model to a task it was not designed for and what are the risks when one tries to predict the operator's activities on the basis of the model?

Adapting the model raises at least two major problems:

(1) A technical problem which is to adjust the model to the new situation (one of controlling the simulator of a nuclear reactor), in particular to build up a knowledge base which corresponds to the knowledge an operator uses to solve the incidents that may occur.

(2) A problem of psychological validity of the model which requires us to evaluate whether the cognitive primitives and the formalized knowledge in the model are the same as those used by the operator in the new situation.

In this study, we will only examine the second problem, that of the psychological validity. More specifically, the objective of the research was to analyze the operator's cognitive mechanisms on the basis of the COSIMO model. We focused in particular on the strategies of observation of information and actions in various situations of problem-solving. The paper will be divided into four parts. In the first one, we will briefly describe the COSIMO model. In the second part, we will explain the situation in which the research was carried out, i.e. the plant, the simulator, and the various incidental scenarios. In the third part, we will give a description of the method used for the experimental investigation. Finally, we will present the results obtained so far.

The COSIMO model uses some concepts of the cognitive under-specification theory developed by Reason (1984, 1990) the cognitive primitives of similarity matching, of frequency gambling and the representation of knowledge through an entity-attribute matrix. The cognitive model presumes a dual information processing architecture comprising a limited, serial but computationally powerful Working Memory (WM), and a virtually limitless Knowledge Base (KB) (Fig. 1). The Knowledge Base is a repository of rules-based frames based on an Entity-Attribute Matrix (EAM). Entities are incidents that can occur and which can be familiar or unfamiliar to the operator; attributes are descriptors of several parameters which can change and vary by their *diagnosticity* accross the several incidents (Reason, 1990). The diagnosticity of attributes may be used to discriminate between one incident and another. The Working Memory (WM) represents a working area for the cognitive processes, including the management and the temporary storage of data. The role of the WM is (1) to assert the external and internal cues to match similarities in the KB; (2) to decide whether the hypothesis is appropriate, and if not (3) to reiterate the search with revised cues. The cognitive architecture has two primitives for selecting stored knowledge units and for bringing them into WM. These primitives, similarity matching (SM) and frequency gambling (FG), operate in a parallel, distributed and automatic fashion within the KB. SM matches perceived cues coming from the work environment with corresponding diagnostic cues described in the KB. FG solves possible conflicts between partially matched hypotheses selected by SM, in favour of the more frequent, already encountered and well known incident scenario. It is assumed that, given the incidental situations in which it is to be applied, no deep reasoning is perfomed and that the operator resorts to previous experience on similar cases. Consequently COSIMO focusses on the rule-based frame implemented in the KB and on the cognitive primitives of SM and FG only.

In consequence, the application of the under-specification theory to the operator's model implies that:

(1) The knowledge used by an operator in incidental situations can be represented by an entity-attributes matrix. This matrix is pre-defined and has a uniform structure, e.g. the elements of knowledge represented have the same properties and are of the same type. Entities are incidents and attributes are parameters of the process which take specific values according to the type of incident. This matrix is supposed to be the heart of the knowledge of the operator, and this knowledge will be activated to recognize the incidental situations.

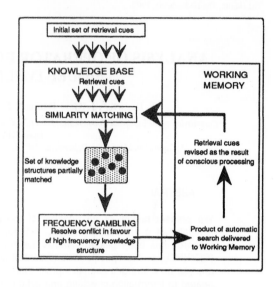

Fig. 1. Dual information processing architecture applied to COSIMO (from Reason, 1990).

(2) In an incidental situation, the operator uses two cognitive primitives to recognize the situation he deals with and on this basis, to decide on which corrective action to take to solve the problem. These two primitives are similarity matching and frequency gambling, which are complementary cognitive processes used to recognize the incident.

It is undoubtedly useful to check whether there are differences between novice and expert operators, in their approach to problem solving. We shall compare the response given by both in specified and under-specified situations. Only one aspect of under-specification will be investigated in this study. The loss of signals on loops of automatic control can be confusing for the operator, but this ambiguity can be avoided once the operator goes back to manual control of the sub-component. In this study, operators were confronted to two types of incidents:

- some incidents have symptoms that are not confusing: they refer directly to the cause of a problem (e.g. incidents 1 &4);

- other incidents are much more confusing because they tend to indicate an incident which in fact does not exist although the process starts to behave as if there had been an incident, because incidents are inferred from faults in instrumentation (e.g. incidents 2,3, and 5).

DESCRIPTION OF THE SITUATION

The SGHWR is a prototype reactor of 100 MWe Located at Winfrith (U.K.). It is a direct-cycle boiling-water reactor, moderated with heavy water. Normal water is used as feedwater. The plant appears similar to a vertical CANDU reactor, cooled with normal water (the presence of a calandria tank and of pressure tubes in zircalloy cause the similarity with CANDU reactors). The reactor operated for 23 years and is presently at the very first stages of decommissioning. In the primary circuit of the SGHWR, the cold feedwater coming from the condenser enters into the two steam drums in which it is mixed with the hot water and extracted from the wet steam. The resulting mixture is fed into the pressure tubes by the primary circulating pumps. In the vertical pressure tubes the water is heated by nuclear fission; the steam water mixture enters into the steam drums were the steam is dried. The steam leaving the steam drums feeds directly the turbine by means of two governor valves (turbine control valves).

The Simulator

The SGHWR simulator interface is a partial replica of the control room instrumentation. The simulation is fully digital at present. The interfaces of the simulator are composed of:
- a control desk with manometers, digital indicators and analogue recorders
- a rear panel also with analogue recorders and digital indicators.

Control Systems

During normal operation, the SGHWR is fully automated by means of three main controllers, regulating respectively: the reactor power, the pressure in the steam generators and the water level in the steam generators. Each one of these automatic controllers may be set on manual, so that the man-machine interaction is mainly related to these three controllers.

The power controller. The power controller regulates the reactor power by acting on the level of the heavy water in the calandria tank. The power is lowered by draining the heavy water under gravity into dump tanks, and it is increased by pumping the heavy water back to the calandria. The power controller may be in two automatic modes, the FLUX mode and the FLOW mode, thus the moderator level is controlled on the error between the demanded neutron flux and the flux measurement or on the error between the demanded steam flow and the steam flow measurement suitably lagged (about 40 s of time lag). Normally the Flux mode is used only at low power, approximately 30 MWe maximum.

The pressure controller. During normal operation, when the turbine is loaded, the pressure in the steam drums is controlled by acting on the governor valves by means of speeder motors. The automatic controller works on the signal error between a measured pressure and a demanded pressure and acts on a variable-speed slow speeder motor (20 % / minute). In manual operation, a fixed-speed fast speeder motor is employed, so that in manual pressure changes should be made by a series of short bursts.

The drum level control system. In order to regulate the water level in the steam generators, each steam generator is provided with 10% and 100% feedwater valve controllers. The 10% controllers act on the 10% valves and are used until the power of 60 MW(t) is reached. At higher power, the 100 % valves and the respective controllers operate. The 10% controllers work only with the level error signal. The 100% controllers run on the difference between the measured steam flow and the feedwater flow, with an additional (proportional+integral) term from the drum level transducers.

METHOD

The elaboration of the data collection method and the first attempt of recording the five incidents were carried out together with a shift manager, a reactor foreman an engineer and a technician. These experiments made it possible to evaluate how difficult the incidents were and the total duration of a session.

Operators

Four operators participated in the experiment, among which were three reactor foremen and one fuel foreman. The age of the operators vary between 30 to 47 years old as shown in Table 1. The average work experience on the reactor vary between four months to six years. The usual hierarchy is the fuel foreman, the reactor foreman and the shift manager. The fuel foreman is thus generally less experienced.

TABLE 1 Personal Data and Work Experience RF = Reactor foreman; FF = Fuel foreman. E1 and E1 : experts operators; N1 and N2 : novices operators.

	AGE	WORK EXPERIENCE (years)	WORK EXPERIENCE ON SGHWR
Reactor foreman E1	31	10	4 (FF) 2 (RF)
Reactor foreman E2	43	27	4 (FF) 2 (RF)
Reactor foreman N1	47	2	4 months (RF)
Fuel foreman N2	30	4	2 (FF)

The 'experts' and the 'novice' operators were distinguished on the basis of their seniority and of the number of positions occupied between the position of fuel foreman, the reactor foreman and the shift manager. The last criterion seems significant as the more an operator has occupied different posts, the more he seems to have a wide experience of the process.

Choice of Incidental Scenarios

Five incidents were chosen for the experiment on the simulator. Each incidental scenario may vary according to its length and its complexity. For each incident, we asked the operator to identify the problem on the simulator and to solve it. Five incidents were thus presented to the four operators. The incidental scenarios were chosen with assistance from three experts: two shift managers and a nuclear engineer external to the enterprise. Incident 1 relates to the response of the plant to an external disturbance. Incident 4 deals with a failure in a plant component. The incidents 2, 3 and 5 correspond to faults of the process control instrumentation. The incidents are (see Fig. 2 for a schematic description) :

(1) A negative step and then a positive ramp in the electric frequency of the external grid.
(2) A sudden loss of the steam flow signal to the power controller. This is a failure of the signal used by the power controller (which uses the steam flow error).

(3) The progressive deterioration, in the form of a negative ramp of the steam flow signal relating only to the feedwater controller of the north drum.
(4) The total loss of one feedwater pump at reduced power of the reactor, i.e. 65 MWe.
(5) The progressive deterioration, in the form of a negative ramp of the steam flow signals relating to both the feedwater controller (North steam drum) and the power controller.

Incidents 1, 2 and 5 have some common element in the initial opening of the governor valve, up to the maximum limit, and in the consequent increase in the pressure of the sensitive oil, working the governor valve. The complexity of the incidents increases with their order: the incidents 3 and 5 are quite difficult since the fault is initially zero so that there are not initial symptoms and the early diagnosis is quite difficult. The probability of occurrence of the scenarios, in a real situation, is approximately decreasing as their number increases. Incidents 2, 3, 4 and 5 may have a delayed common effect of loss of water level in one or both the steam drums. The incident 4 is interesting because the operator may be faced to possible strategies, in particular he may decide to manually operate a Y trip or try to compensate, in order to keep the plant working, with the eventual risk of a X trip, due to very low level in the drums.

Main attributes. The main attributes used to solve the various incidents are shown in Table 2. Those attributes were identified by the three 'experts'.

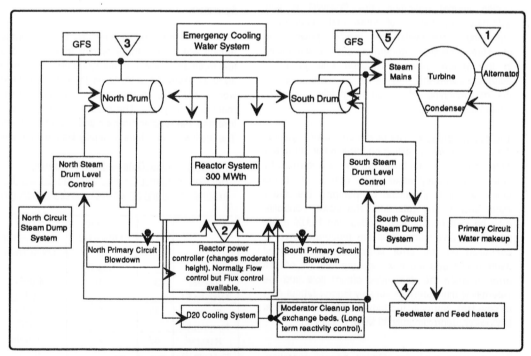

Fig. 2. SGHWR System schema and the five incidents (GFS= Guaranteed Feedwater System).

TABLE 2 Main Attributes Used for Solving the Five Incidents.

Main Attributes\Incidents	1	2	3	4	5
Sensitive oil pressure	x	x	x	x	x
Frequency error	x				
Turbine speed	x				
N. steam drum pressure	x		x	x	x
S. steam drum pressure			x	x	x
North drum level	x	x	x	x	x
South drum level			x	x	x
Governor valve	x	x	x	x	x
Pressure control system error	x	x	x	x	x
Steam main pressure	x	x	x		x
Flux error			x	x		x
Demanded flux			x			x
Flow error		x	x	x	x
D20 level		x			
Ind. position valve moderator		x			
North feedwater flow			x	x	x
South feedwater flow			x		
100/10 N feedw. valve posit. ind.			x	x	x
100/10 S feedw. valve posit. ind.			x	x	x
Alternator output				x	

Description of the incidents. Incident 1. This situation is well known to the operators since it can happen once or twice a day, particularly during the night, when the starting of large pumping stations may provoke stepwise loads on the electric grid, causing a slow down in the turbine speed. Consequentially, the governor valve opens fully and the "governor alarm" may be activated. This happens because once the valve has reached the maximum opening position the sensitive oil pressure can become excessively high. If the pressure continues to increase, in general the operator, realizes that there is some problem on the electric grid and has to decide to intervene, and override the automatic power controller by using his judgement of the sensitive oil pressure. The intervention is not always necessary since the problem can be stabilized by the pressure controller. A possible action is to take manual control of the pressure oil which operates the governor (High pressure= valve open; low pressure= valve closed). The sensitive oil pressure can be increased or decreased automatically or manually by operating with two different motors. The manual motor has a quicker response than the automatic one, and in this way it is sometimes necessary to manually operate on the sensitive oil pressure. This action is done without actually opening the governor, since the TCV (Turbine Control Valves) remains fully open until the pressure of the oil remains above 43 PSI. Once the oil pressure is reduced, it is possible to switch again to automatic control, but not before the pressure error instrument has reached the zero level. Otherwise, the governor opening is also affected.

Incident 2. The second scenario originates from the loss of steam flow signal to the heavy water controller which controls the level in the calandria tank and is thus the power controller of the reactor. The steam flow is measured by two independent instruments feeding separately the feedwater and the power controllers: in this scenario we assume only the failure of the signal to the power controller. This kind of failure is realistic, since it did happen in reality although with a quite low probability; about twice in ten years. However, this is a well known situation for the operators since this case is often presented in the simulator. The power controller can be either in auto FLUX control (usually used at low power) or auto FLOW control. The only other state of the power controller is in manual, where the operator controls the reactor power by raising or lowering the D2O (heavy water) level, using the "dead man's handle". Normally, at power, the power controller is set on FLOW control. The power controller works on the error signal between the measured steam flow and the steam flow demand which may be selected from the control room desk. In this scenario the automatic regulation system can not operate since the flow error is meaningless. The "flow error" instrument goes out of scale since the error is very big. The power rises, since the control on the flow is lost, and the margin to trip instruments (three big analog devices with 2 out 3 logic) show a decreasing margin to the trip level of the reactor. Together with the power, the pressures of the drums rise, so that the pressure controllers open the governor valves and the sensitive oil pressure may become excessive, triggering a warning. Once the warning level is reached, the operator receives a warning (governor alarm) and an acoustic buzzer is triggered. In the absence of any operator action, the margin to trip becomes zero and the reactor is tripped (Y trip). The correct procedure to recover this kind of situation is to switch the power control to manual operation and, eventually, to decrease the reactor power by manually operating on the dead's man handle of the control desk. A reduction in the power occurs due to the operation of the dead man's handle and to the consequent decrease in the heavy water level (this is due to the calandria high level valve and, in some cases, might be enough). Thus the neutron flux is decreased, the margin to trip is increased and the trip is avoided. Once the situation has stabilized, the operator has to switch the control of the reactor to the "flux control" way, by first adjusting the neutron flux to the needed reactor power. Finally, the automatic control can be switched on again.

Incident 3. This kind of situation is extremely rare since it has never happened in the reactor. There is a progressive deterioration of the steam flow signal to the north drum feedwater controller in the form of a negative ramp with a slope of the error of 10 % every minute. Once in twenty years there was a fault like this in the reactor, but the ramp was positive.

Although very infrequent, this kind of progressive deterioration of the steam flow signal to the feedwater controller is very interesting because the operators are faced with a dynamic situation whose effect increases in time: the longer they wait to act, the easier it should be for them to understand that there is something abnormal, but it would also be more difficult to recover the plant state. This scenario might be interesting to assess the time of intervention for operators with different degrees of skills. To better simulate a real situation, some disturbances have been added on the frequency of the electric grid which affects the opening of the governor. Because of the selected negative ramp, the feedwater controller is progressively less able to cope and the controller interprets the signal as a decrease in the feedwater demand. The 100% feedwater valve will then start to close. As the reactor power has not decreased, the net effect results in a steady loss of water in the north steam drum. If the operator does not intervene and selects manual control of the north 100% feedwater valve, the reactor would trip on low drum level (Y trip, 36 inches level). The correct action is to select manual control of the north feedwater controller and open the valve to make up the steam drum level. This has to be done by judgment and expertise.

Incident 4. At the beginning of this scenario the reactor operates at reduced power (65 MWe). There is then a fault in one of the two feedwater pumps, so that only one feedwater pump remains for both the steam drums. Immediately there is a warning on the rear panel for the fault of the pump. Since one pump is lost, the level of the water in the steam drums falls quite rapidly: if no proper actions are taken by the operator a Y trip, due to low level in the steam drums, may be triggered. This incident is very rare since it appeared approximately two times in twenty years in the SGHWR. However the diagnosis of the fault is quite easy because of the warning on the rear panel advising of the pump failure. The scenario is interesting for two reasons, firstly because the action of manual regulation of the of the drum level is quite difficult (operators may under-compensate or over-compensate the feedwater flow), and secondly because there is the possibility of a Y trip which may be manually operated by the operators. Once the pump is failed the drum levels start to lower quite rapidly so that it is necessary to reduce the demanded power to the reactor. This action may be undertaken in automatic control by simply reducing the demanded power. It is further necessary to manually operate on the feedwater valves to increase the water flow to the drums, since the automatic regulation would be too slow. The manual regulation of the valves is a delicate action since, if the drum level is low and the feedwater valves are open slightly, the drum level will initially drop due to the cooler water causing collapse in the voids, and then the level will start to rise.

Incident 5. In this scenario the progressive deterioration of the steam flow signal affects both the feedwater and the power controllers. The slope of the ramp is 5 % per minute. This scenario is a combination of incidents 2 and 3 and presents a very unusual situation for the operators. Due to the fault on both the signals relating to the two controllers there are effects both on the pressure and on the drum level. The position of the governor valve is progressively affected, trying to regulate the pressure of the drums; the power decreases and the drum level progressively drops because of the loss of the power and feedwater controllers. In this scenario the operator is faced with several problems in a relatively long time schedule, where he is given the opportunity to access the situation and to react, dealing with the effects involving all the three main controllers: the pressure, the feedwater and the power controllers. Due to the long time constant of the fault there is no panic from the operator who has time to intervene and do the correct actions.

Data Collection Method

Data were collected as follows:
(1) systematic observation and video recording of the indicators used by the operators and of the actions;
(2) estimate of the subjective occurrence frequency of the incidents on a subjective rating scale.

Observation of the information collecting and of the actions and video recordings. The systematic observation of the operators during incident solving was carried out by recording the observed indicators and the operators'actions on a diagram showing the rear panel and the control desk. A number was given to each newly consulted indicator and to each new action. This allowed the experimenters to keep track of the sequence of information-taking and of the actions and to confront these sequences with the video recording. A video recorder was placed in such a way as to allow the widest possible angle of vision to get information on the operators interacting with the information-supports. Such a device was aimed at getting informations on: - indicators used by the operator while solving the incident; - actions taken (which and when). These data were then transcribed using an observation grid. Those grids include actions taken and their sequence, together with the indicators used and their sequence (when they are used and in which order). If actions are carried out simultaneously with observations of the indicators, this appears on the grid.

Rating scales. At the end of the session, operators were asked to estimate, according to their personal experience, the frequency of occurrence in the reality of each of the five incidents. A descriptive rating scale was used. The degrees of the scale were defined as follows: 'never - rarely - occasionally - commonly - often'.

RESULTS

A first look at the data makes it possible to analyze the subjective frequency of the incidents given by the operators according to their personal experience. We will also look at the effects of specification of the situations on the performance of the operators during the five incidents they experienced on the simulator, in terms of the triggering off of the reactor and of the duration of incidents.

Frequency of occurrence of the incidents. Table 3 shows the answers of the operators to the rating scales of the frequency of occurrence of the five incidents. As shown in the Table, the frequency of occurrence of the incidents is fortunately relatively low. The frequency of occurrence of incident 1 is estimated as relatively high as it is caused by changes of frequency in the electrical network, which occurs daily. This incident can thus hardly be considered as unusual. This situation is the most familiar to the operators and its symptoms are not very confusing.

TABLE 3 Subjective Frequency of Occurrence of the Five Incidents. 1 : never; 2 : rarely; 3 : occasionally; 4 : commonly; 5 : often.

	Incidents	2	3	5	1	4
	Operator					
Novice	1	1	1	1	3	1
	2	2	2	1	4	2
Expert	1	2	2	2	5	1
	2	2.	2	2	5	2
	Median	2	2	1.5	4.5	1.5

Reactor trip. All the operators triggered the reactor (reactor's 'trip') in the fourth incident, i.e. when there is a loss of a feedwater pump.

TABLE 4 Reactor Trip in the Five Incidents

	Incidents	2	3	5	1	4
	Operator					
Novice	1	trip				trip
	2	trip				trip
Expert	1					trip
	2		trip			trip

The second incident, i.e. the loss of steam flow signal, is interesting as novice operators both trip the reactor while the experts do not. In the third incident, one of the expert operator (E2) also triggers the reactor. These results are shown in Table 4.

Duration of the incidents. Table 5 show the time necessary for the operators to solve the incidents. Operators take longer to solve under-specified situations (incidents 2, 3 and 5), except for the novice operators in incident 2. Both novice operators take longer than the two experts to solve the incidents, especially when situations are under specified. Incidents 3 and 5 are particularly long for the novice operators (Table 5). Incident 2 is quick as both novice operators trip the reactor. One of the two novice operator (N2) trips the reactor after 34 seconds. There is no significant difference in the length of time taken by the operators in incidents 1 and 4.

TABLE 5 Duration of the Five Incidents.

	Incidents	2	3	5	1	4
	Operator					
Novice	1	1'08"	12'56"	19'51"	2'32"	2'41"
	2	34"	10'07"	17'09"	2'57"	2'55"
Expert	1	5'25"	9'59"	9'48"	3'25"	1'44"
	2	6'38"	5'32"	16'31"	2'26"	2'39"

Analysis of the Information-Taking

Data obtained allow the analysis of: - the frequency of observation of the operators in the five incidents; - the type of observation made by the operators in the five incidents;

Frequency of observations. It appears from Table 6 and Fig. 3 that the operators, except for operator N2, make more frequent observations in incidents 1 and 4, considered to be specified situations, than in incidents 2, 3 and 5, considered to be under specified situations. Operator N2 makes more frequent observations in under specified situations and less in specified situations. However, this last result is biased by the fact that N2 triggers the reactor in incident 2 after 34 seconds, which gives a frequency of observations of 6.17 per half-minute. The half-minute was used as a unit because it was not relevant to use the minute as incident 2 of novice operator N2 only lasted for 34 seconds.

TABLE 6 Frequency of Observations (per 30 seconds) of the Operators in the Five Incidents.

	Incidents	2	3	5	1	4
	Operator					
Novice	1	2.2	2.2	1.58	3.94	3.54
	2	6.17	2.02	1.83	2.71	1.88
Expert	1	1.47	2.7	2.04	2.92	2.88
	2	2.64	2.53	2.66	1.85	3.39

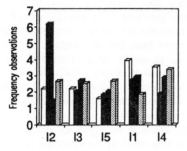

Fig. 3. Frequency of observations in the 5 incidents

Types of Observations. Types of observations correspond to the attributes (indicators of the rear panel and the control desk) observed by the four operators in each incident. As far as the types of observations of the attributes in the five incidents are concerned, it appears from Table 7 that the experts and the novice operators make significantly different observations in incident 2, the null hypothesis that there is no difference between operators in the types of observations can thus be rejected. E1 and N1, $prob<0.01$, E1 and N2, $prob<0.06$, E2 and N1, $prob<0.02$, E2 and N2, $prob<0.01$, and to a lesser degree in incident 3: E1 and N2, $prob<0.07$, E2 and N1, $prob<0.01$. Novice operators between themselves don't make different observations, nor do the experts between themselves (except for incident 5, $prob<0.05$). We cannot talk about a difference between novice operators and experts neither in incident 1, nor in incident 4 (except for E1 and N1, $prob<0.03$), nor in incident 5, except for E1 and E2, ($prob<0.06$). Note that this last result is the only one to show a difference between operators supposed to have the same level of expertise; in any other case novice operators and expert operators don't make different observations.

TABLE 7 Comparison Between Operators Taken by Pairs Concerning the Type of Observations
(«Wilcoxon sign rank test »)

Op.\Incident		2	3	5	1	4
E1 - E2	Sign-r	-9.5	42.5	-50	5.5	-5
	Prob>s	0.57	0.16	0.06	0.65	0.75
E1 - N1	Sign-r	35.5	8.5	-31	-6.5	-21.5
	Prob>s	0.01	0.72	0.11	0.63	0.03
E1 - N2	Sign-r	23.5	42	-30.5	-2	3.5
	Prob>s	0.06	0.07	0.19	0.88	0.79
E2 - N1	Sign-r	26	-32	3	13.5	-21
	Prob>s	0.02	0.01	0.92	0.12	0.24
E2 - N2	Sign-r	24.5	-7	6.5	-15.5	6
	Prob>s	0.01	0.56	0.73	0.23	0.61
N1 - N2	Sign-r	-7.5	24	18	5.5	16
	Prob>s	0.35	0.26	0.4	0.75	0.11

Analysis of the Actions

As for the information-taking, data obtained allow the analysis of: - the frequency of actions of the operators in the five incidents; - the type of actions taken by the operators in the five incidents.

Frequency of actions. As for the observations, and leaving incident 2 apart, expert operators take actions less frequently in under specified situations and more in specified situations (Table 8).

TABLE 8 Frequency of Actions (per 30 seconds) of the Operators in the Five Incidents.

	Incidents	2	3	5	1	4
	Operator					
Novice	1	3.52	1.27	1.73	1.38	2.98
	2	4.41	.99	1.22	1.69	1.37
Expert	1	1.29	.6	1.63	.73	2.59
	2	1.05	.84	1.21	1.85	1.32

As shown on Fig. 4, those differences are more significant than for the frequency of observations shown on Fig. 3.

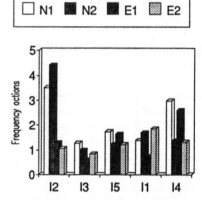

Fig. 4. Frequency of actions in the 5 incidents.

Type of actions. The types of actions correspond to the commands (commands of the control desk) used by the four operators in each of the incidents. Table 9 shows that we cannot talk about a difference between expert operators and novice operators as far as the type of actions is concerned, i.e. there is no difference between the operators in the type of actions taken, except for: E1 and N2 in incident 2, $prob<0.08$, E2 and N1 in incident 3, $prob<0.02$, and E1 and N1 in incident 5, $prob<0.09$.

Incident 2 allows for a closer look at the differences in strategies of the operators according to the expertise as both novice operators trigger the reactor while both expert operators manage to control the situation. In incident 2 a loss of steam flow signal to the heavy water controller which controls the level in the calandria tank and controls the power of the reactor was simulated. A proper action to solve that kind of incident is to take the control of the power in manual. Expert operators do this, while novice operators don't.

TABLE 9 Comparison Between Operators Taken by Pairs Concerning the Types of Actions («Wilcoxon sign rank test »)

Op\Incident		2	3	5	1	4
E1 - E2	Sign-r	0	1.5	-11.5	-3	2
	Prob>s	1	0.83	0.42	0.34	0.79
E1 - N1	Sign-r	10	-17.5	-27	-1.5	-2.5
	Prob>s	0.32	0.13	0.09	0.37	0.59
E1 - N2	Sign-r	15	-5.5	-5.5	-3	1
	Prob>s	0.08	0.17	0.69	0.17	0.93
E2 - N1	Sign-r	5.5	-25.5	-16	1	-3.5
	Prob>s	0.55	0.02	0.16	0.85	0.53
E2 - N2	Sign-r	9	-5	1	-1	-0.5
	Prob>s	0.22	0.44	0.97	0.77	1
N1 - N2	Sign-r	2.5	17.5	19.5	-1.5	5.5
	Prob>s	0.57	0.18	0.13	0.59	0.17

CONCLUSIONS

The first results allow us to analyze the way expert or novice operators behave when confronted with specified or under-specified incidental situations. Those data can then be compared with the answers given by the COSIMO model in incidental situations. When faced with disturbances of the process with which it interacts, the COSIMO model observes a number of cues that reach it, interprets them and acts after finding a solution to the problem. We investigated on the simulator of a nuclear power plant whether operators would act in the same way.

We noticed that the operators take longer to react to under-specified incidental situations than they would for specified ones. In a under-specified incident situation, as incident 2, operators make significantly different observations. Expert and novice operators make different actions in incident 2. Besides, novice operators cannot control the situation.

At this stage, we can already point out two facts. First, it appears clearly that even if they manage to solve the problem and to bring the process to a steady state, operators do not recognize the incidents as the identifiable faults which are simulated. On the contrary, novice and expert operators seem to to proceed step by step by cutting the problem down to sub-problems. This point is important as it is not taken into account by the model. Another important fact is the difficulty of the operators in classifying the attributes according to their importance. Operators have justified this difficulty by insisting on the dynamic character of these attributes: some of them are more important at

the beginning of the incident but then lose their importance at the benefit of others.

Aknowledgements

This work was funded by the United Kingdom Health and Safety Executive as part of the Nuclear Safety Research program and was done by CEC-JRC-ISEI under contract for Control and Dynamics Department of AEA Reactor Services, Winfrith. The authors thank both organizations for permission to publish this paper. We are deeply grateful to Anna Bellorini who participated at the experimental campaign and helped during the analysis of data, and to Frank Kingdom, for his contribution in the selection and in the performance of the sessions at the simulator. Giacomo Cojazzi helped at the selection of the incidents and at the understanding of the nuclear process. Francis Perée, from the University of Liège, contributed to the statistical analysis of the data. Finally many thanks to Carlo Cacciabue for his support and encouragement.

REFERENCES

Amalberti, R., Bataille, M., Deblon, F., Guengant, A., Paignay, J.M., Valot, C. & Menu, J.P. (1989). Développement d'aides intelligentes au pilotage: formalisation psychologique et informatique d'un modèle de comportement du pilote de combat engagé en mission de pénétration. Rapport CERMA 89.09.

Cacciabue, P.C., Decortis, F., Drozdowicz, B., Masson, M. & Nordvik, J.P. (to appear). COSIMO: A Cognitive Simulation Model of human decision making and behaviour in complex work environements. *IEEE Transactions on Systems, Man and Cybernetics.*

Mancini, G., Woods, D.D. & Hollnagel, E. (1987). Cognitive Engineering in Dynamic Worlds. *International Journal of Man-Machine Studies, 27,* 5&6, Londres: Academic Press.

Reason, J.T. (1990). *Human Error.* Cambridge University Press.

Reason, J.T. (1984). Cognitive Under-specification: its varieties and consequences. In B. Baars (Ed), *The psychology of error.* London, UK : Plenum.

Rubin, K. S., Jones, P.M. & Mitchell, C.M. (1988). OFMspert : Inference of Operator Intentions in Supervisory Control using a Blackboard Architecture. *IEEE Transactions on Systems, Man and Cybernetics, 18,* 4.

Woods, D.D., Roth, E.M. & Pople, H. (1987) Cognitive Environment Simulation: An Artificial Intelligence System for Human Performance Assessment. Volume 2. Modeling Human Intention Formation. Nureg-CR-4862.

FAILURE IDENTIFICATION PROCEDURE USING REDUNDANT RULES

T. Kohda and K. Inoue

Department of Aeronautical Engineering, Faculty of Engineering, Kyoto University, Kyoto 606, Japan

Abstract. At the first stage of failure diagnosis, it is very important to confirm the correctness of input data, because the failure diagnosis using such a rule-based expert system identifies the cause of system failure based on them. The operator must check whether the input data are correct or not. This paper develops a simple method to identify an abnormal event using redundant rules, which help the operator understand the current system state. The proposed method requires only a relation table, which shows the values of input data under each abnormal event to be identified. Redundant rules, which are composed of minimal combinations of input data that can identify an abnormal event, can be obtained from the relation table using the concept of "failure distinguishability" criterion. Using redundant rules, the operator can estimate easily not only what state the system is under, but also which input datum may be wrong. The construction of redundant rules can also show which input datum is important and must be improved for its validation. Illustrative examples show the details and merit of the method in the operator's aide.

Keywords. Failure identification; redundant rules; rule-based expert systems; sensor failures; wrong input data; boolean algebra.

INTRODUCTION

In the failure diagnosis, the operator must identify the cause of system failure as quickly as possible to prevent the expansion of its effects. After identifying the cause, the operator can follow the protection or maintenance procedure for it. Since the operator must form a judgement based on values of monitoring sensors and inspection results as well as their knowledge, it is necessary to assure the validity of input data. To help the operator make a correct decision in such an emergent condition, various computer-aided systems have been developed such as disturbance analysis systems and expert systems. Especially the rule-based expert system finds the cause of system failure by comparing the input data with the condition parts of stored rules. A wrong input datum may lead to a wrong conclusion. Thus, the validation of input data is very important before identifying the cause of system failure. The operator must assure the correctness of input data.

Various methods exist to mitigate the effect of sensor failures. Voting systems choose the most common output signal among redundant components as their output signal. A single sensor failure does not affect its output. On the other hand, it increases the number of sensors monitoring each plant state. We must decide which input datum a voting system should be applied to. Another kind of redundancy is the analytical redundancy (Chow and Wilsky 1984), but this also requires redundant sensors. The same sensor allocation problem must be solved. A model of input datum dynamics can be also applied to its validation and verification (Iserman 1984), but its effectiveness depends on its accuracy of modeling the system, and further it needs more sophisticated computations than the previous two method.

This paper proposes another simple redundancy for the rule-based expert systems: redundant rules. The condition parts of rules are composed of minimal combinations of input data to identify an abnormal event, and so redundant rules hold based on different combinations of input data. The proposed method tries to use all input data such as sensor outputs and inspection results as effectively as possible. Whether input data are correct or not can be shown easily as the results of redundant rules, which help the operator verify input data. First, the basic concept of "failure distinguishability" is introduced to obtain redundant rules for identifying abnormal events. Illustrative examples show the merit of redundant rules in failure diagnosis.

CONSTRUCTION OF REDUNDANT RULES

Failure Distinguishability

Assume that a relation table is given which shows values of input data under all abnormal events to be detected. Here, input data mean inspection results, basic conditions and facts in the rule-based expert system, which are usually given by their users and monitoring sensors, while an abnormal event means a final proposition of the expert system, a cause of the system failure. The only necessary condition for input data is that they are represented as basic conditions in the rule-based expert system. The relation between abnormal events and input data can be usually obtained by compiling the knowledge of the expert system. Since the expert system uses a part of the relation table, some knowledge not covered by the expert system must be obtained by asking the system analyst what will happen to input data if a specific abnormal event occurs. In the proposed method, the development of the relation table is the most crucial and important task for the system analyst of the subject plant. During this development process, the system analyst can not only verify the consistency of the rule-based expert system knowledge, but also understand the subjective system more deeply. For a rule-based expert system composed of more than 1000 rules, the consistency check will be difficult and time-consuming. But the check can be performed easily if input data and abnormal events are divided into independent subgroups, which correspond to modules of the system. A module decomposition method can be applied to this case, which is similar to the method in Fault Tree Analysis (Kohda, Henley, and Inoue 1989). Combinations of input data and abnormal events can be reduced so that the system analyst can check them easily.

Based on the given relation table, abnormal events j and k can be distinguished by input datum i, if and only if

$$Y_i^j \cap Y_i^k = \varnothing \qquad (1)$$

where Y_i^j indicates the value of input datum i under abnormal event j, $A \cap B$ denotes the product set of A and B, and $\varnothing$ denotes an empty set.

Eq. (1) requires that input datum i must take completely different values under abnormal events j and k. The occurrence of abnormal event j can be detected by input datum i if and only if the value of input datum i under abnormal event j is different from that under the normal condition.

Failure Identification Rules

To identify abnormal event j, abnormal event j must be distinguished from any other abnormal event. Thus, identifiability of abnormal event j, F(j) can be expressed as:

$$F(j) = \underset{k \neq j}{\wedge} (\underset{i \in D(j,k)}{\vee} X_i) \qquad (2)$$

where X_i is a Boolean variable for input datum i, which takes 1 if and only if input datum i is chosen to be checked, $\wedge_i X_i$ denotes a logical AND combination over binary variables X_i's, which takes 1 if and only if all X_i's are equal to 1, $\vee_i X_i$ denotes a logical OR combination over binary variables X_i's, which takes 1 if and only if at least one X_i is equal to 1, and D(j,k) is a set of input data that can distinguish abnormal events j and k: $D(j,k) \equiv \{ i \mid Y_i^j \cap Y_i^k = \varnothing \}$.

Eq. (2) means that at least one input datum that can distinguish abnormal event j from k is necessary for any other abnormal event k.

Using the property of Boolean variables X_i's: 1) $X_i \wedge X_i = X_i$, 2) $X_i \vee X_i = X_i$, 3) $X_i \vee (X_i \wedge X_j) = X_i$, 4) $X_i \wedge (X_i \vee X_j) = X_i$, Eq. (2) can be transformed into a logical OR of logical AND combinations of input data. Each logical AND combination shows a minimal combination of input data that can identify abnormal event j, i.e., an identifiable set of input data for abnormal event j. If the same values are obtained for all input data of any one logical AND combination, it can be concluded that the system is under abnormal event j. Thus, diagnosis rules to identify an abnormal event can be constructed easily from its identifiable sets. These rules correspond to rules compiled from a relation table to identify an abnormal event.

Redundant Rules

If some identifiable sets are obtained for an abnormal event, different minimal combinations of input data can identify its occurrence. In a sense these rules are considered to be redundant in identifying the abnormal event. Redundant rules can be easily constructed from identifiable sets. If any input datum is wrong, redundant rules can correct it by their vote.

Consider how the effect of wrong input data can be nullified by redundant rules. Two types of wrong input data are considered: one is an input datum which indicates no symptom of system failure when the system is failed (a failed-dangerous one), and the other is an input datum which makes false alarm under the system normal condition (a failed-safe one) (Inoue et al. 1982). When all input data are correct, all redundant rules for a specific abnormal

event hold. The effect of the fail-dangerous failure of an input datum for an abnormal event can be nullified by its other rules. If an input datum appears in all redundant rules for an abnormal event, its failed-dangerous failure produces a situation where no rules hold but input data except it coincide with those for the abnormal event. This situation can indicate that the input datum can be considered as failed-dangerous. The conditional probabilities are compared for two cases: where the other data are under their failed-safe conditions and where the datum are under its failed-dangerous condition. The more probable state can be considered as the current system state. But, this situation is a dangerous state of the failure diagnosis system. In such a case where an input datum appears in all identifiable sets, some validation method should be applied to it.

Consider a case where an input datum is failed-safe. Since rules for an abnormal event are not composed of all input data, one wrong input datum may not affect the rules without it. If the failed-safe input datum appears only in an identifiable set composed of more than 2 input data for an abnormal event, then the other normal data nullify its effect. Otherwise, the datum may produce a situation where some rules hold for more than one abnormal event. The existence of more than one abnormal event occurrence can show errors in input data, because all the rules must hold for the only abnormal event when all input data are correct. Similarly to the failed-dangerous situation, redundant rules can tell that some input datum may be wrong. Comparing these abnormal event occurrence probabilities, the possible wrong input data can be located easily. Thus, the operator can recognize a situation where some input datum is wrong and understand the plant state and the monitoring system state based on the results of redundant rules.

If an abnormal event have only one identifiable set, no redundant rule can be constructed. In this case, the system has to deal with errors of these input data in the set by using an appropriate data validation method. A redundant sensor allocation or analytical redundancy must be applied to these input data, because the expert system may yield a wrong conclusion without it.

The consideration of redundant rules for all abnormal events can point out which input datum needs an additional validation method. The construction of redundant rules can not only minimize the additional cost for the validation of input data, but also use input data as effectively as possible.

ILLUSTRATIVE EXAMPLES

Simple Example

Consider the construction of identification rules for 4 abnormal events using 4 monitoring channels. The relation table is given as Table 1, where N denotes the normal range, H denotes the higher range, and L denotes the lower range for each channel. Abnormal event 1 corresponds to the system normal condition. Comparing the monitored values of channels, set D_{jk} can be obtained as:

$$D_{12}=\{2,4\}, \quad D_{13}=\{1,2,3,4\}, \quad D_{14}=\{2,3\}$$
$$D_{23}=\{1,3,4\}, \quad D_{24}=\{3\}, \quad D_{34}=\{1,4\}$$

Obtain the minimal combinations of channels for identifying each abnormal event. For abnormal event 1, F(1) is expressed as:

$$F(1) = (X_2 \vee X_4) \wedge (X_1 \vee X_2 \vee X_3 \vee X_4) \wedge (X_2 \vee X_3)$$

Using property 4) $X_i \wedge (X_i \vee X_j) = X_i$, F(1) is reduced to:

$$F(1) = (X_2 \vee X_4) \wedge (X_2 \vee X_3)$$

and expanding this into a logical OR of logical AND combinations, F(1) is:

$$F(1) = (X_2 \wedge X_2) \vee (X_2 \wedge X_3) \vee (X_4 \wedge X_2) \vee (X_4 \wedge X_3)$$

Further simplifying this using property 1) $X_i \wedge X_i = X_i$, $F(1)$ is:

$$F(1) = X_2 \vee (X_2 \wedge X_3) \vee (X_4 \wedge X_2) \vee (X_4 \wedge X_3)$$

Finally using property 3) $X_i \vee (X_i \wedge X_j) = X_i$, the simplest form of $F(1)$ is:

$$F(1) = X_2 \vee (X_3 \wedge X_4)$$

Similarly, $F(2)$, $F(3)$, and $F(4)$ can be obtained as:

$$F(2) = (X_2 \wedge X_3) \vee (X_3 \wedge X_4)$$

$$F(3) = X_1 \vee X_4$$

$$F(4) = (X_1 \wedge X_3) \vee (X_3 \wedge X_4)$$

The identifiable sets for each abnormal event are obtained as:

For abnormal event 1: {2} {3,4}
For abnormal event 2: {2,3} {3,4}
For abnormal event 3: {1} {4}
For abnormal event 4: {1,3} {3,4}

Thus, redundant rules can be obtained as:

R1-1: If [Ch 2:N], then abnormal event 1 occurs.
R1-2: If [Ch 3:N]*[Ch 4:N], then abnormal event 1 occurs.
R2-1: If [Ch 2:H]*[Ch 3:N], then abnormal event 2 occurs.
R2-2: If [Ch 3:N]*[Ch 4:H], then abnormal event 2 occurs.
R3-1: If [Ch 1:H], then abnormal event 3 occurs.
R3-2: If [Ch 4:L], then abnormal event 3 occurs.
R4-1: If [Ch 1:N]*[Ch 3:H], then abnormal event 4 occurs.
R4-2: If [Ch 3:H]*[Ch 4:N,H], then abnormal event 4 occurs.

Here, [Ch i:A] indicates that the value of channel i is A, and * denotes a logical AND combination.

Consider a case where channel outputs are [Ch 1:N]*[Ch 2:N]*[Ch 3:N]*[Ch 4:N]. The redundant rules conclude that the system is under abnormal event 1, because only all rules for abnormal event 1 hold. When all the channels are correct, all rules for a specific abnormal event can hold. In this case, even though more than 2 sensors may be failed-dangerous, it is most probable that all the sensors are normal. Consider another case where channel outputs are [Ch 1:N]*[Ch 2:N]*[Ch 3:N]*[Ch 4:H]. In this case R1-1 and R2-2 hold for abnormal events 1 and 2, respectively. Since the system must be under one of 4 abnormal events, this leads to the conclusion that some channel is wrong. Comparing redundant rules for abnormal events 1 and 2, channels 2 and 4 are found to take different values. Thus, it can be estimated that either channel 2 or 4 may be failed. If the system is under abnormal event 1 (i.e., channel 2 is normal), then channel 4 is issuing a spurious alarm. Otherwise, channel 2 is failed-dangerous and the system is under abnormal event 4. Since abnormal event 1 is the normal condition, channel 4 should be examined first because abnormal event 4 requires the maintenance action if it occurs. In this way, the channel to be checked can be determined easily based redundant rules. How to check channels and abnormal events depends on the expected loss caused under each abnormal conditions.

From the viewpoint of failure distinguishability, channels 1, 2, and 3 are sufficient for the failure diagnosis (Kohda and Inoue 1990). In this case, the failure diagnosis rules can be described as:

R1: If [Ch 2:N], then abnormal event 1 occurs.
R2: If [Ch 2:H]*[Ch 3:N], then abnormal event 2 occurs.
R3: If [Ch 1:H], then abnormal event 3 occurs.

R4: If [Ch 1:N]*[Ch 3:H], then abnormal event 4 occurs.

Only one identification rule exists for an abnormal event. This situation is common to the rule-based expert system, where only one rule exists for one conclusion. Consider a case where channel outputs are [Ch 1:H]*[Ch 2:N]*[Ch 3:N]*[Ch 4:N]: channel 1 yields an spurious alarm. Based on Table 1, no abnormal event holds. Diagnosis rules constructed from 3 channels shows that the system is under abnormal event 1 or 3, while the redundant rules conclude that the system is under abnormal event 1 more probably than abnormal event 3, because both of two rules hold for abnormal event 1 while one of two rules holds for abnormal event 3. Assume that more than two failures cannot occur simultaneously, and channel failure 1 can be identified more easily by redundant rules. Thus, redundant rules can increase the certainty of the failure diagnosis.

Redundant Rules for Pump Cooling System

Consider the failure diagnosis for a pump cooling system shown in Fig. 1. Fresh water is pumped from the well and stored in the fresh water tank. The lubricant water pump pushes fresh water from it to the expansion tank. The 2nd cooling water pump pulls up foul water, which is cleaned by the autostrainer. The cleaned water comes into the fresh water cooler through the pump engine, and cooled by the 1st cooling water coming from the expansion tank. The operator must identify 11 abnormal events, based on 15 input data in the failure diagnosis. Table 2 shows abnormal events to be detected, while Table 3 shows input data and their possible results. The expert operator usually identifies the cause of system failure based on the diagnosis rules shown in Table 4, where [*] denotes an input datum, (*) denotes a proposition, and <*> is an abnormal event to be identified. These rules constitute a rule-based expert system for the failure diagnosis of the pump cooling system. The expert system knowledge is constructed in a hierarchical way from system level to component level.

Consider the construction of redundant rules. To obtain the relation table between abnormal events and input data, we first compile the diagnosis rules in Table 4. Table 5 shows the compiled rules, which represent abnormal event occurrence conditions as logical AND combinations of input data. Most of the compiled rules except for abnormal event A2 have only one logical expression. This is due to a straight forward inference of the veteran expert to identify the cause of system failure from system level to component level, which does not include redundant conditions. Since these compiled rules constitute a part of the relation table, the unfilled part must be obtained by asking the system analyst what will happen to input data under each abnormal event. Through this review process, the system analyst can check the consistency of the rules and understand the subject system more deeply. In this example, since A2: loss of main power, a common-cause failure, appears twice as cause in diagnosis rules, two logical AND combinations are obtained. If several logical AND combinations exist for an abnormal event, the system analyst must check the consistency of these situations. In constructing the relation table, we must examine its consistency by considering both failure effect on the system and the system configuration. Thus, the relation table is finally obtained as shown in Table 6, where A0 denotes the system normal condition.

Based on Table 6, the identifiable sets are obtained as shown in Table 7. Since only one identifiable set exists for abnormal events A0, A10, and A11, redundant rules cannot be obtained. Consider a case where only input datum I1 yields wrong alarm (A). Only one rule hold for abnormal event A11, and so it can be concluded that the system is under abnormal event A11 instead of A0. Since input datum I1 is a condition of a non-redundant rule for A0 and A11, its failure can yield an incorrect conclusion. Especially in this case, the system normal condition A0 can be confirmed only by input datum I1, and so its validation is the most important in the failure diagnosis of the pump cooling system. Thus, consideration of redundant rules can show that a special attention must be paid to input data common to non-redundant rules. In this case, the occurrence of abnormal event A11 must be examined first, because no damage will be produced under abnormal event A0.

Consider another case where only input datum I2 indicates abnormal condition (OFF). One rule for A0 holds because I1 is normal, while two rules for A6 hold. Although all rules hold for abnormal events A0 and A6, respectively, some input datum must be wrong since all redundant rule must hold for an abnormal event. Comparing input data I1, I2, I3, I4, and I7 for A0 and A6, it can be estimated that input datum I1 or I2 is wrong. If I1 is wrong, the system state is under abnormal event 6 and the temperature sensor is failed-dangerous. Otherwise, the system is normal and the flow relay is failed-safe. Based on the risk, input datum I2 must be checked first to confirm that the system is under abnormal event A6. By comparing all input data in the relation table with the current state, we cannot conclude what abnormal event the system is under because no list of input data for any abnormal event coincides with the current state. Further, the original diagnosis rules in Table 4 can only show that the system is still under the normal condition, because no rules in Table 4 or 5 hold. On the other hand, redundant rules, which are composed of minimal combinations of input data, can show all possible situation where the system can be. This is considered to be the robustness of redundant rules. The operator can easily identify the current system state by sequentially checking an input datum condition based on the risk under each abnormal event.

CONCLUSIONS

This paper proposes to use redundant rules for identifying an abnormal event. The proposed method requires only a relation table, which shows values of input data under each abnormal event to be identified. Through the provision of the relation table for a subject system, the system analysts can review and verify the present knowledge on the system in the rule-based expert system. Once the relation table is available, failure identification rules can be obtained easily by simple logical calculation. Each rule is composed of a minimal combination of input data that can identify an abnormal event. Using redundant rules, the operator can estimate easily not only what state the system is under, but also which input datum may be wrong. Compared with other validation methods such as voting systems of sensors, the proposed method does not need additional components. As shown in the illustrative examples, the construction of redundant rules can also show what input datum is important and must be improved for its validation. Redundant rules tries to use all input data as effectively as possible. As for the future work of this research, we plant to develop a fast construction method of redundant rules for a large-scale system using a modular decomposition method.

REFERENCES

Chow, E.Y., and A.S. Wilsky (1984) Analytical redundancy and the design of robust failure detection systems, *IEEE Trans. Automatic Control*, AC-29, 603-614.

Inoue, K., T. Kohda, H. Kumamoto, and I. Takami (1982) Optimal structure of sensor systems with two failure modes, *IEEE Trans. Reliability*, R-31, 1, 119-120.

Iserman, R. (1984) Process fault detection based on modeling and estimation methods - a survey, *Automatica*, 20, 4, 387-404.

Kohda, T., E.J. Henley, and K. Inoue (1989) Finding Modules in Fault Trees, *IEEE Trans. Reliability*, R-38, 2, 165-176.

Kohda, T., and K. Inoue (1990) Minimal combination of inspection items for failure diagnosis, *Proc. JAPAN-U.S.A. Symposium on Flexible Automation*, 3, 1025-1030.

TABLE 1 Diagnosis Table

	Ch. 1	Ch. 2	Ch. 3	Ch. 4
Abnormal event 1	[N]	[N]	[N]	[N]
Abnormal event 2	[N]	[H]	[N]	[H]
Abnormal event 3	[H]	[H]	[H]	[L]
Abnormal event 4	[N]	[H]	[H]	[N,H]

TABLE 2 Abnormal Events at Pump Cooling System

A0:	Normal state
A1:	Failure of autostrainer electric system
A2:	Loss of main power
A3:	Failure of autostrainer itself
A4:	Failure of 2nd cooling water pump electric system
A5:	Failure of 2nd cooling water pump itself
A6:	Failure of lubricant cooler
A7:	Failure of lubricant water pump electric system
A8:	Failure of lubricant water pump itself
A9:	Failure of well pump itself
A10:	Failure of 1st cooling water line or expansion tank
A11:	Failure of engine itself

TABLE 3 Input Data and Possible Results

I1:	Temperature of 1st cooling water	Abnormal(A) Normal(N)
I2:	Flow relay state	On(N) Off(F)
I3:	Flow at exit of strainer	Present(P) Absent(A)
I4:	Autostrainer state	Stop(S) Working(W)
I5:	Autostrainer power	Present(P) Absent(A)
I6:	Autostrainer switch and thermal relay	Open(O) Normal(N)
I7:	2nd cooling water pump state	Stop(S) Working(W)
I8:	2nd cooling water pump power	Present(P) Absent(A)
I9:	2nd cooling water pump switch and thermal relay	Open(O) Normal(N)
I10:	Flow of 1st cooling water system	Present(P) Absent(A)
I11:	Lubricant water pump state	Stop(S) Working(W)
I12:	Lubricant water pump power	Present(P) Absent(A)
I13:	Lubricant water pump switch and thermal relay	Open(O) Normal(N)
I14:	Flow at entrance of Lubricant water pump	Present(P) Absent(A)
I15:	Well pump power	Present(P) Absent(A)

TABLE 4 Diagnosis Rules for Pump Cooling System

R1: IF [I1 A] AND [I2 F],
     THEN (Failure of 2nd cooling water system:P1)
R2: IF (P1) AND [I3 A],
     THEN (Failure in autostrainer system:P2)
3:   IF (P2) AND [I4 S],
     THEN (Failure at autostrainer:P3)
R4: IF (P3) AND [I5 A],
     THEN (Power loss at autostrainer:P4)
R5: IF (P4) AND [I6 O],   THEN <A1>
R6: IF (P4) AND [I6 N],   THEN <A2>
R7: IF (P3) AND [I5 N],   THEN <A3>
R8: IF (P2) AND [I7 S],
     THEN (Failure at 2nd cooling water pump:P5)
R9: IF (P5) AND [I8 A]
     THEN (Power loss at 2nd cooling water pump:P6)
R10: IF (P6) AND [I9 O],   THEN <A4>
R11: IF (P6) AND [I9 N],   THEN <A2>
R12: IF (P5) AND [I8 N],   THEN <A5>
R13: IF (P1) AND [I3 P],   THEN <A6>
R14: IF [I1 A] AND [I2 O],
     THEN (2nd cooling water system normal:P7)
R15: IF (P7) AND [I10 A],
     THEN (Failure in 1st cooling water system:P8)
R16: IF (P8) AND [I11 S],
     THEN (Failure at lubricant water pump:P9)
R17: IF (P9) AND [I12 A],
     THEN (Power loss at lubricant water pump:P10)
R18: IF (P10) AND [I13 O],   THEN <A7>
R19: IF (P9) AND [I12 P],   THEN (Mechanical
                    stop at lubricant water pump:P11)
R20: IF (P11) AND [I14 P],   THEN <A8>
R21: IF (P11) AND [I14 A],
     THEN (Well pump stop:P12)
R22: IF (P12) AND [I15 P],   THEN <A9>
R23: IF (P8) AND [I11 N],   THEN <A10>
R24: IF (P7) AND [I10 P],   THEN <A11>

TABLE 6 Relation Table for Pump Cooling System

Event Input	A0	A1	A2	A3	A4	A5	A6	A7	A8	A9	A10	A11
I1	N	A	A	A	A	A	A	A	A	A	A	A
I2	N	F	F	F	F	F	F	N	N	N	N	N
I3	P	A	A	A	A	P	P	P	P	P	P	P
I4	W	S	S	S	W	W	W	W	W	W	W	W
I5	P	A	A	P	P	P	P	P	P	P	P	P
I6	N	O	N	N	N	N	N	N	N	N	N	N
I7	W	W	S	W	S	S	W	W	W	W	W	W
I8	P	P	A	P	A	P	P	P	P	P	P	P
I9	N	N	N	N	O	N	N	N	N	N	N	N
I10	P	P	A	P	P	P	P	A	A	A	A	P
I11	W	W	S	W	W	W	W	S	S	S	W	W
I12	P	P	A	P	P	P	P	A	P	P	P	P
I13	N	N	N	N	N	N	N	O	N	N	N	N
I14	P	P	A	P	P	P	P	P	A	A	P	P
I15	P	P	A	P	P	P	P	P	P	P	P	P

TABLE 5 Compiled Rules for Pump Cooling System Diagnosis

IF [I1 A]AND[I2 F]AND[I3 A]AND[I4 S]AND[I5 A]
   AND[I6 O],   THEN <A1>

IF [I1 A]AND[I2 F]AND[I3 A]AND[I4 S]AND[I5 A]
   AND[I6 N],   THEN <A2>

IF [I1 A]AND[I2 F]AND[I3 A]AND[I7 S]AND[I8 A]
   AND[I9 N],   THEN <A2>

IF [I1 A]AND[I2 F]AND[I3 A]AND[I4 S]AND[I5,N],
                             THEN <A3>

IF [I1 A]AND[I2 F]AND[I3 A]AND[I7 S]AND[I8 A]
   AND[I9 O],   THEN <A4>

IF [I1 A]AND[I2 F]AND[I3 A]AND[I7 S]AND[I8 N],
                             THEN <A5>

IF [I1 A]AND[I2 F]AND[I3 P],   THEN <A6>

IF [I1 A]AND[I2 N]AND[I10 A]AND[I11 S]AND[I12 A]
   AND[I13 O],   THEN <A7>

IF [I1 A]AND[I2 N]AND[I10 A]AND[I11 S]AND[I12 P]
   AND[I14 P],   THEN <A8>

IF [I1 A]AND[I2 N]AND[I10 A]AND[I11 S]AND[I12 P]
   AND[I14 A]AND[I15 P],   THEN <A9>

IF [I1 A]AND[I2 N]AND[I10 A]AND[I11 N],
                             THEN <A10>

IF [I1 A]AND[I2 N]AND[I10 P],   THEN <A11>

TABLE 7 Minimal Identifiable Sets for Abnormal Events

Abnormal event	Minimal identifiable sets (logical AND of {logical OR})
A0	{I1}
A1	{I6}
	{I5}*{I7,I8,I10,I11,I12,I14,I15}
A2	{I15}
	{I12}*{I2,I3,I4,I5,I7,I8,I13,I14}
	{I14}*{I2,I3,I4,I5,I7,I8}
	{I10,I11}*{I2,I3,I4,I5,I7,I8}
	{I5}*{I6,I7,I8}
	{I8}*{I4,I9}
	{I7}*{I4}
A3	{I4}*{I5}
	{I4}*{I6}*{I7,I8,I10,I11,I12,I14,I15}
	{I3}*{I7}*{I5,I6}
A4	{I9}
	{I8}*{I4,I5,I10,I11,I12,I14,I15}
A5	{I7}*{I8}
	{I7}*{I9}*{I4,I5,I10,I11,I12,I14,I15}
	{I4}*{I3}*{I8,I9}
A6	{I2}*{I3}
	{I2}*{I7}*{I4}
A7	{I13}
	{I12}*{I2,I3,I4,I5,I7,I8,I14,I15}
A8	{I11}*{I14}*{I12,I13}
A9	{I14}*{I2,I3,I4,I5,I7,I8,I12,I15}
A10	{I10}*{I11}
A11	{I1}*{I2}*{I10}

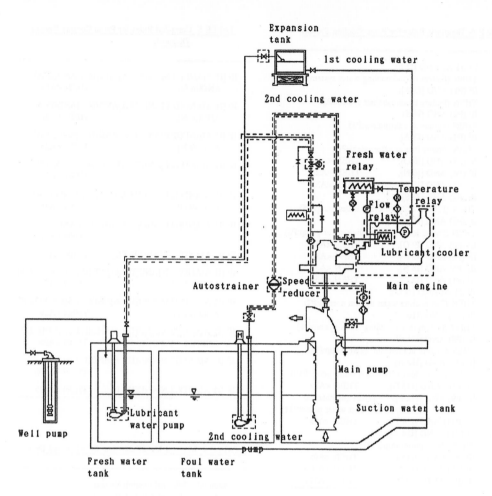

Fig. 1. Pump cooling system

MAN-MACHINE ASPECTS OF COMPUTER-AIDED INTERACTIVE GRASP PLANNING

A.J. Kasiński

Laboratory of Control Engineering and Robotics, Politechnika Poznańska, Poznań, Poland

Abstract. Features of the interactive, graphical simulator for supporting robot fine motion operations are discussed. Some preliminary experiences with using the implemented modules of the simulator are reported. The stress is put on the man-machine aspects of the planning procedure while using this computer-aid. We support the operator planning an action with partly automated, interactive simulation tools based on the simplified models of the scene. The eventual consequences of the decisions taken are evaluated and communicated to the user in graphic form in order to accept or to modify the actual step of the plan. We expect that the adopted approach would help to decompose and structure the decision/action process related to the fine motion of robotics hand and in that way it should enable further studies on error recovery strategies. These experiences might be useful for anyone interested in problems of the tele-operation or of the autonomous fine motion of robots and of dexterous hands in particular.

Keywords. Simulation; heuristic programming; man-machine system; robot fine motion; grasp planning; force control; configuration control; system failure and recovery.

INTRODUCTION.

Dexterous robot hands or smart grippers instrumented with sensors are mainly used or to substitute human activity in assembly processes or to perform some dangerous maintenance functions in hazardous environment. However human dexterity and the ability to operate is still superior in general. This is due not only to the mechanical dexterity of the human hand but even to a greater extent to the implicit involvement of knowledge (i.e. the skill) and to the human heuristic reasoning capabilities. The faculties which are mentioned above are par excellence human abilities and are not easy to be transferred to the machine once the domain of application is complex.

In the following paper we are concerned with the general problem of grasping and mating parts using instrumented grippers or artificial hands. As the most advanced performers we envision the state of art devices actually under development such as University of Bologna Hand [1] or Utah/MIT Hand [2], [3]. These devices are kinematically complex, but on the other hand they are paying back with the remarkable dexterity. Although in their case an explicit motion programming and control could be a

hard job (coordination problems), they do have a potential for performing a large class of grasping tasks included "in-hand" manipulation. Our principal goal is to work out the efficient yet general methods for planning, programming and controlling grasp operations using devices with various dexterity and sensor instrumentation, ranging from the parallel-jaw/force-feedback grippers to the Utah/MIT Hand. Our interest is in supporting a large class of grasping and mating operations normally done by humans.

We begin by proposing a systematic framework for grasp planning, based on deep-knowledge paradigm. The method of planning is founded on the model-based reasoning supported by an appropriate procedural backup (as for locating and classifying contacts, calculating internal forces equilibrium conditions etc.). We advocate for the interactive character of the planning process for the following reasons: at every stage the number of possibilities for decisions to be taken is high (so we would prefer to rely on heuristic reasoning capabilities of the human planner) moreover the premises for the decision rules are heterogeneous by nature (some based on "deep-knowledge" are likely to be involved). The decisions which are in essence the steps of the target program are thus finally taken (accepted) by human operator. Interactive computer support systems with graphical output are of particular interest in that context [4]. As the prototype system under development is actually implemented on the desk top computer we cannot actually guaranty its real-time operational performance in each simulation mode.

Reasoning supported by the system is based mainly on geometrical models, but in the course of task planning some other physical properties of the robot scene are taken into account. Our choice is to begin with the simplest models of the scene involving polyhedral solids represented by the wired-frames and then to upgrade their realism by introducing some relevant "world" characteristics such as curved models, friction models, external forces field, structural elasticity etc. Object oriented programming is an interesting method of building rich models. Normally, at every task-planning stage different aspects of the reality are involved, so the appropriate instance of the model can be called for. The difficult problem of maintaining the consistency between model instances is reduced in that way as well.

The important question is on how to find the planning/programming tools and an expert-interface for human operator to support the transfer of his implicit knowledge related to the task into the explicit robot-hand programs. By assumption some system background activities (purely computational) are not directly visible to the human operator. Only the results of computation are communicated in the indirect graphical form. So the critical factor for the eventual success of our planning method is a good understanding of the job to be done, based on underlying elementary laws of physics, which are relevant to grasping and fitting of solid objects [5], [6], [7].

Our approach is centered on the action, so the most important way of communication with the user-programmer is by the graphical emulation of the scene with animated models, where processes relevant to grasp are simulated and only those aspects of grasp models are rendered visible which are influencing decisions to be taken [8]. Actually we account for the friction and impact forces between the hand and the object of manipulation. Our next step will be to incorporate shape deformation of the contact area effects, with equivalents of the flexibility reactive forces taken into account [9].

GEOMETRICAL MODELS OF THE SCENE.

We are considering simulated scenes where manipulation object, dexterous hand or robot gripper and eventual obstacles are visible. Wired-frame poly-hedral models are displayed in order to get the resulting fast animation. This is our first modeling step aimed at solving the coordinated motion and kinematic analysis problems related to the preliminary preshaping of the hand before grasp. The sole articulated structure on the scene is the grasping device and it is directly manipulated by the user through the 6 d.o.f. sensor-ball manufactured by the ISPRA GmbH Darmstadt. Manipulation object can be pushed and/or affixed to the hand or to the obstacles and it can be relocated in space by direct manipulation. Tree-like kinematical structures of up to 30 d.o.f. can be defined. Geometrical models of the polyhedral envelope of each hand-segment, obstacle or manipulation object are specified by the user in a straightforward way by introducing coordinates of their vertices. Euler test is than automatically performed to check for the correctness of the polyhedral hull definition. Several visibility operators are introduced to ease planning steps by getting better insight into the spatial situation. Namely, a standard axonometric view, orthogonal projection views and observer viewpoint-changing mechanism are available.

On Fig.1. the defined UB Hand model ready for animation can be seen in its axonometry visibility mode. Through the keyboard we get access to every of its 11 joints to program the motion in teach-in mode. Some other operation modes are accessible through the keyboard or the sensor-ball in order to control coordinated motion involving all joints of the selected finger or even few fingers of the hand at-a-time. The sensor-ball provides the user with direct control of the manipulation velocities in translation, rotation and combined (6 d.o.f.) mode. Real-time animation of the model is obtained and it is used to study its manipulability properties or to verify the particular programmed motion before the contacts become active. Complex trajectories of the hand might be assembled by trial and error method without the usual risk.

As the next steps of our work two geometrical representations suitable for more complex volumetric objects are envisioned as candidates for further extensions of the system. We look for good models preserving surface features, which is specially important in our area of application. Compositions of polyhedrons (CPH) and compositions of generalized cylinders (CGC) [10] are taken into account. These particular graphical representations are selected on the the basis of involved computational cost. This economy aspect is important once one would like still to manipulate complex geometrical models in space, especially if some of them are articulated, as it is the case. CPH are more useful to study coordinated motion problems and to solve in a rough sense collision detection problem

[11]. CGC are better adapted for modeling the boundary-surface geometry, an extremely important feature for contact analysis.

PHYSICAL ASPECTS OF GRASP.

The number and final distribution of contact points upon an object depends not only on geometrical features of the interacting solids but also on their mass, inertia and surface related characteristics such as friction coefficient and local flexibility. The initial suggestions for grasp-mode selection, which are based solely on the geometry of the object and on hand kinematics, in most cases must be revised. Considering polyhedral manipulation objects and grasping parts one might have a variety of contact categories. Contacts might be different combinations of vertex-, edge-, face to -vertex,-edge,-face relationships, only four of nine possible being generic. In each of the above situations there is a specific model of the reaction force distribution upon the contact points and the appropriate friction model.

Prior to the computational checking for the internal-forces closure condition a correct classification of contacts must be done. In our modeling system we use a collision detection procedure for moving solids which is based on clipping [11]. Having specified potentially colliding solid segments of the hand and\or the object we are permanently testing for the sign of the inner products of the corresponding subsets of versors, normal to the segments faces. In that way we discover immediately the eventual inter-penetration of polyhedral solids. Moreover the same procedure is able to determine the contact type and to get the location of contact points in space. The procedure is computationally efficient, so despite the concurrent testing for the collision, we get almost natural speed of motion for relatively simple solids, even using an ordinary AT-class computer. On Fig.2. we display a simple situation of the contact detection. Contact analysis is much more involved when curved models of the manipulation object and the hand surface are considered. It is based on relationships between the principal curvatures of the surfaces [12], but its algorithmic version is still to be written.

Given a set of candidate contacts we can start to evaluate the stability of grasp. The analysis is based on checking for the force closure condition. It is worth to be mentioned the way in which the results of that check are communicated to the user. First the arbitrary external force-field active on the scene might be defined, then to each contact point simple friction models, so called friction cones, are associated. All forces are graphically represented as vector arrows. The user is getting access to each force present on the scene and using sensor-ball he can change its direction and/or magnitude. All the time friction cones at contact points are visible. The system treating as the input the value of the force currently under direct manual control, calculates the balance for forces and torques and the responding passive forces, providing the results in graphical form on the screen immediately. Thus the user gets the impression of being able to tune the impact forces one-by-one as to obtain the desired grasp stability margin in some static situation. His activity must be obviously iterative (at the contact-selection level and at the force-equilibrium level), however from our first experiences we would like to point out the efficiency of this grasp-planning method. On Fig.3 we show a simple situation involving grasp stability evaluation.

Local deformation of the contact area is radically changing the assumptions underlying force equilibrium calculus. If moreover the deformation is flexible some extra reaction forces must be accounted for. Actually we are seeking for the adequate yet still simple models for soft and\or flexible contacts. We would prefer to rely on lumped models. Virtual springs are among the concepts which are considered [9].

CONCLUSIONS

As the outcome of the intensive interactive simulation and evaluation of the displayed, simplified, computer-graphic models of the local scene we expect to understand well the necessary and sufficient factors defining successful grasping with grippers or dexterous robotics hands in a broad context of situations. Introducing the interactive, user-controlled animation we expect to support our studies on behavioral aspects of the dexterous manipulation in a quite straightforward way. The above approach can be interpreted as an attempt to transfer indirectly (through the action) of the implicit human (operator) knowledge about the task into the structured scenario of robot actions (i.e. a sequence of ordered commands under various conditions). By the way, it is also a good starting point for studying an autonomous operation of the grasping robot.

We are convinced that our experience might well contribute to the skill-based automation methodology. As to the man-machine aspects of the problem, we see our methodology as the tool-box for the human expert, enabling experiments with more or less complex, computer representations of the real world, providing a mechanism of focusing attention on selected aspects of the reality and of choosing the time-scale for the action (this being impossible in a realistic experiment). It would make possible the emulation of some behavioral patterns for grasp operations based on human understanding of the process. This patterns can then be transformed into robot programs without the human intervention for the subsequent autonomous task execution. In our opinion it is an interesting alternative to the classical idea of the completely automatic generation of robot plans from the specification of the initial and goal situation. In our proposal we expect to get better results in program debugging, as chances for the detection of action-errors using our approach should be improved comparing to the usual task-level programming methods. Moreover, some error recovery procedures could result from the extensive simulation studies.

The described interactive planning system may also be used in order to help choosing the appropriate architecture and displacement of sensors depending on the task requirements. Simulation may help to evaluate the necessary resolution and cooperation modes for tactile and visual sensor systems. Our software tools are developed on our own and run under MS DOS.

Acknowledgements

Author wishes to acknowledge the contribution of Pawel Drapikowski M.S. for his work on computer model of the dexterous hand and of Maciej Witucki M.S. for his work on rigid bodies contact analysis subsystem.

REFERENCES

[1] Bonivento C. and co-authors, "Control System Design of a Dexterous Hand for Industrial Robots" in Robot Control 1988 (Proc. IFAC SYROCO'88 Symp.) U.Rembold Ed., Pergamon Press, Oxford, 1989, pp.389–394.

[2] Jacobson S.C. and co-authors, "The Utah/MIT dexterous hand: Work in progress", The Int. Journal of Robotics Res. no. 3(1984), pp.21–50.

[3] Allen P.K. and co-authors, "An integrated system for dexterous manipulation" in Proc. IEEE Int.Conf on Robotics and Automation, Scottsdale AR, IEEE Washington DC 1989, pp.612–617.

[4] Kasiński A., Bonivento C., Vassura G., "Computer-graphic support system for interactive grasp planning", in Proc. Int. Conf. on Advanced Robotics ICAR'91, Pisa Italy, IEEE Washington DC 1991, v.2, pp.1182–1186.

[5] Akella P., Cutkovsky M., "Manipulating with Soft Fingers: Modelling Contacts and Dynamics" in Proc. IEEE Int.Conf on Robotics and Automation Scottsdale AR, IEEE Washington DC 1989, pp.764–769.

[6] Bicchi A., "Intrinsic Contact Sensing for Soft Fingers" in Proc. IEEE Int.Conf on Robotics and Automation, IEEE Washington DC 1990, pp.968–973.

[7] Yoshikawa T., Nagai K."Manipulating and Grasping Forces in Manipulation by Multifingered Robot Hands", IEEE Trans. on Robotics and Automation, v.7, no.1(1991), pp.67–77

[8] Minsky M. and co-authors,"Feeling and Seeing: Issues in Force Display". Computer Graphics no.5 (1990), pp.235–243.

[9] Van-Duc Nguyen, "Constructing Stable Grasps". The Int. Journal of Robotics Yes. no. 1(1989), pp.26–37.

[10] Besl P.J. "Geometric Modeling and Computer Vision". Proc. of the IEEE, v.26, no.8(1988), pp.936–958.

[11] Moore M., Wilhelms J. "Collision Detection and Response for Computer Animation", Computer Graphics", no.4 (1988), pp.289–298.

[12] Fearing R.S., "Tactile Sensing for Shape Interpretation" in "Dexterous Robot Hands" S.T. Venkatarannan and T. Iberall Eds., Springer 1990, pp.209–238.

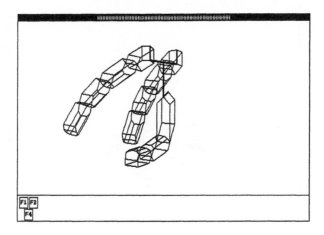

Fig.1. Graphical output to the user programming the 11 d.o.f. dexterous hand through the computer keyboard or sensor-ball. Selected pose of the hand (F1, F2, F4 are used to access the appropriate fingers).

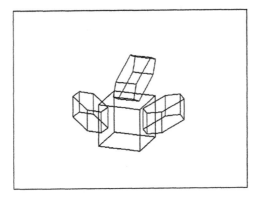

Fig.2. Graphical output showing the situation when the hand fingertips are comming into contact with the manipulated object. Collisions are detected on-line, contact types and locations are returned to the programmer.

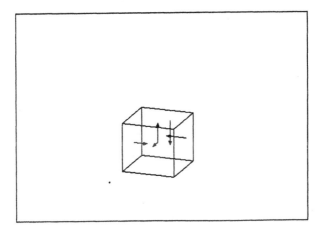

Fig. 3. Checking for the force-closure condition corresponding to the situation from the Fig. 2. with impact forces from every finger set and external force acting on the cube center of mass vertically up.

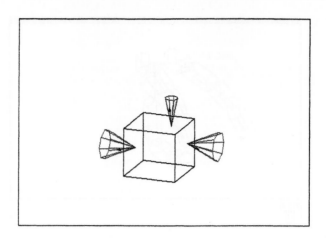

Fig.4. Grasp stability evaluation taking into account the friction at contact points. Note that friction conditions represented as friction cones are different for particular fingertips.

KNOWLEDGE-BASED PATTERN-SUPPORTED
MAN-MACHINE INTERACTION

T. Vámos* and F. Katona**

**Computer and Automation Institute, Hungarian Academy of Sciences, H-1132 Budapest,
Victor Hugo u. 18-22, Hungary*
***Department of Development Neurology and Neurohabilitation, Pediatric Institute Szabadság-hegy,
H1121 Budapest, Mártonhegyi út 6, Hungary*

Abstract. Knowledge base being a key issue in man-machine interfacing, is much more
pattern-like in the human brain's representation than the logic of rule-based systems.
Basic problems of pattern representation, quantification, metric and decision methods
are discussed, having an experience on several problems, one of them a decade's work
of a sophisticated human brain-oriented medical expertise. Some relevant hypotheses
are formed, supporting the practical applicability of patterns combined with logic in
complex man-machine systems.

Keywords. Knowledge-based systems, patterns, representation.

INTRODUCTION

Man-machine interaction is developing into a
complex structure of interface manifold. The
machine as a physical prolongation of human
capabilities suggested the results of ergonomy, a
study of human biological conditions related to
work and a design versatility in matching the
human physical interface to the machine control
devices. Knowledge-based systems, new avenues of
computer technology have to proceed deeper, the
intellectual symbiosis introduces new complexities.
The knowledge, i.e. data, facts, relations, procedures fed into the computer creates a genuine,
sometimes not fully controlled activity and this
should be harmonized with the human brain's
own. Representations, mappings, worlds in a logical sense cooperate and conflict, complexities and
not fully understood brain processes meet in real
time.

The survey given here reports about some selected
cross-sections of the problems related:

- How relevant representation schemes of the
  human brain can be transmitted into a knowledge representation?
- reasoning experience based on these representation methods;

- limitations on both sides (machine and human), matching of capabilities for reaching a
  quasioptimal decision support;
- some results and promising hypotheses.

Basic representation schemes applied in the experiments are patterns, i.e. sets of somehow coherent data. Experiments were conducted in three
very different fields, in medical, legal and manufacturing practice. The ways of approaching the
problems were only partly coherent.

PATTERN REPRESENTATION

Most important lesson was received from cognitive
psychology partly applied in our investigations.
Due to the longeval phylogenetic development of
the neural and especially of the central neural
system the brain's mode of operation can be
modelled as pattern storing, retrieval and
processing activity. The biological system collects
information which looks to be coherent from one
or the other point of view of survival. This
collecting, filtering, ordering activity is organized
by a phylogenetic and ontogenetic learning. In the
first approximation of the understanding these
basic mechanisms the discussion of the role of

inheritance (phylogenetic) and learning (ontogenetic) is irrelevant. Psychology coined several words for these sets of neural representations, Gestalt, scheme are more used; we apply the other similar naming, pattern, because of the further relations to pattern processing of computer science.

In this context we should mention that logic, a seemingly higher level of intellectual activity which is mostly attributed to the humans only, is not more and not less than metapattern, i.e. patterns of pattern relations. Conceptual ways of thinking are clustering procedures and manipulation methods on clustered patterns, static and dynamic relations, interactions of patterns (or pattern clusters formed of these in conceptual representation) are patterns and metapatterns themselves (in the sense as we interpreted the relations represented by logic). These latter static and dynamic behaviour patterns (or patterns of behaviour) are often spiritualized as Laws of Nature, economy, social coexistence etc. The pattern representation view is relevant in several aspects, it looks to be philosophical only but it is very practical as well because by this developmental interpretation it liberates the concepts from their rigid artificial borders and gives us a liberty in computer representations using a much more pragmatic, less ideological approach.

There is nothing new in pattern representation even for computer science. A file is a pattern as well and the efforts in flexible and pragmatic handling of data files, e.g. relational data base systems reflect the conscious relativity of coherence. Different ways of handling uncertainty, fuzzy and possibilistic methods, the acknowledgement of the relativity of logic by modal and intensional variants, different possible worlds' acknowledgement, the tolerance of nonmonotonic logic and the loosening of the closed world assumption are all symptoms of a change in viewing the world as an originally designed, logically structured construct. Connectionist efforts prove the same.

The question to be investigated is the next step: what can we learn from the pattern representation model of the brain's activity and how can we create a quasioptimal interaction between the real mental process and of its computer representations?

QUANTITIES, QUANTIFICATION, METRIC

All popular papers refer to the quantitative differences of the human brain and most advanced computers. This is most probably one of the main reasons of the nonsense of primitive replacements.

Another similarly popularized quantitative relation is more useful for us: the difference between the generally used and as an average-commanded patterns of a speciality and the same possessed by a high level expert. The difference is one order of magnitude or sometimes even more: about 2000 related to 40-100.000. This figure looks to be similar to the linguistic use of words (cf. Zipf's law), to Chinese characters, strategic situations in chess or medical diagnosis, it is supposed to be a general feature of human capabilities.

The lesson for us could be an attempt to reach by our systems the everyday knowledge level characterized by the figure 2000 and buttress it by the high human expertise. The feasibility of this objective is well-supported by our experiments and several other efforts. We come back to this point later.

Problems of quantification of the patterns' (sets') elements and how human judgement does it were and are still a focus of interest both for computer science and for psychology. Estimation of evidence is a subject of all methods related to uncertainty. We discuss these issues elsewhere (Vámos, 1991) where a comprehensive reference to literature can be found collected as well.

Crucial problem is the metric of the pattern space. The metric is built up by the individual evidences related to the pattern elements, it creates a decision space for clustering the coherent patterns and divides them into well-discriminated entities. The task is seemingly the usual pattern recognition one but we encounter here a really nonmetric space, i.e. no uniform distance measure can be defined. The space — according to detailed investigations of cognitive psychologists (Tversky, 1977; Kahneman and others, 1982) and our experience as well — is highly nonlinear, the triangulation relations are mostly invalid, in best cases it is topological, i.e. some ordering of the components is feasible.

In an attempt to reach a viable solution for this special pattern recognition problem we have to investigate the reasons why this decision space is different to the usual metric pattern spaces, what the reason of these subjective looking distortions is. The question is important because the human judgement in spite of its obvious deficiencies is still in many cases better than an objective looking machine one. Looking for an appropriate man-machine interaction we have to analyze these differences: what are the keys of the better performance of the machine and what of the human side, and what are the advisable ways of matching the two?

GRAVITATIONAL DISTORTION OF THE PATTERN SPACE

The main reason of the irregularity likelihoods is a phenomenon which we coin gravitational distortion by its metaphor. The epistemic background is due to the tacit knowledge (Polányi, 1964), i.e. the bulk of background knowledge on (or related to) the much smaller amount of knowledge which is apparently and consciously used in the defined problem solving procedure. This tacit knowledge acts as a feeling, intuition, routine of experience, association, is very pattern-oriented and much less logical than the better definable conscious surface reasoning. The hidden tacit knowledge acts as big hidden mass on a gravitational field which should be uniform due to the surface uniformity.

If all knowledge related to a certain decision (recognition) problem was well defined, it could be represented by well-formed logical formulae in a closed world. In this case we would be able to apply the usual clustering and pattern recognition methods without any further concerns. The statistics-based probabilities transfer us to the world of crisp conceptual-logical methods of the rule-based expert systems.

The tacit knowledge which confuses the picture is mostly not a nuisance. As we have indicated it contains in a hidden form sometimes the most valuable knowledge, the expertise, skill, talent, intuition. Nevertheless, it is confused with biases, psychological influences, prejudices, false or not appropriate dogmas. A relevant task of the man-machine interaction design is a careful continuous separation process of the useful and of the malfunctioning tacit knowledge and effects.

EXPERIMENTAL EXPERIENCE

The longest and most successful experience is received in the medical field during our decade's work. The task is the early diagnosis and habilitation of brain defects in the first few months of life. The procedure is an internationally acknowledged practice of a Hungarian group (Katona, 1988, 1989). The problem is especially interesting because it is related to the partly unknown fast developing process of the human brain, contains lots of deep biological knowledge combined with unknown phenomena of experience. The process is highly dynamic (a fast development period) and this dynamics is one of the most relevant features of recognition. About 2000 patients' data are processed, each patient is checked several times (10 is a usual figure), each check can contain a maximum of about 1000 data. 30-40 basic patterns are identified with several times more subpatterns, each

new check, each newly recognized but earlier fixed pattern is a feedback control for the pattern definition, so the ranking and the quantification of the strength of the phenomenon (data), its relevance related to the pattern concerned and its reliability.

The medical project treated 3 different kinds of patterns. The first sequence (8 patterns) depicts the structural-functional status (see Table 1, Fig. 1).

The second is a mapping of these pattern data by weighted linear combinations into diagnostic patterns (presently 11, see Table 2, Fig. 2).

The third is the combination of these diagnostic patterns to get a general pattern of status. This happens by a complex weighting of developmental features composed especially by those diagnostic patterns which indicate some irregularities. The results can be an estimation of backwardness in development. The system (called KAS-NES as a software product) comprises a statistical modul which enables the user to look after each pattern-combination in the data base of the ward's experience, refine the patterns, tune them according to this experimentation on data. The present scheme is a result of a more than a decade's common work with at least three major redefinitions of the patterns. The new cases fit into this scheme with a nearly 80-90 % accuracy.

The medical system after a decade's experience and several extensions has a vocabulary of 1584 words and only about one third of them are such kind of mostly medical terms which are used as keywords for further contextual search for knowledge query, looking for analog cases, doing statistics. This refinement of expressions has relevant meaning in a double-word term but practically never exceeds 3-4 connected words.

The other project is an analysis of situations and decisions for custody arrangements of children after the divorce of their parents. Rules of Law are combined with estimations of social environment, psychological effects related to the children, i.e. situation patterns. The analysis was carried out on closed files of the cases, without personal contacts. The decision analysis was extended to the experience of the lower and of the higher juridical forums' differences (a hint to the 2000 and an order of magnitude higher tacit knowledge patterns) and to possibilities of using different values, viewpoints, i.e. decision influenced by a liberal or by a conservative view. The legal system's thesaurus contains about 500 terms (maximum 5 words each), these are the basic situation patterns and 35 evaluation patterns as a composition of the situation patterns. The four basic decision patterns (pro/against plaintiff, compromise, other) are derived from these latter ones having more refined versions but not more than 2-3 for each of the four. The hierarchy of these patterns is similar to that of the medical one: factual data (e.g. income, home conditions, school evaluation), evaluation patterns (e.g. new family relations of the parent, personality etc.) and decision patterns (e.g. child's interest).

The number of cases processed by this experiment was in the magnitude of 100, an exact number cannot be given here because the selection of the appropriate cases provided different groups from different points of view of the analysis.

The third set of experience was the result of another group working close to our one. This group has been working for more than a decade in AI application of computer-integrated manufacturing; job scheduling, fixture design are some of their dedicated results (Márkus and others, 1984), they compare human decision with advanced combinatorial solutions, e.g. genetic algorithms.

The three experiments were intended for different purposes. The medical project, the most advanced and extended one supports research of the neurology group, daily practice of specialists working with their results and training of pediatric doctors and nurses. The levels of application is a challenging further study of man-machine symbiosis. The legal project is conducted in the frame of a PhD work, at this phase it has no practical application objective, later it should serve the laymen as a legal advisor, i.e. the lowest application level. The manufacturing project is operating as a design phase, i.e. they build special parts of a complex creative process.

SOME RESULTS AND HYPOTHESES

1/ According to our experience the nonlinearities of the metric are rather well-expressed by the subjective weights of the expert estimations. This is no surprise, the nonlinearity of human sensory perception is a long known fact (Weber-Fechner, Stevens relations). The natural transformation of physical to psychological scaling served just the same objective as our computer procedure: a wide range perception combined with a sharp distinction.

2/ This weighted set of data can be well used in most cases in a linear combination for group estimation and pattern discrimination. This is another interpretation of the Hebb-law of psychology used similarly in connectionist applications.

3/ The above two steps create a viable structural (conceptual) hierarchy which helps in avoiding the combinatorial explosion in the low-medium expertise field.

4/ This level of knowledge can be elicited by the usual but lengthy knowledge acquisition methods, represented in a combination of patterns and rules. The machine knowledge, especially if it is regularly updated by a feedback of experience is about in the usage-centre of the high-level expert's knowledge like the 2000 words truncation of the high brow language (Zipf-Law).

5/ The relations of a machine knowledge and of the high quality human expert's can be somehow similar to the two levels of human expertise, i.e. this relation can be used as a practical metaphor for expert system applications.

6/ Due to the human limitation of parallel processing more variants (channels of thinking) than 5-7, the human decision maker selects decision routes mostly too early (design: Vámos and others, 1991; medical: Elstein and others, 1978; Schwartz and Griffin, 1986). Especially in design an appropriate parallel or quasiparallel process (in our referenced case the genetic algorithm) can select better, less biased solutions than the human expert. Several examples could be quoted for this early commitment in design: the architects start − according to their personal preferences − either with a view on the facade or with a functional ground plan or with an idea of structural embodiment. Design of complex machines (e.g. cars) show similar attitudes, e.g. the major use of earlier details, some existing and accustomed technologies. Our group used genetic algorithms in the design of manufacturing processes for not too complex parts. The number of feasible solutions for the start of selection was more than 30, much more than an average designer is able to handle. An intermediate multitude of feasible process arrangements reached several hundreds.

7/ Another improvement reachable by the pattern method is an improved avoidance of biases due to existing conceptual frames. These conceptual frames can be closely related to the subject or in a loose way as well by the hidden presence of inappropriate metaphors (negative effects of tacit knowledge). The same is due to other kinds of biases, e.g. effects of risk consciousness, etc.

The bias could be experimented by different methods in the different projects. The most obvious way was the comparison of the machine results with the human judgement. The mentioned three major revisions of medical patterns was just a result of this experience. The variations received by the use of genetic algorithms in machining design were also compared with the judgements of average and high authority experts. A systematic experiment was carried on in the legal project where the weight of the data and patterns were changed due to different policies, mostly liberal and conservative (e.g. weighting the present sexual relations of the divorced parents, rigor of discipline and its methods, etc.) The other aspect of comparisons was the practice of the lower and of the higher courts in the mirror of the same cases.

An important concluding remark: As it was stated in the introduction two systems of different complexities and most probably different mechanisms should be matched: the machine and the human brain. Due to this soft interface the resulting statements are even more soft, all major steps on both sides, relating technology and psychology can change the picture and it is changed day by day by the minor advancements or failures of the various projects conducted all over the world. Statistics are weak as well, they are extremely subjective depending on the subject, on conditions of the experience — and most of them belong to the tacit knowledge —, the conscious and unconscious presentation due to selection of cases and circumstances. This is one reason why we preach modesty and advise to take our views with salt and pepper as well.

ACKNOWLEDGEMENT

The authors express their gratitude to their colleagues, Ms. M. Berényi, MD., P. Szabados, MD., P. Somogyi, Mrs. Á. Koczka, P. Danyi for their contributions to the experiments and P. Koch, A. Márkus, J. Váncza for their valuable remarks.

REFERENCES

Boden, M.A. (1988). *Computer Models of Mind*. Cambridge University Press, Cambridge.

Cooke, N.M., McDonald, J.E.: The application of psychological scaling techniques to knowledge elicitation for knowledge-based systems (1988). In Boose and Gaines (Eds.) *Knowledge-Based Systems*, Vol. 2. Academic Press, New York. pp. 65-82.

Csányi, V. (1989). *Evolutionary Systems and Society — a general theory*. Duke University Press, Durham/NC.

Danyi, P. (1989). Advisory expert sytem. *3rd International Conference on Expert Systems in Law*, November 2-5, 1989, Florence/Italy, Vol. 1, 157-173.

Ehrenfels, Ch.(1887). Über Fühlen und Wollen — Eine Psychologische Studie. *Sitzungsberichte der Philosophisch-historischen Klasse der Kaiserlichen Akademie der Wissenschaften*. Gerold, Wien.

Eisler, H. (1965). The connection between magnitude and discrimination scales and direct and indirect scaling methods. *Psychometrika, 30*, , 271-289.

Elstein, A.S., Shulman, L.S., Sprafka, S.A. (1978). *Medical Problem Solving*. Harvard University Press, Cambridge/Mass.

Fodor, J.A. (1983). *The Modularity of Mind: An Essay on Faculty Psychology*. MIT Press, Cambridge/Mass.

Gelsema, E.S., Kanal, L.N. (1988). *Pattern Recognition and Artificial Intelligence Towards an Integration*. Elsevier, Amsterdam.

Hebb, D.O. (1949). *The Organization of Behaviour: A Neuropsychological Theory*. Wiley, New York.

Katona, F. (1988). Early diagnosis and neurohabilitation. In P.M. Vietze and H.G. Vaugham (Eds.) *Early Identification of Infants with Developmental Disabilities*. Grune and Stration, Saunders/Philadelphia. pp. 121-145.

Katona, F. (1989). Clinico-neurodevelopmental diagnosis and treatment. In P.R. Zelazo and R.G. Barr (Eds.) *Challenges to Developmental Paradigms*. Lawrence Erlbaum Ass., Hillsdale/New Jersey, pp. 167-187.

Koch, P. (1987). Some problems of building process control expert systems. In: L.O. Hertzberger (Ed.) *Intelligent Autonomous Systems*. North-Holland, Amsterdam.

Márkus, A., Márkusz, Z., Farka, J., Filemon, J. (1984). Fixture design using PROLOG: an expert system. *Robotics and Computer Integrated Manufacturing, 2*, 67-172.

Mérő, L. (1990). *The Way of Thinking*. World Scientific, Singapore

Piaget, J.(1976). *The Child and Reality: Principles of Genetic Epistemology*. Penguin, New York.

Polányi, M. (1964). *Personal Knowledge: Towards a Post-Critical Philosophy*. Harper, New York.

Pylyshin, Z.W. (1984). *Computation and Cognition: Toward a Foundation for Cognitive Science*. MIT Press, Cambridge/Mass.

Schwartz, S., Griffin, T. (1986). *Medical Thinking* (Contributions to psychology and medicine). Springer, New York.

Tversky, A. (1977). Features of similarity. *Psychological Review, 84*, 324-352.

Tversky, A., Slovic, P., Kahneman, D. (1982). *Judgement under Uncertainty: Heuristics and Biases*. Cambridge University Press, Cambridge. pp. 47-519.

Vámos, T. (1991). *Computer Epistemology*. World Scientific, Singapore.

Vámos, T., Váncza, J., Márkus, A., Somogyi, P. (1991). Learning from nature and Augustine — two experiences. *World Congress on Expert Systems*, Orlando/Florida, December 16-19, 1991, pp. 2759-2766.

Table 1 Structural-Functional Patterns

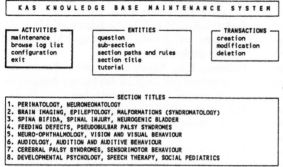

```
 K A S   K N O W L E D G E   B A S E   M A I N T E N A N C E   S Y S T E M
```

```
┌─ ACTIVITIES ─┐   ┌──── ENTITIES ─────┐   ┌─ TRANSACTIONS ─┐
│ maintenance  │   │ question          │   │ creation       │
│ browse log list  │ sub-section       │   │ modification   │
│ configuration │  │ section paths and rules │ deletion      │
│ exit         │   │ section title     │   └────────────────┘
└──────────────┘   │ tutorial          │
                   └───────────────────┘
```

```
┌────────────────────── SECTION TITLES ──────────────────────┐
│ 1. PERINATOLOGY, NEURONEONATOLOGY                           │
│ 2. BRAIN IMAGING, EPILEPTOLOGY, MALFORMATIONS (SYNDROMATOLOGY) │
│ 3. SPINA BIFIDA, SPINAL INJURY, NEUROGENIC BLADDER          │
│ 4. FEEDING DEFECTS, PSEUDOBULBAR PALSY SYNDROMES            │
│ 5. NEURO-OPHTHALMOLOGY, VISION AND VISUAL BEHAVIOUR         │
│ 6. AUDIOLOGY, AUDITION AND AUDITIVE BEHAVIOUR               │
│ 7. CEREBRAL PALSY SYNDROMES, SENSORIMOTOR BEHAVIOUR         │
│ 8. DEVELOPMENTAL PSYCHOLOGY, SPEECH THERAPY, SOCIAL PEDIATRICS │
└─────────────────────────────────────────────────────────────┘
```

```
Select a function, please. ↑↓←┘: Selection    Esc: Abort
```

Table 2 Diagnostic Patterns

Morphological	*Behavioural*
subdural effusion	problem solving
hydrocephalus	control problems
hypertension	spontaneity
hydroencephalia	

Sensory	*Performance*
visual problems	independency
auditory problems	communication
	visuomotory
	mind-thinking

45. Character of exploration
    45.0. adequate
    45.1. inadequate
    45.2. stereotype
    45.3. does not explore

    weight: 1, valid after 4 months

46. Modality of exploration
    46.1. visual, oral, tactile
    46.2. visual, oral
    46.3. only visual
    46.4. visual and tactile
    46.5. oral and tactile
    46.6. tactile
    46.7. only oral
    46.8. does not explore

    weight: 1, valid after 4 months

....
....

52. Pattern recognition
    52.1. certain recognition of complex figures   36
    52.2. several simple patterns but reverse   30
    52.3. several simple patterns similar place   24
    52.4. one simple pattern reverse   18
    52.5. one simple pattern similar   15
    52.6. does not   0

Fig. 2. Small sections of performance patterns
(a part of more than 1000 entries)

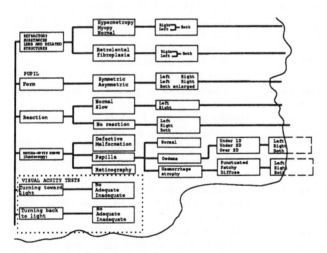

Fig. 1. A small section of the structural pattern
(about 1/3 of visual chart, 3-4 % of the whole)

MODELLING AND ANALYSIS OF HUMAN TASK ALLOCATION IN A BOTTLING HALL

W.A. Blaauboer, P.L. Brinkman and P.A. Wieringa

*Delft University of Technology, Faculty of Mechanical Engineering and Marine Technology,
Laboratory for Measurement and Control, Man-Machine Systems Group, Mekelweg 2, 2628 CD Delft,
The Netherlands*

Abstract. Automation changes the role of a human operator in an industrial production facility. Needed are methods and models to evaluate different task allocation designs with respect to their performance, flexibility and working conditions. A procedure is described for the modelling of human operator teams in a bottling hall. The model of a human operator is based on the assumption that he performs his task in stages. Then a time-sequential network of interacting human operators can be formed, using the theory of Petri Nets. The possibilities for the analysis and simulation of this man-machine model are discussed. The method is applied to three different work organization forms. The model allows the evaluation of the performance and flexibility of different work organization forms. However, it is not possible to determine an optimal form, because this depends on the rate of occurrence of the tasks and other, more subjective, factors.

Keywords. Automation; Coloured Petri Nets; Human Factors; Industrial production systems; Man-machine systems; Modeling.

INTRODUCTION

Most research on the subject of man-machine systems deals with a single human operator. For large and complex processes usually a team of human operators is assigned to the task of (supervisory) control. In that case automation changes not only individual tasks but also the division of tasks among and interactions between team members. Hence the task allocation consists of two phases (Rijnsdorp, Pikaar and Lenior, 1984):
- allocation of system tasks between "machines" and "men";
- allocation of human tasks between men, with the division of responsibilities and communication patterns.
The result of the latter called is work organization.

The aim of this study is to develop methods and models to analyze and evaluate different task allocation designs for an organization. A designer is faced not only with the problem of creating a task allocation design that will meet the desired specifications, but in addition with the problem of assigning meaningful tasks to the members of that organization so no one is under- or overloaded. An organization is defined as a team of human operators that receive and execute tasks, i.e., the organization receives an input from an external environment and produces a response. In recent years studies on this subject have been conducted by Levis and others, e.g. (Andreadakis and Levis, 1987). For this study the task allocation in a bottling hall of a beer brewery was studied. In this paper a procedure is presented for the modelling of human operator teams in such a way that different work organization forms or organization structures can be modelled and evaluated.

HUMAN OPERATOR TASKS IN THE BOTTLING HALL

In a bottling hall of a beer brewery bottles are filled and packed. There are four to six bottling lines working in parallel. A bottling line consists of approximately ten machine operations. The bottles are transported between machines on conveyor belts, which also act as a buffer. Short stops of a machine, approximately two to five minutes, will not affect the throughput of a bottling line. The nominal capacity of a bottling line is 60,000 30 cl bottles per hour. The mean transit time is about seventy-five minutes.

A bottling line is manned by five to eight machine operators. They monitor and, if necessary, adjust or intervene in the process in order to minimize production losses due to machine breakdowns or lack of quality. There is no fixed schedule of activities, hence the operator's activities are event driven. The operator is allocated to a specific part of the bottling line and is not supposed to leave that area unattended for a long period of time. The workload is determined by the rate of occurrence and the nature of breakdowns.

Most breakdowns are easy to correct, but when a complex breakdown occurs the machine operator has to call in maintenance. The maintenance crew is a group of specialized craftsmen to tackle complex breakdowns and to make running repairs, their actions are event driven as well. The maintenance crew is located in a separate area in the bottling hall. Their workload is mostly determined by the frequency and nature of breakdowns that cannot be corrected by the machine operators. One supervisor monitors an entire bottling line and instructs the machine operators, his work is mostly event driven. Finally, one manager monitors the entire bottling hall and instructs the supervisors.

Automation changes the role of the machine operator in a bottling hall. In recent years the number of machine operators in a bottling line is significantly reduced. As a consequence they become more widely spaced apart. Because of the noise, caused by the machines and bottles, people cannot talk intelligibly to one another, which leads to social isolation. Since the frequency and nature of the breakdowns vary and are unpredictable, the workload for the different machine operators is usually unevenly distributed. A higher degree of flexibility of the organization is therefore desirable.

The use of modern electronics makes it possible to distribute on-line information about the machines and to present this information on visual display units, placed further away from the machines. Letting machine operators act as teams of roving field operators, could add to the flexibility of the organization and create better working conditions. Also with the help of an improved information system the machine operator may be able to correct more breakdowns himself, thus taking over parts of the tasks of the maintenance crew.

On the supervisory level the use of modern information systems makes the supervisor's position become ambiguous. His task used to be that of a roving (technical) adviser; now it changes more and more to a control room operator who monitors the process continuously and gives instructions to his operators from a small monitoring room. The situation of a single supervisor in a control room leads to social isolation and, since the supervisor is still sometimes needed on site in the bottling line, this leads to a non optimal use of the information systems in the control room. Letting a team of supervisors work in a central control room, with in part a control room operator's task and in part a field operator's task, creates better working conditions and a more flexible organization. In short, due to the use of modern information systems, a large spectrum of new possibilities arises and makes it necessary to take a closer look at the allocation of human tasks in a bottling hall.

MODELLING AND EVALUATING
TEAMS OF HUMAN OPERATORS

At this stage of the research real life experiments could not be considered so a model was needed to quantitatively evaluate the pro's and con's of different work organization forms. The evaluation should address key questions such as: the performance and degree of flexibility of an organization, the distribution of information among team members, the resulting workloads and the need for decision aids (Blaauboer and Brinkman, 1990).

On the operational level an organization consists of a team of human operators, who react on inputs from the process. These inputs are usually disturbances and require a certain amount of time and expertise to correct. In the bottling hall the human operator's actions are dependent on each other. Each member of the team gives and receives information or instructions by words, in writing and even sometimes by hand signals. The efficient running of a bottling line depends a great deal on these interactions. A different work organization form requires different interactions. The interactions consist of: giving or receiving information from the process or other human operators, receiving commands from or giving commands to other human operators etc.

All the actions of a human operator are event driven; only when an operator receives information can he start to process a response. A systems' theoretic representation, using block diagrams, leads to an ambiguous representation of the internal information structure of an organization (Levis and Boettcher, 1983). Therefore the whole man-machine system will be modelled as a Discrete Event Dynamic System. The organization structure, i.e., the interactions between organization members, is described using Petri Nets, based on ideas introduced by Tabak and Levis (1985). The fundamental assumption of the model is that a human operator processes only one input at a time and that he performs his task in stages. Then a time-sequential network of interacting human operators can be formed. Furthermore, the model can be extended to include hierarchical forms.

To evaluate different work organization forms, first a qualitative and intuitive analysis will be performed to generate feasible work organization alternatives, with, for instance, better or more flexible working conditions. Then a quantitative analysis of the performance, the flexibility and the resulting workloads should take place.

As a measure of performance of an organization, the time interval between the moment the inputs or tasks were received and the execution of a correct response is taken. In the ideal situation this means that the response should be in time, so that the throughput of the bottling line is not affected. Delays in producing a response can consist of external delays, the reaction time of an organization, and internal delays, team members waiting for other member's responses or the time to get the right person at the right place.

Flexibility means that an organization can adapt itself to internal or external changes, thus maintaining an acceptable level of performance and an acceptable distribution of the workloads. An external change is, for instance, an increase in the rate of occurring tasks, an example of an internal change is the temporary absence of a team member. Usually an indication of the flexibility of an organization is the recovery time, that is the time interval between the moment a breakdown occurs and the moment that for all the input places a human operator is again stand-by. A measure of workload is presented in the next section.

A MODEL OF AN INTERACTING
HUMAN OPERATOR

Petri Nets are described by Peterson (1981) and Reisig (1986). A Petri Net defines places, transitions and tokens. Places, represented by circles in diagrams, model signals or conditions. Transitions, represented by bars, model processes or events. Places can only be connected, through arrows, to transitions, and transitions can only be connected to places. The execution of Petri Nets is controlled by tokens in places, denoted by dots in the circles, see Fig. 1. A place can hold zero, one or more tokens. A Petri Net is said to execute when a transition fires. A transition can fire only when it is enabled, i.e., all its input places must contain at least one token. When a transition is enabled it fires immediately; it removes one token from each its input places and puts a new token in each of its output places. If one assigns to each transition a certain amount of time, one can study the performance of a system.

For the actual modelling high-level nets are used. In the Coloured Petri Nets, each token is related to a data set, called colour; this makes it possible to treat tokens differently, depending on their colour. Hence making it possible to handle the description of much larger and more complex systems (Jensen and Rozenberg, 1991). For the practical applications a computer tool Design/CPN (Meta Software, 1991) is used.

The basic model of an interacting human operator A on the operational level is depicted in Fig. 1. The model consists of five stages or transitions: the detection (DETa) stage, the movement (MOVa) stage, the interpretation (INTa) stage, the correction (CORa) stage and the inactive (INAa) stage.

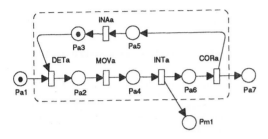

Fig. 1. Petri Net of a human operator

External inputs from the process or other members of the organization are processed by the human operator. A token, representing a disturbance, appears in the input place Pa1, for instance a machine breakdown has occurred. If the operator is stand-by, i.e., the place Pa3 contains a token, the transition DETa fires. A token appears in Pa2 and the operator becomes active. The detection time is the time between the occurrence of the failure and the moment that the operator perceives it, thus the time between the appearance of a token in Pa1 and Pa2. Next the operator moves to the place of disturbance, the movement time, and a token appears in Pa4. During the interpretation stage the operator makes a rudimentary assessment of the situation and either decides to diagnose and correct the failure himself, thus a token appears in Pa6, or decides to call in the help of others, for instance a maintenance man, and a token appears in Pm1. The transition INTa acts as a switch. Having decided to correct the failure himself, the operator moves to the correction stage. After the correction time, a token appears in the output place Pa7 and a token appears in Pa5. It takes a certain amount of time before the operator is ready to attack another problem, represented by the inactive stage. The part of the net, Pa5-Pa3, secures that once an operator is activated, he cannot be activated again, unless he again has passed the inactive stage and is stand-by. In this model the human operator acts as a sequential data processor.

Basic assumptions are:
1. When the human operator is stand-by he is in a fixed location.
2. The mental workload is not taken into account, assuming that the extent of the mental workload will not seriously affect the functioning of the human operator. Hence an indicator for the workload is the time load, the percentage of time the human operator is active.
3. The human operators are well trained for the tasks that they have to perform and there are no learning effects.

With this basic model different work organization forms can be modelled by allowing different interactions. In the Fig. 2 and 3, two different forms of interaction between a machine operator and a maintenance man are modelled, using Coloured Petri Nets. A colour set I, containing a single element or token i, is defined, representing the identity of a human operator. A class of disturbances is defined as the colour set S. The variable x of the colour set S can take either the value t or u. Tokens of the colour t or u, representing machine breakdowns, appear in Pa1. During the interpretation stage the machine operator decides to correct the failure himself, thus if x=t a token t appears in Pa6, or to call in the help of others, meaning if x=u a token u appears in Pm1. Two different situations can be distinguished. In the first, Fig 2, the operator moves to the inactive stage immediately after he has called in for assistance, depicted by the arc with inscription [x=u]%i; a boolean statement meaning that if the variable x has the value u a token i appears in the place Pa5.

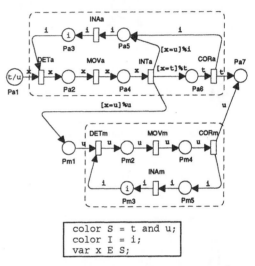

color S = t and u;
color I = i;
var x E S;

Fig. 2. Petri Net of a machine operator and a maintenance man

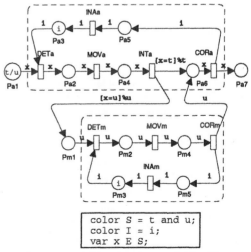

color S = t and u;
color I = i;
var x E S;

Fig. 3. Petri Net of a cooperating machine operator and maintenance man

In the second case, Fig. 3, the machine operator stays involved with correcting the disturbance and moves to the inactive stage after the correction of the breakdown.

APPLICATIONS AND RESULTS

As an example of the application of Petri Nets, three different work organization forms for machine operators will be evaluated. The task of a machine operator consists of correcting machine breakdowns, a fault management task. A part of a bottling line is modelled by four machines: M1, M2, M3 and M4. In the present work organization form A, Fig. 4, two machine operators are working in parallel, there is no information or task overlap.

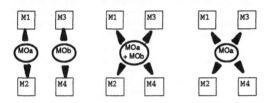

Organization A   Organization B   Organization C

Fig. 4. Three different work organization forms

A possible new work organization form is organization B. The machine operators are placed further away from the machines, but are provided with information about the status of all four machines. They work as a team of roving field operators. In this case there is a complete information and task overlap. The third option, work organization form C, resembles organization form B with the exception that both field operators are replaced by one.

As an assumption these changes in work organization only affect the detection, the movement and the inactive stages. The correction times remain constant. Because we are only interested in the performance and flexibility of the different work organization forms of the machine operators, another simplification could be introduced. In this case the machine operator can correct all the breakdowns himself, thereby, eliminating the interpretation stage. The resulting Petri Nets are given in Fig. 5 and 6.

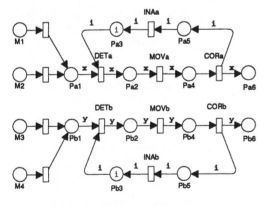

Fig. 5. Petri Net of organization A

Since the two work organization forms B and C, in terms of Petri Nets, differ only in the number of tokens initially present in place Pa3, respectively 2i and i, only the former is shown.

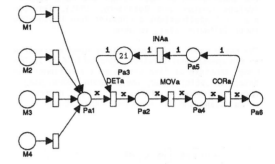

Fig. 6. Petri Net of organization B

The time intervals for the various transitions in Fig. 5 and 6 are shown in Table 1. Transitions with no label have a transition time of zero. Three types of breakdowns are defined: U, V, and W, requiring respectively 30, 60, and 120 seconds to correct, see also Table 1.

Table 1 Transition Time Intervals for Organization Forms A,B and C

	Tdet	Tmov	Tcor	Tcor	Tcor	Tina
			Type of breakdown			
			U	V	W	
	[s]	[s]	[s]	[s]	[s]	[s]
A	10	10	30	60	120	10
B	15	20	30	60	120	20
C	15	20	30	60	120	20

As an assumption each machine has an equal chance to breakdown. Furthermore, the buffer capacity is such that the throughput of the bottling line is not affected if a stop shorter than 180 seconds occurs. Normally for this part of the bottling line only one breakdown occurs at a time. To compare the organization structures, their performance and flexibility are evaluated. As a measure of performance the execution time interval, Texe, is taken. This interval is between the moment a breakdown occurs, i.e., a token appears in Pa1 or Pb1, and the moment the breakdown is corrected and a token appears in Pa6 or Pb6.

Flexibility is evaluated in three ways. Firstly, by determining the ability of an organization to maintain an acceptable level of performance while the frequency of occurring tasks is increased. For each type of breakdown two different scenarios are evaluated, namely the appearance of respectively two and three breakdowns of the same type at the same time. Secondly, the recovery interval, Trec, is measured. This is the time interval between the moment a breakdown occurs and the moment a machine operator is again stand-by for all the input places; meaning that places Pa3 and Pb3 must contain at least one token. Thirdly, a qualitative assessment to evaluate flexibility is made by omitting one machine operator. In the cases described above the results of measuring the machine operator's time load become meaningless, so they are omitted. The results of the simulations, where N is the number of permutations, are presented in Tables 2 up to and including 6.

Table 2 Execution and Recovery Times in case of One Breakdown (N=4)

	Texe			Trec		
	U	V	W	U	V	W
	[s]	[s]	[s]	[s]	[s]	[s]
A	50	80	140	60	90	150
B	65	95	155	0	0	0
C	65	95	155	85	115	175

The values of Texe, for the different work organization forms, show that for low rates of occurring breakdowns organization A produces the best performance. However, organization B has the lowest recovery time, so for higher rates of occurring breakdowns the performance of organization B should be better. All three work organization forms perform the corrections within the allotted time.

Table 3 Execution Times in case of Two Simultaneous Breakdowns (N=6)

	Texe					
	2*U		2*V		2*W	
	mean	max	mean	max	mean	max
	[s]	[s]	[s]	[s]	[s]	[s]
A	70	110	110	170	190	290
B	65	65	95	95	155	155
C	108	150	153	210	223	330

Table 4 Recovery Times in case of Two Simultaneous Breakdowns (N=6)

	Trec					
	2*U		2*V		2*W	
	mean	max	mean	max	mean	max
	[s]	[s]	[s]	[s]	[s]	[s]
A	80	120	120	180	200	300
B	85	85	115	115	175	175
C	170	170	230	230	350	350

As expected, in the case of two simultaneously occurring breakdowns (Table 3), organization B achieves the best performance, and is the only one to perform the tasks within the allotted time. The performance of organization A depends on the distribution of the breakdowns, in some cases large delays occur as can be seen from the maximum values of Texe. The reason is, that in some cases, one operator has to perform both corrections, while the other one remains inactive. Organization C achieves the lowest performance although the maximum values are comparable with the ones of organization A. The recovery times of organization B are slightly better than for organization A (Table 4).

Table 5 Execution Times in case of Three Simultaneous Breakdowns (N=4)

	Texe					
	3*U		3*V		3*W	
	mean	max	mean	max	mean	max
	[s]	[s]	[s]	[s]	[s]	[s]
A	70	110	110	170	190	290
B	93	150	133	210	213	330
C	150	235	210	325	310	505

Table 6 Recovery Times in case of Three Simultaneous Breakdowns (N=4)

	Trec		
	3*U	3*V	3*W
	[s]	[s]	[s]
A	120	180	300
B	85	115	175
C	255	345	525

For three simultaneously occurring breakdowns organization A produces the best performance (Table 5), contrary to the expectations based on the recovery times. In all these cases, one operator has to perform two corrections and the other operator only one. However, the recovery times of organization B remain the lowest so for higher rates of occurring breakdowns organization B should perform better (Table 6). The reason is that once the operator, who has to perform only one correction, is again stand-by, for organization B all the input places are covered. The performance of organization C is dramatically decreased.

The qualitative assessment of the consequences of omitting one machine operator in these cases are rather trivial. In case of organization A this leads to a situation where two machines remain unmanned. Consequently, should a breakdown occur, it will not be noticed by a machine operator. Organization B changes to work organization form C, with all four machines manned. In case of organization C no machines remain manned.

Comparing the timeliness and flexibility of the different work organization forms in this relatively simple case is not straightforward. Organization B is the most flexible organization, but achieves not always the best performance. In case of a low occurring rate of breakdowns the performance of organization A is slightly better. Regarding extreme high occurring rates organization A performs in some cases better, depending on the distribution of breakdowns for the different machines. Usually, the performance of one machine deteriorates, for instance, due to a maladjustment. In case of organization B, the increase of workload will be evenly divided. In this case organization B has also the most favourable working conditions.

In general the structure of the "optimal" organization will depend on the distribution and the rate of occurring breakdowns and on subjective factors such as the social acceptability of a particular work organization forms and the costs of maintaining a human operator.

FUTURE RESEARCH

Now that a basic model for evaluating organization forms has been defined, models on the supervisory level will be developed and a model of the bottling process is needed. To represent a bottling line as a Petri Net a new variant is used: Timed Continuous Petri Nets (Brinkman and Blaauboer, 1990) based on ideas introduced in (Alla and David, 1988). Both kinds of models will be combined into one hybrid model and the combined model will produce results about the organization's performance and the effects on the bottling line's efficiency. The next step will be to validate the modelling approach and to perform several sensitivity analyses.

Coloured Petri Nets are a promising modelling tool. Different organization structures are easy to model by just rearranging the different interactions. Coloured Petri Nets can also be used to verify system properties, for instance with invariant analyses or coverability graphs. Also the use of new theories such as (max,+)-algebra (Cohen and others, 1989; Baccelli and others, 1992) is considered. As an analysis tool Coloured Petri Nets look promising, however, at this moment not all the possibilities for the analytical analysis of systems are available; software is lacking. Therefore Coloured Petri Nets are now mostly used as a simulation tool, but this aspect can hardly compete with other available discrete simulation software. So at this moment, to compare this new modelling approach to more traditional ones is not realistic.

CONCLUSIONS

The aim of the paper was to discuss the methodology used for modelling the bottling hall as a Discrete Event Dynamic System. A quantitative model for evaluating work organization forms at the operational level is presented; Petri Nets are used to model the different processes and interactions. The model makes it possible to evaluate different organization forms with respect to their performance and flexibility.

The method is applied to three different work organization forms. On the operational level, a team structure is the most flexible and also has the most favourable working conditions. However, in case of low occurring rates of tasks and in some cases of extreme high occurring rates, a parallel organization gives the best performance. An

analysis of the results shows that it is not possible to determine an optimal form. Each form has advantages and disadvantages, the choice of a particular one depends on the task arrival rates and on other, more subjective, factors.

ACKNOWLEDGEMENT

The authors are grateful to Dr. A.H. Levis, currently with: Department of ECE, George Mason University, Fairfax, VA, USA, for copies of papers and reports on modelling human operators with Petri Nets.

REFERENCES

Alla, H. and R. David (1988). Modelling of production systems by Continuous Petri Nets. In Proceedings of the 3rd. Int. Conf. on CAD/CAM Robotics and Factories of the Future. Southfield, USA. pp. 344-348.

Andreadakis, S.K. and A.H. Levis (1987). Accuracy and Timeliness in Decision-Making Organizations. In Preprints of the 10th IFAC World Congress. Munich, Germany. Vol. 7. pp. 330-334.

Baccelli, F., G. Cohen, G.J. Olsder and J.P. Quadrat (1992) Synchronization an linearity. Wiley. (To be published).

Blaauboer, W.A. and P.L. Brinkman (1990). Analysis and evaluation of task allocation strategies in a bottling line. In Proceedings Ninth European Annual Conference on Human Decision Making and Manual Control. Varese, Italy. pp. 205-213.

Brinkman, P.L. and W.A. Blaauboer (1990). Timed continuous Petri Nets, a tool for analysis and simulation of discrete event systems. In Proceedings European Simulation Symposium 1990. Ghent, Belgium. pp. 164-168.

Cohen, G.; P. Moller; J.-P. Quadrat and M. Viot (1989). Algebraic Tools for the Performance Evaluation of Discrete Event Systems. In Proceedings of the IEEE. Vol. 77, No. 1. pp. 39-58.

Jensen, K. and G. Rozenberg (Eds.) (1991). High-level Petri Nets. Springer-Verlag, Berlin.

Levis, A.H. and K.L. Boettcher (1983). Information structures in decisionmaking organizations. In Proceedings Mediterranean Electrotechnical Conference (MELECON). Athens, Greece.

Meta Software (1991). DESIGN/CPN: A tool package supporting Colored Petri Nets. User's Manual, vol 1-2. Meta Software Corporation, 150 Cambridge Park Drive, Cambridge MA 02140, USA.

Peterson, J.L. (1981). Petri Net Theory and the Modelling of Systems. Prentice-Hall, Englewood Cliffs, N.J.

Reisig, W. (1986). Petrinetze. Eine Einfuhrung. Zweite, uberarbeitete und erweiterte Auflage. Springer, Berlin.

Rijnsdorp, J.E.; R.M. Pikaar and T.M.J. Lenior (1984). Process Control and Organisation. In Proceedings 9th IFAC Triennial World Congress. Budapest, Hungary. pp. 2627-2633.

Tabak, D. and A.H. Levis (1985). Petri Net representation of decision models. IEEE Transactions on Systems, Man and Cybernetics. Vol. SMC-15, No.6. pp. 812-818.

AN EXPERIMENTAL INVESTIGATION OF DYNAMIC ALLOCATION OF TASKS BETWEEN AIR TRAFFIC CONTROLLER AND A.I. SYSTEMS

S. Debernard, F. Vanderhaegen and P. Millot

Laboratoire d'Automatique Industrielle et Humaine, URA CNRS 1118,
Université de Valenciennes et du Hainaut Cambrésis, Le Mont Houy - BP 311,
59304 Valenciennes Cedex, France

ABSTRACT

The air traffic increase and the air traffic controller workload heaviness lead to provide an assistance to the air traffic controller. The purpose of our research is to propose and validate a new organization of the air traffic control, which allows air traffic controllers to stay active in the control and supervisory loop of the process, in order to maintain the present traffic safety level and to improve the global system performances.

Our research consists in decomposing the problem according to the two levels of the air traffic control organization : a tactical level which aims at regulating the traffic and a strategic level which aims at filtering this traffic. The first step of our research is directed towards an horizontal cooperation that consists in a dynamic allocation of the tactical level control tasks between human air traffic controllers and an assistance tool.

This paper presents the dynamic tasks allocation principles, and then describes the experimental platform for tasks allocation in air traffic control. The experimental protocol used for the experiments with qualified air traffic controllers is described and our first results are presented.

Finally, a global organization involving a multi-level cooperative decision-making is proposed and discussed.

KEYWORDS : Air-traffic Control, Man-Machine Cooperation, Dynamic Tasks Allocation

I. INTRODUCTION

The constant air traffic increase influences the traffic safety and constraints the air traffic controllers to have an irregular and high workload. A solution is to give an active aid to controllers by means of computer tools that allow an optimal control in order to keep the same safety level and to regulate the air traffic controllers workload (Debbache and colleagues, 1990). On the one hand, this study needs the realization of new decision support tools in order to help controllers. On the other hand, a new organization, based on a Man-Machine cooperation, must be defined.

The air space is divided into geographical sectors that are managed by two controllers. The first one is a tactical controller, called "radar controller". Through a radar screen and with paper strips which contain the main parameters of each plane, he/she supervises the traffic and dialogues with aircraft pilots. The supervision consists in detecting possible conflicts between planes that may transgress separation norms, and then in solving them. The dialogue with pilots consists in informing them about traffic and asking them to modify their flight level, cape or speed, in order to avoid a conflict. The second one is a strategic controller, called "organic controller". He/she makes coordination between sectors which consist in determining, and modifying if necessary, the flight parameters of each plane entering or leaving the controller's sector. These coordinations allow controller to filter the traffic for avoiding, firstly unsolvable conflict, and secondly an overload of radar controller. So, the organic controller activities are the input coordinations when planes enter the sector, the output coordinations when planes get out of the sector, the phone calls with the controllers of the other sectors in order to insure these coordinations, and the prediction of the traffic density in order to regulate the radar controller's workload.

Therefore, the organization of the air traffic control tasks shows two levels in which human controllers already cooperate. Then, the integration of support systems must take into account this existing organization, i.e. : these systems must be adapted to the respective objectives of each level.

Because of the complexity of these problems, our approach focused first on the lower level (the radar controller level) with a view to integrate an expert system for air traffic regulation called SAINTEX (Morchelles and colleagues, 1989). Having to regard to the high variations of controller's workload along the day, a dynamic task allocation between the human radar controller and SAINTEX seemed to be a fruitfull way of solution. Indeed, this man-machine cooperation mode should allow the air traffic controller to stay active in the system and to be actively assisted when he/she is overloaded.

Moreover, both the implementation and tests of such a new flexible task allocation cannot be made directly in a real control room, without a sufficient preliminary validation in a simulated context. For these reasons, we have built a simplified but realistic experimental platform for air traffic control, called SPECTRA (french acronym for : Experimental System to Allocate the Air Traffic Control Tasks).

This paper presents first two principles of Man-Machine cooperation, a vertical cooperation and an horizontal one which involves a dynamic task allocation. Then, we focus on the dynamic allocation, and the experimental study for its implementation and evaluation at the tactical level of air traffic control is described. Finally, the results are discussed toward a future global organization integrating both the human controllers according to their respective levels of intervention.

These studies are realized in cooperation with the C.E.N.A. (French Research Center of Air Traffic Control).

II. MAN-MACHINE COOPERATION PRINCIPLES

The concept of Man-Machine cooperation was born as a result of the appearance of Artificial Intelligence tools as assistants to a human decision-maker, and from the necessity to predict, and moreover to prevent the decisional conflicts liable to arise between these two types of decision-makers (human and artificial), (Millot, 1988). The research concerned, in its first stage, single-operator systems, and defined two cooperation modes (Millot, 1990) called respectively "vertical" and "horizontal".

II.1. Vertical cooperation

In a vertical cooperation, the operator is responsible for all the process' variables and if necessary he/she can call upon the decision support tool, which will supply him/her with advice, figure 1. Within this framework, two principles can be distinguished :

- One aims at guiding the operator in his/her problems solving so as to lead him/her to find a solution himself/herself. This principle is very attractive for it lets the operators maintain and enrich his/her operative knowledge of the process. However, it requires a great adaptivity on the part of artificial intelligence system, so as to insure consistance between its own reasoning and the operator's while reckoning with the latter's cognitive level. Moreover, decision time can be fairly long, and this excludes emergency situations !

- The second principle is aimed at emergency situations, wherein the operator can "lose" his capabilities for objective reasoning, through stress, for instance. For this reason, it is advisable to propose him/her solutions, that time pressure could prevent him/her from finding them. In this case, conflicts can arise when the human operator refuses the solutions proposed by the support system; then the designer must provide the support system with an interface for dialogue and justification of its reasoning. Then the human operator will be able to check rapidly who (himself or the support system) is wrong and why. Such an explanation and justification graphical interface has been implemented for a real time expert system integrated in the supervision system of a simulated power plant (Taborin, 1989 ; Taborin, Millot, 1989).

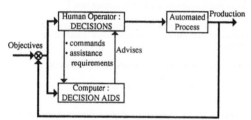

Figure 1 :   Vertical cooperation principles

II.2. Horizontal cooperation principles

In horizontal cooperation, the A.I. decision tool's output is directly connected to the process' control system. This suppose that the A.I. tool reasons in real time, figure 2. From now on, such a decision tool will be called an agent. The two decision-makers, the operator and the agent, are then on the same hierarchical level, and the supervision tasks, as well as the resulting actions, can be distributed dynamically between them, in order to relieve the operator in situation of overload. This cooperation can be implemented according to two principles :

- The first principle is an "explicit" dynamic allocation, controlled by the operator through a dialogue interface. He evaluates his own workload and the process' performances and then allocates some tasks to the agent if he is overloaded. The implementation is fairly easy, but the main drawback lies in the supplementary human workload induced by the task allocation management and the sending of orders to the agent. To that effect, Greenstein and Lam (1985) propose to optimize the dialogue interface's ergonomics in order to minimize the extra load; and in an experimental study on an air control simulator they define and evaluate the criteria to take into account for characterizing this interface.

- The second principle is an "implicit" allocation managed by the computer. The main difficulty then lies in the necessity to define an allocation criterion and in its implementation. To that effect, Greenstein and Revesman (1986) propose to integrate in the allocation control system, a predictive model of human actions and to allocate the tasks that the operator is not supposed to do to the agent, according to this model's predictions. Although the idea is enticing, it is nevertheless hardly feasible at present, given the difficulties encountered in the attempts to model human decision processes. Another method consists in seeking an optimal task allocation between the operator and the agent (Millot, Kamoun, 1988). The principle consists in searching for the optimal performance of

the process controlled by both decision makers, by modifying iteratively the number of variables allocated to each of them. Furthermore, two constraints of respectively maximum and minimum admissible workload, have been introduced in the task allocator controller, in order to avoid either a human overload or an underload. This principle has been validated on a laboratory experimental platform including a simulated continuous process. Additional disturbances were introduced in the process and the decision team (operator and agent) had to compensate the corresponding process errors. The results has shown a near-optimal performance of the process and a well regulated workload for the man, due to the implicit allocation (Kamoun, 1989 ; Kamoun, Debernard, Millot, 1989).

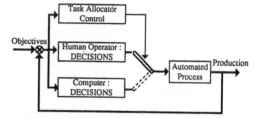

Figure 2 :   Horizontal cooperation principles

These horizontal cooperation principles have been transposed to the air traffic control domain especially for the tactical level involving the radar controller. The experimental platform SPECTRA built for this purpose is now described (Debernard and colleagues, 90).

III. THE EXPERIMENTAL PLATFORM : SPECTRA

III.1. SPECTRA description

The experimental platform SPECTRA, figure 3, simulates a control unit only for radar controller's tasks.

SPECTRA is built around an air traffic simulator, called OLIVE which simulates, in a realistic way, the evolutions of the aircrafts according both to their flight plan and to the instructions given by one of the two decision makers, the radar controller or SAINTEX.

The controller interface is composed of two color graphic screens. The former is a standard radar view which displays on line, the aircraft evolution and enables the human air traffic controller both to supervise the traffic and to detect possible conflicts between aircrafts. Moreover, this module manages alarms when planes taken in charge by the controller are transgressing separation norms. The latter is the operator dialogue screen. It contains an "electronic stripping" and a piloting module. By means of a mouse, the radar controller places the strips on screen as he/she want. Sensitive areas allow controller :

- to send orders (only cape or level changes) for preventing conflicts,
- to modify the radar view (zooms, speed vectors...),
- to allocate tasks to SAINTEX in explicit mode.

The expert system "SAINTEX" (french acronym for : Experimental Automatic System for Night Air Traffic Control) can detect conflicts between two or more planes, but it only can solve those between two planes in rather simple conditions, figure 4. The control context concerns night control which is easier than the daily one because each flight crosses the sector in straight line between two beacons : a sector input point and a sector output one.

The conflict detection is realized by extrapolating planes trajectory. Then, SAINTEX classifies the detected conflicts and determines whether it can solve them or not. The solvable conflicts are then the shareable tasks and SAINTEX calculates the trajectory deviation order to send to the pilot (simulated in OLIVE) so as to solve the conflict. When a conflict is given to SAINTEX, the order is sent and when the conflict is over, SAINTEX replaces the deviated plane on its initial way.

The task allocator informs each decision-maker, radar controller and SAINTEX, about the tasks allocation for preventing decisionnal/command conflicts between them. The functionning of the task allocator control depends on the mode of the dynamic

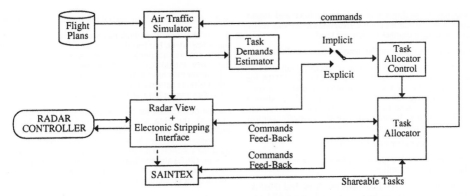

Figure 3 : SPECTRA structure

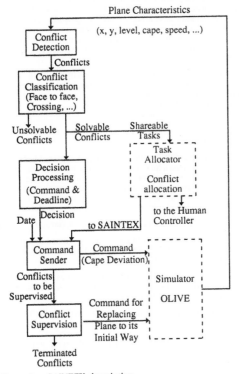

Figure 4 : SAINTEX description

task allocation. In explicit mode, the task allocator is managed by the human air traffic controller. In implicit mode, the task allocator is managed by SPECTRA.

III.2. Dynamic tasks allocation on SPECTRA

III.2.1. Explicit mode

For the explicit mode, the conflict SAINTEX can solve are indicated with a specific color on the operator dialogue screen and on the radar view. The deadline before which the conflict becomes unsolvable by SAINTEX and the deviation order calculated are displayed on the strips of the concerned planes.

Initially, all planes are given to the controller. If he/she feels himself/herself overloaded, he/she can select a conflict and transfer it to SAINTEX. As he/she is the system manager, he/she can take a conflict back. But at that time, if the order deadline given by SAINTEX is over, this conflict can not be solved by SAINTEX anymore .

III.2.2. Implicit mode

For the moment, the control policy of the task allocator is quite simple and is based on the intrinsic abilities of each decision makers and on the human decision maker workload. In such a way, a conflict is allocated to SAINTEX when controller tasks demands are too high and exceed a maximum level. In that case, the conflict with the nearest order deadline is chosen and transfered to SAINTEX. In the contrary case, the conflict is affected to the human controller.

The task demand estimator adds all the tasks demands to which the air traffic controllers are submitted. These demands are only functional and take into account the difficulties encountered when assuming a plane, detecting a conflict, correcting and supervising a conflict and replacing a plane on its initial trajectory. So, it consists in affecting a weight to the demand for each event appearing during experiments.

SPECTRA has been tested and validated by air traffic controllers before the experiments. Then, an experimental protocol has been defined.

IV. THE EXPERIMENTAL PROTOCOL

IV.1. The experimental context

The scenarios used for these experiments have been created in order to overload the air traffic controller, in such a way that a dynamic allocation was useful. Therefore, they involve a great number of planes (between 40 and 50 per hour) that are generating a lot of conflicts (≈ 20) of different natures, into a large geographical sector, (Vanderhaegen, Debernard, Millot, 1991).

One training scenario permits to controllers to get used to SPECTRA interface in explicit mode, during two hours. Other scenarios have been created for three kinds of experiments. In the first type, the controller is not aided. In the second kind, he/she is supported by SAINTEX in an explicit dynamic allocation mode. In the last one, he/she is assisted in an implicit allocation mode.

These experiments have been performed with nine male qualified air traffic controllers. Each controller began with the training scenario. Then three experiments, with three different scenarios prevented him from learning effects, have been realized without assistance, in explicit mode and in implicit mode.

IV.2. The experimental data

IV.2.1. The data recorded on line

Three kinds of parameters have been measured and recorded during each experiment :

- The first class concerns all the data needed to calculate a global performance criterion, i.e. the performance obtained by the decision-makers team. However, this performance is composed of two factors that cannot be aggregated. The former is the traffic security, and the latter is the traffic optimization. Therefore, two criterions have been defined. The security criterion is the ratio between the number of alarms and the number of conflicts appeared during the scenario. The traffic optimization criterion for the controller-SAINTEX team is based on an economic criterion which is the ratio between the real consumption of kerosene of each aircraft and the theoretical one. The theoretical consumption corresponds to the direct way between the input and output beacons. The real consumption corresponds to the way really taken according to changes in cape and level.

- The second class is the task demands measured on line.

- The last one concerns the on line subjective evaluation of controller's workload. A module was added to SPECTRA for that purpose. The air traffic controller estimates his own workload, clicking with a mouse on a small scale, graduated from 1 (lower load) to 10 (higher load). This board appears every five minutes. It cannot disturb the air traffic control because the operator answers when he wants.

IV.2.2. Questionnaires after each experiment

In the final step of each experiment, two questionnaires have been submitted to the controller.

The first questionnaire is the Task Load indeX method (NASA-TLX). TLX is a subjective method developed by the NASA that permits to calculate the global workload. This global workload estimator is based on 6 semantic descriptors : mental demand, physical demand, temporal demand, performance, effort and frustration.

The second questionnaire is a series of questions that allowed us an oriented discussion with the controller after the experiment. These questions concern :

- general information about scenario, radar interface and strip interface, strategies used, etc;
- specific information about workload, estimated performance and alarms;
- specific information about explicit mode, implicit mode and their comparison.

V. THE PRELIMINARY RESULTS

The preliminary results have been analysed for the nine qualified controllers who performed each simulated scenarios. They concern the TLX results, the questions synthesis and the results analysed from recorded data.

V.1. Using the TLX method

From the TLX method, the global workload was calculated for each controller and for each experiment. The figure 5 presents the global workload for each experiment..

All the controllers, except controller number 1, have estimated that the experiment without assistance was more difficult and more overloaded than the others. For four controllers (2, 3, 4 and 5), their workload with the implicit mode was less important than with the explicit mode. For the others (6,7,8 and 9), the explicit mode generates the less important workload.

Therefore, a dynamic task allocation mode (either implicit or explicit) seems to improve the air traffic control task and to reduce the global workload.

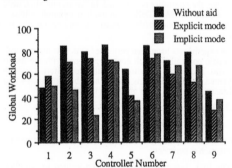

Figure 5 : TLX results for each experiment

V.2. Using the questionnaires

All the answers of the second questionnaire were analysed. They concern general information, workload, explicit and implicit modes, and the comparison between those allocation modes.

V.2.1. Conclusions about general information questionnaire

63% of the controllers think that the number of planes is not too much important, but less than 74% of them consider that the air traffic was not structured. 66% of the controllers estimate that the radar view should be enhanced because :
- the sector they have to supervise is too large,
- the traffic is not structured enough,
- and they cannot have sufficient marks.

On the other hand, for 60% of them, the strip view is correct although the high number of planes involves heavy strips management. Moreover, SPECTRA does not change usual controllers strategies because, as in real control, strip view is used for conflicts detection (74%) and radar view for conflicts solving (96%) and conflicts supervision (100%). As in a loaded daily control, 85% of the controllers solve immediately the conflicts for preventing an overload.

V.2.2. Conclusions about workload questionnaire

When the graduated scale appears on the screen for asking the controllers to evaluate their own workload, 78% of them give higher priority to the traffic management tasks than to the assessment scale answering. Controllers generally carry out their current task and then evaluate their workload before begining the next task treatment. Nevertheless, they are satisfied with their own evaluation (78%). So, we can use these data as a workload index.

Moreover, 70% of the controllers are satisfied with their global performance and the main factors that increase the workload are :
- number of planes (85%),
- number of conflicts (70%),
- strips management (48%).

The results about the other factors have not lead to any conclusion yet.

A lot of alarms (78% of the experiments) appeared because the scenarios involved a lot of planes and the controllers managed the traffic alone. Half of the alarms were due to a bad solving because controllers had not sufficient marks on the radar view. We can notice that the six experiments which have provoked no alarm at all, were all performed with the implicit mode.

V.2.3. Conclusions about explicit mode questionnaire

The explicit mode reduces controllers workload and is a real assistance tool (100%). For 89% of the controllers, it didn't provoke any problem for the air traffic control.

Nevertheless, if the explicit mode is a good assistance tool, an homogeneous policy about the use of SAINTEX by the controllers can hardly be found :

- Two controllers (3 and 6) allocate conflicts to SAINTEX when they are overloaded and when they have too much conflicts.
- Two controllers (1 and 7) use only their workload as allocation criterion.
- One controller (2) uses the number of conflicts as allocation criterion.
- One controller (5) allocates automatically one conflict upon two.
- One controller (4) allocates conflicts to SAINTEX if the delay to send orders to planes is low. In the opposite case, he improves the SAINTEX's solution.
- One controller (8) uses the system as an horizontal cooperation tool : he uses SAINTEX as a conflict detector, and then refines its solutions.
- The last one (9) gives conflicts to SAINTEX if its solution is satisfactory.

V.2.4. Conclusions about implicit mode questionnaire

For 78% of the controllers, the implicit mode reduces controllers workload and provides a real assistance. For 89% of them, like in explicit mode, it didn't provoke any problem for the air traffic control.

The controllers supervise SAINTEX's conflicts as much in implicit mode as in explicit mode although they can't take them back. With this kind of experiment, they have more time to supervise the air traffic control. Moreover, they are less frustrated because they didn't generate a lot of alarms.

V.2.5. Conclusions about the comparison of questionnaires

First, there is no preference between the allocation modes. Nevertheless, the implicit mode seems to be the more efficient (78%) and reduces workload (78%).

Secondly, it is difficult to class the answers about the comparison between the two modes : some controllers prefer the implicit mode but they don't trust SAINTEX or any other future decisional stage. They ask for the possibility to take conflicts back, but they admit they have more time to solve their own conflicts. Other controllers prefer explicit mode because they actually control the air traffic. Nevertheless, they think that this mode is more dangerous because the human controller is the only responsible for the traffic, and the SPECTRA organization is not tolerant to human errors. For instance, when a controller forgets to allocate shareable tasks, a conflict may be not solved. But, we can also notice that this problem already exits in the current real air traffic control context.

V.3. Using recorded data

Figure 6 shows only one example of the tasks demands and subjective workload evolution. Despite the basic difference between the measures, the correlation coefficient between the two variables have been calculated for each experiment. For 63% of the experiments, the correlation coefficient is greater than 0.8, and for 26% it is between 0.6 and 0.8. Therefore, our tasks demands measure reflects fairly well the controller's workload.

The average of the tasks demands corresponding to each of the experiments is presented on the X-axis of the figure 7. We can notice that the values obtained in the contexts without aid are generally greater than those obtained with aid. A dynamic tasks allocation thus actually relieve the work of the controllers. Furthermore, on the same figure, we can see that the more tasks demands increase, the worst the security criterion.

Figure 8 shows the two criteria related to the performance. Here again, we can see that in a context without aid, the performance are not as good as those in a dynamic task allocation. Furthermore, the implicit mode seems to be the most efficient one, with regard to the security criterion.

On the X-axis of the figure 9, is represented the cooperation level defined as the ratio between the number of conflicts allocated to SAINTEX and the total number of conflicts to treat. We can notice that the more SAINTEX is used, the best the security criterion.

Therefore, the dynamic tasks allocation contributes to the increase of the efficiency for the traffic control managment in a high traffic context.

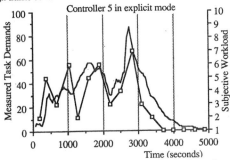

Figure 6 : *Comparison between measured tasks demands and subjective workload*

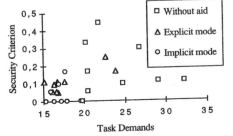

Figure 7 : *Comparison between tasks demands and security criterion*

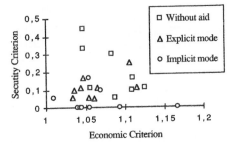

Figure 8 : *Global performances : economic and security criterion*

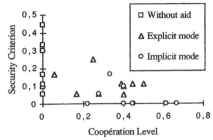

Figure 9 : *Comparison between cooperation level and security criterion*

VI. SYNTHESIS OF THE PRELIMINARY RESULTS AND FUTURE COOPERATIVE ORGANIZATION

VI.I. Synthesis of the preliminary results

There is no doubt on the decrease of controller's workload when they are assisted by SAINTEX either in implicit or in explicit mode. Moreover, it seems that the implicit mode is globally more efficient. Nevertheless, a lot of complementary data, like details on the controller activities during the experiments, have not completely traited.

From a qualitative point of view, a preliminary analysis of the controler activities shows that in explicit mode the controller has to play a double role : 1) a tactical role when solving the conflicts allocated to himself and 2) a strategic role when predicting the possible conflicts and planning the task sharing between SAINTEX and himself for the conflicts solving. Therefore, in this mode, he plays the role of the organic controller as well as the role of the radar controller.

This observation can result in a proposal for the future global organization of the ATC post, including both controllers and support systems.

VI.2. Toward a multi-level cooperative organization in Air Traffic Control

Our future researchs in this way concern, firstly the design of a decision support system at the strategic level, which role consists in predicting future conflicts and planning the task allocation of the tactical level, figure 9. This decision support system is foreseen with 3 modules :

- A module for traffic anticipation. By extrapolating the plane trajectories in a long run, this module permits to know the future evolution of the traffic and to predict more precisely the future problems like conflicts between planes. With this cooperation, the organic controller may regulate his own workload and the radar controller's, because he can make finest planning of all tasks.

- This first assistance can be completed by a task's planning module as the ERATO System (Figarol, 1990), which predicts the tasks and provides their scheduling for avoiding controller's overload.

- The third module can be based on an expert system called PLATONS presently developed by the C.E.N.A. (Ly, 1991). The main goal of this system is to provide the entrance, the cruising and the exit flight levels of the planes.

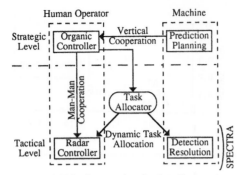

Figure 10 : Multi-level cooperations for air traffic control

At present, these modules are being developped. Therefore, a "combined" cooperation can be defined in the overall organization, figure 10. At the strategic level, a vertical cooperation can be defined between the organic controller and the prediction and planning support system. This cooperation is called vertical cooperation because the human organic controller is the final decision-maker and he can be provided with pieces of relevant information or advices from the support system. Therefore, the interface must cope with possible decisional conflict between these both decision-makers using, for instance the explanation capabilities of the support system. The aim of this strategic level is the management of the dynamic task allocation at the tactical level which already exists.

VII. CONCLUSION

This paper has presented the first results we have obtained with experiments on SPECTRA, the experimental platform for the dynamic tasks allocation study in air traffic control.

After a brief recall of man-machine cooperation modes, the transposition of the dynamic tasks allocation principles to the air traffic control has been detailed. The platform structure and the experimental protocol allowed us to realize experiments with qualified controllers.

Those experiments validated SPECTRA and our dynamic tasks allocation method in the tactical level of air traffic control. The preliminary results are convincing. They show the real help a dynamic task allocation provides to the air traffic controllers. The implicit mode seems to be the most efficient. In the short run, the perspective of this research concerns further analyses in order to generalize the global results and, to find the correspondances between our tasks demands calculation and a workload assessment method. From an organizational point of view, a global structure involving both tactical and strategic levels, has been proposed toward a multi-level cooperative organization in air traffic control. This research will be pursued in this way, with a technical objective concerning the implementation of a decision support system at the strategic level and with an ergonomic objective aiming at implementing a vertical cooperation between the organic controller and this support system.

REFERENCES

Debbache N., Debernard S., Millot P., Planchon P., Angerand L.,(1990). *Toward a New Architecture for ATC using Dynamic Task Allocation between Human and A.I. Systems.* Second International Conference on Human Aspects of Advanced Manufacturing and Hybrid Automation. Honolulu. Hawaii. USA.

Debernard S., Vanderhaegen F., Debbache N., Millot P., (1990). *Dynamic Task Allocation between Controller and A.I. Systems in Air Traffic Control.* 9th European Annual Conference on Human Decision Making and Manual Control. Ispra. Italy.

Figarol D., (1990). *Cahier des charges du banc de test ERATO.* Report of the C.E.N.A. - Ref : CENA/TDV/ERATO-CC

Greenstein J.S., Lam S.T., (1985). *An experimental study of dialogue-based communication for dynamic human-computer task allocation.* International Journal of Man-Machine Studies, n° 23.

Greenstein J.S., Revesman M.E., (1986). *Application of a Mathematical Model of Human Decision Making for a Human-Computer* Communication. IEEE SMC 16. Jan-Feb; N°1 pp 142-147.

Kamoun A., (1989). *Contribution à la Répartition Dynamique des tâches entre Opérateur et Calculateur pour la Supervision des Procédés Automatisés.* Thèse de Doctorat. Université de Valenciennes.

Kamoun A., Debernard S., Millot P., (1989). *Comparison between two dynamic task allocations.* 2nd European Meeting on Cognitive Science Approaches to Process Control. Siena, Italy.

Ly S., (1991). *State of development of PLATONS project.* Report of the C.E.N.A. - Ref : CENA/N91210/S.LY/91.

Millot P., (1988). *Supervision des procédés automatisés et ergonomie.* Edition HERMES. Paris.

Millot P., Kamoun A., (1988). *An Implicit Method for Dynamic Task Allocation between Man and Computer in Supervion Posts of Automated Processes.* 3rd IFAC Congress "Analysis design and Evaluation of Man-Machine Systems" OULU. Finland.

Millot P., (1990). *Coopération Homme-Machine dans les procédés automatisés.* In : J. Leplat et G. de Terssac éditeurs : les facteurs humains de la fiabilité dans les systèmes complexes. Edition Octarés. Marseille. France.

Mordchelles P., Angebault H., Angerand L., (1989). *Système expert pour le contrôle automatique de nuit.* Rapport de stage de fin d'études .

NASA-T.L.X., version 1.0. *Booklet containing the theoritical background of the Task Load indeX.* NASA Ames research center. T.L.X. Human Performance Group.

Taborin V., (1989). *Coopération entre opérateur et système d'aide à la décision pour la conduite des procédés continus : Application à l'interface opérateur-système expert du projet Alliance.* Thèse de Doctorat. Université de Valenciennes. France.

Taborin V., Millot P., (1989). *Cooperation between man and decision aid system in supervisory loop of continuous processes.* 8th European Annual Conference on "Decision Making and Manual Control". Copenhagen. Denmark.

Vanderhaegen V., Debernard S., Millot P., (1991). *Man-Machine Cooperation in Air-Traffic Control.* 10th European Annual Conference on "Human Decision Making and Manual Control". Liège. Belgium.

ROBOTS, WORKING CONDITIONS AND JOB CONTENT:
Opportunities and Hindrances for the Improvement of the Quality of Working Life

J.C.M. Mossink and M.H.H. Peeters

TNO Institute of Preventive Health Care, P.O. Box 124, NL 2300 AC Leiden, The Netherlands

Abstract   The introduction of robots into workplaces may offer opportunities to improve occupational safety, health and job content. In this study 16 case studies of robot work places in 4 companies and a literature survey were performed in order to prepare an instructional publication for the Dutch Labour Inspectorate. It appears that in many cases one fails to fully take advantage of the opportunities robots offer to improve working conditions. With respect to the amelioration of job content it can be concluded that robot related activities offer to little prospects. An approach in which not robot related work is also included offers a better perspective.

Keywords   Robots, Ergonomics, Safety, Health, Job Content

INTRODUCTION

Introduction of robots into workplaces can improve working conditions. Work, done under hazardous or strenuous conditions and monotonous activities can be performed by robots. Improvement of working conditions appears, besides technical and economical criteria, to be a factor in the decision to use robots. For example, it is felt that the use of robots could ease the recruitment of new personnel. The other way around, if robotization is planned, human factors should be taken into account simultaneously.

This study is performed by order of the Dutch Labour Inspectorate. The aim was to prepare an instructional publication on quality of working life in relation to robots. The goal of the publication is to inform participants in the introduction of robots about the influence of robots on occupational safety, health and well-being (in the Dutch Working Environment Act strongly related to job content) and how to deal with it. Attention is also paid to the introduction process of robotic workplaces. This publication fits into the governmental policy to stimulate companies themselves to improve working conditions and job content.

In this study an integral approach is adopted. This implies on one side that robotics, environmental and organizational factors and on the other side effects in terms of working conditions/job content and effectivity/efficiency of the robotical system are seen as highly interrelated, see fig. 1.

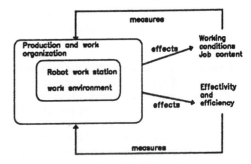

Fig. 1. An integral Approach

Robotization in production departments always gives options in technical, environmental and organizational design. This freedom of choice could well be used to improve working conditions and job content.

Effects of robotization on working conditions and job content are studied by means of a literature survey and by case studies.

In four firms in metal industry 16 robot workplaces are studied. Robots performed tasks like arc welding, spot welding and handling and cleaning of workpieces.

Table 1 gives an overview of the characteristics of the robot workstations.

TABLE 1   Overview of characteristics of robot workstations studied.

Company	case	Robot function	Characteristics
A	1	workpiece cleaning, air hose	stand-alone, fully enclosed; automated workpiece handling by conveyor system
	2	workpiece handling, (un)loading furnaces, quenching hot workpieces;	rail-mounted, workpiece supply by means of a large buffer
	3	workpiece handling, (un)loading furnaces	stand alone, workpiece supply by means of a large buffer
B	4a	spot welding metal plates	part of a 2-robot workstation; workpiece handling with turntable; manual (un)loading after each cycle
	4b	workpiece brushing	part of a 2-robot workstation; workpiece handling with turntable; manual (un)loading after each cycle
	5	spot welding metal plates	stand-alone;workpiece handling with turntable; manual (un)loading after each cycle
	6	arc welding, training	stand alone
	7	arc welding	rail mounted, part of flexible welding unit; manual workpiece handling
	8	laser cutting;	connected with CO_2-laser; fully enclosed; manual workpiece handling.
C	9	arc welding large workpieces	overhead rail mounted; manual workpiece handling
	10a	arc welding large workpieces,	overhead rail mounted; part of a 2-robot work station; manual workpiece handling
	10b	arc welding	overhead rail mounted; part of a 2-robot work station; manual workpiece handling
	11	arc welding	stand alone
D	12	arc welding, part of domestic appliances	stand alone, fully enclosed; use of turntable which is manual loading and unloaded.
	13	arc welding, part of domestic appliances	as case 12
	14	palletizing products	stand alone

In this article section 2 deals with the case studies. Descriptions follow of motives for introduction, safety measures, health hazards, job content and introduction strategies. Information was gathered by interviews (with robot operators, technical staff and management) and observation of the work process. Section 3 gives some general guidelines for the improvement of working conditions and job content, which are derived from literature and from case studies. In section 4 guidelines and results of case studies are discussed. Here a practical evaluation of the guidelines is made. Conclusions of the study are summarized.

CASE STUDIES

Motives for Introduction of Robots
The motives the companies to introduce robots are quite different. Company A gave as the main reason that working conditions were far below standards and it was hard to get personnel for some activities. It was emphasized that robots did not lead to quality and productivity improvement, though a side effect was that costly equipment is

more efficiently used with robot loading and unloading. The cost of robots were high in relation to their return. The pay-out time was about 7 to 8 years. Benefits from reduction in sick rates and turnover are included.

The main objective of Company B to robotize spot welding operations was to obtain a more constant and high product quality, which could not be obtained by manual spot welding.

In company C, introduction of arc welding robots offered better chances to recruit personnel. In this, working conditions were no argument. Welders are merely attracted by the professional opportunities robot welding offers. A positive side effect was a better product quality.

Company D introduced robots to eliminate heavy work (palletizing) and to increase quality and productivity (arc welding).

Safety Measures
The studied workplaces showed much difference in the extend required safety measures had been taken. As can be seen in table 2 many of the work stations suffered from shortcomings. In none of the companies accidents with robots occurred.

TABLE 2 Evaluation of Safety Measures.

Com-pany	case	Robot function	Evaluation of safety measures
A	1	workpiece cleaning, air hose	Completely enclosed. Interlocked entries; automated workpiece supply by conveyor. Enclosure to close to robot
	2	workpiece handling, (un)loading furnaces, quenching hot workpieces;	Safeguarding by fence, interlocked entries; all necessary precautions taken.
	3	workpiece handling, (un)loading furnaces	Safeguarding by fence, interlocked entries All necessary precautions taken.
B	4a	spot welding metal plates	Safeguarding by fence. Turntables with light-curtain. Robot control and welding control unit placed within fence; fence to close to the robot.
	4b	workpiece brushing	As in case 4b.
	5	spot welding metal plates	As in case 4a.
	6	arc welding, training	Safeguarding by fence, interlocked entries; all necessary precautions taken.
	7	arc welding	Safeguarding by fence, interlocked entries; all necessary precautions taken.
	8	laser cutting;	Robot completely enclosed to avoid laser beam to break out of work station.
C	9	arc welding large workpieces	No fence, no detection of presence of operators.
	10a	arc welding large workpieces,	No fence, no detection of presence of operators.
	10b	arc welding	Fence and curtains to absorb UV-radiation, use of turntable
	11	arc welding	Fence and curtains to absorb UV-radiation, use of turntable
D	12	arc welding, part of domestic appliances	Fence and curtains to absorb UV-radiation, use of turntable
	13	arc welding, part of domestic appliances	Installation not complete at time of observation
	14	palletizing products	Fence, automated workpiece supply by conveyor

From interviews it resulted that reasons for omitting safety devices are divers.

In company C, where large workpieces are welded it is very impractical to install a fence. The use of turntables or conveyors turned out to be impossible. Operators said they were very familiar with the movements of the robot, so they felt no danger of being hit by an unexpected movement of the robot. In company A and B the main problem was the lack of space at the work station, so the robot control unit was placed within the enclosure.

All companies provided for the safety devices themselves. Only speed reduction of the robot arm (for programming, optimization and maintenace puposes) is integrated in robot system software. Knowledge was obtained from the Labour Inspectorate or derived from standards.

Occupational Health Measures

In several cases occupational health problems were reduced. However as can be derived from table 3, new unfavourable situations arise. Only in company A, where improvement of working conditions was the main goal, a good situation with respect to health risks was obtained. Here robots performed the tasks autonomously, and only occasional supervision and chages of buffer stock were necessary.

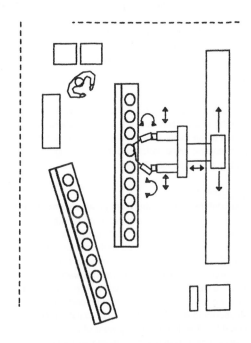

Fig. 2. Lay-out of Robot workstation (case 10a/b)

TABLE 3 Overview of working conditions

Com-pany	case	Robot function	Improvements	Deterioriations
A	1	workpiece cleaning, air hose	No exposure to high noise levels, no need to wear protective clothing, elimination of repetetive tasks	
	2	workpiece handling, (un)-loading furnaces, quenching hot workpieces;	No exposure to high temperatures, reduction in physical load.	
	3	workpiece handling, (un)-loading furnaces	No exposure to high temperatures, reduction in physical load.	
B	4a	spot welding metal plates	Reduction of physical load (handling of spot welding apparatus)	Physical load (mounting parts), repetetive tasks, lifting. Less variation in working postures
	4b	workpiece brushing	reduction of physical load, no exposure to vibrations, less need to wear protective clothing.	As in case 4a.
	5	spot welding metal plates	As in case 4.	As in case 4a.
	6	arc welding, training	Reduction of static muscular load. Reduced exposure to welding fumes and radiation. Less need to wear protective clothing	
	7	arc welding	As in case 6.	
	8	laser cutting;		
C	9	arc welding large workpieces	Reduction of static muscular load. Reduced exposure to welding fumes ar 1 radiation. Less need to waer pro$active clothing	Lifting of occasional heavy parts, repetetive tasks. Unfavourable working positions
	10a	arc welding large workpieces,	As in case 9.	As in case 9.
	10b	arc welding	As in case 9.	As in case 9.
	11	arc welding	As in case 9.	As in case 9.
D	12	arc welding, part of domestic appliances	Reduced exposure to welding fumes and radiation. Less need to wear protective clothing.	Frequent lifting of large workpieces. Less variation in working postures
	13	arc welding, part of domestic appliances	As in case 13.	As in case 13.
	14	palletizing products	Elimination of unfavourable lifting tasks.	

In many of the cases high physical load due to loading and unloading fixtures comes about. In all cases these tasks (sometimes repetitive with high frequencies) were performed in a standing position during a long period of time.

Job Content

Job content of robot operators is largely dependant on both the production and work organization. In table 4 the job content of robot operators is shown. From these results it can be derived that job content of robot operators leaves much to be desired. In company C it was stated that expierenced welders were "degraded" to just loading and unloading the robot after having programmed the robot. Measures to enrich the jobs (task roulation, occasional hand welding, development of new robot applications and addition of maintenance tasks) were taken.

Introduction Strategy, User Involvement

The none of the companies end-users were involved in the robot introduction process. Company A bought robots turn-key. Even initial programming was done by the vendor. In this case there were no problems. The robots were seen as pieces of machinery that should only occasionally be taken care of. Reprogramming and adjustment of programs rarely occurs.

In company B it is recognized that profound knowledge of the robot is a powerful method to avoid excessive idle time and to improve product quality. To become more experienced in robot technology the company started with a small and simple arc welding robot. Now complex robot work stations are partly bought and partly developed by themselves (peripheral equipment, safety devices).

TABLE 4   Overview of operator tasks and job content.

Compa-ny	case	robot function	Operator tasks	Effects of robots on job content
A	1	Workpiece cleaning, air hose	Supervision	Elimination of repetive and isolated work
	2	Workpiece handling, (un)-loading furnaces, quenching hot workpieces;	Supervision. Occasional Loading and unloading buffer stock (about once a day)	Elimination of repetetive work
	3	Workpiece handling, (un)-loading furnaces	Supervision. Occasional Loading and unloading buffer stock (about once a day)	Elimination of repetetive work
B	4a	Spot welding metal plates	Loading, unloading; adjustments and optimizations in programs; supervision	Monotonous tasks, isolated work place. Less opportunities for contacts. Little autonomy.
	4b	Workpiece brushing		
	5	Spot welding metal plates	Loading, unloading, adjustments and optimizations in programs; quality control; performing aftertreatments; supervisions.	Isolated work place. Less opportunities for contacts. Some variation in tasks. Less opportunities for contacts.
	6	Arc welding, training	Not relevant	Not relevant
	7	Arc welding	Loading and unloading four cells (2 operators); supervision	Little autonomy, some possible contacts.
	8	Laser cutting;	Loading, unloading; supervision	Monotonous tasks, isolated work place. Less opportunities for contacts. Little autonomy.
C	9	Arc welding large workpieces	Loading, unloading; manual welding; programming; development of robot applications; supervision; maintenance; supplying of parts.	Fairly complete job, though programming and maintenance take little time. Isolated work-place, due to long robot work cycle opportunity to leave work station for a short time.
	10a	Arc welding large workpieces,	As in case 9	As in case 9
	10b	Arc welding large workpieces		
	11	Arc welding	As in case 9	Fairly complete job, though programming and maintenance take little time. Isolated work-place.
D	12	Arc welding, part of domestic appliances	Loading, unloading	Monotonous tasks, isolated work place. No opportunities for contacts. Little autonomy.
	13	Arc welding, part of domestic appliances	As in case 12	As in case 12
	14	Palletizing products	Occasional removal of pallets; supervision.	Elimination of repetetive work

In company C, users had no influence in the introduction of the first robot. The project failed due to lack of knowledge of robot welding and acceptance of welders. Later welders were actively involved in the introduction. The welders themselves were asked to develop applications. In the beginning, this was a successful approach. Later however, the users felt they had degraded themselves from professional welders to machine loaders.

Company D bought a complete robot system, including additional devices (like turntables) turn-key.

REDUCING RISKS FOR SAFETY, HEALTH AND JOB CONTENT: GUIDELINES

Introduction

In the last few years a massive amount of literature concerning occupational safety and health in robotic workplaces is published . The main issue in these studies is robot-safety (Bonney, 1985; Hartmann, 1989; Helander, 1990; Jiang & Cheng, 1990; Nagamachi, 1986). Occupational health problems and job content are addressed to a much lower extend (Benders, 1991; Hilla & Tiller, 1990; Landau, 1988; Parsons, 1986).

Robots and Safety

Working with robots brings a number of safety hazards to the workplace. The main hazard is being hit by an unexpected moving robot. The strategy to avoid a collision in itself is simple: keep man and moving robot separated and when work has to be done close to a moving robot its speed should be low. Many techniques have been elaborated to protect operators from a moving robot.

Robots and Health

Because of their versatility robots can be used to take over several hazardous tasks. Well known applications are spray painting, arc welding, and handling hot workpieces. However it is important to notice that robots may not fully take away health hazards. In design this situations should be anticipated.

Exposure to unfavourable working conditions may for instance occure at robot down time, when man has to take over robot activities. Often conditions are even worse than at not robotized workplaces. Secondly the introduction of robots may lead to the use of more toxic substances because the argument of health hazards is disappeared. Thirdly it is possible that heath hazards are not adequately reduced. Though the distance between man and the work proces is enlarged, enclosure or ventilation may still be needed. Last, exposure to hazardous conditions may occur at programmnimg, optimizing, testing or repair. Often one has to be close to the robot with the work process in progress.

Robots may also introduce new health hazards. Most important is physical load. Especially robotizing welding may give rise to enlargement of physical load introduction of (most simple) repetitive streneous tasks like loading and unloading. Physical load can be reduced by technical measures like proper workplace design, use of buffers and a balanced cycle time in which rest moments are incorporated. A positive effect can also be obtained with a job design in which heavy and repetetive tasks are alternated by less strenuous tasks.

Robots and Job Content

Improvement of jobcontent in general focusses on a restructuring of the division of labour, especially the production organization and work organization (WEBA-Project Team, 1990). Only tasks like programming, optimizing planning, loading and unloading will be of restricted influence on the content of jobs at robot stations. Technical measures within the robot system itself can contribute to better job content. For example, use of buffers reduces the need of manual loading at the work pace of the robot, thus enlarging autonomy and opportunities to have contacts with colleagues. Some guidelines are summarized in table 5. A distinction is made between measures on three domains: the production organization (departmental structure), the work organization (division of tasks within a department) and the robot system itself. Measures taken on the field of the production organization can be seen as conditional for the improvement of the work organization.

DISCUSSION AND CONCLUSIONS

Motives for Introduction

In only one of the cases working conditions were mentioned as a motive to introduce robots. Because of the small number of cases it is not possible to draw conclusions. But this finding contrasts with the results of questionnaire among 38 robot using companies. It was found by Laurentius (1982) that improvement of working conditions was the most important motive.

TABLE 5 Guidelines for improvement of job content at robot workplaces

---

Production organization
Decentralization of departments like Programming, Maintenance, Technical Assistance, Production planning etc.

Work organization
Integration of daily work planning, after treatment, quality control, administration, exchange activities etc; Building of teamwork.

Robot workstation
Buffers, layout adjustment and noise reduction.

Safety Measures

The reasons for omitting safety precautions in the cases is not clear. From the results of interviews and observations it might be concluded that practical hindrances (like lack of space and the size of workpieces) are a factor. Also a lack of knowledge might play a role. For instance a proximity detection (contact mats, light curtains) which forces a robot to stop is feasable for robot welding of large work pieces (Company C). Yet it was not installed.

Occupational Health Measures

Though technical and organizational measures can lead to reduction of physical load, in none of the companies adequate provisions were present. It was striking that a lot of effort and money was put into the robot and its peripheral devices, but the work place of robot operators left much to be desired.

Job Content Measures

The improvement of job content in robotic and also CNC work stations is sometimes sought in adding programming and adjustment tasks to operator

106

jobs. From observation in several cases it is found that programming and adjusting the robot take only a minor amount of time. Practice shows that, within the robot workplace itself, there are little opportunities for improvement of job content. There are limitations from a number of factors, which are related to:

1. materials processing (processing time, workpiece mounting time, exchange time);
2. planning and product diversity (quantity per series of production, number of different products);
3. present work organization and technical realization of other parts of the manufacturing process (allocation of tasks, way of programming, workplace lay-out and the like).

Even with many different products and small series, programming and optimizing time is in the order of a few percent of the time. So only a minor improvement in job content may be expected.

Also the opportunities to enlarge workers autonomy in choosing their own pace of (e.g. by installing buffers) work are often not taken. In several cases robot operators are almost fully engaged in loacing robot welding stations. Though, technically spoken, the operator is in control of the starting time of the welding process, in many cases operators tried to keep up with the robot. Workers appeared to adjust their pace of work to the possibilities of the robot and the demands of the organization (e.g. high production norms). In one case this situation led to serious quality problems. Here a solution was found in balancing the duration of human and robot tasks. The same goes for creating workplaces that are less isolated.

In our opinion for the improvement of job content it is better not to focus on the robot work station. Following the integral approach it can be pointed out that more positive effects could be expected from an other way of grouping and coupling activities in the surrounding organization, see fig. 3. Nice examples of team work are shown in IG Metall (1986).

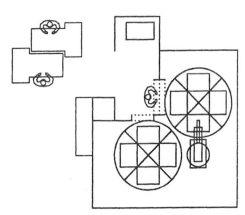

Fig 3. Example of uncoupling activities

In summary an overall opinion could be that in most cases not the full opportunities to ameliorate the quality of working life are seized. In some cases some unfavourable conditions are just replaced by others. Safety hazards are widely known but in some standard applications like enclosure are not practical.

Workplace design needs attention. It is striking that the robot activities are mostly well designed and facilitated but robot operator workplaces are often neglected.

With respect to job content it is concluded that tasks like programming and optimization offer far to little opportunities to design good jobs. Robotic work places, including operator jobs should be designed in the context of the whole organization, which includes both work and company aspects. In that view the robot is just a single and marginal station to be (re)designed.

SELECTED REFERENCES

Arbeidsinspectie.(1991) Werken met robots (CV-20). Arbeidsinspectie, Den Haag. (in Dutch).

Arbeidsinspectie.(1991) Industriële robots (CP-23). Arbeidsinspectie, Den Haag. (in Dutch).

Benders, J.(1991) Dividing labour around industrial robots; A theoretical framework. Paper presented at the WESWA symposium Nijmegen, Department of Business Administration, Tilburg University.

Bonney, M. C., Yong, Y. F. (1985) Robot Safety. Springer Verlag, Berlin.

Hartmann, G.(1989) Erfahrungen mit Sicherheitseinrichtungen an Roboterarbeitsplatzen. Zentralblat Arbeitsmedizin, 39 214-220. (in German).

Helander, M. G.(1990) Ergonomics and safety considerations in the design of robotics workplaces: a review and some priorities for research. International Journal of Industrial Ergonomics, 6 127-149.

Hilla, W., Tiller, R.E.(1990) Automatisierung im Karosseriebau aus arbeitsmedizinischer Sicht. Zentralblatt Arbeitsmedizin 40 94-100. (in German).

IG Metall. (1986) Industrie Roboter, Aktionsmappe. Industriegewerkschaft Metall für die Bundesrepublik Deutschland.(in German)

ILO. (1990) Safety in the use of industrial robots. Occupational safety and health series nr. 60. ILO, Geneva.

Jiang, B.C., Cheng, O.S.H. (1990) A procedure analysis for robot system safety. International Journal of Industrial Ergonomics, 6 95-117.

Landau, K., Brandenburg, U. (1988) Zur Analyse van Anforderungs verschiebungen durch Einsatz van Industrierobotern in Automobilbau. Zeitschrift für Arbeitswissenschaft, 4 201-205.

Laurentius, G., Timmerman, H., Vermeulen, A.A.M. (1982) Flexibele automatisering in Nederland,

ervaringen en opinies. Delftse Universitaire
Pers, Delft. (in Dutch).

Nagamachí, M. (1986) Human factors of industrial
robots and robot safety management in Japan.
Applied Ergonomics, 17.1 9-18.

Parsons, H. M. (1986) Human Factors in industial
robot safety. Journal of Occupational Accidents,
8 25-47.

Rahimi, M., Karwowski, W. (1990) A research paradigm in human-robot interaction. International
Journal of Industrial Ergonomics, 5 59-71.

WEBA-Project Team. (1990) Outlines of the WEBA-
instrument. A conditional approach for the assessment of the quality of work. TNO-Institute
of Preventive Health Care, Department of Work
Research, Leiden

SENSORY SUBSTITUTION FOR FORCE FEEDBACK IN TELEOPERATION

M.J. Massimino and T.B. Sheridan

*Room 3-355, Man-Machine Systems Laboratory, Department of Mechanical Engineering,
Massachusetts Institute of Technology, Cambridge, MA 02139, USA*

Abstract. The objective of this research was to study the capabilities of sensory substitution for force feedback through the tactile and auditory senses for teleoperation tasks, with and without time delay. The motivation and potential benefits of sensory substitution for force feedback with vibrotactile and auditory displays are discussed. Teleoperator experiments that examined the presentation of basic force information through object contact tasks, indicated that operator performance was improved by using the vibrotactile and auditory displays to present force information. Further, the vibrotactile and auditory displays compared favorably to traditional bilateral force feedback. Common manipulation experiments with peg-in-hole tasks of varying complexity were also conducted and showed that when the subjects' view was fully obstructed, the subjects were able to successfully complete the task by using either of the sensory substitution displays. Sensory substitution was also tested in the presence of a three second time delay and significantly improved performance without instabilities.

Keywords: Man-machine systems, force control, teleoperation, sensory substitution, time delay.

INTRODUCTION

Bilateral master-slave force feedback has been shown to be preferable to non-force feedback in many teleoperation studies [Massimino and Sheridan, 1989]. Hill and Salisbury [1977] found in their experiments that with force feedback task completion times were significantly shorter than without force feedback for peg-in-hole tasks. Further, Hill [1979] concluded that difficult tasks were done twice as fast when force feedback was present.

However, providing bilateral force reflection in the form of a force to the operator's arm and hand muscles can have its disadvantages. Systems that provide force feedback are often bulky master/slave manipulators that are impractical in many environments. Further, presenting force feedback to the operator's hand or arm in the presence of even small time delays has been shown to create operator induced instabilities. Ferrell [1966] suggested that the advantages of force sensitivity could be maintained in the presence of a time delay if the force feedback were substituted through the auditory or tactile modalities, and that a tactile display to the active hand might be especially compatible.

The objective of this research was to study the capabilities of sensory substitution of force feedback through the vibrotactile and auditory modalities for teleoperation tasks, with and without time delay. The major contributions of this research are: 1) an experimental study on the use of sensory substitution for force feedback through the tactile and auditory senses for manipulation tasks, encompassing psychophysical and remote manipulation tests, 2) a solution to the instability problem that occurs when force feedback is

presented in the presence of a time delay, and 3) methods by which many remote tasks can be performed when both useful visual feedback and traditional bilateral force feedback are not available.

Bach-y-Rita, Webster, Tompkins, and Crabb [1987] define sensory substitution as "the provision to the brain of information that is usually in one sensory domain (for example visual information via the eyes and visual system) by means of the receptors, pathways and brain projection, integrative and interpretative areas of another sensory system, (for example visual information through the skin and somatosensory system). Some examples include sign language for the deaf, and Braille for the blind." Sensory substitution has been successfully used for many years in helping people who are fully or partially deficient in one or more of their sensory systems, for example, sensory aids for the blind or deaf [Mann, 1974].

The vibrotactile and auditory senses were of interest for several reasons. First, they provided non-reactive representations of force feedback. Non-reactive means sense modalities that do not induce operator movements like bilateral force feedback does when providing force information. Such movements may be undesirable in certain situations, and can cause instabilities in the presence of a time delay. In addition, the ears and the fingertips contain sensitive sensory receptors and are allocated relatively large areas in sensory cortex for information processing [Barlow and Mollon, 1982], suggesting that auditory and vibrotactile displays could be successful in presenting force feedback information. Further, the auditory and vibrotactile modalities presented force information while not placing any extra burden on the operator's visual system which was normally intently viewing the remote task environment via a television monitor. Vibrotactile and auditory displays may also provide cost benefits by reducing the need for expensive bilateral force reflecting manipulators or high quality visual displays.

EXPERIMENTAL SYSTEM

A testbed for experimentation on the use of auditory and vibrotactile feedback displays of force feedback during teleoperation was developed. The auditory display related the loudness of a tone to the magnitude of a force, and the vibrotactile display related the magnitude of vibration intensity to the magnitude of a force. The auditory-force [Verillo and Chamberlain, 1972] and the vibrotactile-force [Scharf and Houtsma, 1986] relationships were developed from sensory magnitude functions. These relationships were controlled through a PC. To present the auditory stimuli, an audio card made by Creative Labs, Inc. was used. Small vibrators (1 inch by 0.5 inches) presented the vibrotactile stimuli. The original vibrotactile display was developed by Patrick [1990]. Force sensing resistors made by Interlink Electronics were interfaced to the PC and served as force sensors when using the sensory substitution displays.

In order to determine the characteristics and capabilities of each of the displays, two initial groups of psychophysical experiments were conducted: 1) experiments to determine the number of absolute judgements, and 2) experiments to determine the number of just noticeable differences (JNDs). These experiments found similar results for the vibrotactile and auditory displays. With the vibrotactile display, over a range of 7 to 32 db (re: 1 micron), there was channel capacity of approximately 1.73 bits which corresponded to 3.3 absolute judgements. For the auditory display, over a range of 46 to 70 db sound pressure level (SPL), a channel capacity of 1.76 bits or 3.4 absolute judgements were possible. In previous studies, auditory loudness (15 to 110 db SPL) was found to have a channel capacity of 2.3 bits or 4.9 absolute judgements by Garner [1953], while vibrotactile intensity was found to exhibit 2 bits of channel capacity or 4 absolute judgements by Geldard [in Miller, 1956]. In order to determine the number of just noticeable differences (JND's) an up and down method was used [Campbell, 1963; Levitt, 1971]. For the auditory loudness display, 14 JND's were determined, while for the vibrotactile display approximately 15 JND's were observed.

A master-slave manipulator was used for the experiments. The manipulator had the capability of operating with direct electronic coupling control both bilaterally with force feedback, and unilaterally without force feedback coming from the slave back to the master arm. Both master and slave arms had six degrees of freedom and were geometrically similar. The slave arm is shown in the remote environment in figure 1. The test subjects were provided with a view of the remote task environment via a television monitor.

OBJECT CONTACT EXPERIMENTS

The object contact experiments tested the representation of basic force information. They were divided into three subgroups of experiments: 1) detecting the presence of a contact force, 2) detecting the magnitude of contact force, and 3) tracking changes in sustained contact force.

Figure 1 displays the basic task board configuration for these experiments.

Figure 1. Task Board for Object Contact Experiments.

Presence of Contact Force

The objective of these experiments was to test the representation of the presence of a contact force through sensory substitution. The forces were presented in on-off mode when using the sensory substitution displays, thus the presence of the force was critical, not the magnitude of the force. Applications for this task include rendezvous and docking when the operator needs to know when he/she comes into contact with the target vehicle, and telemanipulation when the operator needs to perform an operation on a piece of equipment and has to know when contact has been made to prevent further motion and possible damage.

Two identical target objects were placed in the remote environment and projected outwards from a base toward the camera viewpoint (see figure 1). The objects were 0.75 inch squares, were spaced 8 inches apart, and an force sensing resistor was placed on each object. For the presence of contact force experiments, the base and the objects were fixed. This made the magnitude of the force applied to the target irrelevant.

A 0.75 inch square peg was placed in the gripper of the manipulator. At the start of each trial, the manipulator end-effector was in contact with a microswitch. When a subject moved the manipulator away from the microswitch the timer would start. After fifteen seconds a buzzer would sound indicating the end of the trial. The subjects were instructed to alternate tapping between the two objects with the tip of the manipulator gripper as quickly as possible, and the number of taps were recorded. Contact had to be made successfully in order for a tap to be recorded, and therefore knowing when contact was established and reacting quickly was the key to successful performance. More taps indicated better performance. Force information was helpful to the operator by providing a cue as to when contact existed.

Four force feedback conditions were used: 1) no force information, 2) traditional bilateral master-slave force reflection, 3) sensory substitution of force feedback through the vibrotactile display, and 4) sensory substitution of force feedback through the auditory display. A clear view of the remote task was always provided to the subject via the television monitor. Four test subjects were used, providing for a balanced latin square experimental design where each of the four experimental conditions preceded and followed each of the other experimental conditions an equal number of times. Each subject was trained and given warm-up time prior to the experimental trial. Six trials were conducted for each experimental condition by each subject.

The auditory display presented an auditory signal at 1000 Hz to both ears simultaneously, at well above threshold loudness, whenever contact was made with a target object. The vibrotactile display provided a 250 Hz vibratory signal, at well above threshold intensity, using vibrators on the subject's index finger and thumb.

Figure 2 displays the experimental results for the presence of contact force experiments. A series of paired t-tests at the 95% confidence level were performed on the data. The results showed that the auditory (t(23)=4.75,p<0.001) and the vibrotactile (t(23)=3.66,p<0.001) displays as well as traditional force force feedback (t(23)=5.61,p<0.001) provided a significant advantage over the no force feedback condition. The auditory display (t(23)=3.9,p<0.001) and the vibrotactile display (t(23)=2.52,p<0.02) provided a significant performance improvement over traditional force feedback. Further, the auditory display produced significantly better performance than the vibrotactile display (t(23)=3.75,p<0.001).

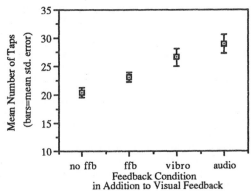

Figure 2. Results for Presence of Contact Force Experiments.

These experimental results were closely linked with reaction time since the subject was required to recognize a stimulus and react as quickly as possible. Recognizing the presence of a contact force and reacting appropriately was quickest with the auditory display, less quick with the vibrotactile display, slower with traditional force feedback, and slowest when using the visual display without any force cues.

Magnitude of Contact Force

The objective of these tasks was to test the representation of the magnitude of a contact force through sensory substitution. Applications for these experiments include a telemanipulation task where a maximum contact force level exists, above which damage and task failure will occur. The task board shown in figure 1 was used, however now the objects were attached to moveable bases as opposed to fixed bases. In addition, weights could be added to the bases to alter the amount of force that would move the objects due to varying frictional resistances.

The subjects were instructed to make contact between the two objects in the same manner as described in the previous section. However, for every 0.25 inch

displacement of an object, a penalty of one tap was subtracted from the total for that trial. Therefore the magnitude of the force being exerted was important. Four trained test subjects completed six trials for each experimental condition. Two different weight levels yielding two different force ranges for movement were used. The lighter level allowed a force range of 0 to 0.5 pounds before movement would occur. The higher level allowed a force range of 0 to 2 pounds before movement would occur. The maximum amount of force that could be exerted with the manipulator was 10 pounds.

The auditory display presented one signal to both ears, and the loudness of the tone was proportional to the magnitude of the force. The vibrotactile display presented a stimulus to the index finger and thumb, with vibration magnitude scaled to force. The experimental results are shown in figure 3.

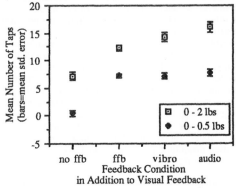

Figure 3. Results for Magnitude of Contact Force Experiments.

A series of paired t-tests were performed. For the higher force range tested (0 - 2.0 lbs.), the auditory display (t(23)=8.42,p<0.001), the vibrotactile display (t(23)=10.71,p<0.001), and traditional force feedback (t(23)=7.42,p<0.001) all produced significant performance advantages over using visual feedback without any force information. The auditory display (t(23)=4.6,p<0.001) and the vibrotactile display (t(23)=3.18,p<0.005) provided significant performance improvements over traditional force feedback. In addition the auditory display produced a significant improvement over the vibrotactile display (t(23)=2.69,p<0.015). These results for the higher force range are similar to the results obtained for the presence of contact force when the object was fixed and the force range was unlimited. It appears that the higher force range was sufficiently high that the operator was able to concentrate on speed and overcome the movement penalty. These results like those for the fixed objects were most closely linked with reaction time.

Different results were obtained for the lower force range. The auditory display (t(23)=7.88,p<0.001), vibrotactile display (t(23)=9.15,p<0.001), and traditional force feedback (t(23)=13.39,p<0.001) again yielded superior performance when compared to manipulation using the television monitor alone. However, there were no significant differences between the vibrotactile display, the auditory display, and traditional force feedback. Therefore, for perceiving the magnitude of the smaller force range, sensory substitution worked as well as traditional force feedback.

Tracking a Sustained Contact Force

These tasks measured the ability of sensory substitution to present information for tracking the magnitude of a force over time. Applications for these experiments include a telemanipulation task where a force must be exerted over a period of time, and has a maximum level above which damage and task failure will occur.

The same task board (figure 1) and force ranges (0 to 0.5 lbs. and 0 to 2.0 lbs.) discussed in the previous section were used for these experiments. However, for these experiments the subjects were instructed to make alternating contact with each object and maintain that contact for three seconds. Force information was important in this case because it indicated to the subject when contact had been established, if it was in the acceptable range so as not to move the object and incur a penalty, and whether the contact was constant so that continuous contact for the full three seconds was possible. Sustained contact for three seconds was indicated to the subject by a contact light. Four consecutive successful sustained contacts constituted a trial. Task time was recorded, with a five second add-on penalty for every 0.25 inches of object displacement. Therefore lower task times indicated better performance. Four trained test subjects completed six trials per experimental condition. The auditory display presented a single tone to both ears with loudness proportional to force, while the vibrotactile display presented vibration to the index finger and thumb simultaneously with vibration magnitude proportional to force. The results are shown in figure 4.

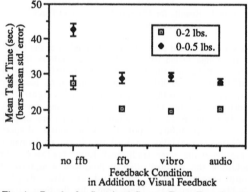

Fig. 4. Results for Sustained Contact Force Experiments

At the lower force range (0 - 0.5 pounds), the auditory display ($t(23)=7.2, p<0.001$), vibrotactile display ($t(23)=6.16, p<0.001$), and traditional force feedback ($t(23)=6.89, p<0.001$) all had significantly lower task times than performing the task without any form of force feedback. The higher force range produced similar results. There were no significant differences between the sensory substitution displays and traditional force feedback for either force range. Therefore, the overall performance results of these experiments indicated that traditional force feedback and sensory substitution of force feedback allowed the operator to track the force changes equally well for both force ranges.

PEG-IN-HOLE EXPERIMENTS

The sensory substitution displays were tested for their effectiveness with two types of peg-in-hole tasks: 1) a

two sided hole (or slot), and 2) a four sided hole. The two sided hole was 1.1 inches wide (x-direction), 7.25 inches high (z-direction) and 2.0 inches deep (y-direction), therefore movement was constrained in the x-direction when inserting the peg into the hole. The four sided hole was 1.2 inches wide, 1.25 inches high, and 1.0 inches deep, constraining movement in both the x and z directions. The peg was fixed in the manipulator gripper, was 1.1 inches in height, 1.05 inches wide, and two inches long. Therefore the two sided hole had a tolerance of 0.05 inches in the x-direction and very large tolerance in the z-direction, while the four sided hole had a tolerance of 0.15 inches in the x-direction and 0.15 inches in the z-direction. The two sided hole also required that the entire length of the peg (2 inches) be inserted to complete the task, while the four sided hole required that only one half of the length of the peg (1 inch) be inserted to complete the task.

The peg-in-hole experiments required more complicated force presentations for the sensory substitution displays since more forces were crucial for successful performance. The key forces to be presented were: 1) the force on the right side of the hole, 2) the force on the left side of the hole, 3) the force on the top side of the hole, 4) the force on the bottom side of the hole, and 5) the force being exerted on the peg. (The forces on the top and bottom sides were not presented when using the two sided hole task board.)

The auditory display presented the force on the right side of the hole as a medium pitch (1000 Hz) tone to the right ear, the force on the left side as a medium pitch (1000 Hz) tone to the left ear, the force on the top part of the hole as a very high pitch (3500 Hz) tone to both ears (which made the tone appear to the subjects as emanating from the middle of the head), and the force on the bottom part of the hole as a low pitch (350 Hz) tone to the middle of the head. The force on the peg was presented as a very low pitch (28 Hz) tone to the middle of the head. The loudness of each tone was related to the magnitude of the appropriate forces.

When using the vibrotactile display, vibrators placed at different sensitive skin locations [Wilska, 1954] represented the different forces. The force on the right side of the hole was presented as a vibration to the index finger, the force on the left side as a vibration on the thumb, the force on the top side of the hole as a vibration to the upper part of the palm, and the force on the bottom side of the hole as a vibration on the lower part of the palm. The force on the peg was presented as a vibration on the wrist. Each vibrator vibrated at 250 Hz, and the magnitude of the vibration was scaled to the magnitude of the associated force.

In order to test these sensory substitution representations of force for the four sided hole, a force direction psychophysical experiment was conducted. A force vector was positioned at sixteen different locations on an imaginary square hole. The subjects were given a stimulus and then had to respond as to which of the sixteen directions was being represented. Four subjects were used, and each direction was presented randomly five times per display for each subject. The results showed that the success rates for recognizing the exact direction of the force, the auditory display (67%) was significantly more accurate than the vibrotactile display (46%) ($t(3)=4.42, p<0.012$). The results for recognizing a dominant force component, i.e. recognizing a gross estimate of the force direction, indicated that there was no significant difference between the success rates of the auditory (98%) and the vibrotactile (94%) displays.

The peg-in-hole tasks started with the test subject moving the manipulator slave arm away from a microswitch which would start a clock. The task ended when the peg was inserted into the hole and made contact with another microswitch which would stop the clock and cause a contact light to go off indicating to the operator that the task was completed. Force sensing resistors were placed on the sides of the holes and on the tip of the peg. Four trained test subjects performed ten trials for each experimental condition, and a counterbalanced experimental design was used. The tasks were conducted with both unobstructed and obstructed views.

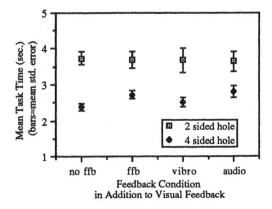

Figure 5. Results for the Peg-In-Hole Experiments with an Unobstructed View

Analyzing the results shown in figure 5, there were no significant performance difference between any of the feedback conditions for the two sided hole. Visual feedback alone appears to have dominated and was sufficient to allow the subjects to perform the task as quickly as possible. However, for the four sided hole, it appeared as though force information could actually impede performance. Using the visual display alone without any force information provided significantly lower mean task times than combining visual feedback with traditional force feedback (t(39)=2.62,p<0.012) or with the auditory sensory substitution display (t(39)=2.67,p<0.011). In the case of traditional force feedback, it seemed as though the subjects were constrained from executing their desired motions. Visual feedback was very clear and gave the subject a clear indication of the desired movement. However force feedback hurt performance by impeding this movement for this particular task. In the case of auditory sensory substitution, the high resolution of the auditory display (that was determined by the psychophysical tests on force direction) overloaded the operator with information and decreased performance.

When using an obstructed view, it was impossible to perform the tasks by relying only on visual feedback, and therefore three experimental conditions were tested as is shown in figure 6. A series of paired t-tests showed that the mean task times for the four sided hole scenario were significantly lower with traditional force feedback than with either the vibrotactile display (t(39)=4.05,p<0.001) or the auditory display

(t(39)=5.0,p<0.001). Similar results were observed for the two sided peg-in-hole task.

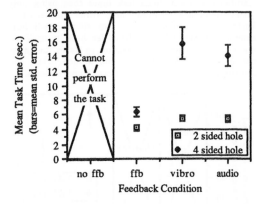

Figure 6. Results for the Peg-In-Hole Experiments with an Obstructed View

The advantage given to traditional force feedback can probably be attributed to the fact that force feedback not only provided a force cue to the operator, but also forced the subjects' hands to proper positions so that insertion was quicker than with sensory substitution which only provided force cues. Therefore this was a case in which the forces on the subjects' hands helped performance unlike the results found for the unobstructed view scenario shown in fig. 4. However, the most interesting result of these experiments was that the task was performed successfully by using either the vibrotactile or auditory display without any useful visual feedback.

TIME DELAY EXPERIMENTS

The peg-in-hole experiments described above were also conducted with a three second time delay. In order to facilitate this delay, the manipulator was interfaced with a Macintosh II PC. The results for the experiments with an unobstructed view are shown in figure 7.

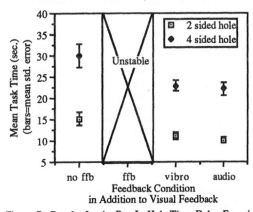

Figure 7. Results for the Peg-In-Hole Time Delay Experiments with an Unobstructed View.

Using traditional force feedback led to instability with a time delay and could not be tested. The experimental results showed that unlike performance without a time delay, performance with a time delay was improved by using sensory substitution in combination with visual feedback. When conducting the four sided peg-in-hole tasks, the vibrotactile display (t(39)=2.32,p<0.025) and the auditory display (t(39)=2.53,p<0.016) both had significantly lower mean task times than using only visual feedback without any force information. Performance results for the two sided peg-in-hole experiments were similar. Visual feedback by itself did not provide sufficient information so as to maximize performance when operating with a time delay, and sensory substitution gave significant improvements.

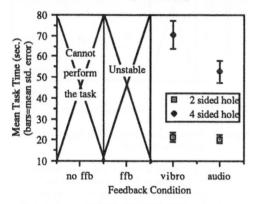

Figure 8. Results for the Peg-In-Hole Time Delay Experiments with an Obstructed View

Figure 8 shows the results of the two sided and four sided peg-in-hole time delay experiments with an obstructed view. Only two feedback conditions were possible. The performance between the vibrotactile and auditory displays was not significantly different for the two sided hole. For the four sided hole the auditory display had significantly lower mean task times (t(39)=2.14, p<0.04) than the vibrotactile display. Combining the time delay with an obstructed view, put the subjects in need of as much information as possible concerning the position of the peg. The psychophysical tests on force direction indicated that the auditory display provided information about force direction with higher resolution than the vibrotactile display. This information apparently was helpful, since performance was significantly better with the auditory display.

The most interesting result from the time delay experiments with an obstructed view, was that it was possible to perform these tasks with a three second time delay and without any useful visual information. This task would be impossible under conventional means since neither visual feedback or traditional force feedback could help the operator perform the task. Thus sensory substitution made it possible to successfully complete a task that would normally be impossible.

CONCLUSIONS

The feasibility of sensory substitution for force feedback through the vibrotactile and auditory displays to present the presence and magnitude of instantaneous and sustained contact forces, and to present multiple force representations for peg-in-hole tasks has been shown through the various experiments conducted. For every object contact task tested, the auditory and vibrotactile displays provided a significant advantage over the no force feedback condition, and compared favorably to traditional force feedback. For the peg-in-hole experiments with an obstructed view, the tasks were performed successfully using either of the sensory

substitution displays alone . The time delay experiments showed that force information could be presented successfully through sensory substitution without the instabilities that would normally occur with traditional force feedback. Further, operator performance was improved when using sensory substitution in the presence of a time delay for both of the peg-in-hole tasks even when clear visual feedback was available. With an obstructed view and time delay, the tasks could still be completed by relying solely on either of the sensory substitution displays. This enabled the tested tasks, which would be impossible to perform even if vision and traditional force feedback were available, to be successfully completed with sensory substitution.

REFERENCES

Bach-y-Rita, P., Webster, J.G., Tompkins, W.J., and Crabb, T. (1987). Sensory Substitution for Space Gloves and for Space Robots. In Proceedings of Workshop on Space Telerobotics, Vol.2.,pp.51-57.

Barlow, H.B. and Mollon, J.D. (1982). The Senses. Cambridge: Cambridge University Press.

Campbell, R.A. (1963). Detection of a Noise of Varying Duration. Journal of the Acoustical Society of America, Vol. 35, No. 11., pp. 1732-1737.

Ferrell, W.R. (1966). Delayed Force Feedback. Human Factors. Vol.8, No.5, October, 1966. pp.449-455.

Garner, W.R. (1953). An Informational Analysis of Absolute Judgments of Loudness. Journal of the Acoustical Society of America, Vol.46,pp.373-380.

Hill, J.W. (1979). Study of Modeling and Evaluation of Remote Manipulation Tasks with Force Feedback. Final Report, JPL Contract 95-5170, March, 1979.

Hill, J.W. and Salisbury, J.K. (1977). Study to Design and Develop Remote Manipulator Systems. Annual Report, NASA Contract NAS2-8652, Nov., 1977.

Levitt, H. (1971). Transformed Up-Down Methods in Psychoacoustics. The Journal of the Acoustical Society of America.Vol.49,No.2(pt.2),pp.467-477.

Mann, R.W.(1974). Technology and Human Rehabilitation: Prostheses for Sensory Rehabilitation and/or Sensory Substitution. Advances in Biomedical Engineering, Vol. 4, 1974, pp. 209-353.

Massimino, M.J., and Sheridan, T.B. (1989). Variable Force and Visual Feedback Effects on Teleoperator Man/Machine Performance. Proceedings of the NASA Conference on Space Telerobotics. Pasadena, CA, January 31-February 2, 1989.

Miller, G.A. (1956). The Magical Number Seven, Plus or Minus Two: Some Limits on our Capacity for Processing Information. The Psychological Review, Vol. 63, pp. 81-97.

Patrick, N.J.M. (1990). Design, Construction, and Testing of a Fingertip Tactile Display for Interaction with Virtual and Remote Environments. MIT Masters Thesis, Dept. of Mech.Engineering.

Scharf,B. and Houtsma, A.J.M. (1986). Audition II: Loudness, Pitch, Localization, Aural Distortion, Pathology. In Handbook of Perception and Human Performance. Chapt.15, Eds. K.R. Boff,L. Kaufman, and J.P. Thomas. New York: John Wiley and Sons.

Verillo, R.T. and Chamberlain, S.C. (1972). The Effect of Neural Density and Contactor Surround on Vibrotactile Sensation Magnitude. Perception and Psychophysics, Vol.11(1B), pp. 117-120.

Wilska, A. (1954). On the Vibrational Sensitivity in Different Regions of the Body Surface. Acta Physiologica Scandinavica. Vol. 31, pp. 285-289.

MANUAL CONTROL USING PREDICTOR DISPLAYS

Chi-Cheng Cheng* and T.B. Sheridan**

*Department of Mechanical Engineering, National Sun Yat-Sen University, Kaohsiung, Taiwan, PRC
**Department of Mechanical Engineering, Massachusetts Institute of Technology, Cambridge, Massachusetts, USA

Abstract. In manual control, predictor displays have been shown that with such predictive information, considerable improvement in control quality can be achieved; however, most of these investigations have been limited to experimental approaches. There still lacks theoretical analysis about how the human utilizes predictive information to accomplish more accurate control. This paper presents a preliminary study of predictor displays in manual control from a view of control theory. The analysis shows that predictor displays decrease the time constant of the controlled process and increase the damping of the control system, which result in reduction of mental loading of human operators and improvement of control performance.

Such a system is applied to a towed submersible, which has no independent propulsion and its positioning can only be achieved through a cable by maneuvering a surface ship. Currently, the maneuvers possible with the towed submersibles are restricted to straight-line strategies. In order to enhance the maneuverability of the towed submersible, a supervisory control system with a predictor display is proposed to compensate for the long time constant of the system. Experiments on manual control demonstrated that predictor displays improved position accuracy for tracking tasks. Experimental results also indicated consistency with the theoretical analysis.

Keywords. Manual control; predictor displays; man-machine systems; supervisory control; display systems.

INTRODUCTION

Although the technology of automatic control has provided humans with an ever-growing mastery over their environment, manual control still plays an important role on controlling many complex systems in variable environment, e.g., driving a car on a curved road, recovering a space satellite with a master-slave manipulator, taking off and landing of an aircraft, etc. The need of human operation stems from a unique gift of human beings, *adaptability*, which cannot be easily replaced by any other machines or control methods. This peculiarity allows humans to have the capability of performing tasks in difficult environment.

Apparently human operators create internally dynamic models of themselves and their environment (Kelly 1968). They seem to perform planning and prediction by means of the internal model. Planning is to choose the goal and the route to a designated goal, and prediction is accomplished by incorporating status information into the model and extrapolating it into the future. Control is normally exercised on the basis of the difference between what is desired or planned and what is predicted. If some changes occur in the system or the environment and the internal model employed for control does not incorporate the phenomenon, the model can be adjusted adaptively to represent the changes.

The principle of a predictor display is to use an *external* model to generate more accurate predictive information to help the human building their own internal models and to improve the performance of manual operation. For high order and nonlinear systems, given any input commands, human operators cannot predict changes in outputs so easily. They may have to wait until perceptible changes of the output occur, and then correct their commands. Predictor displays reveal the future tendency of output and can reduce the waiting time significantly. Thus operators are able to take correct action much earlier.

Many studies have shown that with such predictive information, considerable improvement in control quality can be achieved (e.g., Kelley and Prosin,

115

1971; Poulton, 1974). In addition, Sheridan and colleagues (1964) found that predictor displays can especially be of great assistance to the human operator in controlling higher order systems having little damping. Dey (1971) investigated the human transfer characteristics influenced by predictor displays and concluded that this technique caused an increase of a gain factor and a diminishing of a lag factor of the human transfer function. Kvålseth (1978) compared the effects of input and output predictions on manual control performance. He summarized that the output predictions tended to enhance control performance; however, the input predictions hardly have effects on performance. Multi-input and multi-output systems were examined by van der Veldt and van den Broomgaard (1986). Their experiments indicated that predictions improved the performance of manual control especially in systems with time delay and coupled interaction.

Although the fact that predictor displays are helpful to manual control is already well-known, the knowledge about how and why the predictive information can assist manual control systems is still unclear. Most of early researches were limited to experimental approaches. There still lacks theoretical analysis about how human utilizes predictive information to accomplish more accurate control. In this paper, a preliminary study of manual control systems with predictor displays from a view of control theory is presented.

A supervisory control system with a predictor display was developed for a towed submersible system. The towed submersible is an unmanned underwater vehicle, which has no independent propulsion. Therefore, its positioning is achieved through a cable by maneuvering a surface ship. The cable can be up to 6,000-meter long and causes difficulty in manual control, because of its highly nonlinear characteristics as well as long delay time and long time constant. For the time being, the maneuvers possible with the towed submersible are relatively crude: lining up and making a straight line pass over the area of interest. Experiments on manual control of a towed submersible were conducted to evaluate the performance improvement of manual control with predictor displays and to verify the consistency between experimental results and theoretical analysis.

PREDICTOR DISPLAYS IN MANUAL CONTROL

Two primary techniques exist for generating predictive information. They are *the extrapolation method* and *the fast-time model method*. The extrapolation method is based on the instant outputs and their derivatives with proper Taylor series coefficients to extrapolate the system response into the future. This method has the advantage of easy implementation, but can be in error due to faulty extrapolation. The principle of the fast-time model method is to simulate the actual system with an existing mathematical model in an accelerated time scale to estimate the future response. The first predictor display using this method was introduced by Ziebolz and Paynter (1954). The advantage of using this method apparently is that correct future information can be presented to human operators.

Although the fast-time model method is usually computationally expensive because of the requirement of a complete system model for fast-time simulation. This disadvantage can be overcome by using simplified models and/or high performance computers. Fortunately, rapid improvement on integrated circuit technique makes fast simulation more likely than ever.

Analysis of Manual Control Systems

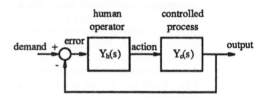

Fig. 1. Block diagram of a simple manual control system.

For simple tracking problems, we assume that what the human does is to minimize the error between the demand and the output variable shown on the display. Hence the manual control system can be illustrated by a block diagram as shown in Fig. 1, where $Y_h(s)$ and $Y_c(s)$ represent the transfer functions of the human operator and the controlled process, respectively. The human is a highly adaptive system and is able to change his dynamics to achieve better control. In other words, the human transfer characteristics will be modified during the controlling period and also vary from one control process to another. Many studies (e.g., Young, 1973) have shown that a simplified human model can be written as :

$$Y_h(s) = \frac{K_h(T_L s+1)e^{-\tau_e}}{T_I s+1}, \qquad (1)$$

where τ_e is the effective time delay comprising the pure time delays associated with the reading and interpreting of the display and deciding the appropriate control action. T_L and T_I are lead and integral time constants, which can be adjusted by the human along with the gain K_h to reach a stable closed-loop system and small errors.

There have been a number of attempts to develop adjustment rules for selecting those parameters in the human transfer function. *The crossover model* is the simplest and the most popular one. The model asserts that human operators will adjust their own transfer functions to compensate the controlled process in order to satisfy the open-loop condition

$Y_h Y_c \approx \omega_c e^{-\tau_e}/s$ near the gain crossover frequency, ω_c (see McRuer and others, 1989).

Assume a task performed by a human operator is to control the position of a vehicle, which has an inertia m and a damping coefficient b with the ground. The controlled element dynamics, $Y_c(s)$, is given by

$$Y_c(s) = \frac{K_c}{s(\tau s+1)}, \qquad (2)$$

with $K_c = 1/b$ and $\tau = m/b$. According to the equalization adjustment rules recommended by McRuer, the human transfer function for this controlled process appears as $Y_h(s) = K_h(\tau s+1)e^{-\tau_e}$, where the value of τ_e is about 0.5 second for the case of $\tau > \tau_e$ and about 0.3 second for the case of $\tau < \tau_e$. The human builds a lead compensator of the same time constant internally to balance the lag of the controlled process. From the root loci of the control system, we found that higher gains reduce damping of the system. The maximum gain for K_h without bringing the system into the unstable region is

$$\max K_h = \frac{\pi}{2\tau_e K_c} \qquad (3)$$

The magnitude of the lead time constant in the human transfer function plays a role in determining the degree of mental loading. When the time constant of the controlled process τ is large, the human operator needs to have a long projection into the future and will experience strong mental loading from the operation. When τ is small, the derivative signal becomes less important. Therefore, the internal calculation for prediction becomes simpler and a low level of mental loading will be experienced.

Manual Control with A Predictor Display

To analyze a manual control system with a fast-time predictor, a deeper understanding of the fast-time model is appropriate. A fast-time model is a duplicate model of a controlled process with a fast-time scale of its response. Through simple algebraic manipulation on Laplace transformation, the fast-time model of a transfer function $G(s)$ can be written as $G(s/\alpha)$. α is a scale factor and should be always greater than 1. Block diagrams with regular-time and fast-time scales are shown in Fig. 2.

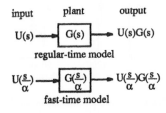

Fig. 2. Regular-time and fast-time models.

When a predictor display is supplied to human operators by the fast-time model method, the operators will pay some attention to controlling the output of the fast-time model instead just the actual model. The block diagram of the control system with a fast-time model prediction is depicted in Fig. 3. ZOH represents a zero order hold, which holds constant values during sampling intervals. Therefore, the predictor display is generated by assuming no further action is taken in the future.

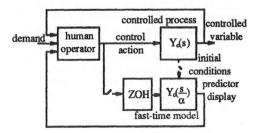

Fig. 3. Block diagram of a manual control system with a fast-time model prediction.

Since the emphasis is on the effect of the fast-time model, only the feedback loop for the predictor display is examined. In addition, the sampling frequency is assumed fast enough so that the controlled process can be approximated in a continuous form. Hence the new controlled process, $Y'(s)$, has the following characteristic:

$$Y'_c(s) = Y_c(\frac{s}{\alpha}) = \frac{K_c\alpha}{s(\frac{\tau}{\alpha}s+1)} \qquad (4)$$

In order to satisfy the crossover model, the human model becomes

$$Y'_h(s) = K'_h(\frac{\tau}{\alpha}s+1)e^{-s\tau'_e}, \qquad (5)$$

with

$$\max K'_h = \frac{\pi}{2\tau'_e K_c\alpha} \qquad (6)$$

The time constant of the pseudo-process model is reduced by a factor of α. The value of α is usually large and in the range of hundreds or thousands. Besides, the lead-time constant in the human transfer function due to the similar effect is correspondingly reduced. Consequently, the prediction span of the internal model decreases and the effective time delay is also shorter. Both effects bring about a reduction of mental loading. Furthermore, the gain is also decreased approximately by a factor of α, because the values of τ_e are within a small range, from 0.3 to 1 second. As a result, the small gain increases the damping to the system response. If controlled errors are in a similar magnitude region, lower levels of control action can be observed in the manual control systems with predictor displays.

APPLICATION TO A TOWED SUBMERSIBLE

During the accelerating exploration of the deep sea, unmanned submersibles have been used extensively. Argo, developed by the Deep Submergence Laboratory of the Woods Hole Oceanographic Institution, is a towed submersible for deep sea survey and inspection (Ballard, 1984). It provides a garage for Jason, a tethered, self-propelled vehicle with three-dimensional mobility. The basic operational scenario is that after Argo has located an area of interest, Jason will be deployed and maneuver to the worksite for detail investigation. Since Argo has no independent propulsion; its positioning can only be achieved through a cable by maneuvering a surface ship. The cable between Argo and the ship is up to 6,000 meters long. Such a long cable brings about difficulty in manual control, because of its highly nonlinear properties as well as long delay time and long time constant. For a 2,500-meter cable, the time lag and the time constant are about 20 seconds and 15 minutes, respectively. Due to the slow and complex dynamics of the cable, current control strategies are only limited to straight-line maneuvers.

In order to enhance the maneuverability of the towed submersible, a control aid of predictor displays for manual operation was implemented. Since a completely dynamic model for a long cable is not possible given limited knowledge, the predictor display is based on a sufficiently good parametric model , which can be run in fast-time relative to the dynamics of the system (see Cheng, 1991). Figure 4 illustrates the structure of the Argo submersible system with the predictor display.

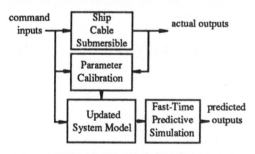

Fig. 4. A predictor display for the Argo submersible system.

EXPERIMENTS ON HUMAN OPERATION

Two experiments, one-dimensional and two dimensional tracking tasks, were conducted to evaluate control performance of human operation for the towed submersible with and without predictor displays. The experimental setup was composed of a personal computer with a graphic terminal. All human subjects were graduate students in mechanical engineering at M.I.T. The subjects were asked to control the trajectory of a submersible at conditions

of different cable lengths and prediction spans in order to follow specified trajectories as shown in TABLE 1. For purposes of experiments, lateral thrusters were disabled in one-dimensional experiments and forward thrusters were fixed in two-dimensional experiments.

TABLE 1 Experimental Conditions

	one-dimensional tracking task	two-dimensional tracking task
trajectory	straight-line vs time	planar trajectory
no. of subjects	8	6
execution time	10 minutes	≈ 10 minutes
cable length (m)	1200, 1600, and 2000	1200 and 2000
prediction span (min)	0^1, 3, 6, and 10	0, 3, 6, and 10
controlled action	forward thruster	lateral thruster

[1] Zero prediction span indicates no predictor display.

Tracking errors measured as root-mean-squares of deviation distance between actual trajectory and the desired trajectory are presented in Fig. 5. Error bars indicate plus and minus one standard deviation from the mean values. Experimental results show that tracking performance can be improved by at least 50% with the aid of predictor displays.

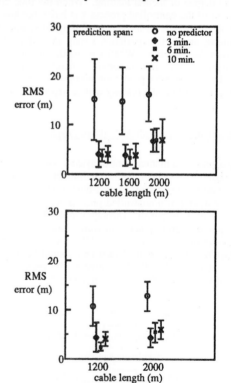

Fig. 5. Tracking errors for one-dimensional (top) and two-dimensional (bottom) tracking experiments.

The results were also compared using the paired *t* test. At over 95% significance the null hypothesis for no difference of means can be rejected. Longer cables, which had long time constants, seemed more difficult to control than short ones, as expected. It is interesting to see that the prediction spans appeared not to have much to do with the control performance, and longer predictions did not always improve the performance. The reason for this phenomenon may due to the form of the predictor displays, which supplies not only future information of the position, but also information about the velocity and the acceleration. If a shorter prediction is provided, longer predictions based on this information could be accomplished through extrapolation by human operators themselves. Thus with both explicit and implicit predictions, control performance seemed to be improved at similar regions.

For the purpose of analyzing control actions of human operators, the thrust was divided into three groups: top 12.5%, bottom 12.5%, and center 75%. All control actions of human subjects for different cable lengths are plotted as 100-percent bar graphs in Fig. 6. The abscissa represents the frequency or the probability density function. It appears that without predictors about 30% of the control actions were devoted to almost full thrust; however, with the help of predictor displays more activity happened at low thrust levels and human operators could apply more sophisticated control strategies. This result matches the theoretical analysis described before. For 2,000-meter cables, this effect is not so obvious. It is because, in order to follow the desired trajectory, operators seemed to have more chance to use large thrust to drive the heavy system.

Typical example trajectories with and without predictors are illustrated in Fig. 7 and Fig. 8 for different experiment tasks. It is apparent that with the aid of predictor displays, the humans could comfortably apply more efficient control and the system response became smooth with high damping. Besides, the trajectories without predictors showed the property of low damping, which again supports the theory presented before.

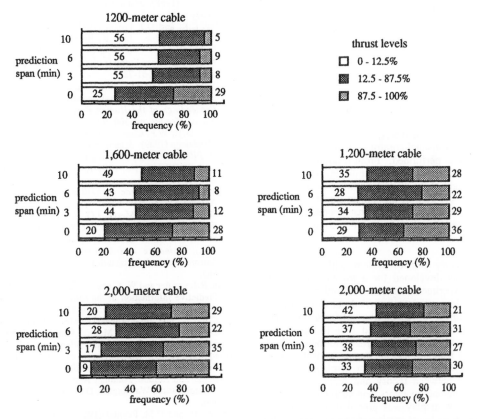

Fig. 6. Control actions of human operators for different cable lengths in one-dimensional (left) and two-dimensional (right) tracking tasks.

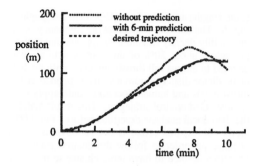

Fig. 7. Example trajectories for a 1,600-meter cable in one-dimensional tracking task.

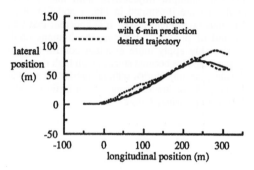

Fig. 8. Example trajectories for a 1,200-meter cable in two-dimensional tracking task.

CONCLUSIONS

Although the technique of predictor displays has been examined to be helpful to manual control, theoretical background is still unclear. A preliminary analysis of a manual control system with predictor displays was performed. From a point of control theory, with the aid of predictor display, the effective time constant of the controlled process equivalently decreases, the gain of the human transfer function and the control action from human operators become smaller, and the damping of the control system increases.

A supervisory control system with a predictor display was developed for a towed submersible system. Experiments on human control of the towed submersible system were carried out to investigate the usefulness of predictive information. The experimental results showed that the predictor display greatly improved control performance. Both the system response and the human control actions, with and without the predictor display, indicated agreement between theory and reality. Control actions of human operators demonstrated that when the predictor display was provided, more control actions were at lower levels and more subtle control strategies could be applied. Besides, the system response with predictor displays showed high damping. Nevertheless, how human operators judge the displays of the controlled variable and its prediction to generate appropriate action is still not fully understood. Further investigations are needed.

ACKNOWLEDGEMENTS

The research described in this paper was funded by the Woods Hole Oceanographic Institution. The authors would gratefully acknowledge the helpful comments and criticism of colleague at Man-Machine Systems Laboratory of M.I.T: Kan-Ping Chin, Jong Park, Mike Massimino, Jie Ren, David Schloerb, and Nick Patrick.

REFERENCES

Ballard, R.D. (1984). ROV development at Woods Hole's Deep Submergence Laboratory. *ROV '84 Technology Update - An International Perspective*. The Marine Technology Society. pp. 82-89.

Cheng, C.C. (1991). *Predictor Displays: Theory Development and Application to Towed Submersibles*. Sc.D. Thesis, Department of Mechanical Engineering, Massachusetts Institute of Technology, Cambridge.

Dey, D. (1971). The influence of a prediction displays on the human transfer characteristics. In R.K. Bernotat and P. Gärtnek (Ed.), *Displays and Controls*, Proceedings of the NATO Advanced Study Institute. Swetz & Zeitlinger, Amsterdam. pp. 483-491.

Kelley, C.R. (1968). *Manual and Automatic Control*. John Wiley and Sons, New York.

Kelley, C.R. and D.J. Prosin (1971). Adaptive display using prediction. In R.K. Bernotat and P.Gärtnek (Ed.), *Displays and Controls*; Proceedings of the NATO Advanced Study Institute. Swetz & Zeitlinger, Amsterdam. pp. 61-73.

Kvålseth, T.O. (1978). Effects of input and output predictions on manual control performance. *Proceedings of the Human Factors Society - 22nd Annual Meeting*. pp. 360-364.

McRuer, D.T., W.F. Clement, P.M. Thompson, and R.E. Magdaleno (1989). *Minimum Flying Qualities Volume II: Pilot Modeling for Flying Qualities Applications*. Technical Report No. 1235-1, Systems Technology, Inc., Hawthorne.

Poulton, E.C. (1974). *Tracking Skill and Manual Control*. Academic Press, New York.

Sheridan, T.B., M.H. Merel, J.G. Kreifeldt, and W.R. Ferrell (1964). Some predictive characteristics of the human controller. In R.C. Langford and C.J. Mundo (Ed.), *Guidance and Control - II*, Progress in Astronautics and Aeronautics - Volume 13. pp. 645-663.

Veldt, R.J. van der and W.A. van den Broomgaard (1986). Predictive information in the control room. In H.P. Willumeit (Ed.), *Human Decision Making and Manual Control*, The 5th European Conference on Human Decision Making and Manual Control. Elsevier Science Publisher B.V. (North-Holland). pp. 249-264.

Young, L.R. (1973). Human control capabilities. In J.F. Parker, Jr. and V.R. West (Ed.), *Bioastronautics Data Book*, 2nd Edition. National Aeronautics and Space Administration, Washington, D.C. Chap. 16, pp. 751-806.

Ziebolz, H. and H.M. Paynter (1954). Possibilities of a two-time scale computing system for control and simulation of dynamics systems. *Proceedings of the National Electronics Conference*, Vol. 9. pp. 215-223.

IDENTIFICATION AND APPLICATION OF NEURAL OPERATOR MODELS IN A CAR DRIVING SITUATION

K.-F. Kraiss* and H. Küttelwesch**

*RWTH Aachen, Germany
**Research Institute for Human Engineering (FGAN-FAT), Neuenahrerstr. 20, 5307 Werthoven, Germany

Abstract.This paper addresses the question of whether neural networks are applicable as operator models in man-machine-systems. For this analysis a two lane car driving task is used as an experimental paradigm. Various network architectures are tested. In particular a combination of functional link and backpropagation is proposed as a novel, rapidly trainable structure. It is shown experimentally, that individual human driving characteristics are indeed identifiable from the input/output relations of the trained networks. Neural nets are therefore candidates for operator models. The applicability of such models as an information source for driver assistant systems is demonstrated.

Keywords. Neural nets; Models; Identification; Teaching; Vehicles; Train control; Learning systems; Adaptive systems

INTRODUCTION

Increased automation in guidance and control systems has changed the role of human operators in such systems from controller to supervisor. This can lead to a loss of expertise, caused by a lack of manual operating time. In order to compensate for such effects, modern automation concepts try to bring the operator back into the loop. One approach to achieving this goal is the provision of adaptive interfaces, which offer on-line assistance to individual operators if and when they happen to need help.

Fig. 1 illustrates schematically how driver assistance can be provided by Neural Operator Models during vehicle control. It is shown that a neural network can be trained to emulate an operator, by observing what he is actually doing in a particular situation. After successful training it may be applied, e.g., as operator substitute (autopilot), as monitor of changes in human operator behavior, as operator assistant, or as trainer for novices.

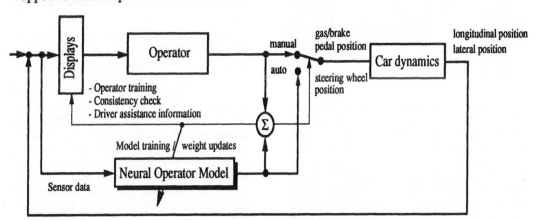

Fig.1: Applications of Neural Operator Models in vehicle control .

Experimental paradigm for manual vehicle control

The feasibility of neural operator modeling has been studied by various authors (Shepanski & Macy, 1987; Kraiss & Küttelwesch, 1990; Fix, 1990). In continuation of this work a more demanding experimental traffic situation has been developed. It is the bird's eye view of a two lane road, with traffic flowing at constant

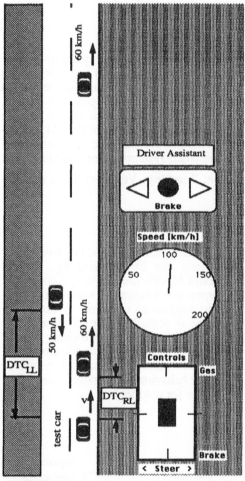

Fig.2: Bird's eye view of a traffic situation used for experimentation

speed, but with random distances between (grey) vehicles, in both directions (fig. 2). The traffic moves relative to a (black) test car. It is the driver's task to close up to and safely overtake preceding cars, while simultaneously trying to maintain a commanded speed of 150 km/h. Gas and brake pedals are operated by longitudinal. Steering is effected by lateral movements with

a mouse. A black area within a 'controls' rectangle (fig.2) indicates what the actual inputs are, while an analog meter indicates speed. An additional indicator, to be explained later, provides the driver with assistant information.

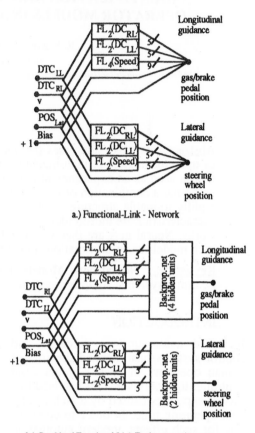

Fig. 3: Two of three network architectures used in this study

Determination of a suitable network architecture

The traffic situation described above can be characterized by at least four parameters, i.e., the distance to collision on the left (DTC_{LL}) and right lane (DTC_{RL}) respectively, the speed (v), and the lateral position (right or left lane) of the test car. All of these have been used as analog net inputs, scaled to a range between +1 and -1. Additional inputs would become necessary, if the surrounding traffic moves at variable speeds (we are in this paper not concerned with the question, of how this information can be picked up by suitable sensors).

Three layer backpropagation nets (BP) of various configurations were used in this study, with 8 hidden units yielding the best results. Alternatively functional link nets (FL) and combined functional link / backpropagation nets (FL/BP) were tested, in order to achieve shorter training times for real time applications (Yoh-Han Pao, 1989). For these again a variety of configurations have been examined. The results given below are concerned with the network architectures outlined in fig. 3, where FL_n (x) is a short notation for the net inputs:

$$(x, (\sin (2^{(i-1)}\pi x), \cos (2^{(i-1)}\pi x): i = 1...n)).$$

Algorithmic controller

Since human drivers behave inconsistently, an algorithmic controller has been developed, in order to make reliable and reproduceable training data for the comparison of different nets available. This controller ensures collision free operation by taking into account the actual and predicted traffic situation, as well as the gas and braking limits of the test car. The strategies applied by the algorithmic controller are illustrated qualitatively in fig. 4.

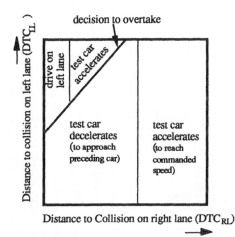

Fig. 4: Strategies applied by the algorithmic controller at 100 km/h

Comparison of competing net architectures:

For the comparison of all network architectures 10 overtaking manoeuvers were performed in a 280 [s] mission, using the algorithmic controller as driver. With data collected at a rate of 5 [Hz], the total training set covered 1400 data points. Since the overtaking manoeuvers were similar in parts, identical data points occurred at multiple times. Therefore filtering was applied, in order to better balance the coverage of the parameter space. To this end all inputs and outputs of a net were quantized into 21 steps. Subsequently data vectors consisting of identical inputs and outputs (within the limits of a quantization step) were eliminated, exept for one. By this measure the training set could be reduced by about 60 [%] without loss of information. It also became obvious that

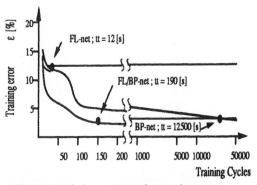

Fig. 5: Training curves for various net architectures. (training times (tt) apply for MacIntosh IIfx)

training results were mainly determined by the last part of the training set, if used as registered, i.e., according to the correct time sequence. To compensate for this effect, a randomization of the data before training turned out to be very efficient.

Learning behavior: As illustrated by fig. 5, the FL-net reaches a residual training error e ≈ 13 [%] after only some 30 cycles. The FL/BP-net levels off at e ≈ 3[%] after about 150 cycles, while the BP-net needs about 30 000 cycles to reach the same training status. This could lead to the conclusion, that FL/BP- and BP-nets are equivalent. A more detailed analysis shows however, that e alone is not a good indicator of control performance. This is demonstrated in fig. 6. Presented are net inputs and car dynamics outputs, which turned out to be more informative than net outputs. Diagrams in column a.) depict for reference the - normative - behavior of the algorithmic controller (compare fig.4).

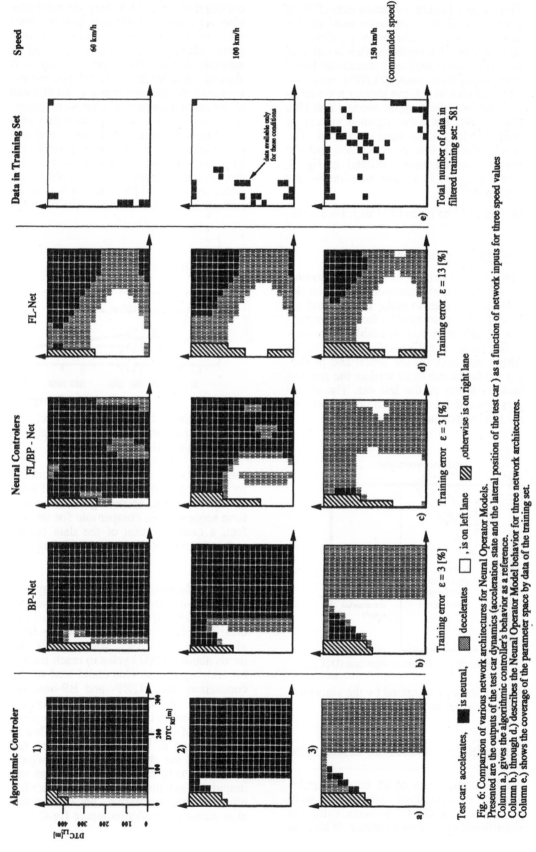

Fig. 6: Comparison of various network architectures for Neural Operator Models.
Presented are the outputs of the test car dynamics (acceleration state and the lateral position of the test car) as a function of network inputs for three speed values
Column a.) gives the algorithmic controller's behavior as a reference.
Column b.) through d.) describes the Neural Operator Model behavior for three network architectures.
Column e.) shows the coverage of the parameter space by data of the training set.

124

As documented by the diagrams in column 6b, the backpropagation net matches 6a behavior almost perfectly and consequently this Neural Operator Model performs fairly identical to the algorithmic controller. The trained BP-net is therefore suitable to identify human driving characteristics, so far, however, not in real time due to the excessive time required for training. The FL/BP-net (6c), in spite of having the same 3 [%] training error, matches 6a much worse. It still works satisfactorily as a neural driver in the sense that no collisions occur, but the driving strategy is different from that of the algorithmic controller. Training, however, is very fast. Therefore this net architecture may be suited to provide real time driver assistant information. The FL-net (6d), finally, is far from emulating 6a correctly. Therefore genuine FL-nets are not suitable for the intended task.

Generalization of performance: Fig 6e illustrates, how little of the parameter space is actually covered by the data in the training set. Most of the data are concentrated around the commanded speed 150 [km/h], very few for other speed values. Nevertheless generalization takes place correctly also in areas where no data exist at all. There is however no guarantee, that generalization covers the whole parameter space correctly. Generalization is also influenced by the composition of the driving mission, from which training data are collected. Both aspects are critical and subject to further analyses.

Identification of individual operator performance

In order to identify whether and to which degree differences in individual operator behavior are reflected by corresponding Neural Operator Models, two drivers performed a 16 minute mission with the experimental setup described in fig. 2. Data collected during these runs were used to train 3-layer backpropagation nets with 8 hidden units. The results are presented in Fig. 7, together with the algorithmic controller's performance.

The algorithmic controller, acting perfectly consistently, is characterized by sharp transitions between phases of car acceleration and deceleration. Since human drivers

perform inconsistently, their results show smoother transitions. Apart from this general effect, other differences can be identified from fig. 7. Driver no. 1, for example, after reaching 150 [km/h], decelerates to approach preceding car just when the car on the left lane comes in sight. In comparison, subject no. 2 appears to be more aggressive in his acceleration/deceleration behavior. Braking happens at shorter distances DTC_{RL} from the preceding car and he decides to overtake even if the distance to collision to the car on the left lane DTC_{LL} is critically small.

From these results it appears that individual driving differences can in fact be derived from trained neural networks. The visualization technique used in fig.7, where diagrams of two net inputs are depicted with all other inputs set constant, turns out to be helpful. So far no experience is available with whether this approach works equally well for more complex nets with many more inputs. It must also be mentioned that the modeling fidelity required for individual driving behavior identification can only be achieved with backpropagation nets. Hence fast and highly parallel hardware would be required, to enable online operation.

Application of a Neural Operator Model as driver assistant

Currently significant European research activities are directed towards improved safety in road traffic (Kramer & Reichart, 1989). One of the goals pursued in this context is concerned with situation and driver specific onboard advice. From the discussion in the previous paragraph it should be clear that Neural Operator Models are a suitable source for operator adaptive information. To this end, a net, trained with data of an individual driver, is operated in parallel with the same driver (see fig. 1). It receives sensor information, which should as far as possible be identical to what the driver sees, and yields steering wheel and gas/brake positions as outputs. These control signals are compared with what the driver actually does. In case of divergence a visual or acoustical alert can be initiated. In fig.2 a proposal for that kind of driver assistant information is presented (head-up projection on the windshield is desirable). Arrows in this display command lane changes and

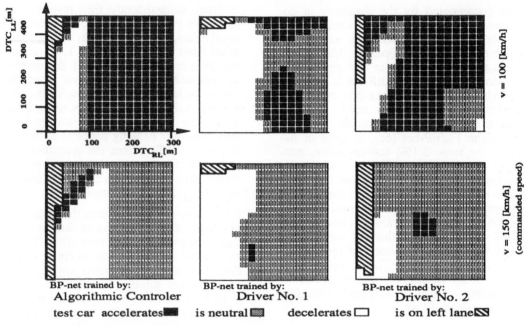

Fig. 7: Identification of individual driver performance from backpropagation nets. Each driver performed a 16 minute mission.

the light in the middle suggests emergency braking during dangerous overtaking manoeuvres. This preliminary installation was favorably accepted by subjects. Further studies concerning the technical implementation are needed however .

CONCLUSION

In this paper possible applications of neural networks in man machine systems have been studied. Various net architectures were tested. From these only three layer backpropagation nets could be taught individual driving performance. Network training and generalization properties were significantly influenced by the training set. Hence training missions must be carefully selected to ensure a reasonable coverage of the network parameter space. In addition the training set must be filtered and randomized to compensate for overrepresentation of data and time sequence effects. Combined functional link/ backpropagation nets could be trained rapidly to drive safely, but are not suited as Neural Operator Models. The application of such networks as information sources for driver assistant systems was successfully demonstrated. Further studies in a more realictic setting are needed however.

REFERENCES

Fix, E. (1990). Modeling Human Performance With Neural Networks. In: Proc. of the IEEE Int. J. Conf. on Neural Networks. S.Diego, I 247-252.

Kraiss, K.-F. & H. Küttelwesch (1990). Teaching Neural Networks to Guide a Vehicle Through an Obstacle Course by Emulating a Human Teacher. In: Proc. of the IEEE Int. J.Conf. on Neural Networks. S.Diego, II 333-337.

Kramer, U. & G. Reichart (1989). Automation and Safety Systems Engineering Aspects of Selected PROMETHEUS Functions, 2nd PROMETHEUS workshop. Stockholm.

Shepanski, J.F.& S.A.Macy (1987). Teaching Artificial Neural Systems to Drive: Manual Training Techniquesfor Autonomous Systems. In: 1st Annual Workshop on Space OperationsAutomation and Robotics. NASA CP 2491, 231-238.

Yoh-Han Pao (1989). Adaptive Recognition and Neural Networks. Addison-Wesley Publ. Comp., New York

126

TEAM RELATIVE OPERATING CHARACTERISTIC: A MEASURE OF TEAM EXPERTISE IN DISTRIBUTED DETECTION TASKS[1]

A. Pete, K.R. Pattipati and D.L. Kleinman

*Department of Electrical and Systems Engineering, University of Connecticut, Storrs,
CT 06269-3157, USA*

Abstract. We consider a generalized distributed binary hypothesis-testing problem within a hierarchical team. In this problem, the subordinate decisionmakers (DMs) transmit their opinions on their own local hypotheses, which are only probabilistically related to the global hypotheses at the primary DM. In a recent paper, we have shown that the normative decision strategies of all DMs are coupled likelihood-ratio tests, but the decision thresholds are also a function of the joint probability distribution of hypotheses at all DMs. In order to assess the discrimination capabilities of a team, we have introduced the concept of a "Team Relative (Receiver) Operating Characteristic (ROC) curve". In this paper, we extend earlier results in the literature, and show that the area under the team ROC curve can be used as a measure of expertise of teams. To predict the expertise of actual teams, the normative model is tested on teams of humans using a hypothetical medical diagnosis task. Potential human biases leading to discrepancies between the normative predictions and experimental results are identified. A normative-descriptive model is developed to capture the cognitive biases of human DMs. The model provides excellent predictions with respect to the individual and team ROC operating points and the confidence estimates.

Keywords. Decision theory; team theory; man-machine systems; human factors; modeling; Kalman filters; inference processes; optimal estimation.

INTRODUCTION

In a certain class of binary (dichotomous) group decision problems, a set of decisionmakers (DMs), using their own private information and expertise, must make a common decision for the group as to which of two possible events has occurred. This paradigm, termed "Distributed Binary Hypothesis Testing" in Signal Detection Theory (SDT), assumes that all the DMs have their own conditionally independent measurements with respect to the task to be judged. In many real-life situations, however, DMs often have to assess a situation based on the knowledge of different events (hypotheses), which are only probabilistically related to the task itself. A typical example is a clinical judgment task (Johnson and others, 1982), wherein a physician is presented a number of stimuli (such as the results of diagnostic tests: X-ray, blood, etc.), which must be interpreted and combined into a categorical classification (e.g., a tumor judged malignant or benign).

In a recent paper (Pete, Pattipati and Kleinman, 1992b), we have extended normative results in distributed detection theory (Tang, Pattipati and Kleinman, 1991; 1992; Pete, Pattipati and Kleinman, 1992a) to a larger class of realistic dichotomous choice problems. Specifically, our formulation allows for different local hypotheses at each DM, which are only probabilistically related to the global hypotheses and to each other. As with the simplified formulation, the normative decision strategies of DMs are likelihood-ratio tests (LRT) with coupled thresholds. However, the decision thresholds of DMs, which correspond to operating points on the individual Relative (Receiver) Operating Characteristic curves (ROC) of DMs, are functions not only of the prior probabilities of events and of the cost structure, but also of the joint probability distribution of hypotheses. Individual ROC curve of a DM (determined by the detection and false alarm probabilities with respect to the *local* hypotheses) can be transformed into a modified ROC curve (termed the *perceived* ROC curve), determined by the detection and false alarm probabilities with respect to the *global* hypotheses. The two ROC curves are linked by means of a transformation matrix, whose entries are the conditional probabilities of the local hypotheses given the global ones. A consequence of this result is that a DM's opinion can become irrelevant to the team decision (irrespective of the quality of the DM's measurement), if the particular local hypothesis of that DM is independent of the global hypothesis to be judged.

Assessment of an individual DM's discrimination capability by the corresponding ROC curve has a wide body of literature in Engineering and in Social Sciences (Swets, 1986; 1988). We extended these results and showed that: (i) the true expertise of a DM (with respect to the global hypotheses) can be assessed by the perceived ROC curve; and (ii) the aggregation process for the team can be characterized by a new ROC curve, termed the team ROC. The team ROC curve is the locus of the detection and false alarm probabilities, associated with the final decision of the team with respect to the global hypotheses, and it assesses discrimination capability of the team, just as the local ROC curve characterizes the quality of an individual DM. The normatively optimal team ROC curve represents the best possible decision strategy for the team, wherein each DM adapts his strategies to the following elements of the organization: (a) the expertise of the other DMs; (b) the probabilistic relationship of the local and global hypotheses; and (c) the goal of collective decisions.

There are several indices of discrimination accuracy associated with the ROC curves, for a survey, see (Swets, 1986). Following the method introduced in (Swets, 1988), we use the area under the ROC curve as a measure of expertise of a team. This measure allows us to evaluate the effect of changes in the team decision environment (such as the information structure at each DM) on team expertise. A major advantage of this method is that it does not require an underlying normative model. For example, in a binary hypothesis-testing experiment, ROC curves of teams can be obtained, and the measures of expertise can be calculated.

To validate the normative model we developed an experimental paradigm using a hypothetical medical diagnosis task. Our experimental results indicate that the normative model does not provide accurate predictions of human decision performance. Hence, following the normative-descriptive approach (Mallubhatla and others, 1991), we constructed a model to provide realistic predictions of performance of actual human teams in the above decision environment, by introducing psychologically interpretable empirical factors. Specifically, we were interested in predicting team decision accuracy, as characterized by the operating points of the team on the ROC space. Our normative-descriptive model uses a sequential

---

[1] Research supported by NSF grant # IRI-8902755 and ONR contract # N00014-90-J-1753.

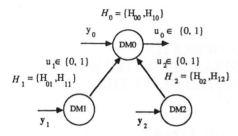

$$H_0 = \{H_{00}, H_{10}\}$$

$y_0 \qquad u_0 \in \{0, 1\}$

DM0

$u_1 \in \{0, 1\} \qquad u_2 \in \{0, 1\}$

$H_1 = \{H_{01}, H_{11}\} \qquad H_2 = \{H_{02}, H_{12}\}$

DM1 $\qquad$ DM2

$y_1 \qquad y_2$

Fig. 1. The team decision scenario.

Kalman filter and incorporates two descriptive parameters: a factor of overweighting more attractive data and a parameter representing conservativeness in the decision. These descriptive parameters are intended to capture the inherent limitations of human DMs. The normative-descriptive model was found to provide excellent match with the experimental data. In this paper, we focus on the experiment and the normative-descriptive model only. For a detailed discussion of the normative model, we refer the reader to (Pete, Pattipati and Kleinman, 1992b).

EXPERIMENT DESIGN

Team structure and the decision environment

In the computer-based experiment we have designed, the team structure and the decision environment are as shown in Fig. 1. The overall objective of the team is to identify infected blood cells. To this end, the subordinate DM1 is provided with measurement y_1, on the amino acid level in the blood. This DM's task is to decide whether the level is normal (H_{01}) or high (H_{11}). Note that the first index of hypothesis H_{qi} represents its type (q=0 or 1), while the second index refers to the DM (j=0, 1 or 2). The measurement at DM1 is a three-dimensional vector corresponding to the normalized levels of three amino acids. Vector [1, 1, 1] represents high amino acid level (hypothesis H_{11}), while [0, 0, 0] corresponds to normal amino acid level (hypothesis H_{01}). Similarly, DM2 is provided with a measurement y_2, on the spectral signals emitted by the DNA of the cell's nucleus. The measurement at DM2 is again a three-dimensional vector corresponding to signal levels at three different frequencies. Spectral components of [1, 1, 1] denote high signal level (hypothesis H_{12}), while levels of [0, 0, 0] correspond to normal signal level (hypothesis H_{02}). The components of measurements of two subordinate DMs are contaminated by independent, identically distributed, zero-mean Gaussian noise with unit variance.

DM0's global hypothesis set corresponds to the global task of identifying whether the blood cells are normal (hypothesis H_{00}) or infected (hypothesis H_{10}). The status of blood cells is also normalized so that a status of 1 denotes hypothesis H_{10}, while a status of 0 corresponds to hypothesis H_{00}. DM0 is provided with a noisy scalar measurement y_0 on the status of blood cells (e.g., physical symptoms of the patient). The conditional probability density functions of this measurement were constructed to be linear: $p(y_0 \mid H_{10})=2y_0$ and $p(y_0 \mid H_{00})=2(1-y_0)$ in the region $y_0 \in [0,1]$.

The local hypotheses at subordinate DMs are assumed to be independent with prior probabilities: $P(H_{11}) = P(H_{12}) = 0.707$; $P(H_{01}) = P(H_{02}) = 0.293$. The local hypotheses are related to the global hypothesis as follows: H_{10} is true if and only if H_{11} and H_{12} are true. Given the priors at the local level and the relationship between the global and local hypotheses, the corresponding prior probabilities of the global hypotheses (those of DM0) can be readily computed as: $P(H_{10}) = P(H_{00}) = 0.5$.

Given the above decision environment, a subordinate DMj makes a decision $u_j \in (0,1)$ on his/her local hypothesis based on the measurement y_j, and associates a confidence c_j with this decision, where $u_j = 1$ denotes DMj's opinion that the local hypothesis is H_{1j}. The confidence estimate is quantized into three levels; that is, $c_j \in (1, 2, 3)$ denoting low, medium and high confidence in the decision made, respectively. DM0 receives the decisions and confidence estimates of the two subordinates, combines them with own measurement y_0, and makes the final decision $u_0 \in (0,1)$ for the team about the status of blood cell, along with an estimate of the confidence c_0 in the decision made. As before, the confidence report is quantized into three levels: low, medium, high.

Independent and dependent variables

Motivated by the normative model predictions (Pete, Pattipati and Kleinman, 1992b), the experiment was designed around three independent variables. The first independent variable is the cost structure imposed on the team. This independent variable was selected to test the prediction that the DMs operate on their ROC curves, and that the DMs adapt their strategies to changes in the team cost structure. A cost, $C_0(u_0=p, H_{q0})$, is imposed only on the final decision - true event pair {($u_0=p$, H_{q0}): p, q = 0 or 1}, but not on the local decisions of the subordinates. The cost of a correct decision is zero; the cost of a missed detection $C_0(u_0=0, H_{10}) = 10$ units; and that of a false alarm $C_0(u_0=1, H_{00})$ is an independent variable taking one of the following three values: 10, 30 and 100. Thus, we considered three different cost structures corresponding to three levels of false alarm cost. The changes in false alarm cost allow us to change the team cost structure in a systematic manner. The second independent variable is the availability of DM0's own measurement y_0 (two possible variations: y_0 available versus y_0 not available). The selection of this independent variable was motivated by the normative model prediction that the subordinate DMs adapt their strategies to the information structure at the primary DM0. The third independent variable is the availability of subordinates' confidence estimates at DM0 (two possible variations: confidence estimates are available versus confidence estimates not available). This independent variable was selected to test the normative result that with confidence estimates available, the decision strategy of DM0 corresponds to an optimal centralized likelihood ratio test, resulting in improved team performance. The dependent variables that were recorded were the actual decisions and confidence estimates of all team members, the resulting detection and false alarm probabilities for a particular set of independent variables, and the team cost associated with DM0's final decision.

Experimental procedure and data analysis

Four teams of three DMs each were formed with paid subjects from the University of Connecticut. The teams were offered bonuses as incentives to strive for best team performance. Each team was subjected to three experimental sessions, corresponding to each of the three cost structures. During each session, the subjects were asked to identify tasks in eight sets of trials, each set containing sixteen individual tasks. The eight sets represented four possible variations of independent variables for a given cost structure (two variations on the availability of y_0 at DM0 x two variations on the availability of confidence estimates at DM0) with one replication. The total number of collected data points (decisions made) was 1536 for DM0 and 3072 for subordinate DMs.

Detection and false alarm probabilities were computed for each set of sixteen decision tasks, and used in a statistical analysis (MANOVA) to determine the statistical significance of the influence of each independent variable on them. Reported confidence estimates with respect to decisions u = 0 were converted into ones with respect to decisions u = 1, averaged over the eight subordinate subjects and the four leaders, respectively, and plotted as a function of the log-likelihood ratio. To make plots of confidence versus the log-likelihood ratio smoother, the abscissa was divided into 24 regions; for each region, the points lying in each region were replaced by their average value. The averaged false alarm and detection probabilities and the averaged confidence estimates were used in the model-data comparisons.

Normative predictions and experimental results

Figures 2-4 show the experimental operating points of the subordinate DMs and those of the teams along with the corresponding normative predictions. Experimental ROC curves were fitted on the three experimental data points, using least squares criterion. The applied analytical ROC curve was that of two Gaussian variables with equal variances and different means for the subordinate DMs (Fig. 2), and the power ROC curve for the team cases (Figs. 3-4). We use these ROC curves to demonstrate the method of assessing individual and team expertise in the next Subsection.

Analyzing the influence of independent variables on the operating points, we can see that our subjects adapted their decision strategies to the team cost structure: there are significant shifts in the operating points on the ROC curve due to changes in the team cost structure (p < 0.05 for subordinate DMs and

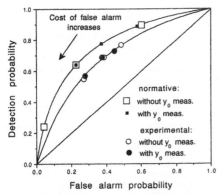

Fig. 2. Operating points of subordinate DMs.

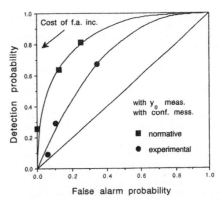

Fig. 4. Operating points of the team with y_0 measurement.

p<0.01 for the team decisions), and the directions of shifts correspond to the predictions of the normative model. However, actual changes in the applied decision thresholds are different from those predicted. Operating points of subordinate DMs are in a cluster at the intersection of the main (NW-SE) diagonal and the experimental ROC curve (Fig. 2). We can attribute this phenomenon to the conservative behavior of DMs. For the team ROC curves, the applied thresholds are not so apparent, because a DM0 is supposed to apply different thresholds for different sets of incoming messages, and the team ROC points are linear combinations of these particular operating points (Pete, Pattipati and Kleinman, 1992b).

The prediction of normative model that subordinate DMs adopt different strategies for different information structures at DM0 was disconfirmed; there was no significant change in the decision strategies of the subordinate DMs with respect to the availability of own measurement y_0 and confidence estimates at DM0 (the corresponding p values are p>0.5 and p>0.16, respectively). As the availability of confidence estimates at DM0 did not improve the team performance, our normative prediction that a primary DM with confidence estimates from subordinate DMs acts as a centralized detector was also disconfirmed.

Clearly, the normative model provides poor predictions with respect to the operating points both for individual DMs and for teams of humans. Furthermore, comparing the predicted and reported confidence messages (a typical pair of curves is shown in Fig. 5), non-monotonicity of experimental confidence curves indicate <u>structural</u> problems in the normative model. Specifically, the optimal batch processing method results in monotone confidence curves with respect to the likelihood ratio value. Non-monotone confidence curves are typically results of sequential data aggregation method, wherein data are ordered and weighted in a non-optimal manner (Mallubhatla and others, 1991). This phenomenon can not be captured by parameter selection in the normative model. Instead, we needed to develop a normative-descriptive model on a different basis.

ROC curve: measure of expertise

Performance of a diagnostic system, faced with binary hypothesis-testing problems (defined in terms of percentage of correct answers) depends: (i) on the inherent discrimination capability ("true expertise") of the system; (ii) on the prior probabilities of the events; and (iii) on the decision criterion applied. The problem of measuring the true expertise was extensively studied by Swets (1988). He assumed that an appropriate measure of expertise of diagnostic systems should be independent of the decision environment, that is, of the last two parameters. He used the area below the ROC curve to measure the expertise of a DM in binary discrimination tasks and illustrated the method on various practical examples. The method is extremely powerful because ROC curves of systems can be obtained in a purely experimental manner, with no further theoretical considerations.

The maximum possible expertise in *any* task corresponds to the area above the minor (SW-NE) diagonal of the ROC space. Let this area be equal to 100 units of absolute expertise. With reference to Fig. 2, for a subordinate DM the normative model predicts an expertise of 54 units, while the experimental value is only 40. Having introduced the team ROC curve, we may extend the above method to assess the discrimination capability of the whole team. Specifically, contribution of a given information source can be quantitatively evaluated. For example, considering Figs. 3-4, we can see that availability of the direct measurement y_0 at DM0 resulted in a normative improvement of 31 units. The effect of different local hypotheses is also apparent. Although in Fig. 3, DM0 relies on <u>two</u> local measurements (availability of the confidence estimates represents sufficient statistics), the normative team expertise is <u>less</u> then the measure of the local expertise of <u>one</u> DM (43 versus 54 units). The reason is that even a noiseless measurement at the local level (with the maximum expertise of 100 units) contains ambiguity with respect to the team task, due to the non-deterministic relationship between the local and global hypothesis-sets.

Non-optimal methods of information aggregation had a serious impact on the expertise of the actual teams: it is consistently less than the normatively optimal expertise. The actual drop in the discrimination capability of the team (as opposed to the

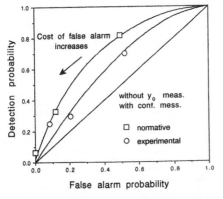

Fig. 3. Operating points of the team without y_0 measurement.

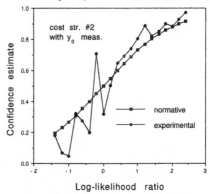

Fig. 5. Confidence curves of subordinate DMs.

129

normatively optimal value) is 17 units in Fig. 3, and 30 units in Fig. 4. The team did not utilize the extra measurement at DM0 properly (the improvement is only 18 units, as opposed to the normative value of 31).

NORMATIVE-DESCRIPTIVE MODEL

Normative-descriptive Model of the Subordinate DMs

Following the earlier studies in human decisionmaking (Pattipati, Kleinman and Ephrath, 1983), we hypothesize that a human DM's decision process can be separated into two sequential phases: (i) information processing (task attribute estimation and hypothesis evaluation); and (ii) decisionmaking (option or alternative evaluation and selection). Our normative-descriptive model assumes that the first phase is independent of the second. Specifically, in phase (i), the posterior probabilities of hypotheses $P(H_{1j} | x_{fj}) = 1 - P(H_{0j} | x_{fj})$ are evaluated, where x_{fj} is an estimate of task attributes. This is a standard estimation problem, where Kalman-filtering techniques have found several successful applications. For references, see (Pete, Pattipati and Kleinman, 1992b).

It has also been shown (Mallubhatla and others, 1991) that human DMs tend to misaggregate the available data. Misaggregation is the nonoptimal sequential revision of subjective probabilities, and can be modeled as a nonoptimal weighting of data in the estimation process. Sequential data processing may be due to sequential presentation of data or human's limited ability to process multiple data in a batch. In our experiment, all the measurements were presented at the same time. In order to gain insight into the method of ordering of data used by humans, we analyzed the experimental confidence curves of subjects. A typical curve of subordinate DMs (such as one in Fig. 5) has sharp peaks and deep valleys at certain values of the log-likelihood function. The positions of these deviations do not change with the independent variables. This means that measurement vectors having the same support to the decision u=1 in the normative model are perceived differently by a human DM.

Normatively, a high measurement value, e.g., y = 1.9, supports hypothesis $H_{1j} = [1, 1, 1]$ more than y = 0.9. However, humans perceived it differently: the closer was a measurement y to one of the possible true values (0 or 1), the more was its subjective validity to a DM, and the more weight was given to it in the sequential estimation process.

The above phenomenon was first discovered by (Kahneman and Tversky, 1973). In a series of experiments, they showed that people predict by "representativeness", that is, they select (or order) outcomes by the degree to which the outcomes represent the essential features of the evidence. Moreover, in the case of multiattribute inputs, the more consistent is the input, the greater is the confidence in that prediction. A sequential data-processing experiment, where multiple inputs of different representativeness ("salience") were applied is reported in (Wallsten and Barton, 1982). It has been shown that data with more representativeness are given more weight in developing a final confidence estimate. Thus, our normative-descriptive model assumes that subjects processed data triplets in the order of their representativeness, wherein the measurement processed earlier is given more weight. We select a descriptive factor $\alpha_s>1$ that accounts for this misweighting of data by subordinate DMs.

The recursive equations of the filter are as follows:

$$\hat{x}_{k+1} = \left[\alpha_s P_k^{-1} + V_{k+1}^{-1}\right]^{-1}\left[\alpha_s P_k^{-1} \hat{x}_k + V_{k+1}^{-1} y_{k+1}\right] \quad (1a)$$

$$P_{k+1}^{-1} = \left[\alpha_s P_k^{-1} + V_{k+1}^{-1}\right] \quad (1b)$$

where P_k is the updated variance of estimate $\hat{x}_k$ and V_k is the measurement noise variance. The filter starts with an initial state estimate x_0, that corresponds to the prior probability distributions of the scalar measurement y. At the end of the estimation process, the output of the filter is a scalar attribute estimate x_f and its estimated variance P_f. The estimate x_f serves as an effective measurement for the decision stage with P_f as the measurement noise variance. This is a standard binary hypothesis-testing problem with different means, but with the same noise variance under the two hypotheses (VanTrees, 1968). A LRT on the "effective" measurement x_f yields the decision of the DM. The confidence estimate c is computed as the ratio of weighted priorities (Pete, Pattipati and Kleinman, 1992b):

$$c(u=1) = \frac{\dfrac{\Lambda(\hat{x}_f)}{\lambda}}{1 + \dfrac{\Lambda(\hat{x}_f)}{\lambda}} = 1 - c(u=0) \quad (2)$$

where $\Lambda(\hat{x}_f)$ is the likelihood-ratio, associated with the effective measurement x_f and λ is the applied decision threshold of the DM.

Normative-descriptive Model of the Primary DM

In the information-processing phase, DM0 combines his/her prior knowledge (i) with the confidence estimate, $c(y_0)$, based on the direct measurement y_0 (if available); and (ii) with the confidence reports c_1 and c_2 of the subordinate DMs. However, there are two major difficulties: (a) the confidence messages of subordinates relate to the local hypotheses H_{q1} and H_{q2}, but not to H_{q0}; and (b) they are not independent given the true global hypothesis H_{q0}, q=0,1. Hence, sequential Kalman filter model may not be applied directly.

Theoretically, the set $\{c_1, c_2\}$ represents sufficient statistic for the measurements $\{y_1, y_2\}$. However, from (2), the evaluation of $\{c_1, c_2\}$ requires DM0 to estimate the applied thresholds λ_j, or the perceived risks, $R_j = \lambda_j * P(H_{1j})/P(H_{0j})$, j=1,2. It can be shown that if subordinate DMs apply unit perceived risk: $R_j=1$, j=1,2, then DM0's confidence in global hypothesis H_{10} (based on the reports of the subordinate DMs) can be obtained as $c_1 * c_2$ (this is a consequence of the AND relationship among the local and global hypotheses). In the normative-descriptive model we decouple the strategies of DMs by assuming $R_j=1$. The resulting confidence estimate: $c_1 * c_2$, contains all the information at the two subordinates and is assumed to be independent of y_0. Hence, it can be applied as an input to the filter.

The confidence update equations, treating the confidence reports as "measurements" (z_k), are:

$$c_0^{(k+1)} = \left[\alpha_0 C_k^{-1} + Q_{k+1}^{-1}\right]^{-1}\left[\alpha_0 C_k^{-1} c_0^{(k)} + Q_{k+1}^{-1} z_{k+1}\right] \quad (3a)$$

$$C_{(k+1)} = \left[\alpha_0 C_k^{-1} + Q_{k+1}^{-1}\right]^{-1} \quad (3b)$$

where $c_0^{(k+1)}$ is the updated confidence estimate at stage k+1 based on $c_0^{(k)}$ and the measurement z_{k+1}; C_{k+1} and Q_{k+1} are the variances of updated confidence estimate and that of the "measurement" noise at stage k+1. As the data pieces for DM0 come in a natural order (first the prior knowledge, then y_0, and finally the messages), the descriptive factor α_0 here refers to primacy ($\alpha_0 > 1$) or recency ($\alpha_0 < 1$), associated with the sequential data processing. Given the conditional probability distribution functions of the measurement y_0, the associated confidence level can be computed via (2) as: $c(y_0) = y_0$. The measurement noise variances Q_k, that is, the variances of y_0 and of the product $c_1 * c_2$, can also be estimated.

Model to capture human conservativeness

Previous work in the literature of behavioral sciences suggest that humans are conservative in their decisions. Healy and Kubovy (1981) found a linear correspondence between the logarithm of the computed optimal decision threshold $\hat{\lambda}$, determined by the prior probabilities of hypotheses and the cost structure, and the logarithm of the decision criterion λ, employed by subjects. The model uses $\lambda = \phi(\hat{\lambda})$ as the applied threshold, where $\phi(\cdot)$ is expected to be linear on a logarithmic scale. We select the slope of this linear function as the second descriptive parameter.

For a DM0 the cost function and the prior probabilities of global hypotheses are known, implying that $\hat{\lambda}$ can be determined. For a subordinate DMj lack of a direct cost imposed on the local decision u_j represents an additional uncertainty. As demonstrated in (Tversky and Kahneman, 1974), in complex decision situations, especially when there are no clear-cut rules to adhere to, human DMs try to simplify their tasks. We hypothesize that the subordinate DMs use a heuristic rule in finding an appropriate cost structure for their decisions. Specifically, a subordinate DM, instead of perceiving the cost structure optimally (as implied by the normative model), simply accepts the global cost structure of DM0, as if it were his own (rule of availability). Hence, the computed threshold is a product of the true risk factor of the global cost structure and the ratio of the local prior probabilities, $\hat{\lambda}_j = R_0 * P(H_{0j}) / P(H_{1j})$, with $R_0 = C_0(u_0=1, H_{00}) / C_0(u_0=0, H_{10})$.

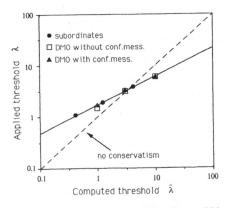

Fig. 6. Conservative decision thresholds of human DMs.

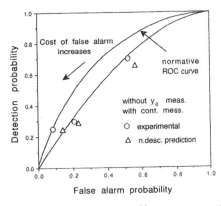

Fig. 8. Operating points of the team without y_0 measurement.

Normative-descriptive model validation

The normative-descriptive model was implemented for each of the 24 sets of trials. Values of the descriptive parameters α_s, α_0, and the applied decision threshold, λ_j in a particular scenario were chosen by trial and error to fit the ROC points and the computed confidence curves to their counterparts from the experiment. The criterion applied was to minimize the mean square error (MSE) between the experimental data and the predictions. Due to symmetry of team structure, we assumed that $\alpha_{s1} = \alpha_{s2}$ throughout. Then, using the least squares method, we constructed the transformation curve ϕ.

Descriptive parameter α_s was found 1.3 for all the three cost structures. It means that subjects, presented with three independent measurements consistently overweighted data with higher representativeness. We could not measure recency (or primacy) effect in the decision behavior of DM0, because the confidence message information (c_1*c_2) had relatively low variance in comparison with that of y_0, hence the decision was dominated by the incoming messages. The reason for this is that the former was based on six independent measurements at the subordinate level. In Fig. 6 we plotted the decision thresholds applied by the DMs versus the computed 'optimal' thresholds. Our results are similar to those in (Healy and Kubovy, 1981), the transformation $\phi(\cdot)$ is linear on a logarithmic scale. The slope of the line is 0.55, suggesting that in this experiment subjects were less conservative than in (Healy and Kubovy, 1981), where the corresponding value is in the range (0.25, 0.35). An interesting feature of Fig. 6 is the fact that the same transformation applies to both the subordinate and primary decisionmakers. Moreover, subjects were biased in their decisions. The predicted value of the criterion applied for a symmetric problem is $\phi(\lambda=1) = 1.7$. As subjects were not confronted with decision tasks, where the cost of a false alarm would have been less than that of a missed detection, they developed an anxiety toward the decision u=1. This behavior manifested itself as a bias in their decision thresholds and as a relatively low level of conservatism.

Comparing the normative results in Figs. 2-4 with those of the normative-descriptive model in Figs. 7-10, it is apparent that the latter matches the data better than the normative model. In Fig. 7 the model captured the characteristics of the subordinate DMs: it exhibits less expertise than that of an optimal processor (the predicted points lie on an ROC curve below the normative ROC curve); and the operating points are clustered around the intersection of the ROC curve with the main (NW-SE) diagonal of the ROC space. In Figs. 8-9, we provide predictions of the decision behavior of the primary DM. As DM0 makes the final decision for the team, these operating points are the team ROC points. The normative - descriptive model yielded good predictions for the team operating points, as well. As in the case of subordinate DMs, the team has less expertise than the normatively optimal centralized detector (corresponding to the normative predictions in Figs. 3-4). Sequential data processing in the model resulted in non-monotonic confidence curves as functions of the log-likelihood ratio. The predicted confidence estimate curve in Fig. 10 has peaks and valleys at the same log-likelihood ratio values as the reported ones. This means that the model also captured human bias of misaggregation of data.

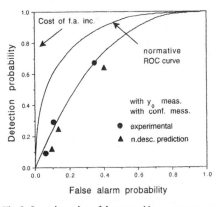

Fig. 9. Operating points of the team with y_0 measurement.

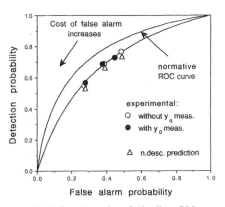

Fig. 7. Operating points of subordinate DMs.

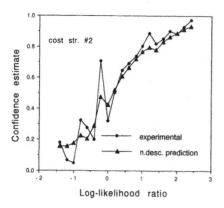

Fig. 10. Confidence curves of subordinate DMs.

DISCUSSION

In this paper, we have studied a generalized distributed binary hypothesis-testing problem, wherein the DMs may have different views of the decision environment. Introducing the concept of "team ROC", we extended the method of characterizing discrimination capabilities of individuals to those of teams. Using the area under the team ROC curve as the measure of team expertise, we assessed actual expertise of human teams along with the impact of changes in the decision environment.

The ROC method of characterizing expertise of teams can be applied both on the normative and on the experimental ROC curves. To construct an experimental ROC curve, however, is time consuming and requires a special testing environment. Therefore, we were interested in constructing a mathematical model to predict ROC points of teams in a class of binary decision tasks. A hypothetical medical diagnosis experiment was designed to test the predictions of the normative model. Results showed that the normative model had structural deficiencies in predicting the actual human decision behavior. Although the optimal data processing is in the form of batch processing, humans apply a non-optimal sequential data aggregation method. As a consequence, reported confidence curves are non-monotonic (as opposed to the monotonic predictions of the normative model), and actual expertise of a team is less than predicted.

To capture the inherent limitations and cognitive biases of actual human DMs, a normative-descriptive model was developed, wherein psychologically interpretable (descriptive) factors are quantified and incorporated into a more suitable model to produce better predictions of the experimental data. Misaggregation of the data is modeled by a factor representing the nonoptimal weighting of independent measurements. Specifically, humans put more weight on data with higher representativeness (being closer to one of the possible true values). Following earlier results in human decisionmaking literature, the normative-descriptive model assumes that in cases of uncertainty, human DMs use heuristic rules instead of complicated computations; subordinate DMs consider the team cost structure as if it was their own. The total decision risk is assessed in terms of a decision threshold, while human DMs' conservatism manifests itself in a transformation between the optimal and actual thresholds. Having determined this function, the normative-descriptive model is capable of predicting the ROC operating points of human teams in distributed binary hypothesis-testing tasks.

REFERENCES

Healy, A.F., and M. Kubovy (1981). Probability matching and the formation of conservative decision rules in a numerical analog of signal detection. *J. Exp. Psych. Learn. Mem. Cogn.*, 7, 344-354.

Johnson, P.E., F. Hassebrock, A.S. Duran, and J.H. Moller (1982). Multimethod study of clinical judgment. *Org. Behav. Hum. Perf.*, 30, 201-230.

Kahneman, D., and A. Tversky (1973). On the psychology of prediction. *Psych. Rev.*, 80, 237-251.

Mallubhatla, R., K.R. Pattipati, D.L. Kleinman, and Z.B. Tang (1991). A normative - descriptive model for a team in a distributed detection environment. *IEEE Trans. Sys. Man Cybern.*, 21, 713-725.

Pattipati, K,R., D.L. Kleinman, and A.R. Ephrath (1983). A dynamic decision model of human task selection performance. *IEEE Trans. Sys. Man Cybern.*, 13, 145-166.

Pete, A., K.R. Pattipati, and D.L. Kleinman (1992a). Optimal team and individual decision rules in uncertain dichotomous situations. *Public Choice*, in press.

Pete, A., K.R. Pattipati, and D.L. Kleinman (1992b). Team relative operating characteristic: a normative-descriptive model of team decisionmaking. Submitted to *IEEE Trans. Sys. Man Cybern*.

Swets, J.A. (1986). Form of empirical ROCs in discrimination and diagnostic tasks: implications for theory and measurement of performance. *Psych. Bull.*, 99(2), 181-198.

Swets, J.A. (1988). Measuring the accuracy of diagnostic systems. *Science*, 240(3), 1285-1293.

Tang, Z.B., K.R. Pattipati, and D.L. Kleinman (1991). Optimization of distributed detection networks: Part I - Tandem Structures. *IEEE Trans. Sys. Man Cyb.*, 21, 1044-1059.

Tang, Z.B., K.R. Pattipati, and D.L. Kleinman (1992). Optimization of distributed detection networks: Part II. Generalized tree structures. To appear in *IEEE Trans. Sys. Man Cyb.*, Nov./Dec., 1992.

Tversky, A., and D. Kahneman (1974). Judgment under uncertainty: heuristics and biases. *Science*, 185, 1124-1131.

VanTrees, H.L. (1968). *Detection, Estimation, and Modulation Theory*, Vol. I. John Wiley & Sons, New York.

Wallsten, T.S., and C. Barton (1982). Processing probabilistic multidimensional information for decisions. *J. Exp. Psych. Learn. Mem. Cogn.*, 8, 361-384.

ADAPTATION TO UNEXPECTED VARIATIONS OF AN INERTIAL LOAD IN GOAL DIRECTED MOVEMENTS

R. Happee

Man-Machine Systems Group, Delft University of Technology, Faculty of Mechanical Engineering and Marine Technology, Mekelweg 2, 2628 CD Delft, The Netherlands

Abstract. Human motor behavior has been studied experimentally in goal directed movements. At the movement onset the inertial load to be displaced was changed unexpectedly. The adaptation of movement and muscular activity have been described with a moving average model. Significant adaptation effects were demonstrated in the first two or three movements in the new condition. Adaptation only partly compensated the mass effects: A higher mass led to a persistent reduction of movement velocity. Amplitudes of muscular activity showed no adaptation of muscular effort, only activation durations were modified. Thus the hypothesis that adaptation pursues a certain movement trajectory as a function of time had to be rejected. However, after scaling towards peak velocity, a global shape invariance was demonstrated in the movement trajectory.

In the first movements after a change of mass effective and substantial modifications of muscular activity appeared about 90 milliseconds after movement onset. Earlier modifications suggest a force feedback enlarging the disturbance. Such force feedback may however increase system bandwidth as it will allow increased position/velocity feedback gains.

Keywords. feedback, adaptive control, human movements.

INTRODUCTION

This paper concerns human motor behavior in changing conditions. We will study both feedback and adaptation. Feedback will be defined as a direct effect of sensory information on neural inputs while adaptation has an indirect effect also affecting behavior in future movements. (This definition of adaptation overlaps common definitions of learning.) A plausible interpretation of adaptation is that an internal representation exists in the central nervous system containing knowledge of the task, the environment and the body (Stassen and co-workers, 1990). Sensory information will be used to adapt the internal representation. This internal representation will then be used to adjust the control strategy.

That proprioceptive information affects muscular activity in fast goal directed movements is demonstrated in perturbation experiments (Wadman and co-workers, 1979; Hallet & Marsden, 1979; Smeets and co-workers, 1990, 1991) and by tendon vibration experiments (Hannaford, Cheron & Stark, 1985; Inglis & Frank, 1990). Jeanerod and Prablanc (1983) state that the presence of visual information contributes to the accuracy in movements lasting longer than 250 ms. These observations indicate that feedback is effective also in fast movements. However, due to neural delays, the first phase of fast movements is unaffected by sensory information and is thus under feedforward control. This feedforward phase is about 100 ms (Wadman and co-workers, 1979). This is substantial as movement durations can be as low as 150 ms.

Contrary to current robots, the human arm can accurately control objects of much larger weight than the arm itself. It is obvious that a feedforward control should be adapted to the mass to be displaced. Also, feedback parameters should be adapted to maintain stability and a reasonable performance. In daily life often objects have to be displaced of which the mass is not accurately known. So, the displacement of an unknown mass is a realistic movement condition in which adaptation and feedback effects can be expected.

Movement kinematics are often found to have certain invariant characteristics or a shape invariance (Soechting & Laquaniti, 1981; Ruitenbeek, 1984, 1985; Gielen and co-workers, 1985). Position, velocity and acceleration traces as a function of time are reported to have an identical shape for movements of a different amplitude and/or velocity or even direction (Flash, 1990). The shape is invariant; but duration and amplitude are scaled to match the movement objective. This concept has also been linked to actual control schemes; it is then assumed that this standard trajectory is a time-dependent desired position/velocity which is pursued by low level feedback eventually combined with feedforward control (Ruitenbeek, 1985). In changing conditions the goal of adaptation can be to retain the desired trajectory. The current paper focuses on the goal of adaptation in terms of shape-invariance. A second aspect studied is the temporal course of adaptation and whether feedback and adaptation can be separated in time. Both maximally fast and relatively slow movements will be considered. In maximally fast movements, inputs are (near-) maximal. Due to this constraint, certain adaptation goals may not be reached. In slower movements, a more complete adaptation is to be expected.

METHODS

Experiments. Four healthy male subjects (average age 23 years) volunteered to participate in the study after being fully informed about the experiment. All, except one subject showed right-side dominance.

With their right hand, the subjects controlled a manipulator with one degree of freedom; a forward or backward movement of the hand. The arm was in the sagittal plane with a roughly horizontal forearm. Both the displacement of the manipulator and a target position were presented on a display. The manipulator has been built as a linear electro-hydraulic servomotor which has the characteristics of a damped mass-spring system. The parameters of the manipulator can be varied over a large range. A more elaborate description of the setup can be found in Ruitenbeek and Jansen (1984). In the experiment, the target jumped to another position after random time intervals. In this way, a step tracking task was executed, resembling a goal directed movement.

A stepsize of 0.2 meter was chosen and both forward and backward movements were recorded. Visual feedback was continuously present during movements which yields a natural movement condition to test adaptation. Unexpectedly, at the onset of some movements the virtual mass to be displaced was changed from 0.6 to 5.6 kg or back to 0.6 kg. Thus two conditions and two transitions (a rise or drop of mass) were studied. The sequence of mass as a function of movement number was generated with a random generator. The sequence was checked to have a zero correlation between mass and movement direction. The sequence was also checked to have a negligible mass auto-correlation over three movements. The random generator was run a few times until the sequence met these checks. Between one and five movements were performed in the same condition. This is acceptable as previous research showed that adaptation processes are observable only in a few movements after the change of the movement condition. This procedure has the advantage that a large number of transitions is studied in one experimental run. This is very important as adaptation effects are of the same order of magnitude as the natural variability (Ruitenbeek, 1984).

First, subjects were trained by verbal feedback to make movements with a varying mass with a peak velocity of about 0.8 ms^{-1}. Then, hundred movements with varying mass were recorded at this moderate velocity. Thereafter a similar run was performed with the instruction to move maximally fast. Finally thirty movements with a constant low mass and thirty movements with a high mass were recorded with a maximal movement velocity.

Range effects cannot be avoided in such experiments; subjects can execute their task in a way adapted to intermediate values in the range of expected conditions although in this experiment a task with intermediate conditions is never executed. The experiments with a constant mass were executed to determine in how far the behavior depended on the mass uncertainty. In this task, the possibility exists that subjects recognize the two mass conditions and switch to a prepared program thus reducing the adaptation task to a recognition and switching task. However, the subjects appeared not to be aware of the fact that only two conditions occurred.

Data collection. The surface electromyogram (EMG) was recorded of twelve shoulder muscles of which the biceps also spans the elbow (Table 1). This large number of muscles was recorded to gain insight in the coordinated control of shoulder muscles. However, in this paper we will not treat muscular activity in great detail. Most muscles were studied in all subjects, however some muscles were not recorded in all subjects or the data were discarded because of movement artifacts. The resulting number of subjects per muscle is shown in Table 1. Bi-polar EMG was recorded with disposable Ag/AgCl electrodes with an inter-electrode distance of 23 mm directly attached to a pre-amplifier. The signal was bandpass filtered from 20-1500 Hz, rectified and low pass filtered (100 Hz). These EMG signals were sampled at 250 Hz together with the position and velocity of the hand and the force exerted by the hand.

TABLE 1. Muscles and functions as apparent from EMG patterns:

+ contributing to a forward force, or acceleration
− contributing to a backward force, or acceleration
* only clear as agonist, not as antagonist in other movement direction
b both directions, agonist
c continuing activity during agonist and antagonist phase
? function not clear

muscle	(subjects)	EMG	remarks
trapezius descendens	(3)	+ *?	scapular muscle
trapezius transversalis	(3)	− *?	scapular muscle
trapezius ascendens	(3)	− *?	scapular muscle
serratus anterior	(4)	+	scapular muscle
infraspinatus	(3)	b	rotator cuff muscle
anterior deltoid	(4)	+	prime mover
medial deltoid	(3)	?	abductor
posterior deltoid	(4)	−	prime mover
latissimus dorsi	(4)	−	prime mover
pectoralis major clav.	(4)	+	prime mover
pectoralis major thor.	(4)	+	prime mover
biceps	(3)	− c	bi-articular; shoulder elbow

Data analysis. In the EMG of goal directed movements, three relevant phases can be found (Wadman and co-workers, 1979; Gielen and co-workers, 1984, 1985; Marsden and co-workers, 1983; Flament and co-workers, 1984; Happee & Daanen, 1991, 1992). Agonist activity in the first phase causes acceleration of the limb. Antagonist activity in the second phase serves to decelerate the limb. The agonist activity in the third phase is shown to be relevant in fast movements (Hannaford and Stark, 1985, 1987; Wierzbickca and co-workers, 1986). In Happee (1992a) it is shown that a third phase is required for a time optimal movement: The third phase serves to compensate the slowly decaying antagonist force.

From the EMG, timing and average amplitudes of the three phases were determined. This was done per movement and per muscle with a statistical detection method (Happee & Daanen, 1992). Parameters are defined in Table 2.

In Happee (1992b) a method has been presented to analyze movement kinematics (position, velocity trace). From the kinematics, five parameters can be estimated. Four parameters describe a triphasic input pattern. These parameters correlate highly to the relevant EMG parameters. A fifth parameter represents neuromuscular dynamics. These five parameters are defined in Table 2 where also parameters scaled to movement velocity are defined. The scaled parameters provide a means to compare the shape of movement traces under different experimental conditions.

Thus the EMG and the movement kinematics are described globally by two sets of parameters (Table 2). For these parameters, with a linear regression, a dynamical model is made of the response to a changing mass. The regressors used are given in Table 3. A static effect of mass is described with the regressor M which is the mass in the current movement. Adaptation is modelled with moving average regressors for the change of mass relative to the previous three movements DM_i. To describe asymmetric responses for a rise and a drop of mass, the regressors $Rise_i$ and $Drop_i$ are used. If these asymmetric regressors were significant it was tested whether they remained significant when the symmetric regressor DM_i was forced into the equation. Only if the asymmetric regressor remained significant it was assumed to be relevant and was presented. Besides these mass related regressors some other regressors were used: the movement direction (dir), the movement number (trend), and per subject a parameter which is one for this subject and zero for other subjects. These regressors mainly served to reduce the variance and thus enhance the significance of the mass regressors. Only the mass regressors and the movement direction (dir) will be presented. Regressors were entered stepwise until a significance level of 1% was reached.

The parameters in Table 2, describe the EMG and the movement kinematics. However, this description is too global to reveal the latency and the nature of the first effects of feedback. These aspects will be studied from plots of average EMG traces. This averaging was performed after alignment in time to the onset of the EMG of agonists which showed a clearly detectable activity in all subjects. For anteflexion, the average onset of the anterior deltoid and the pectoralis major clavicularis was used and for retroflexion the posterior deltoid and the latissimus dorsi.

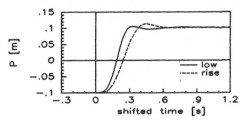

Fig. 1a. Fast movements with a stationary low mass (-) and first movements after a rise of mass (..).

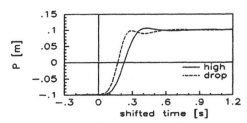

Fig. 1b. Fast movements with a stationary high mass (-) and first movements after a drop of mass (..).

RESULTS

Already the first disturbed movements, arrived at the desired position within reasonable time, be it with a movement trajectory different from previous movements (Fig. 1).

Kinematic parameters. In the linear regression it appeared that most kinematic parameters were significantly and relevantly affected by the mass to be displaced (Tables 4a,b). The coefficients for M indicate a persistent dependence on the mass. Thus these coefficients describe the final result of adaptation. In both fast and slow movements, a higher mass led to increased durations of the different movement phases and to a subsequent reduction of velocity. The coefficients for DM, $Rise_i$ and $Drop_i$ indicate temporary effects after the change of mass. Thus these coefficients describe adaptation dynamics. Most of the dynamic parameters in Table 4 concern the first and second movement after a change of mass. Only for the velocity of slow movements, adaptation seems to extend to the third movement. For a number of parameters, the regressors $Rise_i$ and/or $Drop_i$ are significant (Table 4). This demonstrates a difference in the temporal course of the adaptation process after a rise of mass and a drop of mass. The regression parameters for the peak velocity V_{max} are depicted in Fig. 2. Here it can be seen that V_{max} does not return fully to the original value. This is not surprising for maximally fast movements where probably bounds on input amplitudes limit movement velocity. However, in the experiment at submaximal velocity, a velocity independent of the mass could have been obtained. Thus no support is found for the hypothesis that adaptation pursues a certain desired peak velocity. The last five parameters in Table 4 describe the shape of the movement trajectory after scaling to the movement velocity. For these dimensionless parameters, adaptation yields static values independent of the mass to displace. This follows from the fact that none of these parameters depends significantly on the regressor M (current mass). Thus it can be concluded that the adaptation process contributes to a shape invariance rather than a certain movement velocity.

TABLE 2. Parameters evaluated from EMG and kinematics

EMG, evaluated for each muscle separately	
T_d	delay from stimulus to first activity
T_a	duration first phase (agonist)
T_b	duration second phase (antagonist)
T_c	duration third phase (agonist)
A_a	amplitude first phase (agonist)
A_b	amplitude second phase (antagonist)
A_c	amplitude third phase (agonist)
kinematic parameters, index: k=kinematic, s=scaled	
V_{max}	peak velocity
T_{dk}	delay from stimulus to movement
T_{ak}	duration first phase (agonist)
T_{bk}	duration second phase (antagonist)
T_{1k}	neuromuscular time constant
R_c	magnitude of the third phase relative to the first phase
T_{aks}	$T_{ak}*V_{max}*0.2/D$ (D=stepsize which is 0.2 in this paper)
T_{bks}	$T_{bk}*V_{max}*0.2/D$
T_{1ks}	$T_{bk}*V_{max}*0.2/D$

TABLE 3. Regressors, n=movement number.

dir	movement direction, 1=anteflexion (forward), -1=retroflexion	
M	M(n), the current mass, 0=low, 1=high mass	
DM_i	M(n)-M(n-i), mass change	with i=1,2,3
$Drop_i$	MAX($-DM_i$,0)	with i=1,2,3
$Rise_i$	$Rise_i$=MAX(DM_i,0)	with i=1,2,3

TABLE 4a. Slow movements, regression coefficients of dynamic response to mass change. Paramameters defined in Table 2, regressors in Table 3.

parameter	regression coefficients significant with p < .01 for t-test with separate variance analysis
V_{max} [m/s]	.913 -.127M -.028DM_1 -.033DM_2 -.025DM_3 +.023dir
T_{dk} [ms]	337 +25$Drop_1$
T_{ak} [ms]	149 +29M +37$Rise_1$ +25$Rise_2$
T_{bk} [ms]	214 +54M -6dir
T_{1k} [ms]	150
R_c	.29
T_{ak}/T_{bk}	.77 +.16$Rise_1$ +.08$Drop_1$
T_{aks}	153 +26$Rise_1$
T_{bks}	213
T_{1ks}	133 + 19dir

TABLE 4b. Maximally fast movements, further as in Table 4a.

V_{max} [m/s]	1.590 -.373M -.201$Rise_1$ +.098$Drop_1$ +.0421dir -.070DM_2
T_{dk} [ms]	315 +37$Drop_1$
T_{ak} [ms]	104 +29M +4dir
T_{bk} [ms]	152 +52M +9DM_2 -14dir
T_{1k} [ms]	91 +68$Rise_2$
R_c	.33 +.05dir
T_{ak}/T_{bk}	.73 -.05DM_2 +.08dir
T_{aks}	161 -14DM_2 +10dir
T_{bks}	236 -6DM_1 -12dir
T_{1ks}	140 +27dir

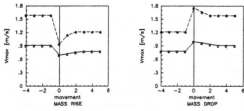

Fig. 2. Response of peak velocity to change of mass, movement 0 is the first movement with a new mass.

These results for an unexpectedly varying mass were compared with parameters from experiments with a constant mass. This comparison was only made for the well defined task of maximally fast movements. The static behavior following from Table 4b was in close accordance with parameters estimated for movements made with a constant mass. Only for a constant mass the ratio T_{ak}/T_{bk} was .77 for a low mass and .68 for a high mass (p < .007). Similar values were found in the experiment with varying mass but there the difference was not significant. The ratio T_{ak}/T_{bk} is dimensionless and thus must be constant with an invariant shape. The significant difference in the constant mass experiment therefore contradicts the shape invariance indicated by the experiment with varying mass. Still this difference is rather small compared to the effects on movement velocity and duration.

EMG parameters. From the EMG, timing and amplitude parameters were computed for twelve muscles. These parameters have been analyzed, except for those of the trapezius and the medial deltoid in which no consistent pattern was observed in the subjects studied now.

EMG timing showed qualitatively the same behavior as the parameters T_{dk}, T_{ak}, T_{bk} in Table 4 (which were estimated from movement kinematics). Agonist and antagonist durations increased with a higher mass. The static dependence of EMG timing on the mass was nearly always significant but the dynamic regressors DM_i, $Rise_i$, $Drop_i$ mostly did not meet the 1% significance level.

EMG amplitudes were largely affected by the instruction regarding movement velocity. Compared to fast movements, in slow movements agonist amplitudes were about forty percent and antagonist amplitudes were about fifty percent. The mass variations had far smaller effects on EMG amplitudes; effects on the agonist amplitudes were insignificant and very small for all prime movers.

Plot of EMG and kinematics. From the parametric analysis of the EMG, a variation of timing rather than amplitude emerged. However, these parameters describe durations of about 60-150 ms and average amplitudes over these durations. A more detailed analysis of the first phase of the EMG will be given hereafter to assess the first effects of the mass variations.

Figure 3 describes the activity of two agonist muscles for a high mass, just after a drop of mass and for a low mass. It can be seen that, after a drop of mass, activity is first similar to the undisturbed activity. Between t=30 and 70 ms, activity is enlarged (p < .007 for two-sided t-test). After t=70 ms, activity is reduced (p < .001).

After a drop of mass, displacement and velocity will be larger than expected. Thus feedback of displacement and/or velocity would naturally lead to a reduction of inputs. This explains the later reduction of agonist activity. However, the early enlargement is quite unexpected. This finding unfortunately is not fully consistent over muscles and subjects.

After a rise of mass a similar reaction was found; first a reduction of agonist activity and later an enlargement. Again, the first reaction is unexpected and the second reaction is expected. Unfortunately again this finding was not fully consistent; only in the two parts of the pectoralis it was

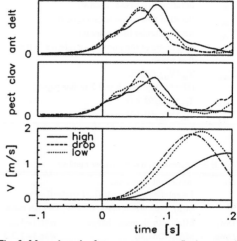

Fig. 3. Mass drop in fast movement, anteflexion, agonist EMG (upper traces), velocity (lower trace), stationary low mass (—), first movement after mass drop (--), stationary high mass (..).

136

observed in all four subjects. This unexpected activity takes place as little as 25 ms after movement onset.

While these early feedback effects are unexpected and not fully consistent, on average about 90 milliseconds after movement onset, effective and substantial modifications of muscular activity appear which are largely consistent over subjects and muscles.

DISCUSSION

At the onset of goal directed movements the inertial load to be displaced was changed unexpectedly. Already the first disturbed movements, arrived at the desired position within reasonable time, be it with a movement trajectory different from previous movements (Fig. 1). Significant differences were found between the first two or three movements after a change of mass (Table 4, Fig. 2). These differences prove that adaptation takes place which can be argued as follows. After a change of mass the dynamics of the arm plus load remain constant. The fact that movement parameters change must therefore be caused by an adaptation of the control strategy.

It can be argued that the successful completion of the first disturbed movement is (mainly) due to feedback. This implies relatively fast feedback combined with slower adaptation which is common in control theory. Such an adaptation could even be a discrete process in which information from one movement is used to adapt the control strategy for following movements. However, alternatively, adaptation may well be effective already within the first disturbed movement. An argument for such a fast adaptation is given by Smeets and co-workers (1990) who concluded that modifications of muscular activity in the first movements with a changed mass could not be explained by linear position or velocity feedback. Smeets concludes on a more complex use of sensory information already in the first disturbed movement. This he calls reprogramming which well falls within our definition of adaptation.

As to the nature of the adaptation process we have the following observations: Subjects can increase movement impedance by co-activation of antagonist muscles and enlarging of feedback gains. This is a well documented adaptation mechanism to uncertain conditions (Winters and co-workers, 1988, Hogan, 1990). Yet in our experiment co-activation is negligible. We therefore, consider only the following aspects to be possibly adapted:
1) a load parameter; the mass,
2) feedback gains,
3) feedforward terms (motorprograms).
Of course, adaptation of a load parameter is only useful if serving to adapt feedback or feedforward controls. Adaptation of a load parameter is the most powerful option: with one parameter the control of all degrees of freedom in the arm can be adapted. Using the "knowledge" that only the load changes only this global parameter has to be estimated in some way which can be done in a relatively short time. The controls can be adjusted using knowledge of how the system can be controlled given a certain mass. This knowledge can be the result of a much longer learning process. Such an adaptation of the control strategy based on adapted load parameters is a process which requires an internal representation of some kind.

Alternatively, adaptation can directly affect feedback or feedforward control. Direct adaptation of feedback gains could be performed in a global manner. Yet, a global adaptation of all gains is presumably sub-optimal. Direct adaptation of feedback and/or feedforward terms will be rather complex as feedforward terms for several muscles would have to be adapted. It is to be doubted whether such an adaptation could be completed in only a few movements.

Our experiment contained both forward and backward movements. Thus information is provided on in how far adaptation for these directions is a separate process. Finding a separate process would indicate that direction-specific information like a motor program is adapted. In case of a separate process, information from a first disturbed movement would not be used to adapt the second movement which has an opposite direction. Thus behavior would only depend on the current mass M and the mass two movements ago $M(i-2)$ and eventually four movements ago (etcetera). In the regression (Table 4) such a behavior would lead to dominance of the parameters DM_2, $Rise_2$ and $Drop_2$ and absence of DM_1, $Rise_1$, $Drop_1$ and DM_3, $Rise_3$ and $Drop_3$. Table 4 does not show such a dominance, instead a rather continuous adaptation is observed. This indicates a combined adaptation for forward/backward movements.

Concerning the goals of adaptation we have the following observations assuming that adaptation is successful and that thereby the obtained movements reflect the goals of the adaption. We found that after adaptation, the movement velocity still strongly depended on the mass. So, adaptation does not retain the old trajectory. After scaling towards velocity, a shape invariance was demonstrated. This indicates that the adaptation process contributes to a shape invariance rather than a certain movement velocity or duration. Apparently the desired movement velocity is adapted to the load. Here it must be noted that in the experiment with constant low or high mass a slight but significant deviation from the shape invariance was demonstrated. Still our results globally support Ruitenbeek (1984) who concluded on a shape-invariance in various conditions in relatively slow movements. Still we only found the shape to be practically independent of mass. However, in fast movements, the shape does depend strongly on movement direction (Table 4b).

The EMG reflects muscular activity as it results from neural inputs. EMG timing was strongly affected both by the instruction regarding movement velocity and by the mass variations. EMG amplitudes practically only depended on the instruction. Gottlieb and co-workers (1990) relate EMG amplitudes (and acceleration) to instructions regarding movement velocity and not to the desired displacement. Such task dependent input levels may actually be determined by the size of the recruited motoneuron populations. Pre-determined input levels would largely constrain possibilities of adaptation. Trajectories after adaptation may follow from these constraints rather than from goals like a shape-invariance. As stated in the introduction such predetermined input levels were not expected in submaximally fast movements.

In the first movements after a change of mass effective and substantial modifications of muscular activity appear about 90 milliseconds after movement onset. Subtracting a delay between EMG and detectable movement of 20 ms this indicates feedback latencies below 70 ms. This latency allows relatively complex processing by multi-synaptic pathways and thus does not contradict adaptation within the agonist burst of the first movement.

One would expect that with an increased mass, feedback enlarges muscular activity to obtain the desired displacement. With a decreased mass a decrease of activity would be expected. Such effects would follow naturally from a position (and velocity) feedback using for instance information from the muscle spindles. (At least when assuming negative closed loop gains as is required for stabilizing feedback of position and velocity.) However, suggestive evidence was presented that a first effect of feedback has an effect opposite to what could be expected (Fig. 3). Only later the expected effect of feedback was observed. This finding can theoretically be explained by a velocity or acceleration feedback with a (unusual) positive closed loop gain. A more reasonable explanation is the presence of force-feedback with a (usual) negative closed loop gain. After a decreased mass, the resisting force in the hand will be lower than expected. This will lead to an increased acceleration and velocity. Due to the force-velocity relation this will lead to reduced muscle forces. Negative feedback of either the tactile force in the hand or the muscle force could thus explain our findings. Especially the tactile force is strongly and early affected by the load which would also explain the early occurrence of this feedback component. Yet, it has to be noted that this finding is not fully consistent over subjects and muscles.

Such a force feedback seems not to be useful. However, it would help stabilizing the system as has been verified in a simulation study. A model was used with one degree of freedom, two muscles (non-linear, third order) with a sensory delay of 50 ms. For this model feedback was optimized assuming feedback of muscle length and velocity. Adding feedback of muscle force allowed increased position and velocity feedback gains. Thus adding force feedback yielded a doubling of system bandwidth (unpublished results).

EMG was recorded of 12 muscles to gain insight in the coordinated control of shoulder muscles. Data not presented in detail suggest differences between synergist muscles in the adaptation process. However, more experimental data are required to test the significance of these effects.

Manual control. This paper concerns step tracking tasks, where the system to be controlled was a unity gain. Adaptation was studied with changing manipulator dynamics. Changed was the manipulator impedance; the force required to displace the manipulator. In manual control tasks, such changes will occur when using active manipulators. When states of the system to be controlled are fed back on the manipulator, the operator will feel a changing system dynamics as a changing manipulator impedance. In our experiments, subjects had no problem stabilizing the system which is remarkable as the inertia of limb plus manipulator changed with about a factor three. Adaptation was found to be incomplete, manipulator impedance had a persistent effect on the movement velocity (Fig. 2). This incomplete adaptation contrasts the full adaptation to changing system gains as predicted by the cross-over model (McRuer & Jex, 1967).

REFERENCES

Flament D., Hore J. & Vilis T. (1984). Braking of fast and accurate elbow flexions in the monkey, *J. Physiology*, 349: 195-203.

Gielen C.C.A.M., Van den Heuvel P.J.M., Denier van der Gon J.J. (1984). Modification of muscle activation patterns in fast goal-directed arm movements, *J. Motor Behavior*, 16-1: 2-19.

Gielen C.C.A.M., Van den Heuvel K. & Pul ter Gunne F. (1985). Relation between EMG activation patterns and kinematic properties of aimed arm movements, *J. Motor Behavior*, 17- 4: 421-442.

Gottlieb G.L., Corcos D.M., Agarwal G.C. & Latash M.L. (1990). Organizing principles for single joint movements III. Speed-insensitive strategy as default. *J. Neurophysiology*, 63-3: 625-636.

Hallet M. & Marsden C.D. (1979). Ballistic flexion movements of the human thumb. *J. Physiology*, 294: 33-50.

Hannaford B., Cheron G. & Stark L. (1985). Effects of applied vibration on triphasic electromyographic patterns in neurologically ballistic head movements. *Experimental Neurology*, 88: 447-460.

Hannaford B. & Stark L. (1985). Roles of the elements of the triphasic control signal. *Experimental Neurology*, 90: 619-634.

Hannaford B. & Stark L. (1987). Late agonist activation burst (PC) required for optimal head movement: A simulation study. *Biological Cybernetics*, 57: 321-330.

Happee R. (1992a). Time Optimality in the Control of Human Movements. *Biological Cybernetics*, 66: 357-366.

Happee R. (1992b). Goal directed movements II, Kinematic analysis and analysis of the relation between kinematics and EMG. Submitted for publication.

Happee R. & Daanen H.A.M. (1991). Activation patterns of shoulder muscles in goal directed movements. In: Anderson P.A., Hobart D.J & Danoff J.V. (eds). *Proceedings of the 8th Congress of the International Society of Electrophysiological Kinesiology, 12-16 August 1990, Baltimore USA.*

Happee R. & Daanen H.A.M. (1992). Goal directed movements I, Analysis of the EMG in shoulder and elbow muscles. Submitted for publication.

Inglis J.T. & Frank J.S. (1990). The effect of agonist/antagonist muscle vibration on human position sense. *Experimental Brain Research*, 81: 573-580.

Jeanerod M. and Prablanc C. (1983). Visual control of reaching movements in man, *Motor control mechanisms in health an disease*, edited by J.E. Desmedt, Raven Press, New York.

Marsden C.D., Obeso J.A. & Rothwell J.C. (1983). The function of the antagonist muscle during fast limb movements in man. *J. Physiology*, 335: 1-13.

McRuer D.T. & H.R. Jex (1967). A review of quasi-linear pilot models, *IEEE Transactions HFE*, 8-3: 231-249.

Ruitenbeek J.C. and Jansen R.J. (1984). Computer controlled manipulator display system for human movement studies. *Medical & Biological Engineering & Computing*, 22: 304-308.

Ruitenbeek J.C. (1984). Invariants in loaded goal-directed movements. *Biological Cybernetics*, 51: 11-20.

Ruitenbeek J.C. (1985). Visual and proprioceptive information in goal directed movements, a system theoretical approach, *Ph.D. thesis*, Delft, ISBN 90 70879 23 9.

Smeets J.B.J., Erkelens C.J. & Denier van der Gon J.J. (1990). Adjustment of fast goal-directed movements in response to an unexpected inertial load. *Experimental Brain Research*, 81: 303-312.

Smeets J.B.J., Erkelens C.J. & Denier van der Gon J.J. (1991). Perturbations of fast goal-directed arm movements: different behaviour of early and late EMG-responses. *J. Neurophysiology*, submitted.

Soechting J.F. & Laquaniti F. (1981). Invariant characteristics of a pointing movement in man. *J. Neuroscience*, 1-7: 710-720.

Stassen H.G., Johannsen G., Moray N. (1990). Internal representation, internal model, human performance model and mental workload. *Automatica*, 26-4: 811-820.

Wadman W.B., Denier van der Gon J.J., Geuze R.H. & Mol C.R. (1979). Control of fast goal directed arm movements, *J. Human Movement Studies*, 5: 3-17.

Winters J.M., Stark L. & Seif Naraghi A.H. (1988). An Analysis of musculoskeletal system impedance. *J. Biomechanics*, 21- 12: 1011-1025.

MENTAL SET AND COMPLEXITY EFFECTS IN THE DIAGNOSIS OF SIMULTANEOUS FAULTS IN A SIMPLE BINARY ADDER

P.M. Sanderson, T. Appeddu and D.V. Reising

*Department of Mechanical and Industrial Engineering, University of Illinois at Urbana-Champaign,
1206 West Green Street, Urbana, IL 61801, USA*

ABSTRACT

This paper focuses on the challenge to human diagnostic reasoning posed by multiple faults. Two sources of difficulty were tested; mental set and multiple fault complexity. Mental set was manipulated by training subjects on single or multiple faults and then transferring them to multiple fault diagnosis. Half the subjects were transferred to a multiple fault with simple easily separable symptoms and the other half to one with more complex highly interactive symptoms. Results suggest that mental set and multiple fault complexity affect the correctness, latency and subjective workload of transfer multiple fault diagnosis, but not the number or range of tests made before the final diagnosis. Inability to diagnose the transfer multiple fault was wholly due to (1) failure to explore the symptom space so as to see the sufficient and necessary evidence to diagnose the multiple fault (2) failure to consider the possibility of multiple faults, or (3) both.

INTRODUCTION

Over the last 15 years there has been a growing amount of interest in the issue of what makes faults in systems difficult to diagnose (Gitomer, 1988; Rasmussen and Rouse, 1981; Sanderson and Murtagh, 1990). One likely reason is the cognitive complexity of diagnosing some types of faults. To pursue this, we seek to establish how much more difficult two or more simultaneous faults in a system ("multiple faults") are to diagnose than non-simultaneous faults ("single faults"). With some multiple faults the evidence for the component faults exists as independent, non-overlapping sets of symptoms ("separated fault symptoms"), whereas for other multiple faults the evidence is interdependent and overlapping ("integrated fault symptoms"). We hypothesize that fault diagnosis will be considerably easier in the first case than the second case (Mendenhall, 1990).

Another possible source of difficulty in diagnosing faults is having to break a mental set in order to successfully diagnose a fault (Luchins and Luchins, 1991). We examined subjects' diagnostic reasoning when encountering multiple faults after experiencing either a series of single faults or a series of multiple faults. We hypothesized that subjects transferring from a series of single faults to a multiple fault problem would find it more difficult to diagnose the multiple fault at transfer (Appeddu, 1991). However subjects transferring from a series of multiple faults to another multiple fault would experience no extra difficulty. We wished to explore the source of this difficulty and to see how might it interact with the type of multiple fault (separated or integrated symptoms).

EXPERIMENTAL TOOLS

Binary Adder Task

These hypotheses were tested using an interactive simulation of a binary adder. The binary adder was chosen because a complete analysis of all possible faults, and all symptoms of these faults, could be identified, making it possible to identify the "problem space" in which subjects worked, and to judge the quality of their performance against certain logical standards. Subjects could decide upon inputs to the adder and view the resulting output (see Figure 1). The normality, or abnormality, of these input-output relations served as clues to the location of a break, or breaks, in the binary adder itself. Subjects were required to make test inputs to the adder and to register their diagnoses of the symptoms they saw as each problem solving session progressed.

Ideal Problem Solver and Diagnosis Model

In order to judge the quality of subjects' fault diagnosis performance, a computer-based "Ideal Problem Solver" (IPS) was programmed which could take each human subject's data file and work out which possible diagnosis solutions had been eliminated, and which were still possible, after each test of input-output relations made by the subject (Holden, 1990). A model of fault diagnosis was developed which divided diagnostic reasoning into three phases and is shown in Figure 2. Many of the data analyses performed were specific to individual phases and so it is necessary to understand the logic behind them.

In this model there are three major milestones after the beginning of a fault diagnosis episode: the logical isolation of the fault solution through testing (as revealed by the IPS), the selection of a final fault diagnosis solution, and the decision to terminate the fault diagnosis episode. These milestones divide the fault diagnosis trial into three phases: the period of testing and diagnosis before the fault's location is logically isolated (A), the period between fault isolation and the final fault diagnosis, during which further testing and diagnosis occur (B), and the period after final fault diagnosis during which further testing but no further diagnosis is performed (C). Figure 2 shows some of the pathways to incorrect diagnoses.

METHOD

Design and Procedure

The experimental design is shown in Table 1. Six groups of five different subjects were allocated to six experimental conditions created by crossing learning group and multiple fault at transfer. Short (SS) and long (SL) single fault conditions were used as ways of handling experimental control problems arising from the relative scarcity of unambiguous single faults. In the SS condition, subjects experienced four unique single faults on Day 1 and a different four on Day 2. In the SL condition, subjects completed as many unique single faults as they could on both days. The initial faults on Day 2 were the ones least likely to have been completed on Day 1, but the possibility of repeats was present. In the ML condition, subjects completed as many unique multiple faults as possible, and there were no repeats.

At the end of Day 2, all subjects transferred to either a hard or an easy multiple fault. The multiple fault FS was chosen as the easier fault. (See Figure 1 for location of F and S faults). Here, the subject sees two states symptomatic of the F fault, six that are symptomatic of the S fault, and two that combine these symptoms separately. Thus the effect of F and of S are quite separate,

reproducing local symptoms similar to those that would be seen if F alone or S alone were the fault.

PS was chosen as the harder fault (see Figure 1 for location of P and S faults). With PS, the subject will see three states which could be P or S and four states which could only be S. There is only one state which gives evidence for both the P and S faults in this circuit. Because this state cannot be due to S alone and because other states have indicated the possible involvement of P, the subject must infer that there must be faults both at P and at S. Clearly, the deduction required by the diagnostician is far more complex in this case than the former.

All subjects experienced two one-hour training sessions on successive days on either single or multiple faults before transferral to the multiple fault problem at the end of Day 2. After each trial on Day 2, subjects were asked to rate their mental workload, using the NASA TLX workload scales (Task Load IndeX: Hart and Staveland, 1989). After the last trial, subjects were given a questionnaire in which they were asked how many faults they had assumed there could be on each trial, followed by a series of other questions about their fault diagnosis strategies.

Subjects

Thirty Electrical Engineering undergraduate students who had taken basic courses in digital logic and digital circuit design were used as subjects and were paid $4 per hour for their participation. They were not given any predisposition about whether the experiment would present them with single or multiple faults.

RESULTS

Subjects who trained with single faults were significantly less likely than those who trained with multiple faults to correctly diagnose the multiple fault at transfer (see Table 2). All cases of failure to correctly diagnose the multiple fault at transfer could be attributed to one of the following:

(1) the subject's failure to provide themselves with the necessary evidence during testing to diagnose the fault ("lack evidence")
(2) the subject's belief (using retrospective questionnaire) that there could only be single faults in the adder ("lack hypothesis")
(3) both the above ("both").

When the multiple fault at transfer had integrated fault symptoms (PS), subjects were generally less likely to diagnose it correctly than when it had separated fault symptoms (FS). This effect was stronger for subjects trained on single faults than for those trained on multiple faults. These results support our initial hypotheses.

However Table 2 makes it clear that further practice on single faults helps a couple of subjects in the SL group diagnose the PS fault.

In many cases where subjects failed to correctly diagnose the multiple fault they provided a single fault diagnosis instead. In the FS fault, for subjects trained in single fault diagnosis the temptation was to stop testing after seeing the F fault in the half adder at the right. One of the two FS subjects who diagnosed incorrectly failed to see the evidence for S. In contrast, with the PS fault as many as five of the eight subjects who failed to diagnose the PS fault actually saw the evidence for it, but because of the fault's complex, integrated nature the subjects failed to recognize that this evidence meant there were two faults present.

The factors of type of training and type of transfer fault also affected subjective measures of diagnosis difficulty at transfer. Subjects were asked to rate their mental workload, using the NASA TLX workload scales (Task Load IndeX: Hart and Staveland, 1989). Results revealed that subjects showing the highest level of mental workload were those who had trained with single faults and who had transferred to the multiple fault with integrated fault symptoms (see Figure 3).

Another important indicator of fault difficulty at transfer is time taken to diagnose the transfer fault. Figure 4 shows the time to complete each trial for the S1 and ML groups on Day 2 training and transfer trials. It is clear that SL subjects in both groups show an increase in time. The ML/PS subjects shows a very slight increase and the ML/FS subjects show a decrease in time.

We also analyzed the time subjects took in each phase of fault diagnosis (A, B, and C) according to the model shown in Figure 2. Results for all subjects who *correctly diagnosed* the transfer fault are shown in Figure 5 (results for SS group are not shown because an insufficient number of subjects correctly diagnosed the transfer fault). The results show that ML subjects tend to diagnose faster overall and that isolation time (A) takes longest, diagnosis time (B) somewhat less time and checking time (C) the least time of all. The effect of multiple fault training upon transfer fault performance (ML group) was to decrease isolation and diagnosis time. Moreover, there was a trend for the advantage to be particularly pronounced for isolation time for ML/FS over SL/FS subjects, but for diagnosis time for ML/PS over SL/PS subjects.

Further results also show that ML/PS subjects perform fewer tests in phases A and B and more in phase C (after diagnosis) than subjects in other groups, and a greater percentage of their phase C tests are logically redundant with prior information. Thus the ML/PS subjects appear to sift through the existing evidence again after making their final diagnosis. In contrast, the SS/PS and SL/PS subjects perform fewer tests after their final diagnosis and those tests are novel. Whatever final diagnosis the SS/PS and SL/PS subjects give, they clearly feel less need to review the critical evidence for their final diagnosis than the ML/PS subjects, because they are more likely to have ignored the evidence for the PS fault. Interestingly, when results are analyzed including only the SL subjects who correctly diagnose the transfer multiple fault, their percent test redundancy approaches that of the ML subjects.

Further tests have been run on the relative difficulty of multiple faults by taking faults diagnosed by ML subjects during Day 2 training rather than by concentrating solely on the transfer faults. Results show significant differences between faults on various performance measures. These differences appear to be related to the cognitive complexity of disentangling evidence, as hypothesized. For example SV and EF prove to take much longer to diagnose than FP and DK. (Similarly, for this group, the transfer fault PS takes much longer to diagnose than FS.) The results show a significant interaction between the time taken in each phase of diagnosis (A, B, or C) and the particular fault being diagnosed.

CONCLUSIONS AND IMPLICATIONS

This research indicates that both mental set and the inherent complexity of multiple fault symptoms can affect subjects' success at diagnosing a transfer multiple fault. Higher degrees of workload and longer fault diagnosis times are seen in subjects who have training only in single fault diagnosis (SS and SL) and who encounter a more complex multiple fault with integrated symptoms at transfer (PS). However, these performance measures show that subjects with a greater amount of practice with single fault diagnosis (SL) perform somewhat better than those with less single fault practice (SS), suggesting that greater familiarity with single faults in the binary adder can help eventual multiple fault diagnosis, although not as much as familiarity with a range of other multiple faults. The variables manipulated affect mental workload and time to solution, but have less clear-cut effects on gross measures of strategy such as the number, range and quality of tests performed.

The failure of some subjects in the single fault training conditions to diagnose the transfer multiple fault can be attributed to their failure to examine system states holding the critical evidence for the fault, their failure to entertain multiple fault hypotheses, or both. Seeing the evidence and considering multiple fault hypotheses were both necessary *and sufficient* to lead subjects in the present sample to the correct diagnosis. It is logically possible that subjects could fail despite having the evidence and the correct mental set, but no such cases were seen in this sample.

The three-phase model of fault diagnosis provides insight into subjects' strategies. We find that SL subjects trained on single faults who nonetheless manage to correctly diagnose the difficult transfer fault (SL/PS), spend more time in phases A and B, before the final diagnosis is made, than do ML/PS subjects, and there is some suggestion that the greater relative time increase occurs for phase B, the diagnosis phase..

In many cases where subjects failed to correctly diagnose the multiple fault they provided a single fault diagnosis instead. In the FS fault, for subjects trained in single fault diagnosis the temptation was to stop testing after seeing the F fault in the half adder at the right. However only one of the two FS subjects who diagnosed incorrectly failed to see the evidence for S. Not only did the FS subjects see this evidence, but they also recognized that it meant there were two faults present. In contrast, with the PS fault, as many as five of the eight subjects who failed to diagnose the PS fault saw the evidence for it, but because of the fault's logically complex and integrated nature the subjects failed to recognize that it meant there were two faults present.

Further analyses currently being undertaken examine the trajectories that subjects take through a "space" defined by (1) beliefs (rows) and (2) states of evidence (columns). The goal is to visualize each fault diagnosis history with a view to discerning systematic differences between subjects in different conditions. For example, in what order to subjects see the evidence, and is there a tendency to "get stuck" in certain beliefs? Figures 6 through 9 show typical trajectories on the transfer fault. For each multiple fault, 'X' refers to evidence for, or a belief about, the fault represented by the first letter (e.g., 'P' of 'PS'), 'Y' the second letter ('S' of 'PS') and N represents some incorrect diagnosis. Figures 6 and 7 are good and poor subjects, respectively, on PS. Figures 8 and 9 are good and poor subjects on FS. More details about the particular cases illustrated are given in the figure captions

Overall, it seems clear that multiple faults differ in the potential difficulty they impose on a troubleshooter, and this is related to how separated or integrated the symptoms of the contributing faults are. The present research has provided a conceptual and operational framework for more detailed analyses and studies of when and why the diagnosis of multiple faults can be complex.

REFERENCES:

Appeddu, T.A. (1991). *Mental set and complexity effects in human diagnosis of multiple faults in a binary adder.* Unpublished M.S.I.E. thesis, Department of Mechanical and Industrial Engineering, Urbana, IL. [Also Engineering Psychology Research Laboratory technical report EPRL-91-03, Department of Mechanical and Industrial Engineering, Urbana, IL].

Gitomer, D.H. (1988). Individual differences in technical troubleshooting. *Human Performance, 1*, 111-131.

Hart, S., and Staveland, L.E. (1989). Development of a multi-dimensional workload rating scale: Results of empirical and theoretical research. In P.A. Hancock & N. Meshkati (Eds.), *Human Mental Workload.* Amsterdam: North Holland.

Holden, J.A. (1990). *Understanding subject performance in fault diagnosis: The Ideal Problem Solver.* Unpublished manuscript, University of Illinois, Department of Mechanical and Industrial Engineering, Urbana, IL.

Luchins, A.S., and Luchins, E.H. (1991). Task complexity and order effects in computer presentation of water jasr problems. *Journal of General Psychology, 42,* 279-297/

Mendenhall, J.H. (1990). *Multiple fault diagnosis and human behavior: Diagnosing a two-bit binary adder.* Unpublished B.A. Honors thesis, Department of Psychology, University of Illinois, Champaign, IL.

Rasmussen, J. and Rouse, W.B. (1981). *Human Detection and Diagnosis of System Failures.* New York: Plenum Press.

Sanderson, P.M., and Murtagh, J.M. (1990). Predicting fault diagnosis performance: Why are some bugs hard to find? *IEEE Transactions on Systems, Man, and Cybernetics, 20,* 274-283.

Learning Group	DAY 1 Training Session	DAY 2 Training Session	DAY 2 Transfer to Mult. Fault
Single Short (SS)	single faults (no repeats)	single faults (no repeats)	easy (FS) hard (PS)
Single Long (SL)	single faults (no repeats)	single faults (poss repeats)	easy (FS) hard (PS)
Multiple Long (ML)	mult. faults (no repeats)	mult. faults (no repeats)	easy (FS) hard (PS)

Table 1:
Experimental design. Training condition (SS, SL and ML) and type of multiple fault (FS or PS) were between-subjects variables.

	Easy Transfer Fault (FS)					Hard Transfer Fault (PS)				
	Correct (Total)	Incorrect... Total	Lack Evid	Lack Hyp	Both	Correct (Total)	Incorrect... Total	Lack Evid	Lack Hyp	Both
SS	4	1	0	0	1	0	5	1	3	1
SL	4	1	0	1	0	2	3	0	2	1
ML	5	0	0	0	0	5	0	0	0	0

Table 2:
Performance of subjects in SS, SL and ML conditions when diagnosing either the FS or PS multiple fault at transfer at the end of Day 2.

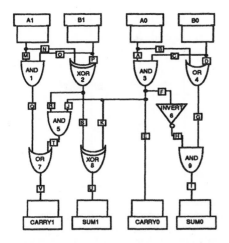

Figure 1:
Binary adder circuit with labeled fault locations, A through I in the half adder (right side) and K through V in the full adder (left side). Faults were breaks in the circuit along the connection indicated. Subjects entered either 0 or 1 in A1, B1, A0 and B0, and saw the resulting sum and carry outpus (either normal or abmornal) in the boxes at the bottom.

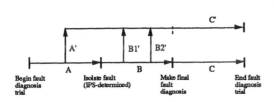

Figure 2:
Model of fault diagnosis phases. Path A' represents failure to see the critical evidence for (isolate) the fault, Path B1' represents a failure to entertain a multiple-fault diagnosis and Path B2' represents an inability to reason through to the correct multiple-fault diagnosis, despite having avoided A' and B1'.

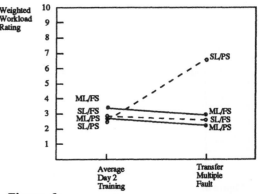

Figure 3:
Results for Weighted Workload, using the NASA TLX workload scales and analyzing results for the SL and ML subjects who correctly diagnosed the transfer multiple fault.

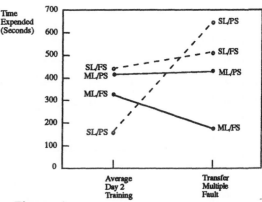

Figure 4.
Average time before final diagnosis trial for Day 2 training trials and transfer multiple fault, by Learning Group and Transfer Multiple Fault type. Analysis is confined to subjects who correctly diagnosed the transfer fault.

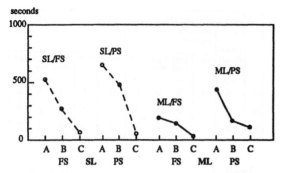

Figure 5:
Time spent in each phase of diagnosis for SL and ML subjects who correctly diagnosed fault. A is phase before isolation, B is phase after isolation and before final diagnosis, and C is phase after final diagnosis.

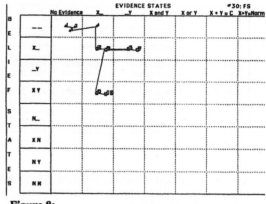

Figure 8:
Trajectory of a good subject through the belief/evidence state space during diagnosis of FS.

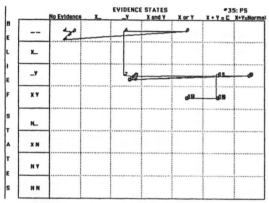

Figure 6:
Trajectory of a good subject through the belief/evidence state space to a successful final diagnosis of the PS transfer fault (final node is in XY row). Normal numerals are no beliefs or beliefs the subject marked as exploratory diagnoses only, whereas outlined numerals are ones subject marked as being definite diagnoses.

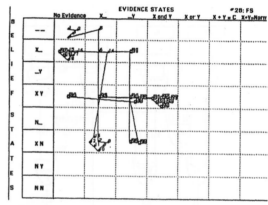

Figure 9:
Trajectory of a poor subject through the belief/evidence state space during diagnosis of FS. Subject finally diagnoses FS correctly, but makes incorrect diagnoses along the way (XN)

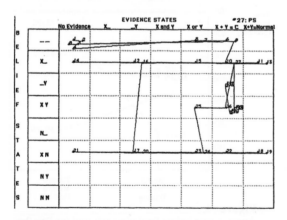

Figure 7:
Trajectory of a poor subject through the belief/evidence state space during diagnosis of PS. Subject correctly diagnoses PS, but then abandons P and makes only S the final definite diagnosis.

THE DESIGN, IMPLEMENTATION AND EVALUATION OF A NOVEL TRAINING ENVIRONMENT FOR CONTROL OPERATORS

N.V. Findler

Department of Computer Science and Engineering, and Artificial Intelligence Laboratory,
Arizona State University, Tempe, AZ 85287-5406, USA

Abstract. After a brief discussion of decision support systems and the problems of Air Traffic Control (ATC), we describe the design and implementation of a Predictive Man-Machine Environment (PMME) that can be used in the automatic training and evaluation of control operators as well as serve as the basis of their routine planning and decision making activity. PMME works with two connected workstations, one of which displays the essential features of the Current World and the other those of the Extrapolated World. The latter predicts the future consequences of the control operator's tentative decisions made in responding to the Current World. If the operator is satisfied with these, he finalizes his decisions; otherwise, he modifies the tentative decision and, time permitting, goes through this cycle as long as necessary.

We have also implemented a simulated air traffic control environment for the PMME and performed two sets of experiments with tasks of equal complexity — one with and the other without the extrapolation facility. We have shown a statistically significant improvement in the ATC operations that rely on the extrapolation facility. It can be said in general that the PMME increases the effectiveness and efficiency of human judgmental processes by extending their range, capabilities and speed.

Keywords. Automatic training and evaluation of control operators; air traffic control; tentative and final decisions; predictive man-machine system;

INTRODUCTION

The increasingly high rate of automation of a multitude of operations has raised the demands on the human element left in the control loop. The difficulties of the human controller are due to many factors, such as

• The knowledge base needed for a decision is too large for a human to access or even to make conceptual use of.
• To arrive at a satisfactory solution requires a large amount of processing of the available information.
• There is a time pressure to perform the computations (because the environment may continually change) and/or to obtain the solution (because certain actions must be performed *before* a certain point of time).
• Delicate judgements are needed as to which environmental and control variables are relevant and what their current values are. *Decision making under uncertainty* refers to the fact that the knowledge available about the world can be partial, imprecise and even inconsistent. [See Chapters 1-3 of (Findler, 1990).]

It is, therefore, important to devise techniques that would enhance the quality and shorten the length of the training process of control operators. It would also be necessary to implement some objective measures for evaluating their performance and use these as feedback for self-correcting control actions.

ON DECISION SUPPORT SYSTEMS

We now briefly discuss decision support systems (DSSs). A DSS is a computer-based information system that helps a user make decisions by providing him with the necessary information in an easily understandable form. The user is in an interactive environment and usually faces a menu-driven front-end. The system accesses a database to locate the necessary data, applies appropriate mathematical and/or statistical models, and displays the information needed at the user's terminal. Based on the information provided, the user can explore different alternatives in order to arrive at a decision.

The important DSS factors are accessibility, flexibility, facilitation, learning, interaction, and ease of use. Some additional issues to note are:

• Some DSSs focus on supporting decision making in effective planning and control for managerial problem finding and problem solving. The purpose is to assist users in fulfilling their organizational duties and responsibilities.
• DSSs give support to the user but usually do not replace him. Humans' qualitative models and other non-tangible considerations enhance the whole decision making process.
• DSSs encompass the user and the computer as well as various models and the languages and databases of modelling and simulation — which all contribute to the decision making process.

• A DSS is an important tool when a decision making process references some uncertain factors or involves many conditions and constraints to be considered simultaneously.

ON AIR TRAFFIC CONTROL

Air traffic control (ATC) has evolved from flag waving, when the major concern was to avoid collision with the ground and trees, to modern computer-based systems aiding the control of thousands of commercial, military and small private planes. Current systems have the capabilities of detecting the airplane's ID, speed, location and predicting potential traffic conflicts before their occurrence. Future generations of ATC systems will be built on space and satellite technology. Communication, navigation, and surveillance will be accomplished from earth orbit and will cover the entire globe.

ATC is a complex topic and we cannot give even a brief outline of its major aspects in this paper. It suffices to say that ATC is performed in areas designated to contain instrument-flight-rules operations during portions of the terminal operation and while transiting between the terminal and the *en route* environments. (The other operations are under visual-flight-rules when the pilot, and not the controller, is responsible for the adherence to the rules). Low altitude airways are between 1,200 feet above ground level up to 17,999 feet. The high altitude ones are designated at or above 18,000 feet.

Separation is the essence of air traffic safety. It is provided by establishing approved minimum longitudinal, lateral, and vertical distances between adjacent aircraft. Longitudinal separation is the spacing between two aircraft at the same altitude and on the same route. Lateral separation is also between aircraft at the same altitude but on different routes. On low altitude airways, the minimum horizontal separation is 3 miles if the airplane is within 40 miles of the radar antenna and 5 miles if it is outside the 40 miles range. In the high altitude case, the separation ranges from 5 to 10 miles according to various factors such as speed and weather condition. Vertical separation is established by assignment to different flight levels. Minimum vertical separation is 1,000 feet for airplanes of altitudes less than 29,000 feet, and 2,000 feet above 29,000 feet. Concerning separation between planes and obstructions, if the airplane is at least 3 miles from an obstruction and at least 1,000 feet above it, a minimum 3 miles distance from it must be maintained until the obstruction has been passed. If the airplane is less than 3 miles from the obstruction and its altitude is at least 1,000 feet above it, the minimum lateral distance of 3 miles must be maintained until it passes the obstruction.

An *event* is the product of interactions among aircraft, airspace, facilities, and ATC operations. The types of events associated with a single aircraft include conflict, violation, clearance request, and flight status. The event among multiple aircraft is *conflict*, a situation in which two or more planes get too close to each other or one plane gets too close to geographical obstructions. *Violation* involves only one aircraft violating some flight parameters, such as the landing angle. *Clearance request* events may refer to takeoff, approach or landing. Initial contact, filing flight plans and missing the approach are events concerning the *flight status*.

Activities are defined as top-level sequences of man-machine interactions responding to a group of closely related events. A typical flight goes through several terminal and *en route* activity steps: a jet is waiting on the runway and the captain informs the controller in the control tower about being ready for takeoff; he then gets the takeoff clearance and departure instructions; the jet leaves the ground and an electronically generated "data tag" (containing the jet's flight number, ground speed and altitude) is shown on the radar scope; at the boundary of the *en route* control sector, the terminal controller hands the flight off to the *en route* controller; close to the destination, the pilot asks for a landing clearance; the flight is in a waiting pattern until the clearance is received; the data tag disappears from the radar scope after the plane has landed.

We have been interested in applying AI methodology to this intellectually and economically challenging area for some years (Findler 1984, 1987; Findler and Lo 1986, 1990).

THE DESIGN OF THE PREDICTIVE MAN-MACHINE ENVIRONMENT (PMME)

Relying on our previous work concerning a simpler environment[1] (Findler 1985, 1988; Chapter 8 of Findler 1990), the design of the PMME was done according to the flow of information and control shown in Fig. 1.

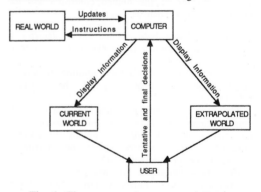

Fig. 1. The conceptual structure of PMME

The *Current World* displays the essential features of the *Real World*. Events from the Real World (or, as in our case, from the Simulated Air Traffic Control Environment) *update* the status of

---

[1]In this early work, the user was not able to define the ATC· environment — the location and the number of airports, airways, radio beacons, obstructions, etc. were fixed. Also, the environment included only *en route* operations and no near-terminal ones. Consequently, the performance measures of the control operations did not cover those that are relevant near an airport.

the affected objects. The updating can be controlled in three different ways:

- at regular points of time, according to a user-defined frequency (*time-driven updating*);
- whenever one of the set of user-specified events takes place, an update is triggered (*event-driven updating*);
- on user's command, regardless of how long time has elapsed since the last update or whether certain conditions in the environment are satisfied (*user-driven updating*).

The user makes *tentative decisions* on the basis of the Current World display, say, at time t_0 and transmits them to the computer. He then specifies a time point $t_1 > t_0$ at which he wants to see the expected consequences of the tentative decisions. This extrapolation of the world status is computed — according to the computer-resident model — by periodic (user-specified) time increments, up to the final time point t_1. However, if between t_0 and t_1, a *conflict* occurs, as defined by the user again, the calculation is interrupted, and the conflict situation with the corresponding time is displayed on the second unit called *Extrapolated World*. If no conflict has occurred between the time points in question, the relevant features of the world will reflect the permissible consequences of the tentative decision on the second screen.

If the user is satisfied with the status of the resulting world, he finalizes the tentative decisions and informs the computer accordingly. Otherwise, he makes a different set of tentative decisions and goes through the above cycle as many times as necessary, time permitting.

THE SIMULATED AIR TRAFFIC CONTROL ENVIRONMENT

First, some terminology is needed. To set up any simulated environment for the PMME, we have to define a set of *objects* inhabiting it. The *conditions* that affect the objects can be either constant over time (e.g., the location of a mountain range) or subject to change (e.g., current weather). Objects can also affect each other (e.g., two planes with an intersecting flight path). An object passes through different *phases* in its interaction with the environment. Each can be considered a distinct stage in the lifetime of the object in the environment. During a phase, the object may perform certain *functions* toward achieving a subgoal and, finally, the goal. Such act places the object into another phase. (A plane passes through eight major phases in its flight: preparation for take-off, take-off, climb, cruise, descent, approach, landing and taxiing). The overall goal is, of course, the safe and timely flight between the origin and the destination.

Associated with each object is a set of *attributes* which determine the constant and varying properties of the object — including the phase it is in. The number and the type of the attributes, in general, depend on the domain of application. The user must define the functions that can alter the values of the attributes every time a user-

specified *incremental parameter* (usually time) value elapses. There is a trade-off here between "too large" and "too small" triggering values for the incremental parameter. In the former case, more computation has to be performed but there is a lesser chance that a critical conflict can get by unnoticed in between two attribute updatings. In the latter case, the opposite is true.

It is important to point out that the attributes of the objects change independently from each other in the Current and in the Extrapolated World. This separation of *attribute scopes* is a basic requirement when causal relations between decisions and consequences are sought.

A large knowledge base needs to be prepared for any non-trivial domain of application. For example, in case of the Simulated Air Traffic Control Environment (SATCE), it must include

- the physical characteristics of several airports (length and location of airstrips; height and location of obstructions, such as mountains and towers; direction of radio beacons; etc.)
- the physical characteristics of the planes (symbolic notation for the manufacture/model, maximum initial landing speed, approach angle and speed, descent angle and speed, maximum and average cruising speed, climb angle, take-off speed, turning radius, fuel capacity, fuel consumption in different phases, etc.);
- dictionary and syntax for communicating between the pilots and the air traffic controller;
- a program to interpret for the communication;
- links between controller instructions and the functions changing attribute values accordingly.

THE IMPLEMENTATION OF THE PMME

The PMME runs on two workstations, each with a separate control unit, user interface, communication unit and graphics unit. We have assigned Workstation I to display the Current World containing simulated air traffic scenarios. Workstation II displays the Extrapolated World obtained through the Extrapolator to predict future situations. Fig. 2 illustrates the structure of the different components.

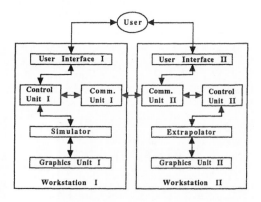

Fig. 2. The implementation structure of PMME

The simulated radar scopes are similar to those in use at the TRACON centers of the FAA. The display presents a map, about 100 miles in radius, and contains the "control section" area for which the controller is responsible. It is centered over the location of the radar dish generating the display. The circular grid lines are shown five miles apart. The location of the airways, radio beacons, mountain peaks and other landmarks are indicated in a standard symbolic format.

In real-life, ATC displays produce only a blip on the screen for each aircraft and the transponder on the aircraft continually sends messages giving additional display information as needed. This includes the aircraft's identification, ground speed, altitude and other data, put in a block connected with a line to the exact location of the aircraft on the screen. Our display simulates this final result (see Fig. 3).

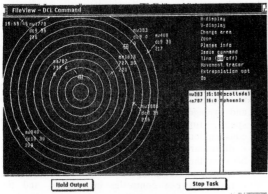

Fig. 3. Current World, horizontal projection

We have made the simulated environment fairly realistic and enhanced the displays with certain facilities that are not yet available in real-life. For example, we have made use of the *color capabilities* of our display units, to make the identification of various symbols easier. The user can *center the display* over any part of the map, and *zoom* it in and out as necessary to obtain a more detailed view of potential problem areas. (see Fig. 4). There is a *system clock*, a *text window* for presenting the status of the system and the airplanes, and a *menu-display* to make PMME more powerful and flexible.

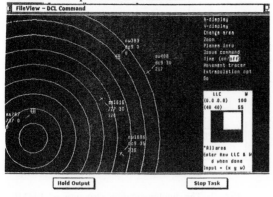

Fig. 4. Changing an area of interest, horizontal projection

Through the menus, the user can select or change any simulation parameter not associated with airplane behavior. Also, he can vary the time parameters of the simulation, change the color of a group of symbols, or alter the center and scale of the map. The vertical profile of an airplane's movement can be viewed in the second *display mode*. It is centered on the aircraft and the viewing plane is rotated so that the target is moving perpendicular to the controller's line of sight. Although the current ATC radar scopes do not yet have this capability either, the facility was provided to enhance the man-machine interface in certain critical situations. The user can request in our system a *trace of an aircraft's trajectory* in either projection (see Fig. 5).

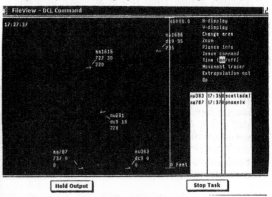

Fig. 5. Trace of an aircraft's flight, vertical projection

The program handles each aircraft as an independent object, with its own performance characteristics. Airplane objects carry data about their flight plan, commands previously issued to them, a description of their current status, and a pointer to a generic block of information containing the performance characteristics of a class of airplanes (e.g., DC-9, Boeing 747, etc.).

The user to manipulate *system time* easily. To update the environment to a new time, a set of operations has to be performed on the representation of each airplane. Extrapolation is accomplished by taking the current state of the environment and then repeatedly updating it by the time increment until the specified time is reached and the display is triggered. (If there are conflicts, the extrapolation process is interrupted, warning sound and light signals are given, and the conflict in question is displayed.) Also, the environments can be saved to allow the controller the possibility of an *"instant replay"*.

Another addition is a versatile *reminder system*. The controller can describe a variety of anticipated situations using the menu system. When the program senses that a predefined situation is about to occur, it alerts the controller with a tone and a message with reference to the triggering condition. This facility may save the controller valuable time by providing retrieval cues for previously generated plans. In setting up the PMME, the user can define the geographical properties of the environment (the control sector, the location of the airports, waiting stacks, airways, ranges of the mountains, towers, and radio beacons. The user can also adjust the level of task difficulty — that is, the range[2] of traffic density (upper and lower limits of the number of aircraft) and the ratio between simulated and real time (to suit the needs of student controllers). The geographical properties can be stored in a file to be retrieved at later training sessions. One can thus evaluate the improvement in the user's performance while the level of difficulty is kept constant.

EXPERIMENTS AND THEIR EVALUATION

To evaluate the PMME concept, two groups of experiments were designed and performed. The objective was to compare the efficiency and effectiveness of human planning and decision making in a complex, real-time problem domain — with and without the extrapolation facility. The subjects were six graduate students in Computer Science.

The task specified for the subjects was *en route* and near-terminal ATC under light, moderate and heavy traffic conditions. The performance of the subjects was automatically evaluated in each experiment. In addition to the primary quality measures, the number of airspace conflicts and violations occurred during the session, there were control efficiency-related, secondary ones:

• The number of instructions issued — the smaller it is, the better the control strategy is (pilots prefer fewer and effective instructions).
• The number of extrapolations requested — same as above (extrapolation takes time away from concentrating on the situation, particularly under time-stressed conditions.)
• The number of conflicts — for all aircraft, updated after every internal time increment.
• The average "off-distance" — the difference in miles between the lengths of each aircraft's requested (usually straight-line) route and the sum of the distance actually travelled through the air traffic controller's sector plus the distance from the point it left the sector to its destination.
• The average "climb/descend distance" per aircraft — the total distance in feet each aircraft was told to climb and descend during its flight through the sector, plus the difference between its requested and final altitude.
• The average delay of departures and arrivals per aircraft — the difference between actual and scheduled time points.
• The ratio between the simulated and real time. The faster this user-selected number approached one, the more effective the controller's learning process was.

The program randomly generated the number of participating aircraft, their flight plans and certain other characteristics. For the benefit of objective comparisons, task complexity was the

---

[2] The actual number and the location of the planes participating in the experiment at hand are within the ranges specified and are determined by means of random number generators.

same for the corresponding experiments — 30 with and 30 without the extrapolation facility. Each session lasted 20 minutes of simulated time. The time increment was 60 seconds in the Current World and 30 seconds in the Extrapolated World. Each group of 30 sessions was further divided into three sets with light, moderate and heavy air traffic density — about 10, 15 and 20 airplanes, respectively.

The results are summarized in Table 1 and the effect of the extrapolation is shown in Table 2.

TABLE 1 Summary of experimental results

Air Traffic Density	With Extrapolation			Without		
	Light	Moderate	Heavy	Light	Moderate	Heavy
Av. No. of Planes	10.9	15.2	20.5	10.3	15.4	20.5
Av. No. of Commands	8.5	11.2	16.1	7.6	10.4	14.3
Av. No. of Extrapols.	5.0	7.5	10.2	0.0	0.0	0.0
Av. No. of Conflicts	1.9	4.0	9.1	3.2	9.4	12.7
Av. No. of Violations	1.4	0.8	2.6	2.2	2.0	3.9
Av. Off-Distance	0.13	0.10	0.22	0.19	0.26	0.31
Av. Climb/Desc.-Dist.	0.27	0.12	0.31	0.39	0.41	0.52
Av. Dep./Arr. Delay	1.0	1.4	2.1	1.2	1.9	2.3
Av. Sim./Real Time	1.21	1.39	2.42	0.82	0.97	1.41

TABLE 2 The improvements due to extrapolation

Air Traffic Density	Light	Moderate	Heavy
Av. No. of Planes	—	—	—
Av. No. of Commands	+12%	+8%	+13%
Av. No. of Extrapols.	—	—	—
Av. No. of Conflicts	-41%	-57%	-28%
Av. No. of Violations	-36%	-60%	-33%
Av. Off-Distance	-32%	-62%	-29%
Av. Climb/Desc.-Dist.	-31%	-71%	-40%
Av. Dep./Arr. Delay	-17%	-26%	-9%
Av. Sim./Real Time	+32%	+30%	+42%

CONCLUSIONS

The experimental results over time clearly indicated that in the task environment chosen, the PMME is an effective teaching tool, and the subjects' performance with the extrapolation facility was significantly better than without it. (There is also an extraneous and small contributory factor to note. Since the two display units are connected to the same hard disk memory, the Current World environment runs somewhat faster in real-time when the extrapolator is disabled, which fact makes the controller's task a little more difficult. However, the difference in speed was small enough to let us state that we have obtained an accurate evaluation of the usefulness of the PMME.)

The results were especially impressive in reducing the number of conflicts when the air traffic density is moderate (57% less conflicts and 60% less violations with the extrapolator enabled). In general, the number of conflicts, violations, average off-distance, average climb/descend distance, average arrival/departure delay were all reduced significantly. However, the number of commands used increased to some extent. This was expected because some of the commands were made to avoid the potential conflicts warned about by the extrapolator. The simulated and real time ratios also increased when the extrapolator was used, particularly under heavy traffic conditions. (This, to a large extent, was due to the time-consuming way — via the keyboard — in which the subjects communicated with the simulated pilots.)

Although the use of the extrapolation facility improved the subjects' performance, we have found that it can also be abused and lead to undesirable outcomes as follows.

• After having gained some experience in the use of the extrapolator, it is easy to become complacent about the control task when no conflicts occur within the time interval between the current and the extrapolated time.
• If extrapolations are continually requested for short periods, the major part of one's attention is drawn away from the Current World to the Extrapolated World — resulting in a possible oversight of conflicts in the immediate future.
• In "crisis" situations in which many decisions must be made in a short period of time (e.g., several conflicts are imminent), extrapolation can often be a hindrance. When facing such a confusing situation, a seemingly safe thing to do is to run the extrapolation into the near future. However, this action may result in wasting time during which effective directives should have been issued to resolve the conflicts.

In spite of the above caveats, PMME is a useful, general-purpose environment for real-time decision making and planning. It should, however, be relied upon only as an auxiliary source of information — the planner and decision maker should always focus his attention on the unfolding events in the Current World.

The PMME can be improved in several aspects. Low-level mouse programming utilities would strengthen the graphics user interface. The student controller should be able to select different viewing areas and retrieve information about an aircraft via mouse positioning instead of keyboard typing. SATCE should be able to generate more realistic scenarios, such as those based on Poisson rather than uniform distributions and incorporating also unexpected events (emergency landing, hijacking attempts, sudden changes in the weather, etc.).

Summing up, the PMME can in general be used to improve trainees' analytical skills, long-range planning ability, diagnostic capabilities so that they can balance objectives, associate methods with results, convert objectives — specified as the desired state of the environment — to dynamic process models, utilize resources to minimize risks, and contrast costs with benefits.

Several other applications seem also possible, both for automated training and evaluation as well as for routine control operations in different conflict situations.

ACKNOWLEDGMENTS

I am grateful for the programming efforts and many useful ideas to C. Luo, M.-H. Kuo T. W. Bickmore, R. F. Cromp and N. Mazur, and for our experimental subjects' efforts and enthusiasm.

REFERENCES

Findler, N. V. (1984). Some Artificial Intelligence contributions to air traffic control. *Proc. Fourth Jerusalem Conf. on Info. Technology*, 470-475.

Findler, N. V. (1987). Air traffic control: A challenge for Artificial Intelligence. *AI Expert Magazine*, 2, 59-66.

Findler, N. V. (1988). A prototype of a man-machine environment for the study of air traffic control. In A. Kent and J. G. Williams (Eds.). *Encyclopedia of Computer Science and Technology*, Vol. 19. Marcel Dekker, NY. pp. 3-21.

Findler, N. V. (1990). *Contributions to a Computer-Based Theory of Strategies*. Springer-Verlag, New York, Berlin and Heidelberg.

Findler, N. V., T. W. Bickmore, and R. F. Cromp, (1985). A general-purpose man-machine environment with special reference to air traffic control. *International Journal of Man-Machine Studies*, 23, 587-603.

Findler, N. V., and R. Lo. (1986). An examination of distributed planning in the world of air traffic control. *Journal of Distributed and Parallel Processing*, 3, 411-431. Also in A. Bond, and L. Gasser (Eds.). *Readings in Distributed Artificial Intelligence*. Morgan Kaufmann. Los Altos, CA.

Findler, N. V. and R. Lo. (1991). A distributed artificial intelligence approach to air traffic control. *IEE Proceedings Part D, Control Theory and Applications*, 138, 515-524.

MODELING AND ANALYSIS OF MANNED ROBOTIC SYSTEMS

P.H. Wewerinke

Department of Applied Mathematics, Systems and Control Group and Mechatronics Research Centre
Twente, University of Twente, P.O. Box 217, 7500 AE Enschede, The Netherlands

Abstract

In this paper a modelling approach is presented to describe complex manned robotic systems. The robotic system is modelled as a (highly) nonlinear, possibly time-varying dynamic system including any time delays in terms of optimal estimation-, control- and decision theory. The role of the human operator(s) is modelled varying from supervisor of the automated (part of the) system to controller in terms of the various functions involved to perform goal-oriented tasks.

In the paper it is indicated how the model differs from previous optimal control models. Furthermore, the use of the model is illustrated in discussing typical manned robotic tasks and applications.

It may be expected that the model is capable of answering questions related to reliability and efficiency, design alternatives, function allocation, automation, etc.

1. INTRODUCTION

Robotic systems are more and more applied in many areas. Examples of industrial operations are part assembly, material transfer, repair of parts and inspections. In addition, robotic systems play an important role in many teleoperations, e.g. in space applications (space stations, serviceable sattelites, material processing platforms) and operations in a risky or unaccessible environment.

Autonomous systems can meet the safety, reliability and especially economic requirements for specific tasks, but many operations involve the interacting contribution of both human operator(s) and robotic system(s). This concerns especially complex, non-standard operations in an unstructured environment. One example is shared compliant control, especially in applications with tele-manipulators where time delays are considerable.
The role of the human operator(s) may vary from direct controller to supervisor of the automated (part of the) system. This depends on the goals to be achieved and the related functions to be fulfilled.

In the next section manned robotic systems are discussed in more specific terms. One approach to design and analyze a manned robotic system is based on mathematical models of this complex man-machine system. This is contained in section 3. The paper is concluded with some remarks about how the model can be utilized.

2. MANNED ROBOTIC SYSTEMS

In general the task to be performed with a manned robotic system can be described in terms of the various components involved. This is indicated in the block diagram of Fig. 2.1.

The first important aspect of the task are the goals to be achieved under given boundary conditions. Realistic operations may involve a complex goal hierarchy (interrelated, or even conflicting goals, subgoals, procedures, etc.). These goals dictate the tasks to be performed. The complexity of the task hierarchy will correspond with the complexity of the goal hierarchy. The defined task will be affected by the operational environment.
Next, the functions can be derived to perform the defined task. The motives to fulfill these functions originate from the goals to be achieved. The human operator (HO) will perform these functions utilizing the available resourses, being separate items or elements of the system. These functions are the result of a function allocation procedure to the man and the machine, taking into account the human capabilities and limitations and the possibilities of the system.

Finally, the result of HO functions are actions, taken on the basis of drives (derived from his motives, etc.). Simple stimulus response behavior takes place at this level only. The actions affect the system resulting in a certain system behavior. Based on performance criteria and measures of this behavior, total system behavior can be evaluated with respect to the goals to be achieved.

More specifically, a manned robotic system can be analyzed or designed based on the assumption that a goal-oriented operation can be defined, e.g. controlling the robot from A to B, with given constraints (due to robot dynamics, control limits, environment, etc.) in a given (disturbance) environment. Furthermore, the robotic system is described as a (highly) nonlinear, possibly time-varying dynamic system including any transport or communication time delays.
The role of the HO may vary from supervisor of the automated (part of the) system to controller, or combinations. This involves HO functions such as perception of sytem outputs provided by (e.g.) displays and /or the visual scene, information processing to assess the task- and system variables of interest, decision making involved in monitoring the autonomous (sub)system and involved in intermittent control, and control. In this context, control is used in a broad sense including planning and compensatory actions.

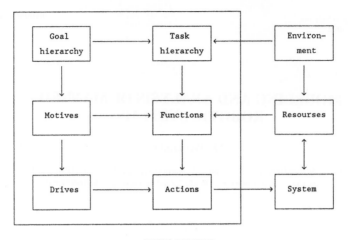

HUMAN OPERATOR

Fig.2.1   Manned robotic system components

There are various ways to analyze a manned robotic system.
One approach utilizes a variety of system performance and HO measures obtained in an experimental situation. The drawback of an experimental approach is that it involves per se comlex man-in-the-loop simulation. This implies that experiments mainly serve to verify system performance rather than to predict the performance of new systems. This limits their utility, especially early in the design stage.
An alternative method to describe manned robotic system behavior is the use of mathematical models. The main advantage of mathematical models is that they provide a precise (quantitative) formulation of task aspects such as goals and system dymanics and basic concepts of HO functioning. Therefore, models potentially have a predictive capability, rendering a basis for selecting design alternatives. Furthermore, models provide a powerful tool to analyze operational problems of existing robotic systems systematically.
In the next section a model structure is presented describing the foregoing manned robotic system.

3. MODEL STRUCTURE

3.1 General

In this section a model of the manned robotic systems discussed in section 2 is presented. For details the reader is referred to Ref.1. Basically, the task considered is to control the robotic system from an initial state X_0 at time $k = 0$ to a final state X_n at time $k = N$ following a given desired trajectory $X_d(k)$.
Strictly speaking, the time aspect involved in this formulation is absent in most of the control tasks of interest. For example, the task is to arrive at a certain point in the state space, irrespective of time, but following a given sequence (e.g. a spatial trajectory). This is known in control theory as a space constrained control problem. Theoretically this is a difficult problem, certainly for robotic (i.e. nonlinear) systems. Thus, it is attractive to prescribe to desired trajectory for each time step and solve for the corresponding optimal control at each time step. This seems for many control tasks a meaningful approach.

As indicated before, the role of the HO may vary from supervisor of the automated (part of the) robotic system to controller. In this paper the manual control task will be assumed. The human functions involved include state estimation and decision making about (ab)normal system behavior (supervising). So the supervisor task is part of the manual control task. This will be discussed in the following.

The following elements are included in the model as indicated in the block diagram of the manned robotic system model shown in Fig. 3.1. The HO perceives (e.g. via a TV-camera/monitor system) the outputs of the system state (characteristic system variables such as position, velocity, attitude, etc.), which is perturbed by external disturbances. The HO utilizes these (inaccurate) observations of the state to estimate the state recursively. Based on this estimate and the task specifications a control input is selected and executed, resulting in a closed loop manual control.
In addition, the future desired state is observed (e.g. from pictorial information) and estimated. Based on this estimate and the system dynamics a pre-programmed control is selected. The resulting open loop control is combined with the feedback control to compensate for deviations from the planned trajectory due to random disturbances.

3.2 System dynamics

A robotic system may be represented as a nonlinear, time-varying dynamic system by

$$X(k) = f(X(k-1),U(k-1),W(k-1),k) \qquad (1a)$$

$$Y(k) = g(X(k),U(k),k) \qquad (1b)$$

where $X(k)$ is the n-dimensional state vector at time k, f is the n- dimensional vector function, U is the r-dimensional control vector, W is the l-dimensional disturbance vector, Y is the m-dimensional system output vector and g is the m-dimensional vector function.

The standard procedure is followed to describe the nonlinear system behavior X in terms of a state reference X_0 and a "small" perturbation x around this reference; thus $X = X_0 + x$, $U = U_0 + u$, etc. This lineari- zation scheme yields a time varying reference model

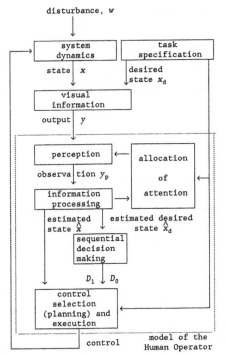

disturbance, w

```
┌─────────────┐   ┌─────────────┐
│   system    │   │    task     │
│  dynamics   │   │specification│
└─────────────┘   └─────────────┘
   state  x        desired
                  │state $x_d$
┌─────────────┐
│   visual    │
│ information │
└─────────────┘
  output  y
```

model of the Human Operator:
- perception → allocation
- observation y_p
- information processing → of attention
- estimated state $\hat{x}$ / estimated desired state $\hat{x}_d$
- sequential decision making
- D_1 D_0
- control selection (planning) and execution
- control

Fig. 3.1   Block diagram of the Manned Robotic System model

$$X_0(k) = f(X_0(k-1), U_0(k-1), W_0(k-1), k) \qquad (2a)$$

$$Y_0(k) = g(X_0(k), U_0(k), k) \qquad (2b)$$

and a time-varying **linear** system description

$$x(k) = \Phi x(k-1) + \Psi u(k-1) + \Gamma w(k-1) \qquad (3a)$$

$$y(k) = Hx(k) \qquad (3b)$$

with $\Phi = \Phi(k, k-1)$ being the Jacobian matrix of f with respect to X, etc.; w is assumed to be a zero mean, Gaussian, purely random sequence, with covariance matrix R, representing disturbances and other system uncertainty.

This linearization scheme holds for relatively small x, u and w. This fact dictates the update rate of the reference and, therefore, of the various Jacobian matrices.

3.3 Control task

The task considered is to control the system state x of eq (3a) over some fixed interval of time $[0, N]$, so as to follow the desired state x_d, by realizing a control sequence $\{u(k), k = 0, 1, \ldots, N-1\}$ that minimizes the performance index

$$J_N(u) = E\{ \sum_{i=1}^{N} (x(i) - x_d(i))' Q_x(i)(x(i) - x_d(i))$$

$$+ u'(i-1)Q_u(i-1)u(i-1)\} \qquad (4)$$

where Q_x and Q_u are weighting matrices.

The solution of this optimal control problem is discussed in the following subsection dealing with human control behavior.

3.4 Human observer and controller

The model of the HO comprises various functional aspects, which are shown in the block diagram of Fig. 3.1.

Perception

The perceptual model describes how the system outputs y are related to the perceived variables y_p. It is assumed that the HO perceives these system outputs with a certain inaccuracy and with a certain time delay

$$y_p(k) = y(k-i) + v(k-i) \qquad (5)$$

Here v is an independent, Gaussian, purely random observation noise sequence, representing the various sourses of human randomness (unpredictable in other than a statistical sense). Each element v_j, therefore, is specified by its covariance V_j. This covariance is functionally related to the signal level, human attention allocation and threshold phenomena.

The lumped time delay i can be associated with the HO's internal time delays related to perceptual, central processing, and neuromotor pathways. For systems with relatively large time constants, these delays can be neglected. Also system-related delays, for instance the communication or transport delays of remotely controlled systems may be modelled as a lumped equivalent "perceptual" time delay involved in eq(5).

Information processing

Based on the perceived data y_p up to time k (corresponding to data y up to $k-i$) and the known (learned, thus assuming that the HO is well-trained) dynamics of the system, the best (minimum variance) estimate $\hat{x}$ of the system state x can be made corresponding to time $k - i$. The resulting Kalman-Bucy filter equations are given by

$$\hat{x}(k-i) = \hat{x}(k-i/k-i-1) + K(k-i)n(k) \qquad (6)$$

with

$$\hat{x}(k-i/k-i-1) = \Phi\hat{x}(k-i-1) + \Psi u(k-i-1) \qquad (7)$$

$$K(k-i) = P(k-i)H'V^{-1}(k-i) \qquad (8)$$

and

$$n(k) = y_p(k) - H\hat{x}(k-i/(k-i-1)). \qquad (9)$$

Hence $\hat{x}(k-i/(k-i-1))$ indicates the estimate of x at time $k - i$ based on the data y up to time $k - i - 1$, K represents the optimal trade-off between system uncertainty (in terms of the estimation error covariance P) and reliability of the data (in terms of the observation noise covariance V), and the innovation sequence n represents the new information.

The best estimate of x at time k, $\hat{x}(k)$, is obtained on the basis of $\hat{x}(k-i)$ and the known system dynamics by means of an optimal linear prediction process. The resulting prediction equation becomes

$$\hat{x}(k/k-i) = \Phi^i\hat{x}(k-i) + \sum_{\ell=0}^{i-1} \Phi^{i-1-\ell}\Psi u(k-i+\ell) \qquad (10)$$

153

which corresponds to eq(7) for the one step prediction.

Perception and estimation of the future desired state x_d is described in a similar way. It is assumed that visual cues y_0 can be observed that are related to the difference between the present state and the future desired state. Estimation of the future desired state is depending on the assumed a priori knowledge that the HO may have about x_d. Three cases may be considered: no prior knowledge of x_d, only statistical knowledge of x_d and imperfect knowledge of x_d. The latter case results in the same filter equations as above. For the other cases the reader is referred to Ref. 1.

The HO's allocation of attention among all the system outputs y and y_0 is described by a visual scanning model. The basic assumption is that the HO divides his attention among the visual cues such that the performance index of eq (4) is minimized (Ref. 1).

Sequential decision making

After the finite time interval for which the control task defined in subsection 3.3 has been performed, the decision has to be made about what to do next.
This amounts to the binary decision as to whether the system behaves according to the small perturbation model of eq(3), corresponding to a given state reference, or a systematic discrepancy between both necessitates a correcting action of the HO and an update of the system model. In the first case the HO continues to control this system steady-state. In the second case, the HO initiates another maneuver to track the systematic deviation (x_d) over some fixed interval of time, after which the HO updates his system model.
It might be necessary to update the sytem model more often. The extreme is to update the system model each time step. The reference is adjusted based on the estimated state (deviation from the old reference). This is known as the extended Kalman filter.

The comparison of system behavior as observed by the HO in terms of y_0 and the expected behavior on the basis of the present system model is made by the HO in terms of the innovation sequence n_0. A systematic deviation of the zero mean sequence x due to a change in the desired state (with respect to the present state reference) results in a non-zero mean innovation sequence (Ref. 1).
This can be tested by comparing (the log of) a generalized likelihood ratio with a threshold T according to

$$L(k) \underset{D_0}{\overset{D_1}{\gtrless}} T \qquad (11)$$

with

$$L(k) = L(k-1) + \frac{1}{2} \bar{n}_0'(k) N_0^{-1}(k) \bar{n}_0(k) \qquad (12)$$

and

$$T = \ell n((1-P_M)/P_F) \qquad (13)$$

where $\bar{n}_0$ is a moving average of n_0, N_0 is the covariance of n_0 and P_M and P_F are the decision error probabilities ("miss" and "false alarm").

Human control behavior

The control task discussed in subsection 3.3 is defined in terms of $J_N(u)$ of eq(4). Optimal human control corresponds to the minimal J_N. The resulting optimal control sequence $\{u(k), k = 0,1,\ldots,N-1\}$ is derived in Ref. 1. The result is given by

$$u(k) = S(k)\hat{x}(k) + S_m(k)z(k+1) \qquad (14)$$

with

$$S(k) = S_m(k)W(k+1)\Phi \qquad (15a)$$

$$S_m(k) = -(\Psi'W(k+1)\Psi + Q_u)^{-1}\Psi' \qquad (15b)$$

$$W(k) = Q_x(k) + \Phi'W(k+1)[\Phi + \Psi S(k)] \qquad (15c)$$

and

$$z(k) = [\Phi' + S'(k)\Psi']z(k+1) - Q_x(k)\hat{x}_d(k) \qquad (16)$$

with

$$z(N) = -Q_x(N)\hat{x}_d(N), \qquad k = N-1,\ldots,0. \qquad (17)$$

Thus, the control is composed of two parts: a feedback control operating on the state estimate and a feedforward (open loop) control operating on the estimate of the desired state x_d and computed recursively backwards in time.

3.5 Constraints

So far, the estimation and control problem of the nonlinear system is solved by linearizing the nonlinear system around an estimated reference yielding a linear estimation and control problem. Control comprises permanent feedback control and intermittent open loop control based on sequential decision making.
Several constraints may play a role in a given task. Control is, in principle, constrained because of limited control authority, hardware limits, etc. Generally these constraints have to be included a priori in formulating and solving the optimal control problem.
Secondly, it may be necessary to consider (hard) constraints in the state space. Examples are the requirement to realize precisely a final destination of the system state (apart from stochastic effects), the limited space available to go from A to B (partly due to fixed obstacles), and limited space because of moving obstacles.

Both types of constraints can be handled in a similar way. Conceptually, the procedure is (as discussed in Ref. 2) to "adjoin" the constraints to the performance index J_N, given by eq(4), by a set of (socalled Lagrange) multipliers, which are chosen in such a way, that an optimal control is obtained (corresponding with an minimal J_N) given the constraints. Such optimization problems with constraints are conceptually straightforward. Numerical solutions, however, can require considerable computational cost.
This can be aggrevated if the control problem is stochastic by nature. In that case the state has to be estimated requiring the solution of a nonlinear estimation problem (e.g. based on an extended Kalman filter).

For example, the solution of the optimal control task inclusive collision avoidance of moving obstacles requires:
- estimation of own system state and the state of relevant obstacles;
- definition of the constraints that have to be met in terms of the estimated states (e.g. the estimated distance to obstacles);
- computation of the optimal control while meeting the constraints.
A more simple (engineering) approach to solve this and similar problems is given in Ref. 3.

3.6 Comparison with previous optimal control models

An important class of man-machine system models may be referred to as control theoretic models. These models are based on the modeling principles described previously. An important result of this approach is the so-called optimal control model (Ref. 4). The basic optimal control model structure has been extended in several directions (Ref. 1). Several of these extensions are combined in a model (PROCRU) for analyzing flight crew procedures in the approach to landing (Ref. 5).

Instead of trying to review and evaluate the control theoretic modeling literature, it is indicated which aspects of the model of this paper are unique.

First of all the model considers the general task to control a system from an initial state to a final state, following a given desired trajectory. Such a tracking task implies not only regulating control (to minimize deviations from a given reference) but also an open loop control to track the desired state. In the present model this desired state has to be estimated based on outside world cues possibly in combination with an internal model of the desired state.

Secondly the model deals with nonlinear, time-varying system dynamics by assuming a state reference and "small" perturbations around this reference. In case the reference is unknown it can be estimated by means of an extended Kalman filter. Both state and control constraints will generally be involved in robotic tasks. These are accounted for in the model by adjoining the constraints to the performance index resulting in an optimal control while meeting the constraints.

Furthermore, the model assumes finite time intervals for which the control (sub)task is specified. After each interval the decision has to be made about what to do next. This sequential binary decision process is modelled in terms of a generalized likelihood ratio test. The result is intermittent open loop control (apart from the standard continuous feedback control).

Finally, a model is included for visual scanning. It provides an optimal strategy to allocate the human attention among all system outputs y (related to the state x) and y_0 (related to the desired state x_d).

3.7 Model applications

The model presented in this paper can be applied to a number of manned robotic systems. Industrial applications tend to autonomous operations, such as part assembly, material transfer, spot welding and inspection, but for many operations the role of the human operator is (still) required (e.g. shared compliant control).

An important area concerns teleoperations. Applications in space concern (e.g.) space stations, serviceable sattelites and material processing platforms. Also important are operations in a risky or unaccessible environment (deep water, dangerous radiation, etc.).

For all these applications the model can be used to simulate the task (generally to control the system from state A to state B), the system dynamics and the environment. In addition HO functioning is included to adequately perform the task.

For analytic purposes, tasks can be simulated fast time, i.e. including the HO model. This allows a straightforward investigation of all task variables of interest, e.g. the effect of time delays in teleoperations, the visual environment, the use of graphic-enhanced video images and additional cues or complete graphic substitution, system dynamics and HO related aspects.

For the linearized model structure the model outputs are obtained in terms of means and variances of the system variables providing all statistical information of the process (assuming Gaussian disturbances). In case nonlinear system dynamics or non-Gaussian distributed disturbances are assumed, Monte-Carlo simulations can be performed resulting in time sequences of the system variables.

Real time simulations can be carried out using the model to assist the HO in performing the task. For example, the model can provide an improved (graphic-enhanced) visual environment or additional visual information. In addition, the model dicision and control outputs can be used for a decision support system.

In this case of real time operations, the model structure must be evaluated (and possibly simplified) to check the numerical requirements for real-time use.

4. CONCLUDING REMARKS

In this paper a model structure is presented of manned robotic systems in terms of optimal estimation-, control- and decision theory. The resulting model can be utilized to analyze, design and evaluate these systems by establishing the effect of task variables of interest on model outputs.

In case the model is used in a time simulation mode, the results are in terms of time histories (sequences) of interesting system- and HO- related variables. For the linearized model version, statistical measures can be obtained (i.e. ensemble mean values, covariances and probabilities) of all variables of interest. In the latter case the model provides a very cost-effective tool to assess the performance and reliability of manned robotic systems.

It may be expected that the model is capable of answering questions related to design alternatives, function allocation, automation, system reliability and efficiency, etc.

5. REFERENCES

1. Wewerinke, P.H. Models of the human observer and controller of a dynamic system. Ph.D. thesis, University of Twente, the Netherlands, 1989.

2. Bryson, A.E. and HO Y.C. Applied optimal control. Halsted Press, 1975.

3. Hoogland, M. Modeling vessel traffic (in Dutch). Thesis, University of Twente, the Netherlands, 1991.

4. Kleinman, D.L., Baron, S. and Levison, W.H. An optimal control model of human response. Part I: Theory and validation, Automatica, Vol. 6, 1970.

5. Baron, S. A control theoretic approach to modelling human supervisory control of dynamic systems. In: Advances in Man-Machine Systems Research (ed. Rouse), Vol. 1, JAI Press, 1989.

CONTROL OF A FISHING TRAWL:
A MULTI-INSTRUMENT PROCESS CONTROL SITUATION

J.L. Minguy*,** and P. Rabardel*

*Laboratoire d'Ergonomie et de Neurosciences du Travail, Conservatoire National des Arts et Métiers,
41 rue Gay-Lussac, 75005 Paris, France
**Laboratoire Sécurité et Conditions de Travail à la Pêche Maritime, Institut Universitaire de Technologie
de Lorient (Université de Bretagne Occidentale), 10 rue Jean-Zay, 56100 Lorient, France

Abstract. Current technological developments in the field of sea fishing are based on approaches which are independent from analyses in real situations and do not take real user requirements into account. A "nautical chart" is established by a skilled fishing boat skipper on the basis of his own requirements in a fishing trawl process situation. This instrument participates in the constitution of operational representations for control of the process and we study the conditions of development and use by its designer. On the basis of the result obtained after analysis of some of the content of this work tool, pertinent study elements are proposed for the definition of design criteria for the new instruments for the bridge of a fishing boat.

Keywords. computer-aided system design; inference processes; marine system; navigation; process control; ship control; undersea systems; expert systems; activity analysis.

INTRODUCTION

Technological developments for fishing boat bridges (computerization, digitization and automation) are part of a plan to optimize the fishing process (digital depth sounders with crossed beams and multi-beams), the management of navigation and fishing tasks (slaving of direction commands to positioning systems), management of stocks and tides (electronic fishing books, satellite telecommunication systems). They partly satisfy certain user requirements (Barnouin, 1988; Benoish, 1988), in terms of functionality, profitability and efficiency.

At present, state of the art in the field of the design of fishing boat bridge instrumentation has partly followed electronic and data processing developments in the field of the merchant navy as presented by Larsen and Istance (1984) for example, but in particular economic and social context (postulate of the "One-Man Bridge"). Then, various new types of apparatus are proposed (Mross, 1988) according to the principles of centralization and multi-functions after the development of a cybernetics type approach.

One of them enables real-time follow-up of the catch on the screen indicating the ship and a vertical cross-section of the sea-bed. Another cumulates several functions (positioning and catching) and gives a flat view of the route followed by the ship, the field scanned by the trawler and the position of schools of fish. All this is shown on a radar type screen. The definition of these new apparatus satisfies three questions which, according to the author, are the main questions asked by the fishing boat skipper :
- in which zone will I find the most fish?
- where will I place my trawl the next time?
- what route should I follow in order to catch the right amount of fish?

We think that an approach which gives preference to analysis of the real activity on the basis of collection of data in situation is able to provide pertinent elements for definition of the criteria for designing fishing boat bridge instruments.

Global analysis of the activity on the bridge of a fishing boat highlights two separate types of tasks (Minguy, 1990) :
- those relative to steering the ship,

- those relative to fishing operations.

They also highlight certain particular uses, depending on the amount and type of information presented and the trust placed in the instrument. In the case of navigation on the basis of double positioning with automatic correction of position errors, the user is not only concerned by the validity of the boat's position but also the validity of the correction made.

Finally, they show that certain instruments are considerably under-used (the route plotter for example) while the functions available are designed to satisfy the user's requirements : memory storage of the ship's trajectories, recording of special events, possibilities of enlargement, coupling with a positioning system, etc.

The question we ask is : "On what basis can we design a tool for assistance with the control of the trawling process?". The idea is that, on the basis of analysis of the content and the nature of properties of instruments designed by users in situation, it is possible to highlight certain functional characteristics of future instrument systems onboard fishing boats. From the viewpoint of improving working conditions, we consider that the concepts of adaptability and flexible use are vital both in terms of offering experts the means of appropriating a working tool and giving trainees the possibility of individualized learning.

Here we shall study the elaboration and use of a special assistance tool (a sea-chart) designed by an expert operator on the basis of his own requirements and a collection of data from a multiple instruments on the bridge. We shall discuss the functions of the instrument produced as such in regard to the "route plotter" instrument and the pertinence of such an approach for the design of new fishing boat bridge instrument systems.

ANALYSIS OF THE INSTRUMENT PRODUCED BY THE SKILLED SKIPPER

The sea chart is an instrument which is produced by the fishing boat skipper on the basis of pre-existing "official" documents (topographical survey maps, temperature readings, navigation chart) provided by national and international organizations. It is actually the enlargement ("zoom") of a specific zone chosen from the criteria linked to

fishing (presence and types of species of fish, accessibility of waters, etc.) and navigation (community fishing zone, territorial waters, special weather conditions - swell, temperatures, currents, iceberg, etc.). In the rest of the communication, we shall concentrate our analysis on a special fishing zone which represents an area of around 150 square miles.

Two main families of data, which are encoded and symbolized in a special way, are shown on this map : those relative to the environment of the sea-bed which present a feature of stability (geographical positions, sounders values, special events) and those relative to actions linked to control of the fishing tackle on the sea-bed (trajectories, indications for manoeuvres, etc.).

Then, first of all, we shall talk about pertinent data concerning control of the fishing itself : type of trajectories and pertinent identifications for control. After that, we shall take a look at the data relative to the pertinent environment for trawling operations (intrinsic characteristics of the sea-bed, characteristics and properties of certain elements of the sea-bed).

The figures shown hereafter represent simplified extracts from the map established by the fishing boat skipper. To make it easier to read, we have used a "grid" (squaring) in which each element can be referenced by a couple associating a letter and a figure.

Content Relative to Control of the Trawling Process (see Fig.1)

Various indications shown on the chart concern control of the fishing tackle on the sea-bed in relation with steering the ship on the surface and the special characteristics of the environment. Movement of the fishing tackle along the sea-bed through control of the ship's movement on the surface : the trajectory of the trawl on the sea-bed is not identical to that of the boat on the surface since it is drawn at a certain distance from the boat and is therefore subject to influences other than those of the ship.

So the purpose of the indications shown on the chart is to enable anticipations about the changes in the respectives trajectories in terms of actions performed on control of the boat and its speed, on the cables linking the boat to the trawl and also knowledge of the characteristics of the environment (intrinsic properties of the sea-bed).

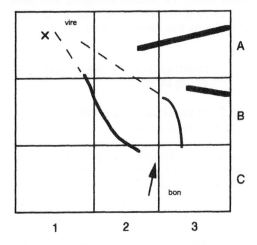

Fig. 1. Indications relative to the trawling process

Sea-bed trajectories. They indicate the trajectory of the trawl on the sea-bed and appear in the form of portions of curves (**B2**) or straight-line segments (**A3**) which are coloured yellow or blue.
The thickness of the length gives an indication of the portion of space taken up by the trawl on the sea-bed (horizontal and vertical openings) as well as the acceptable limits of changes in the trawl relative to the characterics of the sea-bed. This thickness may be interpreted as intended to define the "acceptable bracket" inside wich the trawl is drawn.

Surface-trajectories They take the form of black dotted lines (**A1**, **A2**). The length of the lines on the chart indicates the time necessary to perform certain manoeuvring operations, knowing the speed of the boat and the type of operations to be done.

These two types of trajectories are connected (**A2/B3**, **A1/B2**) and drawn on the chart if control is done in an unstable environment : presence of different types of current, possible changes in the relief, obstacles. In this environment which is not known a priori (on the basis of information on the sounder), the control of the situation is simply done from a representation constructed around empirical knowledge and procedures in relation with technical data.
- openings in the trawl,
- inclination of the boards on the ground,
- résistance to traction of the rig elements,
- gyration properties of the rig.

References for "critical actions" are shown on the map in the form of literal or symbolic indications (**A1, C3**). They concern direct actions to be carried out on the conditions of development of the process (gyration movement, turning[1] or paying out[2]) or indicate the conditions for triggering actions or a special feature of the environment (current) which is at the origin of the modifications to be made in the process control (**C2**).

When the environment is static, which means that the conditions under which indications are obtained do not depend on a given state of the environment at a given time, construction of the instrument, in particular the chart, shows the indications relative to the trajectory of the trawl on the sea-bed (**A3, B3**). The environment is sufficiently determined to be able to accept, in a reliable way, a movement bracket for the trawl on the sea-bed that is represented on the chart. On the basis of determination of this movement bracket, the conditions of actions to be performed of the boat/rig interaction are easily anticipated and take place in real time.

Contents Relative to the Pertinent Environment for Fishing (see Fig.2)

Geographical data. They refer to the history of the construction of the map in regard to evolutions of the positioning instruments. The "map background" is formed on the basis of a grid of coordinates established with LORAN C. They are the indications shown on each side and on the top of the map (see on the top of the figure 2).

Special events inscribed in the form of a circle (**E2**) refer to a DECCA positioning : it is a position indicator which is no longer used but from which some information is transferred to this fishing chart. Square form (**E1**) refers to a geographical position, must often to LORAN C. This symbolic encoding system is complemented by indications shown in the form of a triangle (**D1, D2**), which refers to events that are actually recorded on the image shown on the echo-sounder. The user knows that this

---

1 turning means all manoeuvring actions which enable the rig to be brought on the deck
2 paying out means all manoeuvring operations which enable the rig to be put into the water

information refers to a considerable deformation of the ground. The positioning system associates these identifiable events with the sounder and the system being used (LORAN C or GPS) at the time of detection.

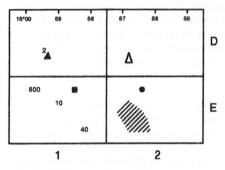

Fig. 2. Indications relative to the pertinent environment for fishing

Topographical data. These are either whole or partial sounder values (E1) or data on the type and quality of the sea-bed. Detailed indications are sometimes associated (D1) with symbolic indications (circle, square form and triangle) which are intended to give a more comprehensive idea of the event indicated (hazards[3] , wrecks). If a special event has been identified on the echo sounder or simply inferred during the run (cable vibrations, boards[4] sinking into the mud, etc.) or after the run (loss of part of the rig, tearing of canvas, etc.) codification will be different (colour and form).

As such, red shaded areas (E2) indicate a zone with identical characteristics over a large area (hard or soft ground, special type of sea-bed - sponge, coral, etc.) which should be avoided. The determination of this zone results from the production of inferences on the basis of a collection of values noted in situation or a posteriori on the rig, according to a generalization process based on the accumulation of these values, as and when the rig passes through this zone.

Also, a red triangle with an indication of a height of two metres (D1) means that a "hard" hazard ("piece of metal") which has provoked considerable damage has been identified by the

3 hazards may be defined by particularities of the sea-bed likely to interfere with the rig. These are natural particularities ("rock summit", etc.) or artificial particularities

4 these are two heavy metal otter with a divergent effect

sounder and its characteristics (position, height and dangerousness) has been inferred, partly during the process (snagging took place) and partly a posteriori (evaluation of damage after the rig has been pulled in).

FUNCTIONS OF THE SKILLED SKIPPER'S INSTRUMENT

The Chart as a Place for Integration of Data from Multiple Sources

This graphic medium is a place where all the information from multiple sources is integrated and converged to form a consistent whole. The user of the chart takes into account the relative value of data in the subsequent use which he makes thereof. Thus, the systems of signs used to characterize certain properties of the environment implicitly bear a "validity coefficient" depending on whether they have been collected in real time or a posteriori from instruments on the bridge or from values. These data are systematically compared with their origin in order to characterize their intrinsic value and to relativize the significance of indications shown on the chart.

The Chart as a Place of Generalization and Production of Inferred Information

The processing of data which come from multiple sources and are collected at different times during trawling is aimed at producing generalizations about the characteristics of the environment (determination of a potentially dangerous zone, a good fishing zone, etc.). This processing also leads to an operationnal representation for the fishing process. For example, on the basis of extrapolation of the sounding lines values, it is possible to construct a level curve which is pertinent for the control of the fishing tackle and the hunting of fish.

The Chart as a Place of Data Processing

Thanks to different types of graphic operators placed on the chart, it is possible to convert the indications which already exist or which concern features that are pertinent for the fishing process :
* a rule instantly converts nautical miles into kilometers. On the basis of a report of the length of cables measured in meters, it can estimate the distance (in miles) separating the boat from the trawl and thus infers the trawl's

geographical position. Due to this, it can estimate the time when fish detected directly under the boat will be liable to be caught, in the knowledge of the scrolling speed of the image or the advance speed of paper on the sounders as well as the speed of the ship.
* An algorithm developed by the expert, represented in the form of a vector, makes corrections of the positions and trajectories in order to take account of the instrumental error specific to each positioning system (GPS or LORAN C).

The Chart as a Medium for Mental Representations which are the Sources of Actions in the Control of Fishing Tackle

The production of a sea-chart mainly takes place during action. The graphic operational representation for control of the fishing process constitutes a medium for the production of mental representations by the operator relative to :
- the state of the environment at the time when the fishing process takes place (characterization of the static environment and the dynamic environment)
- the methods of process management in relation with the complexity of the situation.

The comparison of data about the elements collected during action with the data already known, indicated on charts (and possibility reprocessed) enables the constitution of these representations which guide the actions, including mentals actions, concerning the process itself and knowledge of development of the process (LEROY, Y. and MINGUY, J.L., 1992). These representations are used to anticipate pertinent action procedures for control of the fishing tackle.

DISCUSSION

Therefore, when the instruments produced by the expert skipper are analyzed, it is possible to identify properties and functions which are desirable from the viewpoint of the design of new generations of instruments that are better suited to the requirements of users.

As regards the route plotter, we note that :
* at no time is there any question of the sea-bed trajectory. It is only the trajectory of the ship on the surface which is recorded and may be recalled,
* the indications of the sounding line values are correlated with the ship's trajectory :

what is obtained is a simple topographical survey (values of the sounding lines positionned) on the basis of a surface trajectory and not a sea-bed trajectory, and yet this is what the skipper is trying to manage,
* no functions are provided for producing the representations of the variables calculated (level curve for example) while this would appear to be one of the pertinent elements of the work environment for our expert fishing boat skipper,
* the symbolic encoding elements do not enable the creation of "instrument systems" containing precise significances for the user.

The consequences of this are that :
1- The indications concerning manoeuvring actions, for example, cannot be compared with the boat/rig and trawl/sea-bed interactions since most of the information given by the "route plotter" instrument only refers to the boat (flashing "spot" indicating the position of the boat on a trajectory that is pre-recorded or not) and has very little to do with the boat-rig relation (boat-trawl distance, knowing the length of cables, and the position of the trawl, knowing the angle of the cables compared to the surface and the boat).
2- The construction of new routes on the basis of modifications made to the information present and in terms of targets to be reached (passing as close as possible to a hazard without touching it, going against a current, hunting a school of fish, etc.) requires special processing of this information which is not taken into account in the plotter's functions : direct knowledge of the entire trajectory followed, modification of the positioning of particular events, etc. on request.
3- The choice of the number of characteristics of variables chosen to produce this graphic representation (imagery) is reduced, cancelling the processing intended to make it operational (generalizations, inferences, links with their origins, recognition of associated significants, etc.).

The multi-functional character of this instrument and the centralization of information about this sea chart inform the user of both the process and its dynamics, about the tool (method of construction and use) and about the conditions of the work environment. The use of symbolic differentiated encodings (shapes, colours, thicknesses, size, scales, etc.) combined with open-ended graphic representation, which is directly accessible at all points, gives this instrument multiple properties :

* it is an instrument for the integration and concatenation of past and present data according to processes of correction, validation and determination level,
* it is a medium for the construction of operational mental representations for actions involved in controlling fishing tackle,
* it is a space whose properties enable the simple, real-time production of data processing as and when required (example of correction of positions and trajectories according to the type of instrument which is the information source).

CONCLUSIONS

The production of a work instrument by the user himself, using information which he has collected because it is pertinent for management of the process, is a sign of the relative adaptation of the instruments at his disposal. The "sea-chart" instrument of this fishing boat skipper show essential characteristics of the environment of the trawl on the sea-bed, the pertinent properties of interactions between the ship and the rig and between the trawl and the sea-bed as well as indications about procedures. This assistance tool is perfectly suited to his activity since it represents the translation of his information requirements, which are personalized, retranscribed and easily referenced on a medium which evolves along with the tides.

So, thanks to an analysis of the instruments produced by the expert skipper, it is possible to put forward hypotheses about the reasons for under-use of existing instruments and also identifies the properties and functions that are desired for the design of new generations of instruments better suiter to user requirements.

These analyses must now be backed up by other analyses carried out on board which would help understand the conditions of production and use of this chart in regard to characteristic situations, paying particular attention to the collection and processing of information by the skipper on the basis of all the instruments on the bridge.

Over and above an approach based on the expertise of a user, the problematics of the design of bridge instrumentation for a fishing boat also concern the activity of control of the fishing tackle by less-experienced operators (executive officer, second mate) on the basis

of the use of such an instrument. Here, we shall study this from the viewpoint of analysis of the collective work activity which is sharing experience.

REFERENCES

Barnouin, B. (1988). Les enjeux, les besoins de la pêche maritime; le contexte économique, les actions en cours. Bulletin du Centre de Culture Scientifique, Technique, Industrielle et Maritime, n°1 juillet 1989, actes du Colloque électronique et pêche maritime, Lorient November 17-18th, 1988. Publisher : Maison de la Mer de Lorient.

Benoish, M. (1988). Le point de vue de l'utilisateur à bord. Bulletin du Centre de Culture Scientifique, Technique, Industrielle et Maritime, n°1 juillet 1989, actes du Colloque électronique et pêche maritime, Lorient, November 17-18th, 1988. Publisher : Maison de la Mer de Lorient.

Larsen, P., and H. Istance (1984). Application of computer technology and system integration in ship operation. Second International Bridge Design and Operation Forum. Hamburg, 25-26 september 1984. Hamburg Maritime Research : Forschungsstelle

Leroy, Y., and J.L. Minguy (1992) (to be published). La conception des passerelles de navires de pêche : approche globale du travail et apport de l'expertise d'un patron pêcheur à la définition de nouveaux critères de conception. 1st European Ergonomics Symposium, Estoril, 27-29th February.

Minguy, J.L.(1990). Analyse et modélisation de situations de travail en passerelle de navire de pêche, orientations en vue d'une conception ergonomique de l'instrumentation. Mémoire de D.E.A. d'Ergonomie du Conservatoire National des Arts et Métiers, Paris, September 1990.

Mross, S. (1988). Fishing electronics : new trends in industry and training. Proceedings of the world symposium on fishing gear and fishing vessel design. Published by The Newfoundland and Labrador Institute of Fisheries and Marine Technology, St. John's, Newfoundland, Canada : Marine Institute. pp.249-253.

SEPARATION OF USER INTERFACE AND APPLICATION WITH THE DELFT DIRECT MANIPULATION MANAGER (D2M2)

C. van der Mast and J. Versendaal

Delft University of Technology, Delft, The Netherlands

Abstract. In this paper a user interface management system is presented based on the same object-oriented approach to the design of both the user interface and the application semantics. However, data for user interface and application semantics are stored and processed apart. The outstanding feature of the system is that the modelling technique is identical for both, using the same tool for semantic data modelling. The modelling technique supports most characteristics of the object oriented paradigm: objects, classes and inheritance. The diagrams used can describe among others the concepts classification, aggregation, decomposition, generalization. Attributes of objects can be described in property forms. Transition diagrams are applied to specify the interaction possible to the end user. From these specifications a run time version of the whole application can be generated. A protocol is used for the communication between application and user interface manager during run time. Some features of Delft Direct Manipulation Manager are demonstrated with the design and implementation of a small application with direct manipulation.

Keywords. Human factors, Man-machine systems, Software development, User interface design, Visual programming.

INTRODUCTION

The need to develop and maintain complex information systems has triggered new approaches to system design and development. The object-oriented paradigm addresses some of the main problems with the traditional life cycle model with distinct phases. Most references (Korson et al., 1990) eliminate the boundaries between phases of the traditional activities analysis, design and implementation. This offers alternatives to the waterfall model because all phases can be integrated within the same conceptual framework. Explicit classification structures developed during the analysis phase can be the base for a seamless transfer to the implementation phase with object-oriented languages. Moreover, it seems natural that design pieces are closely identified with real-world concepts which they model. This holds both for the development of the semantics and the user interface of an application.

Therefore we emphasize user interface development as a natural part of the application development, when done according to the object-oriented paradigm. One of the consequences of this approach is that user interface design can be done in parallel with information analysis and application design. This approach also provides a way to cope with complexity and to reuse objects and methods. Another consequence is that user interface design should start, like system analysis, top-down using logic diagrams. When more details concerning *objects* become available during the design, independent interaction *tasks* could be specified to be performed on a predefined subset of those objects by the end user. These tasks provide entries for bottom-up rapid cut-and-past-prototyping, maintaing the overall top-down strategy. One main assumption within the research described in this paper is to support this top-down approach as the base of a formal description of the design.

Separation of user interface *code* and application *code* is an important approach to support maintainable complex information systems. Many generations of solutions have been proposed and implemented for this goal (Hix, 1990). The separation of the user interface and application can also be considered during the *development* of the software. The user interface can be designed more or less separately. This means that human factors specialists are able to bring up their experience within a multi-disciplinary design team.

To establish the separation of the application and its user interface User Interface Management Systems (UIMS) were built. The purpose is to let the UIMS provide an environment for the management of user interfaces, during development and execution. During execution, the application simply sends requests to the UIMS and the UIMS takes care of the interaction process, checks syntax and returns the results of the user's actions to the application. The application does not know how the information is obtained from the user or how its (feedback) information is presented to the user. Therefore the user interface can be changed in appearance as well as in behavior without the application noticing it.

The Delft Direct Manipulation Manager (D2M2) is a UIMS under development at Delft University of Technology. This UIMS manages graphical user interfaces which allow direct manipulation on Unix workstations. D2M2 consists among others of D2M2edit, the development part for non-programmers and programmers, and D2M2run, the run-time part. D2M2 can be positioned within the "4th and the future generations" of Hix (1990). It supports maximal separation between both data and processes of user interface and application. At the same time D2M2 tries to integrate the methodologies and concepts of the user interface designer and the designer of the application semantics. This integration is the most outstanding feature of D2M2.

In section 2 related research is discussed. In section 3 the main concepts of D2M2 are presented. In section 4 the architecture of the tool is presented. In section 5 the use of D2M2 is illustrated with a small example: a filemanager. The final section summarizes the paper and indicates other research being done with D2M2.

RELATED WORK

UIDE, Foley's User Interface Design Environment, is a complete UI design platform, founded on a knowledge base, offering user interface design and building tools, run-time, evaluation and documentation facilities and a run-time help system (Foley et al., 1990a). Strong emphasis is placed on the user interface; the application is allowed to modify the UI object rule base, but the UI determines which part of the application gets invoked. Control lies with the UI, while the application routines can actively partake in the management of the user interface.

Szekely (1989) attempts to decrease the dependance between application and UI building blocks by standardizing the procedures used in the UI-application interface, based on the representation of operations/procedures as objects, with attributes like preconditions, cancelable, preview and with input, type validation, transformation etc.

The Serpent UIMS (Hardy and Klein, 1990) is constructed upon the Seeheim model (Green, 1985). It encourages the separation of concerns between the UI and the semantics of an application. Serpent contains the standard layers presentation, dialogue and application. Serpent uses a special-purpose language called Slang to specify user dialogues. With a special structure editor Slang definitions can be created without the burden of knowing all the syntactical details of this language. The Slang code has to be compiled to create the dialogue layer. The presentation layer and the application layer have to be programmed according software interface definitions prescribed by Serpent.

VUIMS is an object-oriented UIMS (Pittman and Kitrick, 1990). Objects implement primitive presentation and interaction functions. A token language controls interaction and visual style. It is independent of a particular presentation and interaction style. Complete separation of application from UI is claimed. The presentation and the dialogue component are embodied in tokens that are stored in external files during run-time. VUIMS currently provides minimal support for interactive design tools.

Evaluating separation between UI and application we can classify UIDE as specifying the UI fully separate from the application; the UI is not derived from any application semantics description. Szekely sees UI and application as two completely separated parts, communicating through a well defined interface. He does not yet offer any tools to design or build either side. VUIMS has only tools for the UI component. Serpent is offering also both separation and an environment for the design and prototyping of the UI component. But it does not support the application development.

If we compare the systems mentioned here we can conclude that those having simple specification techniques have little separation. Others do support some or good separation of UI and application. However, none of the described systems does pay attention to the similarities of UI and application data, although the object-oriented approach offers possibilities to do so (Korson et al., 1990).

So, as a goal for this research has been chosen to develop concepts and an architecture which do separate the components UI and application completely, both for design and for run-time. And, for run-time a communication protocol has to be defined between both components which supports also semantic feedback. The chosen approach is to use the same semantic data modeling techniques and diagrams to design both the data model for the application and the user interface, so aiming that all disciplines in the design team can communicate in the same "language", see figure 1 (Versendaal, 91).

CONCEPTS

Applications are designed to help users in performing tasks; therefore human-computer interaction should never be an independent goal, but a means to trigger and evaluate computer actions. We define a computer action as the execution of a *computer task*; the triggering of a computer task is performed by an *interaction task*; the user can evaluate the results of the computer action from the visible results of a *feedback task*.

We are now able to describe the sequencing in human-computer interaction for a great deal of applications in terms of interaction tasks, feedback tasks and computer tasks. A user executes an interaction task; the interaction task will trigger a computer task; during and/or

just after computer task execution feedback tasks can be executed. This sequencing holds not only for command language interaction styles, but also for Direct Manipulation and others (Versendaal, 91).

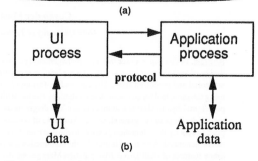

Figure 1: Separation of UI and application during runtime (b), while using an conceptually and physically integrated design environment (a).

During interaction task execution so called user interface data (UI data) is consulted and updated. Examples of UI data are windows, icons, scroll bars, text fields, etc. Also during feedback task execution UI data is consulted and updated. During computer task execution application semantics data (AS data) is consulted and updated. In order to achieve independency of user interface and application semantics, UI data is separated from AS data and stored independently. If AS data must be consulted *during* interaction task execution in order to update the UI data, we say that *semantic feedback* must be accomplished.

UI data is a collection of instances of *UI data types*. We specify UI data types by identifying classifications and generalizations: this specification process is identical to the process of semantic data modelling see e.g. (Smith et al., 1977; King et al., 1985; Ter Bekke, 1991). We use the concepts and the notation technique for semantic data modelling of Ter Bekke (1991). In the next section diagrams of this notation technique will be introduced. We note that other notation techniques like entity-relationship diagrams could be used as well.

Besides UI data types we distinguish *interaction sets* and *feedback sets*, both consisting of some UI data types needed to execute a single interaction or feedback task. During run-time, an instance of an interaction set allows for both user interaction in its rectangular region as well as output of feedback tasks; an instance of a feedback set does not allow user interaction in its rectangular region: only output of feedback tasks can be directed to the rectangular region.

We specify *interaction tasks* by transition diagrams (Green, 1986), in which transitions can be traversed when a certain premise, a boolean function which consults UI data, holds. On traversing a transition UI data can be updated, so realizing lexical and syntactic feedback (Foley et al., 1990b). We specify so called *data transfer* with an interaction task in order to send the result of the user interaction to the application. Interaction tasks belong to an *entry*. Not the name of an interaction task, but its entry name is known to the application. If different interaction tasks produce the same data to be sent by the data transfer, we can group these tasks in the same entry. As a consequence the application could receive, during run-time, data from the user, not knowing from what interaction task it came from; only knowing that it came from a certain entry. Obviously, this construction supports the emphasis on the separation of user interface and application. Interaction tasks are considered to be "small" at human scale, i.e. taking seconds to be executed physically.

We specify *feedback tasks* by functions which consult and update feedback sets.

We specify *AS data* by identifying AS data types, generalizations, aggregations and attribute identification like in UI data specification. We specify *computer tasks* by application code which consults and updates AS data. We support semantic feedback by *semantic feedback functions* which consult AS data. See for an extensive discussion of the concepts described here (Versendaal, 1991).

ARCHITECTURE

D2M2 is a tool which allows for designing, implementing, interpreting and maintaining applications with a separated user interface. The interpreting part of D2M2 is D2M2run; it executes the user interface by interpreting the UI data and takes care for the communication with the application semantics. The implementing and maintaining part of D2M2 is D2M2edit; it allows for defining and editing of both the user interface (consisting of UI data, sets, feedback tasks and interaction tasks) as well as the application semantics (consisting of AS data, computer tasks and semantic feedback functions). In figure 2 the relations between D2M2edit, D2M2run and their environment are depicted. In (Versendaal, 1991) D2M2edit and D2M2run are described.

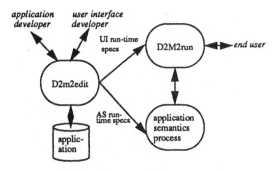

Figure 2: Main components of D2M2 and its relations. The stored application is the definition form for editing.

D2M2edit

The main functions of D2M2edit are shown in the editor window after starting up. Figure 3 depicts the editor window with the main pull down menus to activate the main functions of D2M2edit.

One can choose between library maintenance and application (project) management. Only the latter will be presented, for this is the prime function of the development environment. Selecting application management brings up its menu, offering:

* Application file management, which comprises retrieval, storage, cataloguing, creation and deletion of applications under implementation.
* Application editing facilities, the creation and manipulation of the user interface and application elements through the use of 'views' (see below).
* Printing of documentation of the application being edited.
* Generation of run-time specifications from the user interface and application specifications.

Of these, the application editing best shows the graphical interactive

character of D2M2edit. As explained before, a set of specifications under D2M2 consists of user interface elements (the UI data types, interaction sets, feedback sets, interaction tasks and feedback tasks) and application elements (AS data types, application code and semantic feedback functions), together with the parts required for communication (entries to group interaction tasks, data transfer code to return results of interactions and creation code for instantiation of an interaction or feedback set). All these can be edited, to which end several *views* are available. A view displays a small portion of the specifications, of one type of element, and allows easy and manageable creation and manipulation of user interface and application, giving all information required and retaining consistency without bothering the designer with irrelevant details or distracting relationships.

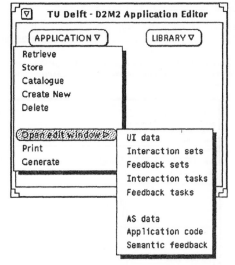

Figure 3: Main command window of D2M2edit with submenu containing several "views" on the application.

An example of such a view is the 'UI data type view'. Through such a view, the designer can specify the hierarchy of data types comprising the interface, and their behavior. The basic UI data types, offered by the relevant window manager and available through the built-in library, are always present and called 'native' types. A special UI data type 'native' is ever present in an application: every UI data type directly related to the window manager is directly or indirectly derived from 'native' by specialization (being the complement of generalization), inheriting its basic properties and methods, but adding some of its own. UI data types not related to the window manager are specializations of the ever present UI data type 'own'.

Graphically, in this view two objects can be distinguished: UI data types, represented by labeled rectangles, and generalization relationships, shown as lines, connecting the UI data types. In figure 4 the way UI data types and their relationships are represented in D2M2edit is depicted.

Buttons in the UI data view window (figure 4a) are used to *Create* new types, *Show* types created earlier, but not visible in this view and *Set* a subtype (i.e. create a generalization between types). These are all functions that place data types in the view. All other functions to be performed on these graphical objects are invoked through manipulation of the objects themselves, as dictated by the concept of direct manipulation interaction. They fall into two categories: functions which merely change the appearance of the view and functions changing the UI data type specifications.

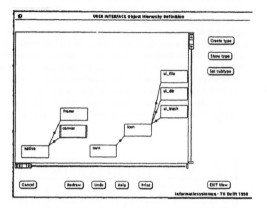

Figure 4a: The UI data view window of D2M2edit. The lines between the corners of the rectangles represent generalization relations.

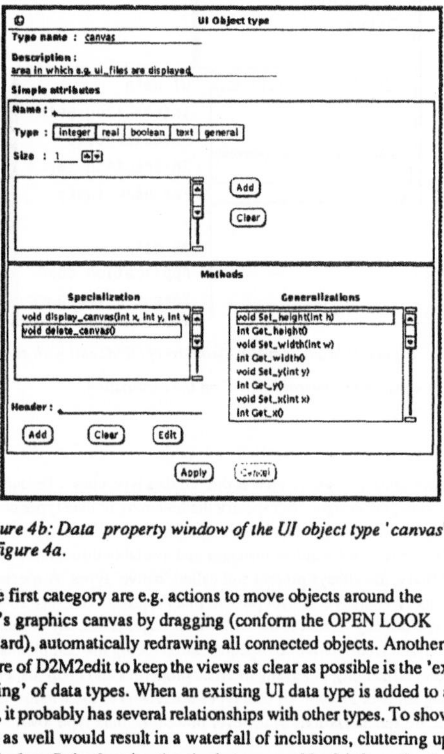

Figure 4b: Data property window of the UI object type 'canvas' in figure 4a.

In the first category are e.g. actions to move objects around the view's graphics canvas by dragging (conform the OPEN LOOK standard), automatically redrawing all connected objects. Another feature of D2M2edit to keep the views as clear as possible is the 'expanding' of data types. When an existing UI data type is added to a view, it probably has several relationships with other types. To show these as well would result in a waterfall of inclusions, cluttering up the display. Only showing the single type would misinform the designer. D2M2edit displays such a type as a double-edged rectangle (see UI data type 'canvas' in figure 4a). When the designer requires more information he 'expands' the type, causing D2M2edit to arrange all directly related types in the view as well.

The second category of functions actually edit the specifications of the UI data types. Examples are: copying and deleting types (with their connected relationships) and editing the properties of data types and generalizations. A special property window is placed on screen when the designer clicks on the object, displaying all properties and allowing them to be edited. E.g. in the case of UI data types there are the name, description, attributes (each with a name, data type and size) and methods (each with a return type, parameter list

and body). In figure 4b a snapshot of a UI data property window is depicted.

Analogous to the previous view, there are the (interaction and feedback) 'set views'. These allow the designer to create or edit interaction sets and feedback sets. A set consists of a collection of UI data types created earlier on or retrieved from a library, and parent-child relationships between some of these types, indicating their hierarchy when used together in an application. A child is always displayed *inside* its parent.

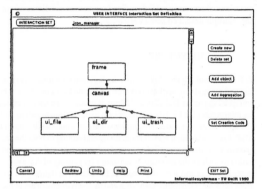

Figure 5 : Interaction set view window showing the interaction set of the example file manager

Again, the types are displayed as rectangles, while the relationships are represented by connecting lines. Within a set, no changes can be made to the actual data types used; this must be done in UI data type views. In figure 5 an interaction set view window is depicted.

The third view of interest is the 'interaction task view'. After having selected an interaction set, the designer can create or edit an interaction task. Such a task is graphically displayed as a state-transition diagram, where one creates and places the states (circles) and the connecting transitions (directed arcs). Conditions and actions for the transitions are regarded as their properties, and so can be edited through a property window. The layout of the diagram can simply be changed by moving the graphical objects. Functions are furthermore provided to specify data transfer code and the entry to which this task belongs, and to check the validity of the transition diagram (determinism, islands, initial/final states, etc.). In figure 6 an interaction task view window is depicted.

For the specification of the application data a slightly different version of the UI data view editor (see figure 4) is provided.
The above represents a brief impression of part of the facilities offered by D2M2edit. By consistent use of graphical user interface D2M2 can be used both by programmers and non-programmers. Currently D2M2edit is only being used by programmers in research projects in several countries.

D2M2run

D2M2run manages the user interface for the application during runtime. We recall that the static part of the user interface is implemented as a semantic data model in D2M2edit. The dynamic part of the user interface is implemented as state transition diagrams. The static part is translated by D2M2edit into C++-classes, which can be interpreted by D2M2run directly; state transition diagrams are processed directly from the specications by D2M2run, according to the user input and pre- and post-conditions within the application.

During run-time D2M2run and the application communicate with each other using a special protocol. This protocol allows the application to specify the type of user interface by enabling entries. After this enabling it tells D2M2run to activate the user interface and starts waiting for D2M2run to send its user input. In order to support real time applications the application component has priority over the interaction component.

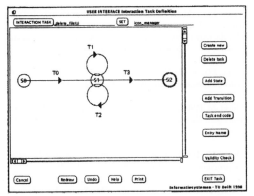

Figure 6: Interaction task view window showing the interaction task 'delete a group of selected files' of the example (file manager). T0: button clicked on UI_file, T1: button clicked on UI_file, T2: button pressed on selected UI_file and move T3: button released on UI_trash. These conditions and belonging methods can be specified on property windows.

During the human-computer interaction D2M2run handles all input and output until the user completes a command or data input sequence. D2M2run takes care of lexical, syntactic and semantic feedback (Foley et al., 1990b). However, when, during run-time, D2M2run needs information from the application semantics in order to produce semantic feedback, it sends a request to the application to give it the required information. It cannot access the application semantics data itself.

When the user completes an input sequence, D2M2run notifies the application by telling it what type of input is retrieved from the user, via the entry name, along with the appropriate data. The application interprets the input sequence and executes a relevant computer task; during and after computer task execution requests for feedback task execution can be sent. At no time either the application semantics process or D2M2run has access to each other's process data.

So, the application requests D2M2run to manage its user interface of which it only knows its functionality. D2M2run only manages the user interface as it was told to do by the specifications in D2M2edit, sending complete input sequences, commands and/or input data, leaving the interpretation of it to the application. The application does not know HOW the input was retrieved from the user and HOW its output and feedback is presented to the user while D2M2run does not know what the semantic meaning of the data is it sends and what the objects it manages mean to the user or the application. The only thing they know is how to communicate with one another.

Presuming the communication channel can vary from a local message passing mechanism to a satellite connection, D2M2run and the application semantics process may run on any machine. As long as they use the protocol they can communicate, allowing D2M2run to handle the application's user interface. The current version is supporting only one application at a time on the local processor.

EXAMPLE

The design of an application roughly consists of three (not independent) steps. The first step is to determine the functionality of the application and the application data types. The second step is to determine what functional regions an application needs and what commands the user can give. The third step is to precisely determine the user interface data types and lay-outs. As an example of these steps we present the design of a simple file manager.

The file manager must perform the following commands:

1) a group of selected files may be deleted;
2) a file may be copied to another file or directory;
3) a group of selected files may be copied to another directory;
4) the working directory may be changed;
5) the file manager can be quitted.

These requirements result among others in the following specifications which must be created in D2M2edit.

Application semantics data types (figure 7):

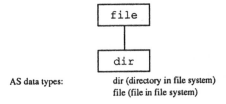

AS data types: dir (directory in file system)
file (file in file system)

Figure 7: application semantics data types of file manager

The only interaction set of the file manager is already presented in figure 5.

The user can e.g. delete a group of selected files (requirement 1). The transition diagram for this interaction task is depicted in figure 6.

In figure 8 the lay-out of the file manager is depicted. This picture is produced by D2M2run according definitions made with D2M2edit. The code which D2M2edit generated had to be changed by hand in the current prototype of D2M2.

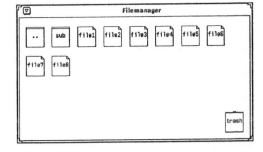

Figure 8: lay-out of the file manager as produced by D2M2run

The presented implementation results in the following entry identifications, which are the identifiers with which the application can activate the commands mentioned above and with which D2M2run can indicate what command the user has given.

* FM_ENTRY_FILES_DELETE
* FM_ENTRY_FILES_COPY
* FM_ENTRY_DIR_CHANGE
* FM_ENTRY_QUIT

Notice that the second and third requirement (see above) are joined into one entry identification, since they are semantically the same. In the second requirement the group simply consists of one file.
The application must have the possibility to add instances of user interface data types to the functional region, delete them and tell D2M2run it is terminating. To establish that the following feedback task identifiers in D2M2run can be distinguished:

* FM_CREATE_UI_FILE
* FM_CREATE_UI_DIR
* FM_DELETE_UI_FILE
* FM_DELETE_UI_DIR
* FM_DELETE_ALL
* FM_QUIT

Of course the file manager must have a functional region in which the instances of user interface data types representing files and directories are placed and manipulated by the user. The functional region is placed within a frame. The identifier in D2M2run for this functional region is:

* FILEMANAGER

These are all identifiers needed for the communication between the file manager application and D2M2run.

CONCLUSIONS

D2M2 aims to separate the UI data and the AS data. This separation proved possible during run-time and during design. An advantage of separation during design is, that human factors specialists and information engineers can focus on their own view and expertise. By using the same object-oriented modelling paradigm the understanding and the co-operation within a multidisciplinary design team may be improved. This will be investigated further in usability experiments. Moreover, D2M2 provides a basis to alter the user interface in a rather easy way, because the user interface is not highly intermingled with the application semantics. As long as the protocol between user interface and application will be obeyed, user interface changes do not affect the application semantics.

Currently some realistic applications with graphical user interface were designed in order to explore these advantages (Versendaal, 1991). These applications are a graphics manager to present raw data on request in tables, bar-, dot-, line- and area diagrams. The second is a computer assisted instruction-application which has the purpose to teach children to add and subtract using a combination of coins and 3D spatial cubes. From these two real applications the first practical experiences with D2M2 are produced. Later, controlled experiments are needed to prove this hypothesis on object-oriented design in different application domains. The current version of D2M2 is supporting only some of the trajectory during the design and maintenance of applications and their user interface. Recently D2M2edit has been integrated successfully with the interactive user interface design tool Devguide (De Baar et al., 1992). This approach provides generating an interface directly from the data model as designed with D2M2 itself. A single data model of the application semantics can be mapped onto multiple graphical user interfaces. Future research is planned for prototyping the lay-out and generating better run-time code. D2M2 is being used in several research intsitutes to explore the semantic data modeling techniques as implemented in D2M2edit.

REFERENCES

Baar, D.J.M.J. de, J.D. Foley, K.E. Mullet (1992). *Coupling Application Design and User Interface Design*, Conference Proceedings ACM CHI'92, May 3-7, 1992, Monterey, California.

Foley, J., C. Gibbs, W. Kim and S. Kovacevic (1990). "UIDE, An Intelligent User Interface Design Environment", in: J. Sullivan and S. Tyler (eds.) *Architectures for Intelligent Interfaces: Elements and Prototypes*, Addison-Wesley Publishing Company.

Foley, J.D., A. van Dam, S.K. Feiner and J.F. Hughes (1990). *Computer Graphics*, Addison-Wesley, Reading, MA.

Green, M. (1985). Report on Dialogue Specification Tools, in: G. Pfaff (ed.) *User Interface Management Systems*, Springer Verlag, Berlin.

Green, M. (1986). "A Survey of Three Dialogue Models", *ACM Transactions on Graphics*, vol 5, no 3, pp 244-275.

Hardy, E.J. and Klein, D.V. (1990). "The Serpent UIMS", in: *Proceedings EUUG Autumn '90*, Nice, 22-26 October, pp. 219-225.

Hix, D., (1990). "Generations of User Interface Management Systems", *IEEE Software*, pp 77-87, September, 1990.

King, R. and D. McLeod (1985). "Semantic data models", in: S. Bing Yao (Ed.) *Principles of Database Design*, Vol. 1, Logical organizations, 1985, pp. 115-150.

Korson, T and McGregor, J.D. (1990). "Understanding Object-Oriented: a Unifying Paradigm", *Communications of the ACM*, vol 33, no 9, pp 40-60, September 1990.

Pittman, J.H. and Kitrick, C.J. (1990). "VUIMS: a Visual Interface Management System", in: *Proceedings of the ACM Siggraph Symposium on User Interface Software and Technology (UIST90)*, October 3-5, pp. 36-46.

Smith, J.M. and Smith, D.C.P. (1977). "Database Abstractions: Aggregation and Generalization", *ACM Transactions on Database Systems*, vol 2, no 2, pp 105-133, June, 1977.

Szekely, P. (1989). "Standardizing the Interface Between Applications and UIMS's, in: *Proceedings of the ACM Siggraph Symposium on User Interface Software and Technology (UIST89)*, November 13-15, pp. 34-42.

Ter Bekke, J.H. (1991). *Semantic Data Modeling in Relational Environments*, PhD dissertation, Delft University of Technology.

Versendaal, J.M. *(1991)*. *Separation of the User Interface and Application*, PhD dissertation, Delft University of Technology.

DESIGNING FOR TELEPRESENCE:
THE INTERDEPENDENCE OF MOVEMENT AND VISUAL
PERCEPTION IMPLEMENTED

G.J.F. Smets

Department of Industrial Design Engineering, Delft University of Technology, The Netherlands

Abstract. In this paper an overview of the Delft depth television system is given, referring to
definitions of concepts from robotics and perception psychology that are crucial in order to explain
the working principle of the system, as well as the system's performance measurements. The
second section refers to the theoretical background. What is the relevant visual input in optic flow
to allow for depth perception and, therefore, for telemanipulation and for enhanced telepresence?
The third section discusses telepresence applications using the Delft depth television system.

The paper intends to emphasize the critical human-machine interaction research problems from
the perspective of the Gibsonian theory of perception. It is parochial in that it deals with research
and design engineering problems tackled at our laboratory. Our research effort concerns the role
of *active* (observer-generated) movement in perception and its application in technology.

Keywords. Telecommunication; Telecontrol; Robots; Control systems; Cognitive systems; Human
Factors; Artificial Intelligence.

INTRODUCTION

The question of how we come to see a world containing
objects which have stable sizes and shapes and
positions, is of technological relevance for the design of
workstations that can carry out physical tasks safely in
a hostile environment, e.g. the hold of an oil tanker or a
nuclear hot cell. However, this question is also of high
theoretical interest: the psychology of perception has
been dominated by it. We want to show how the
answers to this question offered by perception
psychology, and especially the answer given by the
Gibsonian theory of perception, might be of help in
optimising the design of teleoperators. We will limit
ourselves to the visual aspects of it: television. The
whole world has been exposed to television, and the
technology is inexpensive and highly developed.
However, there remain some problems which, while not
critical in ordinary television programming, are critical
for teleoperation. Foremost among these is depth
perception, which continues to be a major reason why
direct manipulation performance is not matched by
telemanipulation (Sheridan,1989a). In analogy to color
television we will speak of depth television when
referring to television allowing for telemanipulation. In
this paper we will present a new working principle for
such a depth television system, referring to its
performance measurements, its theoretical background
and applications.

Sheridan (1989a 1989b) gives an excellent overview of
the history of teleoperating and telepresence including
a survey of current applications. Vertut and Coiffet
(1984) provide a review of the mechanical technology
that still links up to recent developments and current
problems. As background for the psychology of
perception we refer to Bruce and Green (1990) since
they consider perceptual theory from the point of view
of its relevance for daily life as well as of its scientific
merit.

DEFINITIONS AND SYSTEM OVERVIEW
OF THE DELFT DEPTH TELEVISION
SYSTEM

Definitions

The concept of teleoperation is discussed along with the
related concepts of robotics and telepresence. Those
definitions are adapted from Sheridan (1989a).

Robotics is the science and art of performing, by means
of an automatic apparatus or device, functions
ordinarily ascribed to human beings, or operating with
what appears to be almost human intelligence. A
teleoperator includes at the minimum artificial sensors,
arms and hands, a vehicle for carrying these, and
communication channels to and from the human
operator. The term *teleoperation* refers to direct and
continuous human control of the teleoperator.
Teleoperation problems can be distinguished in four
broad categories: (1) telesensing, including vision,
resolved force, touch, kinesthesis, proprioception, and
proximity, (2) teleactuating, combining motor actuation
with sensing and decision making, (3) computer aiding
in human supervision of a teleoperator, and (4)
meta-analysis of the human-computer-teleoperator-task
interaction. This paper is limited to telesensing, and
more specifically to depth television, from a
meta-analytical point of view. *Telepresence*, then, is the
ideal of sensing sufficient information about the
teleoperator and task environment, and communicating
this to the human operator in a sufficiently natural way,
that the operator feels physically present at the remote
site.

When talking of *depth television* we are talking of
television allowing for telemanipulation and providing
telepresence. The television screen forms a virtual
window where what lies behind it forms a rigid whole
with the world before it, just like a window in daily life.

The present work is embedded in the search to account
for the perceptual consequences of active movement
generated by an observer as opposed to merely passive
motions presented to a passive onlooker. Therefore we
define the concepts of optic array, optic flow, motion
and movement along with the related concepts of

movement parallax and motion parallax, proprioception, exteroception and exproprioception. Since it has not always been considered relevant to make a sharp distinction between movement and motion, lexical discussions about those concepts are rather complicated (see, a.o. Owen, 1990). Experimental data illustrating the relevance of this distinction will be discussed later (in the second section).

For an understanding of the Gibsonian approach to perception, the concept of reciprocity between individual and environment is fundamental. Perceiving is an act *of* an individual *in* an environment, not an activity in the nervous system (Mace, 1977; Gibson, 1979, p. 239–240; Owen, 1990). The visual system has a dual role, being at once both *exteroceptive* (giving information about extrinsic events in the environment) and *proprioceptive* (giving information about the actions and reactions of the individual) (Gibson, 1966). Following Lee (1976) we also mention *exproprioceptive* information about the position of the body or parts of it relative to the environment.

Gibson (1966) defined the *ambient optic array* as the arrangement of variations of the light intensities (structured light) available at an observation point, a point in space where an observer can sample light. For each observation point a unique optic array exists. The *optic flow* is the optic array at a moving observation point. Displacements of this point can be generated by the observer or can be imposed on the observer. Following Cutting (1986) we reserve *motion* for imposed displacements[1], resulting in a purely exteroceptive experience, and *movement* for observer-generated displacements, occuring when they behave or perform with any of the motor systems of their body, giving potential information for proprioception and exteroception as well as exproprioception. In accordance with the distinction between motion and movement we differentiate between *motion parallax* and *movement parallax*. This difference is comparable to Sedgwick's (1986) distinction between relative and absolute motion parallax. Parallax refers to the fact that when two objects are at different distance from a moving observation point, the objects seem to shift relative to each other. We speak of movement parallax when the movement of the observation point is self-generated, and of motion parallax when this motion is imposed. Suppose that you are walking in the countryside and looking at the landscape. Let us also suppose that your gaze direction is perpendicular to your moving direction, that your direction of movement is from right to left, and that you are gazing at the fixation point. Under these conditions all of the objects closer to you than the fixation point would appear to move in a direction opposite to your movement. On the other hand objects that are farther away will appear to move in the same direction you are moving. Not only the direction, but also the speed of movement varies with the object's proximity to your fixation point and your movement speed and direction. This is movement parallax. Motion parallax contains the part of it used by a cinematographer. He uses the apparent movements around the fixation point as registered by camera movements, without them being linked to the spectators' movements. With movement parallax potential information for the absolute distance of the object from the observer would be available if the observer were able to register both the translatory component of the eye's movement relative to the object and the visual direction of the object. With motion parallax there is no potential information for absolute distance, the distance between ourselves and the perceived object. Both movement and motion parallax offer potential information for shape perception, a distance order within a limited cluster of coherent features (defined by Koenderink, 1990).

---

[1]   Imposed displacements occur when members of the body are moved, when the head is accelerated or turned, when the whole individual is passively transported and the eyes are stimulated by motion perspective, or when the observation point is passively moved as when looking at motion pictures.

With the help of those definitions an overview of the Delft depth television system is given, with reference to performance measurements.

System Overview and Performance Measurements of the Delft Depth Television System

If perception and movement are tightly interlocked, as is the case in movement parallax, then exproprioceptive information must be provided. Therefore we contend that depth television, where the screen is perceived as a window with a virtual reality behind it, allowing for telemanipulation, can be created by continuously updating the display output to match the observer's movements in front of the screen. This is accomplished by sensing the observer's head position and moving the camera in the remote site accordingly (Fig. 1). Patents on the working principle have been obtained for Europe and the U.S. (Smets, Stratmann and Overbeeke, 1988, 1990). The original version of the system used a simple video-camera and an observer watching with one eye only. Other versions used stereo and computer generated images, but already the monocular version provides a compelling and reliable spatial impression contrary to what most textbooks on perception and the majority of existing spatial display systems (reviewed in Overbeeke and Stratmann, 1988) suggest, in both of which space perception is often equated with the use of two eyes.

Fig. 1.   The Delft depth television system operates by sensing the observer's head position and moving the camera in the remote site accordingly. The screen then doesn't act as a screen any more, but as a window, where what lies behind it forms a rigid whole (and therefore a virtual reality) with what is before it. The sensor at the observer's head is only symbolic. With existing technology, head position can be registered without a sensor attached at the head.

Exproprioception is lost when the visual input is considered without the observer's movements as is the case with ordinary television programming. In such a case the visual input cannot be scaled relative to the observer's body movements and this means that the perception of a rigid world, allowing for adequate spatial behavior and, hence, telepresence, is impossible. Shape perception however, as we all know, remains possible.

Experiments with the Delft depth television system (Smets, Overbeeke and Stratmann, 1987; Overbeeke, Smets and Stratmann, 1987; Overbeeke and Stratman, 1988) indicate a perceptual advantage of the active observer, whose head movements steer the camera, over the passive onlooker, whose movements are not coupled to the display output (Fig. 2). Both active and passive observers receive identical output on their monitor screens, yet their perceptual input differs. The former receives movement parallax information, the latter motion parallax (Fig. 3). The task consists of aligning wedges by means of a remote control.

As shown in Fig. 2 the experimental set-up consisted of a frame with a central rod and three wedges. The camera and the rotation of the camera (as steered by the active observer) were focused on the central rod. Yet another wedge was mounted before the monitor. The lower rear wedge and the upper frontal wedges were reference wedges. They could be placed in one of three positions (0, –2.5 and +2.5 cm). The upper rear wedge and the wedge mounted before the screen were the aligning wedges. They were randomly placed within a range of 15cm forward or backward from the 0 position of the corresponding reference wedges.

Both observers were required to align the wedges 'in' or 'in front of' the screen. In half of the conditions the *active* observer aligned the wedge '*in front of*' the screen (*condition 1*) while the *passive* observer at the same time was aligning the wedges '*in*' the screen (*condition 2*). In the other trials the reverse was done: the *active* observer aligned the wedges '*in*' the screen (*condition 3*), whereas the *passive* one aligned those '*in front of*' (*condition 4*). A *control condition* was added to assert the maximum level of reliability of depth estimates for a comparable task in a real-life condition (Fig. 4).

Given reports in the literature of a learning effect in depth estimation based on motion parallax (Ferris, 1972) each subject went through a training phase in the real-life set-up. The learning phase was terminated once the deviation of the alignment wedge from the reference wedge did not increase more than five percent in three successive trials. The reliability of the depth estimates in the real-life condition was then measured in ten alignment trials per subject, without feedback. When this was over the subject was passed on to either the active or the passive condition according to preset random order.

Forty subjects took part in the experiment, ten in each condition, each of whom carried out ten depth estimates. All forty were tested in the control condition. Were more subjects necessary to obtain significance, the working principle under test would appear not to lead to useful teleoperation applications. For the same reason we used a somewhat insensitive test: we calculated the general mean and variance of depth estimates for each condition in the experimental design without eliminating the variance among subjects. Since we wanted to estimate the reliability of depth estimates the variances of the depth estimations are more important than their means: an unreliable depth estimation cannot be corrected whereas a systematic over- or underestimation of depth can. Therefore F-tests were used.

Telemanipulations of the active operator almost match direct manipulation performance: the average over- or underestimations are consistently small and their variance does not differ significantly from the variance of errors in real life condition ($F = 1.24$, with ten subjects in each condition). The teleperformance of the passive onlooker, on the contrary, is not consistent, although the average error size remains small. The variance of his error size is significantly greater than that of the active observer ($F = 3.25$, $p < 0.01$, ten subjects in each condition) and than that obtained in the real life set-up ($F = 2.25$, $p < 0.01$, ten subjects in each condition). Data indicate that the active condition, where movement parallax information was provided on the display, even allows for things apparently leaping out of the screen. In this case perceived depth is somewhat less compelling, however, than in the real life set-up. There is a systematic underestimation in aligning a wedge virtually leaping out of the screen with a real wedge mounted in front of the display, yet the variance remains relatively small. Movement parallax allows for telemanipulation, motion parallax does not.

How can one account for those data?

Fig. 2.   Performance measurements of the Delft depth television system: a sketch of the experimental design. Two observers receive the same display output, yet the head movements of the left observer steer the camera. His or her visual input contains movement parallax information, whereas the observer on the right hand disposes of motion parallax information. Although the displays are the same, the perceptual input differs (Fig. 3). This proves to have a significant effect on spatial performances.

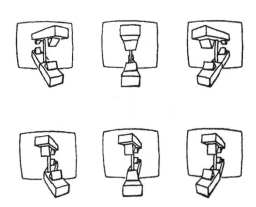

Fig. 3.   Views on the Delft depth television system display. Like a hologram, the Delft depth television system implements a stable link between the visible layouts of the remote (or virtual) and real environments. If, e.g., the observer whoms movements are coupled to camera places his finger 'visually on' a remote object, it will appear there even when he moves his head (top). Passive observers, whose movements do not drive the camera, do not experience this overlay (bottom).

Fig. 4. Control condition with real wedges

PERCEPTUAL THEORY AND TELEPRESENCE

In order to describe the importance of motion and movement for telemanipulation we begin by contrasting two positions, which we will call traditional and Gibsonian, although there is much diversity of opinion within each viewpoint. Although the latter approach indicated the importance of a dynamic approach to perception, the difference between motion and movement as to the perceptual input was neglected. We will describe how this came about and what the consequences are for telemanipulation and telepresence.

By traditional approach we mean psychologists and computational theorists of the tradition which describe the visual input in terms of point intensities of the light at the retina. By the Gibsonian approach to perception we mean those theorists who argue that the input for vision can be described in terms of the structure of the light. The theory is rooted in the research of James Gibson and his followers. We regard the two approaches as complementary.

The traditional approach

Often the eye is thought of as a camera acting to focus light onto a mosaic of retinal receptors. At any instance of time therefore one can conceive the pattern of excitation of retinal receptors as a picture curved around the back of an eyeball (Bruce and Green, 1990, p.142). Though curved, the image is two-dimensional, and yet our perception is of a three-dimensional world. How might depth be recovered? Without invoking learning or memory, pictorial cues and binocular disparity are the most popular answers, with motion parallax being a good third. The movement of the observer is thought to be irrelevant for visual perception. In fact experiments about visual perception, even those about parallax effects, are often designed so that the subject cannot move, his head being fixed by a chin rest, for example.

As a consequence the adherents of this approach create telepresence by reproducing a high resolution stereoscopic image of the teleenvironment. As far as motion is concerned it is restricted to presenting parallax shifts on the stereo display as a consequence of performances in the teleenvironment. Movement of the human operator is important to control the mobility of the effectors of the workstation, but it is not considered important for visual perception as such. The human operator provides largely symbolic commands (concatenations of typed symbols or specialized key presses) to the computer. However, some fraction of these commands still are analogical (body control movements isomorphic to the space-time-force continuum of the physical task) since "they are difficult for the operator to put into symbols" (Sheridan, 1989a, p. 488), yet this analogy is not considered important to enhance teleperformance nor telepresence.

A rather unpleasant human factor these teleoperators' designers must contend with is simulator sickness, a euphemism for heavy nausea if there is a conflict between the operator's perception of motion (shifts on the stereo display of his head-mounted system) and his own movements (which is strengthened by the stereo display often producing pseudo parallax and pseudo convergence).

The Gibsonian approach

In the traditional approach shape and depth perception are expressible in essentially static and pictorial terms. However, they can also be defined, in accordance with Gibson, as the detection of formless invariants *over time*. For a moving observer the retinal projection undergoes a continuous serial transformation on the retina, due to movement parallax, that is unique for this movement pattern in this environment. We believe that movement parallax is sufficient to creating (tele)presence, by rigidly linking the observer's proprioceptive experiences to his exteroception, providing for exteroproprioceptive information.

About the reduction of movement parallax to motion parallax. Movement parallax causing space perception was first described by von Helmholtz (Gibson, Gibson, Smith and Flock, 1959). He postulated that a difference in angular velocity between the projections of objects on the retina is an indication of the perception of distances between objects and between the observer and an object. From that time on typical experiments were designed so that the subject cannot move, thus reducing movement parallax to motion parallax. This has been done partly for reasons of convenience — it is sometimes easier to simulate movement of the display than to allow the head to move — and partly to eliminate questions about the role of proprioceptive information concerning head movements (Sedgwick, 1986, p.45).

At the same time a distinction has been drawn between the role of movement parallax in absolute and relative distance estimation, the emphasis being placed on estimation of relative distances. It was shown that motion parallax provides information about the relative distances of points or surfaces around the subject, but it cannot specify absolute distances or surfaces. A human or an animal may be able to solve this problem by scaling its visual input through head or body movement of a fixed velocity (Bruce and Green, 1990, p. 259–260), thus using movement parallax instead of motion parallax. It has been proven that animals do this.

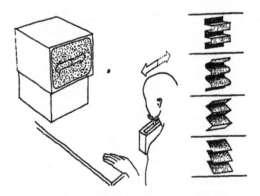

Fig. 5. Experimental set-up of the Rogers and Graham (1979) experiment (for explanation see text).

Head movements by human beings might be useful to their visual perception as well. Rogers and Graham (1979; Graham and Rogers, 1982) built a set-up in which movements of the subject's head caused shifts on a screen (Fig. 6), corresponding to the projection of different wave patterns (square wave, sine wave, triangular wave or sawtooth wave) on the screen. When the subject kept his head still he saw a random dot pattern. However, if he moved he had a convincing and unambiguous shape impression of the different patterns, seen from different angles. The phenomenon only occured when the pattern on the screen was shifted according to the movements of the observer. Perceived depth was more pronounced than in an analogous experiment where motion parallax was used. In this experiment the observer's head remained fixed and the oscilloscope moved back and forth with internal depth in the display specified by motion gradients tied to the movement of the oscilloscope. This demonstrates the effectiveness of movement parallax alone in producing vivid perceptions of surface layout in depth and suggests that previous failures to find such pronounced effects have been due to limitations in experimental displays used to simulate movement parallax rather than to limitations in the visual system's ability to make use of it (also Sedgwick, 1986, p. 21–46).

Evidence for the relevance of movement — and therefore also movement parallax — for visual perception has also come from studies about the development of perceptual skills (Held and Hein, 1963). The applicability of the principle of active control for perception and, hence, movement parallax, is witnessed by the successful experiments in sensory substitution, like Bach-y-Rita's (1972) tactile visual substitution system and Bower's sonar visual substitution system (1977, 1978; Aitken and Bower, 1982). Stassen (personal communication) mentions that thalidomide babies, born without arms, did not show adequate spatial behavior when they started walking. Yet they did when they received elementary (rigid) arm prostheses from their first months on. The fact that they encounter obstacles with their armprostheses seems to be of help in developing their depth perception.

About the importance of the task. Whether motion is sufficient for adequate spatial behavior, or movement is required depends on the visual task.

For some perceptual tasks shifts in the optic array can be reduced to shifts on the retina. This goes for those tasks that do not imply relative shifts in the optic array, or, in perceptual terms, where two-dimensional form perception is at stake without shape nor depth perception being necessary. For this kind of task the movement of the observer is irrelevant. Think of the recognition of a bird high in the sky (as in Fig. 4), or the recognition of a landmark deep down when you look out of an airplane, or a graphic task . This type of task can easily be simulated on a computer display, with standard animation software and videoprocessor hardware. The screen is perceived as a flat screen.

Yet, most perceptual behavior, all manipulation tasks for instance, implies shifts in the optic array. Different parts in the optic array shift relative to each other as well as to the observer. Those kinds of tasks are concerned with shape and depth perception, or, in other words, with the perception of a rigid world. For them the visual input cannot be simulated in the same way as for the former tasks since, here, the exploratory movements of the observer causing the serial transformation in optic flow, are essential. Simulating a virtual world of which the visual input forms a rigid whole with the perception-action system of the observer, and that allows for telemanipulation, requires the movements of the observer to be coupled to the shifts in his optic array. This can be realized on an ordinary television screen if the video-camera in the teleenvironment is steered by the head movements of the observer. This creates a monocular virtual window display: the television screen does not look like a flat display any more, but like a window with a virtual reality behind it, forming a rigid whole with what is before it, allowing for reliable and convincing depth and shape experience. There is no need for stereo-images. As long as movement (head movements of the operator) and motion (shifts on the display) are consistent there is no simulator sickness.

The optic flow of those tasks cannot be simulated on a two-dimensional display without the movements of the observer and the motion on the screen being coupled as described above. In that case simulation of flow in the

ambient optic array is confounded with and reduced to the simulation of retinal flow. Flow on the retina or in the image does not necessarily copy flow in the array (Torrey, 1985). How the failure to distinguish flow in the optic array from flow on the retina can result in confusion is clearly stated by Owen (1990, p. 44–47).

A second implication when studying the second kind of task is that we think it might be necessary to extend the models of optic flow so as to include the movements of the observer, thus pointing to a kinetic concept of optic flow rather than a merely kinematic one. This is explained in Smets (1992), Stappers (1992) en Smets, Overbeeke en Stappers (1991).

APPLICATIONS: NEW PERCEPTUAL AIDS USING TELEPRESENCE

A history of applications of teleoperation — extension of human sensing and manipulating capability by coupling to (remote) artificial sensors and actuators — is given by Sheridan (1989a). This review of relevant historical developments shows that teleoperation is primarily used for manipulation in hazardous or inaccessible environments, e.g. outer space, undersea or nuclear 'hot laboratories', or to compensate for motor deficiencies, e.g. for the motorically handicapped or, more general, when forces have to be exerted that are far too large for a human operator. Until now teleoperation has not been used to overcome visual (or, more general, perceptual) deficiencies. Through the recent attainment of the sense of telepresence in telemanipulation, however, we are convinced that this is a promising, as yet virgin territory for applications. Two applications illustrate how we faced this challenge. We will demonstrate the advantage of simple human-supervised teleoperation for visual tasks, using the Delft depth television working principle.

Needless to say that, although we will not elaborate on them, those designs are subjected to extended performance measures and assessment techniques. We are conducting studies on considering the specific mission requirements of both industrial design engineering projects described below. This is being done together with industry. At the moment there are essentially no accepted standards for asserting the accuracy, repeatability, linearity, etc., of teleoperating systems in general.

Helping the Visually Impaired

Leifer's (1983) review of the use of telerobotics for the disabled illustrates that most applications are designed for motor disabilities. One (famous) exception is the development of a teleoperated electromechanical guide dog for the blind (Tachi, 1981).

Poorly sighted people can be helped by an optical eye correction by means of spectacles or eye surgery. If this is not possible they often can read with the aid of a so called video-magnifier. This is a system consisting of a camera that registers a text and displays it extremely enlarged and eventually with enhanced contrast. This allows the poorly sighted to read and write. Yet other tasks, such as sewing, tasks requiring depth perception and eye-hand coordination, remain unassisted. However we developed a hybrid of the video-magnifier and the Delft depth television system as depicted in Fig. 6. Head movements steer the camera. The images on the monitor are displayed enlarged and with enhanced contrast. Exploratory research with a working prototype proved that this new device allows for reliable depth perception and manipulation as compared to a real life set-up as well as a comparable set-up with a static camera, although it is based on monovision. Despite its limited range of action, it is seen of as a striking and relevant perceptual aid by the visually impaired. More elaborated validity studies and product development are under way.

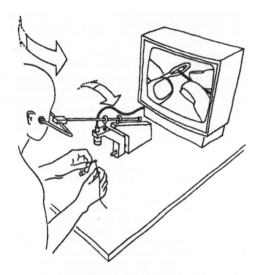

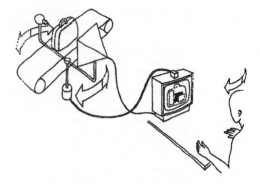

Fig. 7.   Working principle for a luggage
          inspection system allowing for full depth
          perception.

Fig. 6.   Working principle for a video-magnifier
          facilitating spatial tasks for visually
          handicapped people. The product
          developed out of this working principle
          works 'head free'.

Luggage Inspection

An especially exciting challenge is to develop a luggage
inspection teleoperator. For security reasons the
inspection of personal and freight luggage becomes
more and more important. Part of this control consists
of inspecting luggage with an X-ray camera. Luggage
conveyed through such security devices can only be
inspected from one viewpoint: that of the camera.
Nevertheless the safety attendant often would like to
look in and around every hole and corner to solve
perceptual ambiguities. A pistol for instance can be
easily confounded with a lot of other objects when only
looked at from one point of view. If the X-ray camera is
steered by the head movements of the attendant,
perceptual ambiguities disappear and depth perception
is provided for. The working principle of this luggage
inspection device, that can be used for freight as well as
for personal luggage, is sketched in Fig. 7. The security
attendant can do the inspection task nearby or at a safe
distance.

The advantages of this security system as compared to
the existing security systems are greatly acclaimed by
Amsterdam Airport's security service. Product
development is under way.

CONCLUSIONS

Research with and on teleoperators persuades us that
the principles of the Gibsonian theory of perception
offer a powerful engineering paradigm and also
demonstrates that this applied research can help to
clarify the nature of telepresence itself. Since
telepresence is commonly claimed to be important for
direct telemanipulation a theory of
presence/telepresence is sorely needed (Sheridan,
1989b). We think that the ecological theory of
perception might be of help here, by explaining the
difference between motion versus movement parallax as
to their perceptual meaning, that is, as to their
importance in allowing for adequate spatial behavior.

ACKNOWLEDGEMENTS

Work mentioned in this paper was supported by almost
the entire staff of the laboratory of Form Theory at the
Department of Industrial Design Engineering at Delft
University of Technology. This is a multidisciplinary
team with a.o. industrial design engineers, psychologists
and physicists working together to build an
interdisciplinary approach to perception. I specifically
want to mention and thank C.J. Overbeeke, P.J.
Stappers, O.A. van Nierop, R. Wormgoor, A.M.
Willemen, G.W.H.A. Mansveld, H. Subroto, A.C.M.
Blankendaal, J.P. Claessen and C. Kooman.

SELECTED REFERENCES[2]

Bruce, V., and Green, P. (1990). *Visual perception:
    physiology, psychology, and ecology.* London:
    Lawrence Erlbaum Associates.

Cutting J.E. (1986). *Perception with an eye for motion.*
    Cambridge, MA: MIT.

Gibson, J.J. (1962). Observations on active touch.
    Psychological Review, 69, 477– 491.

Gibson, J.J. (1966). *The senses considered as
    perceptual systems.* Boston, MA: Houghton-Mifflin.

Gibson, J.J. (1979). *The ecological approach to visual
    perception.* Hillsdale, NJ: Lawrence Erlbaum.

Gibson, E.J., Gibson, J.J., Smith, O.W., and Flock,
    H. (1959). Motion parallax as a determinant of

_____

[2]   An extended bibliography can be requested at
the author's address.

perceived depth. *Journal of Experimental Psychology*, 58, 40–51.

Graham, M., and Rogers, B. (1982). Simultanuous and successive contrast effects of depth from motion-parallax and stereoscopic information. *Perception*, 11, 247–262.

Mace, W.M. (1977). James J. Gibson's strategy for perceiving: Ask not what what's inside your head, but what your head's inside of. In R. Shaw and J. Bransford (Eds.), *Perceiving, acting, and knowing*. Hillsdale, NJ: Erlbaum. 43–65.

Marr, D. (1982). *Vision*. San Francisco: Freeman.

Overbeeke, C.J., and Stratmann, M.H. (1988). *Space through movement*. Ph.D. thesis, Delft University of Technology.

Overbeeke C.J., G.J.F. Smets, and M.H. Stratmann (1987). Depth on a flat screen II. *Perceptual & Motor Skills*, 65, 120.

Owen, D.H. (1990). Lexicon of terms for the perception and control of self-motion and orientation. In R. Warren and A.H. Wertheim (Eds.), *Perception and Control of Self-Motion*. Hillsdale, NJ: Erlbaum. 33–50.

Rogers, B., and Graham, M. (1979). Motion parallax as an independent cue for depth perception. *Perception*, 6, 125–134.

Sheridan, T.B. (1989). Telerobotics. *Automatica*, 25, 4, 487–507.

Smets, G.J.F. (accepted). Designing for telepresence. In: J.M. Flach (Ed.), *The Ecology of Human-Machine Systems*. Hillsdale, NJ: Erlbaum.

Smets, G.J.F., and Overbeeke, C.J. (in press). Trading off spatial against temporal resolution: the importance of active exploration. *Proceedings of the second international conference on visual search*. Durham.

Smets, G.J.F., Overbeeke, C.J., and Stappers, P.J. (1991). The ecology of human-machine systems III. The importance of actively controlled movement: toward richer operationalizations of optic array & optic flow. Delft University of Technology: Internal rapport.

Smets, G.J.F., Overbeeke, C.J., and Stratmann, M.H. (1987). Depth on a flat screen. *Perceptual & Motor Skills*, 64, 1023–1024.

Smets, G.J.F., Stratmann, M.H., and Overbeeke, C.J. (1988). Method of causing an observer to get a three-dimensional impression from a two-dimensional representation. *U.S. Patent 4757, 380*.

Smets, G.J.F., Stratmann, M.H., and Overbeeke, C.J. (1990). Method of causing an observer to get a three-dimensional impression from a two-dimensional representation. *European Patent 0189233*.

Wertheim, A.H. (1990). Visual, vestibular, and oculomotor interactions in the perception of object motion during egomotion. In R. Warren and A.H. Wertheim (Eds.), *Perception and Control of Self-Motion*. Hillsdale, NJ: Erlbaum. 171–217.

USER ORIENTED DESIGN OF MAN MACHINE INTERFACES:
The Design of Man Machine Interfaces for a
Processing Line; A Case Study

G. van Hal and I.E.M. Wendel

Planning and Engineering Ergonomics Group, Hoogovens IJmuiden, NB WOC 2H.13, P.O. Box 10.000,
1970 CA IJmuiden, The Netherlands

Abstract. In this paper the contributions of ergonomists to a large engineering project are described. After some global remarks concerning the design process, a more detailed description of contributions is given. Major attention is paid to the design and evaluation of the Man Machine Interfaces. Not only implemented ergonomic know-how is discussed, but also organisational aspects of the design process. In the conclusions remarks are given for coping with comparable problems in future engineering projects.

Keywords. Computer interfaces, Design process, Ergonomics, Man-machine interfaces, Process control, Steel industry, User participation.

1. INTRODUCTION

This paper discusses the role of ergonomics in engineering projects. The Engineering Department at Hoogovens IJmuiden steel works has had an Ergonomics Group for some 25 years. It is thus well positioned to know about and participate in engineering projects from the very beginning. The fact that an Ergonomics Group is part of the Engineering Department illustrates the commitment of the Engineering Department management to ergonomics.

Notwithstanding this advantageous position the main problem for the ergonomists is how to organize the design process in such a way that basic ergonomic principles are accepted and applied by the design team and the suppliers of the installation. In other words: "How to make a technically-oriented design process user-oriented".

As a case study the contributions of the Ergonomics Group in a processing line engineering project are presented, with ergonomic know-how implementation problems highlighted.

Fig. 1. The building of the processing line for cold rolled strip.

177

At Hoogovens IJmuiden a new processing line for cold rolled strip has recently been built (see fig. 1). Engineering began in 1985, and production started in March 1991. Project costs amounted to some HFL 280 million (= $ 150 million). The total length of the line is 220 m, the building accommodating it is 300 m long, 30 m high and 25 m wide. Coils to be processed can have a mass up to 16 tons and a strip length up to about 10,000 m.

The processing line is highly automated. The main tasks of the operators consist of monitoring and optimizing. During normal running the line is supervised from two control rooms: one near the entry and one near the exit of the line. The operators use several types of computer systems, each with their own interface (VDUs with alphanumeric and function keyboards). Some switches have been provided for safety cutouts as well as for fast interventions where closed loop control could not be implemented. In addition to the control rooms, about 75 local operator stations have been placed near the line to be used in faulted conditions and during maintenance. The installation is operated by five shifts each consisting of eight operators.

The Ergonomics Group participated in the design team from the very beginning. Having no responsibility for any hardware budget, the task of the ergonomists was to support the other members of the design team in problems concerning workplace design and design of the Man Machine Interfaces.

The contributions of the Ergonomics Group in the different stages of the design process will be described in more detail in section 2. Section 3 deals with a description of the specification stage, section 4, 5 and 6 with the design of control rooms, computer interfaces and local operator stations and section 7 with the conclusions.

2. THE DESIGN PROCESS

The design process comprised several phases. The first phase was the analysis, which started by developing an operational philosophy. This was done in cooperation with engineers and operation experts to gain understanding of the intended operations. Based on this philosophy the ergonomists wrote parts of the specification. The next phase was the synthesis; resulting in the design of the Man Machine Interfaces. In this project the ergonomists designed the control rooms, the computer interfaces and the local operator stations, based on the operational philosophy and the specifications. Part of the task they carried out alone, part was done in cooperation with other engineers of Hoogovens, future operators and the various suppliers.

The last phase, evaluation of the Man Machine Interface was carried out during commissioning of the line. Also, after some months of production, eight interviews were held concerning workplaces and Man Machine Interfaces and the role of ergonomists in engineering projects. The ergonomists had interviews with two members of the engineering department, two members of the production staff and four operators. Two of these operators had taken part in the development of the Man Machine Interfaces, the other two operators had not. The results of the

evaluation will be described in the following sections. We are well aware of the fact that the results of the interviews only give a faint impression of the problems and benefits of the workplaces and Man Machine Interfaces. Because the operators and staff can get used to some incompleteness of the functioning of the line, observation of people operating the line is also necessary. Furthermore, it can be seen that the way of operation of the line varies between operators of different shifts. Firstly, because the training method differs, secondly, because the operators have more or less experience with computers.

A main feature of the design process is the iterative character. This is shown, for example, by the fact that some process computer screens had to be re-designed three or four times. The future operators participated in the design process because of their experience with similar installations. Participation of operators is also a proper aid for training purposes, and helps to achieve better acceptance of the design. To give the future operators an idea of the functioning of the installation and of the design of the Man Machine Interfaces, scale models and three dimensional CAD programs have been used together with two dimensional (technical) drawings. However, the contribution of operators should be jugded with care. It is difficult to give the future operators a good idea of how the installation will work. They are not used to reading two dimensional drawings and technical specifications. Another problem is that experienced operators easily get stuck to their way of operating older installations. Also, layouts of process computer screens on paper are hard to understand, because most of them are not used to operate installations by means of computers.

3. SPECIFICATION

The first activity of the ergonomists in the project was the development of an operational philosophy. The operations experts in this stage were the manager of the production department and a very experienced foreman of an installation similar to the processing line to be built.

The number of operators was determined and their tasks were described. Based on this philosophy the ergonomists wrote specifications for the suppliers concerning the Man Machine Interfaces, noise, climate, lighting and maintainability. A large part of the specification deals with VDU-usage. Directives were specified for system performance and presentation and handling of messages, as well as for design of screens, such as use of colours, symbols, coding and screen lay-out.

Finally system-oriented minimum requirements with respect to the application of VDU-screens were specified, like minimum number of available colours, minimum number of available characters and screen format, but also upper limits of response times.

The ergonomists evaluated the proposals of various potential suppliers, which were subsequently discussed during several meetings. The main purpose of this evaluation was to identify which aspects would require major attention by the Ergonomics Group during the detail engineering.

Based on this evaluation the suppliers adjusted their

proposals; in one of the proposals, for example, the number of process computers had to be doubled in view of requested response times. The final contract included the complete specification for Man Machine Communication, as well as stipulations on the early involvement of future operators and production staff. The complete specification for Man Machine Communication was much longer and more detailed than it used to be in former design projects. The ergonomists were very pleased with this way of specifying ergonomic aspects, and were convinced that the design would proceed as they wanted and as had been laid down in the contract. It was the first time at Hoogovens that ergonomic aspects formed quite an extensive part of the contract and had been specified in so much detail. At that moment no problems were expected. The supplier and Hoogovens were convinced that both parties would understand the ergonomic problems and their consequences, and that appropriate agreements had been laid down in the contract concerning the design approach.

However, at the beginning of the design process it turned out that the supplier expected to implement ergonomic aspects at more final stages of the design of Man Machine Interfaces. Apparently, it was not totally clear for either the designers of the supplier nor those of Hoogovens which ergonomic aspects had to be implemented at what time. Furthermore, on the supplier's side nobody had an overall responsibility for the Man Machine Interface aspects and so it was hard to make agreements concerning Man Machine Interfaces. The specifications were written as directives, an ergonomics coursebook, to try to show our view of Man Machine Interaction. The intentions of the contract were good, but it appeared that the specification was not clear enough with respect to contents as well as to scheduling of activities. Hereafter the consequences for the design of control rooms, VDU-screens and local operator stations will be outlined.

Fig. 2. The main control room.

4. CONTROL ROOMS

In cooperation with civil, electrical and mechanical engineers, as well as operations experts, the Ergonomics Group designed the control rooms. Overall dimensions and a location were determined, the latter with special emphasis on direct visibility of installation parts and accessibility of the control rooms and the installation. Due to their limited number, the operators have less time to go to other places, so toilets and a kitchen have been provided in the control rooms. In the main control room at the exit side of the installation, training facilities and a room for the foreman have also been provided. In the event of problems, the operators can quickly consult the foreman. The training facilities are situated in the main control room to improve contact between operators and trainees. In fig. 2 the main control room is shown. From the control room the strip surfaces in the inspection stand can be seen. The inspection stand is used to detect surface damages caused by the processing line and allow for quick response on the part of the operators. The position of the inspection stand, as well as the location and dimensions of the mirrors and the lighting requirements, have been determined by the Ergonomics Group.

The positioning and lay-out of the control rooms has been worked out on paper. For the entry control room a three dimensional CAD program has been used to judge the visibility of the installation and the strip. Based on the results of this program, the height of the control room has been fixed. For the exit control room a scale model of the control room has been made. It turned out that the visibility of the strip surface and installation parts where strip transport can cause problems is good. The same observation applies to the workplaces in the control rooms.

Once the line was in operation it turned out that not all rooms of the exit control room were being used as expected.

Because of financial problems the exit control room had become smaller than originally designed. At the upper level only the control desks and the room for the foreman could be accommodated. The training facilities, social room, kitchen and a control desk for one operator had to be placed one level lower. Because this operator frequently has to go to the line, his workplace had been situated at floor level, allowing quick access to the line. Very soon, the social contacts between the operators proved to be more important than the accessibility of the line. The operator concerned prefers to stay with the other operators upstairs; for him it was worthwhile to climb the stairs.

The same tendency could be observed with respect to the social room and kitchen. Because the operators have no common lunchbreak, they prefer to eat and remain upstairs with their colleagues, rather than resting alone. Some adjustments will be made to overcome the problems mentioned here.

5. COMPUTER INTERFACES

A major part of the effort of the ergonomists concerned the design and evalution of screens, especially mimics. Mimics can be seen as control desks at which control actions can be performed. The necessary information for the actions is also presented on the mimics. It will be clear that each mimic should be designed for a special purpose.

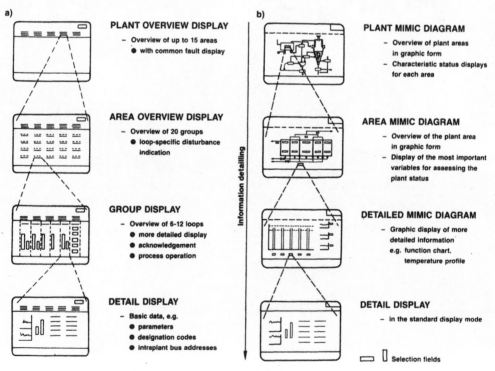

Fig. 3. Display hierarchy (ABB, Publication No. D IA 1309 86E).

The lay-out of the screens is defined by the information that is necessary at a certain place to perform actions which depend on a given situation. The total amount of information is considerable, and there are several operating modes to be distinguished. Therefore, a decision has to be made on how the information is to be distributed among the mimics and how these mimics connect to each other. This connection is based on the way of operation and the structure of the installation and the control system. In designing mimics the way of operation is fixed. This means that the way of operation has to be known before designing the mimics.

Beside mimics, which are designed for special purposes, the control system contains a lot of standard screens. E.g. for each pump, valve, temperature controller, etc., a single loop display is available. To find these single loop displays the most important information of a number of connected loops is shown in group displays. These group displays are condensed in area displays. So the control system has a standard hierarchy, built up of three types of screens: single loop displays, group displays and area displays (see fig. 3). Additionally, the first few lines of each screen contain an overview of different areas of the line. Screen selection can take place via hierarchy, screen number or frame selection fields.

The period after assigning the contract has been used to give more detailed design directives (see fig. 4). Use of symbols, colours and coding etc. were agreed upon with the supplier. Colours are only used as a second, redundant coding parameter. Symbols for pumps, valves, drives, etc. were agreed upon and agreements were also made about the way of operating and how to make clear, by presentation, which elements are operatable.

For example, every parameter which may be subject to operator action is surrounded by a frame, different types of operation being distinguished by form and colour of the frame. Thus the operators can see at a glance which parameters are adjustable.

The contract specified that the supplier of the process control system had to make the first proposals for the screens. For reasons mentioned, these proposals came a year later than expected. The proposals were evaluated in cooperation with operators and other engineers. It turned out that they all had to be fully redesigned in a very short period of time.

The redesign and design of additional screens were necessary because of the following kind of problems:
- missing screens for special purposes like starting up the furnace process
- screens were too full and confusing
- unnecessary information, for example, the graphic representation of installation parts was designed too much in detail
- misinterpretation of agreements concerning the use of symbols and colours
- missing information, for example, installation codes which are used for unique identification of installation parts
- technical faults, for example, the screen showing four pumps for an installation part comprising five pumps.

For simulation of operation by screens and development of training scenarios the contract specified hardware and software. Due to delivery problems the simulator was not available during the design stage. Later on, during commissioning of the line, it became available.

Dynamic aspects of screens are hard to judge on paper only. It appeared, for example, that the movement of the cursor in the mimics was neither logical nor predictable: the direction of the cursor keys on the keyboard did not match the cursor movement on the screen.

However, another consequence of the delay was that very little time was left to train the operators who had to help redesigning the screens. Training was also hampered by the fact that training materials only consisted of descriptions; no working simulations were available.

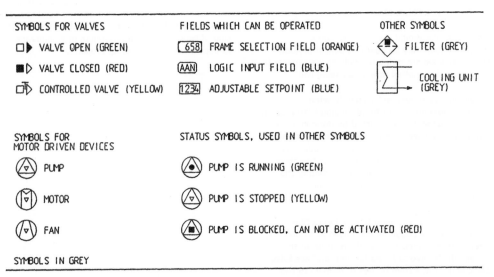

Fig. 4. Examples of some design directives

So we redesigned the mimics in cooperation with future operators. At that time we did not look at the standard screens. They were taken care of later, to match them to the mimics' structure.
After redesigning, the screens worked well: no major problems arose.

A specific problem was the language of the descriptions on the screen. The language of the supplier was German, although we had obliged the supplier to communicate in English. So they made descriptions in English.
However, the production staff decided that all descriptions and alarms on the screens had to be in Dutch. But the descriptions and alarms could only be translated after the line was running well, as the supplier's personnel was to test the line. So during the first 7 months or so the descriptions and alarms remained in English. The operators were almost used to it; now they have to learn the Dutch descriptions.

6. LOCAL OPERATOR STATIONS

A supplier designed the local operator stations. Local operator stations are situated all along the line. Usually these stations are used in disturbed situations and for maintenance. At the outset design directives for lay-out had been given. The corresponding proposals were evaluated by the ergonomists together with future operating staff (experienced foreman and five future operators). It then turned out that most of the panels had to be completely redesigned for the following reasons:
- functions were missing, for example functions to transport the strip after strip break
- the lay-out of panels did not always match the way of operation, for example, the sequence of operation
- the type of components did not always match the way of operation.
The local operator stations are functioning well now. In the evaluation no specific remarks concerning these situations have been made.

7. CONCLUSIONS

In conclusion, we would like to make some statements which may be useful in similar projects. At Hoogovens we are trying to apply these aspects in a new project for another strip processing line.

The ergonomists should firstly specify what functional requirements are to be fulfilled with respect to ergonomic aspects. This should be done in such a way that the supplier understands these requirements and can make an estimation of the consequences for the design, such as money and time. It is therefore important that the ergonomists verify whether the supplier has the same understanding of the requirements.

Next, agreements concerning the design of Man Machine Interfaces shoud be laid down in a contract. This refers to a description and distribution of the activities which have to be performed, and sceduling of these activities. Attention has to be paid to the fact that suppliers of process control systems do not have detailed knowledge of the different installation parts. Also, they do not have a good idea about our desired way of operation.
Therefore it is necessary to design Man Machine Interfaces in cooperation with mechanical and electrical engineers, future operators, production staff and ergonomists.
Also, it is necessary to start designing the Man Machine Interface very early, due to possible implications for the installations and computer systems. One example is that the standard screens are implemented and used for commissioning purposes prior to the operator mimics. So it is necessary to design the mimics of process computer systems at a very early stage, because the standard screens must match the structure of the mimics.
To ensure the fulfillment of the agreements concerning Man Machine Interfaces, overall responsibility must be given to one person on the supplier's side and one person on the customer's side, and this from the very beginning of the design project.

Another point is that the ergonomists do not have time to pay attention to all Man Machine Interface and other ergonomic aspects. The ergonomist should make the other engineers aware of the ergonomic aspects in their design. Consequently, the ergonomists should be kept informed about the way the design process of the various suppliers and engineering departments is proceeding. This view must be supported by the project manager and accepted by the various engineers. A long term approach is to make engineers of all disciplines more aware of a user-oriented approach. In our engineering department we try to achieve this by teaching an ergonomics course. On the other hand, the technical universities and colleges should emphasize the need for user oriented design wherever this applies.

Final Remark

The problems discussed in this paper are certainly not strictly related to the suppliers of the processing line considered in this paper. In other engineering projects with other suppliers, similar problems occurred. The case described in this paper is used to illustrate the kind of problems the ergonomists of Hoogovens have to cope with in engineering projects.

ALARM FILTERING vs. FAILURE PREDICTION: HOW BEST TO REDUCE OPERATOR OVERLOAD

P.F. Elzer*, C. Weisang** and K. Zinser**

*Technical University of Clausthal, Institute for Process- and Production Control Technique (IPP),
Leibnizstrasse 28, D-3392 Clausthal-Zellerfeld, Germany
**ABB, Corporate Research Center, DECRC/L, POB 10 13 32, D-6900 Heidelberg 1, Germany

Abstract: The paper discusses he usual approach of 'alarm filtering' as a means to reduce operator over-load in critical situations and proposes a 'cognitive viewpoint' that takes into account the stress level of an operator in critical situations. It then describes a new technique for dynamic failure prediction in technical processes and discusses the advantages of the described mechanism as compared to 'static', pre-planned 'alarm filtering'. The new method is implemented in the form of a model-based expert system that does not describe the behaviour of a technical process as a whole, but propagates a failure of any individual component through the system, starting with the affected component and checking for the influence of that failure on each of the other components that are technologically connected with the first one. The flexibility of that approach even allows to predict consequences of failures that were not foreseen at the time of the installation of the S&C system. It is shown that this technique bears the potential to significantly reduce the stress level of operators and therefore improve the overall performance and safety of the man-machine system.

Keywords: Alarm Systems, Artificial Intelligence, Cognitive Systems, Computer Control, Control Systems, Ergonomics, Failure Detection, Man-machine Systems, Power Station Control, Supervisory Control.

ALARM FILTERING
AS A MEASURE AGAINST OPERATOR OVERLOAD

'Cognitive overload' is a common buzzword in today's discussion about the design of man-machine interfaces for dynamic systems. For the practitioner its meaning seems to be quite clear: too many measurements taken, too many values displayed, too many alarms presented to the operator, etc. This is especially true in emergency situations, when the operator anyway faces the problem of deciding what to do in order to manage the situation. In many cases the S&C system is therefore regarded more as an additional nuisance than as a support.

The reason for this problem is mainly that a failure in one component of a technical process usually triggers malfunctions in other components as well, which, in turn, may raise more alarms, etc. Fig.1 illustrates this effect.

One countermeasure that is mostly proposed is 'alarm filtering'. Its goal is evident: to reduce the information flow that has to be processed by the operator by filtering out the 'unnecessary' or 'irrelevant' alarms, i.e. mainly those that are 'just consequences' of the alarm that indicated the 'original' failure. This is usually implemented in such a way that during system design it is decided which alarms should have precedence over others in which situation, which ones can be completely neglected, because they are consequences of others, etc. Several such systems have been implemented and reportedly function to the satisfaction of their users. One has e.g. been described by Sill and others (1989)

But there are some drawbacks of that method:

• Firstly, it is expensive to plan and to implement such alarm filtering mechanisms, because every possible fault has to be anticipated and all its foreseeable consequences assessed beforehand with respect to their criticality and necessary operator reactions. Besides, it appears to be impossible to really plan everything in advance.

Of course, this drawback is common to nearly all computerized S&C systems, because they have to be programmed on the basis of previous knowledge of the designer about the process and therefore can only reproduce this static knowledge. Certainly, if the answer the system produces is based on solid grounds, it is better than a 'near miss' by a human. But, as it is not possible to pre-plan everything, such systems will always have to contain features that support the handling of unforeseen failures or combinations of alarms, e.g. diagnosis expert systems.

• Secondly, it can happen that failure messages are suppressed which normally should be consequences of others, but in that particular case have a different cause - a situation which clearly can lead to disastrous consequences.

• Finally, the suppressed 'follow-up alarms' may contain time-relevant information which is implicitly contained in the sequence and the type of the simple alarm messages. Actually the operator would be able to (re)construct a complete and comprehensive picture of the course of the failure situation, if s/he would be able to recognize all of the alarm messages. But as s/he is unable to follow all the alarms, s/he normally skips all of them in order to prevent being stressed and overloaded.

Thus, the information about the typical profile of the disturbance over time, which is unique for each failure situation, gets lost in systems where alarm filtering is applied. From an 'information processing viewpoint' it is appropriate to reduce the high number of alarm messages presented to the operator by only showing those that are connected to the origin of the disturbance. But from a 'cognitive viewpoint' it would be necessary to apply techniques that concentrate the high number of simple messages into a few 'intelligent' messages without losing e.g. this typical profile of events over time.

So, what is really needed, is a reduction of the number of alarms (or other messages to the operator) without reducing the meaning. Fig.2 illustrates this desired state. In this particular case it is assumed that the operator is notified of an ongoing failure (e.g. in an appropriate high level graphical fashion) and then requests further information.

THE 'COGNITIVE VIEWPOINT'

The justification for this is that human cognition works in clusters ('chunks') of information that are recognized and 'understood' as a whole. If information is presented to a person with the right size of such 'chunks', a much more complex situation can be understood with the same amount of 'information processing' than if the information is presented as simple isolated facts.

Further analysis shows that 'cognitive overload' is a relative term as well. Firstly, it depends on the constant properties of the individual, secondly, of her/his momentary conditions, like e.g. stress (or even panic). Cognitive Science teaches us that an individual 'under normal circumstances', i.e. healthy, not tired, relaxed, can accept and process a fairly higher amount of information than under stress. Additionally, under stress conditions a 'focussing effect' develops, blinding the individual against all information that does not seem to be relevant for the solution of the problem at hand. In 'panic mode' a person is even prone to 'blocking', i.e. becomes unable to act properly. This has e.g. been shown by Prodaska (1981) or Popper and Eccles (1984). It should therefore be an important goal of MMI (and even S&C-System design) to avoid situations in which the operator is subject to high degrees of stress or, worse, panic. Fig.3 shall illustrate this situation.

This basically means that in the system under supervision nothing must happen unexpectedly. Of course there is a class of events - those caused by external influences - that can not be predicted (like e.g. the notorious lightning in an electrical grid) by any means. But nearly every event that originates from within the process can - in principle - be predicted. Even the behaviour of a process after some (catastrophic) external event can be predicted to a certain degree if the exact nature and impact of that event can be identified soon and exactly enough.

So, what does an operator really need in a critical situation:

- early warnings in order not to be caught by surprise;
- an 'impression', i.e. a rough but correct overview over the actual status of the system (i.e. information grouped in appropriate chunks);
- sound advice, including the limits of its validity (i.e. information about the real causes of a malfunction or a message from the system that it cannot help);
- advice concerning the options open and the time left for action;
- an impression of the consequences of possible actions or omissions;
- not to be overloaded !

Or, in still more condensed form:

- a system should present information in appropriate chunks and
- avoid unexpected events in order to reduce stress.

However, nearly none of today's S&C systems provides this kind of help. The reason is that traditionally S&C-systems present a 'temporal cross-section' of the behaviour of the process to the operator and that trends are very rarely monitored and indicated.

It is interesting to note that problems with this traditional way of looking at all kinds of dynamic systems - highly detailed, but extremely limited in scope and time - have been observed on a much more general level as well. In a recent article, De Greene (1991) refers to it as the 'cross-sectional paradigm'. A problem that can arise from this 'snapshot-like' way of looking at dynamic systems is illustrated by Fig.4. It shows a process that e.g. has been oscillating for a while. This was not noticed by the operator, because all values were still within the tolerances ('in the green band'). But at a certain point in time one (or several) values exceed an alarm limit and the operator has to make a decision. But s/he cannot know whether this alarm is nearly meaningless, because the process will stabilize itself again (curve a), whether e.g. some slight modification in the setpoints will be sufficient to stabilize the process (curve b) or whether some drastic action has to be taken, because the process is getting out of control (curve c).

Had the operator been informed about these oscillations early enough and been able to observe their shape, s/he might have been able to deduce from this information the nature of the disturbancy and predict the behaviour of the process in advance. Surprise and stress would have been avoided.

But in order to properly assess the situation after an alarm has occured, an operator should also be able to know about the dynamic behaviour of the process in the time frame under consideration about the consequences of possible actions or omissions (the 'longitudinal paradigm' according to De Greene). This would allow the operator to make informed decisions and greatly reduce the risk of panic actions.

OVERVIEW OVER THE GRADIENT SYSTEM

In the GRADIENT system, developed within ESPRIT project P857 and described e.g. by Elzer (1989) or Elzer, Weisang, Zinser (1989), the authors have therefore tried to reduce the surprise effect as well as the uncertainty level with respect to possible consequences of failures by various methods. A GRADIENT System consists of a set of cooperating expert systems that provide the operator with advice in real time. The information is presented mainly in graphical form.

The system provides warnings and decision support on various levels of the 'Rasmussen triangle' (as e.g. explained by Elzer, Siebert, Zinser, 1988) or - in the light of the above - in various phases of the 'decision window' (see below).

By means of a special type of expert system - the QRES (quick response expert system) - the operator is warned in advance of upcoming problems before these lead to an abnormal situation that - in turn - can lead to an alarm. S/he can then take appropriate measures in time and does therefore (normally) not need to react under pressure and stress. In addition to this warning the operator is - if possible - provided with advice concerning appropriate immediate countermeasures.

The GES (Graphical Expert System) and the PRES (Presentation System) then take care of the proper presentation of the situation by e.g. configuring on-line a P&I diagram of the affected subsystem on the VDU screen with proper contents, size and scale and by automatically showing the trend curves of all relevant values in a special window.

The necessary information for decisions to be made by the operator with respect to the next actions to be taken is provided by another component in the GRADIENT system - the 'Support Expert System' (SES) - which consists of three submodules for three different classes of additional operator support:

1) SES-CAUSES, a system that explains the deeper reason for a disturbancy to the operator, using state based diagnosis,

2) PSES, a support expert system giving 'procedural support', i.e. advising the operator (on request) about possible actions or action sequences for routine tasks or failure compensation, and

3) CP-SES with the task to predict possible consequences of failures. By this, the operator can decide which one of several alarms represents the most serious disturbance and therefore has to be dealt with first.

The prediction mechanism of CP-SES works on the basis of a model of the technological structure of the technical process being supervised, together with the properties, behaviour, and failure responses of its components. The model also includes a description of the properties and the behaviour of the supervision and control equipment and takes care of the response of the control system to disturbances in the process.

On request of the operator a dynamic check is performed as to which components might be affected by a failure of the one under consideration. All components detected by this algorithm are then displayed in a special window of the GRADIENT screen system. If several failures happen within a time interval too short for reacting to all of them, the operator can thus decide which failure will have the worst consequences and deal with it first. In case that a single failure has impacts on several components or subsystems the operator can also decide which impact has the most serious consequences on the overall behaviour of the plant and therefore has to be dealt with first. Fig.5 shows the approximate regions of applicability (usefulness) of the various components of a GRADIENT system with respect to the operator's 'decision window' as shown in Fig.4.

THE MECHANISM OF THE CP-SES

As described above, the task of CP-SES is to detect those components and systems of a power plant that may be affected by disturbances or malfunctions of other components. This is done by propagating the failure condition within a model of the plant and its S&C system, starting at the subsystem where it originated and identifying its effect on other subsystems and components. The object that represents the failure condition on its way through the model is called a 'Token'.

First of all, the token contains the pattern of a disturbance. For all known failure situations these patterns can be made available in a knowledge base. For disturbancies, especially classes of disturbances, CP-SES holds the description of those effects that change the process states within the component in which the failure occurred. E.g. the temperature, the flow or the pressure can decrease or increase dangerously or high transients of these process variables can arise. Alarms can be triggered by the effects of a disturbance. When the operator asks for the consequences of a special failure, the Token receives the pattern of this failure and starts the propagation through the power plant. This is illustrated in Fig.6.

On its way from one component or subsystem to the next one, the Token has to fulfil different demands. The main task is to decide whether the component or subsystem will be affected by the failure or not. For this purpose, the Token must compare the failure pattern with a pattern describing the process conditions under which the components or subsystems are stressed (for instance by a high pressure or a high temperature transient). By representing this knowledge in an object-oriented or component-related way the drawbacks of a completely 'pre-planned' mechanism can be avoided.

To channel the Token through the plant, CP-SES uses a model-based description of the power plant. For each component, e.g. a pipe or a valve, predecessors and successors are defined, as well as a unique identifier, like e.g. the Power Plant Designation Number ('KKS-No.'). The predecessor/successor relation is used to describe the 'flow' of a disturbance through a system or the entire plant. The model used by CP-SES currently consists of about 500 objects that represent the feedwater-system of a coal-fired power plant highly detailed as connected components and the rest of the plant less detailed as connected subsystems.

The power plant process is a complex system consisting of cycles with different media or the same medium in different states of aggregation. The incremental qualitative description of physical correlations within subsystems is based on the deviation of the process variables pressure, flow and temperature. For some systems or connected cycles there exist such transfer descriptions. If e.g. the flow in the heat-exchanger cycle 1 is going up and the temperature in cycle 1 is lower than the temperature in cycle 2, the temperature in cycle 2 is going down. The knowledge source 'transfer-descriptions' represents the knowledge about such changes of the process conditions from one system to another or within a complex system. In such components or subsystems the Token must create a new token and/or change its own failure pattern. The new or the changed failure pattern is derived from the old pattern via the transfer description. These tasks of the Token are also shown in Fig. 6. As mentioned above, the power plant process, like many other technical systems, consists of several closed loops with intersections. To avoid loops, the Token must remember the components it has already passed. On its way through the plant model the information about the affected systems will be gathered and finally presented to the operator in a comprehensive form.

DISCUSSION

Series of investigations using a realistic powerplant simulator and a first pilot application have shown that the described combination of support systems within a GRADIENT system provides a great potential for reducing the 'surprise-effect' and, consequently, the stress level in case of alarms that originate from within the supervised process (and/or the related S&C system) and therefore can be predicted on the basis of system knowledge. But, in addition, it is even possible to reduce the 'surprise effect' with respect to external events, because CP-SES can also be used for training. This can e.g. be achieved by inducing hypothetical, though possible, failures during idle time of the operators in order to be prepared for their occurrence that can not be predicted on the basis of system knowledge.

Certainly - in the light of the above considerations concerning the optimal size of the 'chunks', in which process information is presented to the operator - more work will have to be done in order to 'fine tune' the mechanism as well as the representations with respect to fast and correct comprehensibility of the information presented. It also has to be made sure that an operator spends less time in asking for the possible consequences of failures, understanding the answers of the system and making decisions than s/he wins by the widening of the decision window.

It is also not claimed that another common problem has been solved that plagues all computer-based systems: how to avoid that an operator gets overly confident in the advice they give and thus becomes too dependent on their support. But the techniques, by which GRADIENT eases the access to process information also help to avoid another disadvantage of today's highly structured and stiffly organized S&C systems: 'the loss of serendipity', that has also been criticized by De Greene (1991). To the contrary, the presentation system of GRADIENT provides an extremely flexible way of accessing process information, the 'picture pyramid', as e.g. described by Elzer, Siebert, Zinser (1988), that in principle allows a completely free 'navigation in information space' and thus contributes to the solution of another well known ergonomic problem: that a person under stress has to have several options open for solving the problem at hand, lest s/he will feel unable to solve the problem at all.

ACKNOWLEDGEMENTS

The work described in this paper was funded in part by the Commission of the European Communities in the framework of the ESPRIT I programme (Project P857: 'GRADIENT'). The authors further want to thank Dipl.Ing.B.Boussoffara and Dr.M.Reuter from the Institute for Process and Production Techniques at the Technical University of Clausthal for helpful discussions and valuable hints.

REFERENCES

Sill, U., Elzer, P., Roghmanns, H. (1989). Informationsverarbeitung und -darstellung zur Stoerungsfrueherkennung und -behebung; in: Rechnergestuetzte Prozeß- und Betriebsfuehrung im Kraftwerk. ETG/GMA/VGB-Symposium ('Fachtagung'), Saarbruecken, ETG-Report ('Fachbericht') 28, pp. 147-172, VDE-Verlag

Prodaska, J. (1981). Problem solving under stress, Medical Tribune, No. 23.

Popper, K.A., Eccles, J.C. (1984). The self and its brain - an argument for interactionism, Springer, New York

De Greene, K.B. (1991). Emergent Complexity and Person-Machine Systems. Intntl. Journal on Man-Machine Studies, Vol.35, pp.219-234

Elzer, P., Weisang, C., Zinser, K. (1989) Knowledge-based System Support for Operator Tasks in S&C Environments, IEEE Conf. on Systems, Man and Cybernetics, Boston, MA.

Elzer, P. (1989). GRADIENT - ein Schritt in Richtung intelligenter Prozessueberwachung; in: Wissensbasierte Systeme, 3rd Intern. GI-Congress, Muenchen, Informatik Fachberichte 227, Springer Verlag, pp. 453 - 463.

Elzer, P., Siebert, H., Zinser, K. (1988). New Possibilities for the Presentation of Process Information in Industrial Control. in: Selected Papers from the Third IFAC/IFIP/IEA/IFORS Conference, Oulu, Finland, pp.139 - 143.

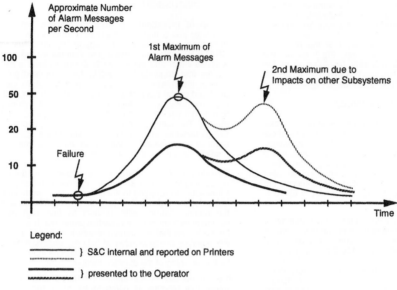

Fig.1: Typical Alarm Message Profile of today's S&C Systems

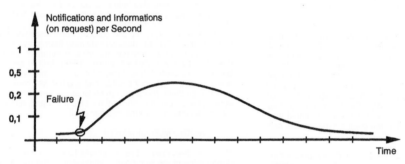

Fig.2: The Desired Flow of Messages from the Process

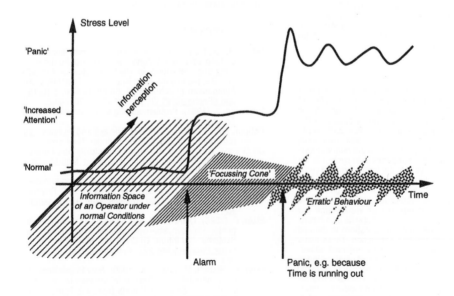

Fig.3: The Effect of Adverse Conditions on Human Information Processing

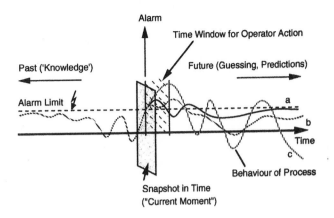

Fig.4: The Subjective Time Window

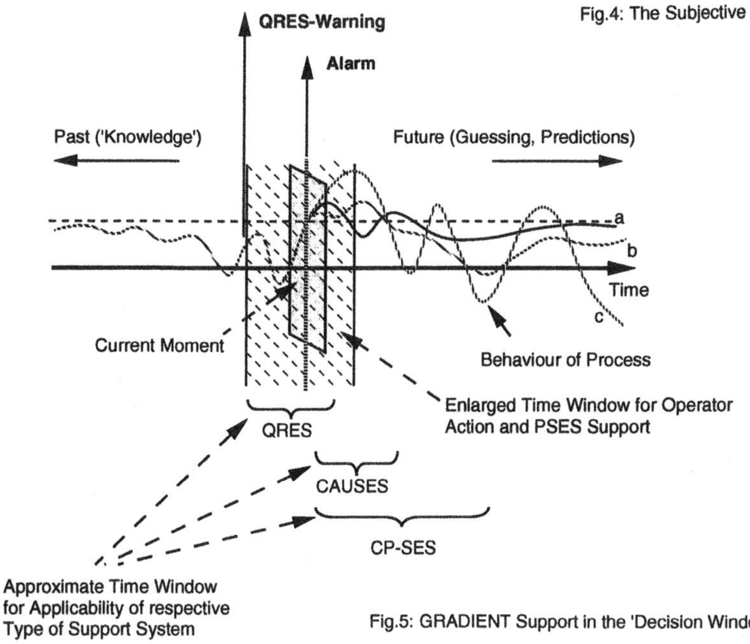

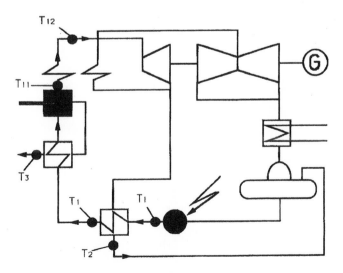

Fig.5: GRADIENT Support in the 'Decision Window'

Fig. 6: A Failure in the Feedwater Pumping System 'generates' a Token that propagates
through the Plant

Fig. 5. GRSAD first Scenario in the Reaction Window

Fig. 6. A Feature in the Feedwater Pumping System (pressure & level) that changes during the Plant

DAISY, A KNOWLEDGEABLE MONITORING AND WARNING AID FOR THE DRIVER ON GERMAN MOTORWAYS

M. Kopf and R. Onken

Universität der Bundeswehr München, Fakultät für Luft- und Raumfahrttechnik, D-8014 Neubiberg, Germany

Abstract. Design and implementation aspects of the driver monitoring and warning system DAISY (Driver Assisting SYstem)[1] are presented. The system is designed to work similar to the human driver except his "shortfalls" with respect to traffic safety. A mathematical formulation of danger is given based upon a parameter space description of traffic situations in terms of the time reserve with respect to a danger avoiding action.

There are three main modules of the system. First, the reference driver module where a situation analysis and evaluation is accomplished by instantiation of several knowledge representation structures. Second, the model of the actual driver where knowledge about the actual driver's normal behaviour is acquired and processed. Third, the discrepancy interpretation where the output of the other modules are combined in order to warn the driver in dangerous situations. In a learning phase DAISY extracts the actual driver's driving characteristics in order to adapt the warnings later on.

At the time being the system is implemented in a well equipped fixed-base driving simulator. It will be reported about some experimental results derived from simulator trials.

Parts of the system will be ported to an experimental car where machine vision is the main source of environmental measurement information.

Keywords. Man-machine systems; driver support; driver modeling; situation analysis; haptic displays; time reserve.

1. INTRODUCTION

The DAISY system (Driver Assisting SYstem) has been designed for driver support in a German motorway environment. With the implementation is in its final stage it is just in time in a phase of a general shift of interest from route guidance support systems to driver monitoring and warning systems.

DAISY is designed to meet both of the goals "increase of traffic safety" and "driver acceptance". The basic idea in order to meet the second goal is the "situative approach" paradigm from traffic psychology (cf. Fastenmeier, 1988). This means that the driver's behaviour is strongly depending on the actual traffic situation. Earlier attempts in developing driver support systems suffered mainly from disregarding this aspect (cf. Thomas e.a., 1989, p. 13). In consequence, the basis for the DAISY design is the study of driver's behaviour in different situations.

Since the system users' driving behaviour varies within a wide range of driving styles the goal "driver acceptance" can only be met, if the system has got the capability for adaptation to the individual driver.

Concerning the goal "increase of traffic safety" DAISY should be able to detect the driver's actual status with re-

spect to degraded performance caused by under/overload, fatigue, intoxication or other causes of declined or misled attention.

DAISY has been implemented in a fixed-base driving simulator. In the current stage of implementation the environmental measurements are assumed to be complete and precise.

Parts of the system will be adapted to real measurements, as they are available mainly based on machine vision, and ported to an experimental car.

2. THE HUMAN DRIVER

It is useful to study the driver's cognitive process as a basis for the functional design of a monitoring and warning system. From existing driver behaviour models (e.g. van der Molen & Bötticher, 1988 or Näätänen & Summala, 1976) can be extracted two principal modules: situation analysis / prediction and situation evaluation / intent.

Concepts for the representation and instantiation of situational elements are contained in the situation analysis module establishing an internal model of the situation. From this the future situation is predicted.

In the module situation evaluation / intent the situation is evaluated mainly with respect to "hindering", "danger" and "observance of traffic rules". These terms have to be formulated mathematically for use in the system.

---

[1] Sponsored by BMFT and Daimler Benz under contract TM 8900 A

The term "danger" is the most important one. Here it is used in accordance to Ross, 1974 cit. from von Benda e.a., 1983 where the danger is construed to be caused from both the environment (situation) and the driver himself. It is necessary to find a description which allows for computing a danger measure from physical measurements. In addition it should be possible to describe the individual behaviour within such a framework. The concept used is based on a situation specific parameter space description (Kopf & Onken, 1991). For every traffic situation this parameter space can be split up in two categories of regions. In the Controllable Region (CR) the driver can avoid an accident by an appropriate action (i.e. braking or steering wheel movement). In the Uncontrollable Region (UR) an accident will occur undoubtedly unless there are some benign circumstances not making the accident happen. These circumstances cannot be controlled by any driver actions. The boundary between these two regions, called Actual Danger Boundary (ADB), can be calculated from the situation parameters. The actual driver's behaviour can be described in terms of surfaces in this parameter space. The region used normally by the individual driver is called the Normal Region (NR). An example taken from a car following situation with two drivers in the driving simulator is given in Fig. 1.

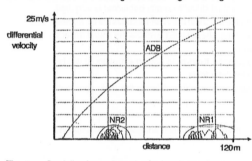

Fig. 1.   Car following trajectories in the phase plane

As pointed out in Kopf & Onken, 1990 the distance from the actual trajectory point to the ADB expressed in terms of time (-reserve) is used as an objective measure. Two situation specific driving styles can be compared with respect to objective dangerousness evaluating the Danger Integral DI defined by

$$DI = \frac{1}{T} \int_0^T \frac{k}{(TR - T_{rea})^n} dt$$

with   T   = time elapsed within the considered situation
        TR   = time reserve
        T_{rea} = minimum reaction time
        k   = scale factor
        n   = power for "curve sharpening"

Although it can be useful to implement human traits in a technical system it is not at all desirable to implement together with these traits the human "shortfalls" like emotions, degraded vigilance, limited processing capacity or improper internal models. It is nevertheless necessary to model these "shortfalls" for the purpose of detection of the driver's status, the design of driver interface and the experimental system evaluation. A model relating task demands, mental properties, fatigue and performance is given for instance in Warburton, 1978 cit. from Gstalter & Fastenmeier, 1987.

Multi capacity theories or the stimulus /central processing/ response compatibility principle (e.g. Wickens e.a., 1983) are also useful particularly for the design of the driver interface.

3. SYSTEM ARCHITECTURE

DAISY consists of four basic modules: The "Reference Driver" (RD), the "Model of Actual Driver" (MOAD), the "Discrepancy Interpretation" (DI) and the "Warning Device" (WD). The interaction and the internal structure of these modules are depicted in Fig. 2. The principal structure is adapted from a pilot supporting system being developed in the same institute (Dudek, 1990). Moreover, there is some resemblance to the structure of GIDS (Generic Intelligent Driver Support) being developed within the DRIVE project (Smiley & Michon, 1989). The RD extracts from the incoming data stream a symbolic description of the actual situation. In addition it computes situation specific action limitations. The DI judges the driver's actions with respect to these limitations considering also the output of the MOAD. In case of actual danger the WD will be triggered to issue an appropriate warning. The MOAD contains all available information about the actual driver's individual driving behaviour.

4. OPERATIONAL MODES

DAISY has to be operated in a learning phase if it has no driving experience with the actual driver. In this phase the driver's actions are judged only with respect to average/normative driving behaviour and objective danger measures (see par. 5). The system is not yet able to adapt the warnings to the individual behaviour but it is already able to warn in dangerous situations. In this phase the system learns the actual driver's individual traits.

Problems could arise because the driver is already influenced by the activity of the system. This has to be taken carefully into account in learning algorithm design. With the system having learned the relevant features of the individual driving style it is able to adjust the warnings to the indiviual driving style and to detect deviations from normal behaviour possibly caused by over/underload, fatigue or intoxication in order to determine the driver's state.

5. IMPLEMENTATION

5.1 Experimental environment
DAISY is integrated in a fixed-base driving simulator. A Silicon Graphics SkyWriter 320 IG 2 is the main computer. There, a detailed mechanical model of the own vehicle, simple models of environmental vehicles, the picture generation algorithms and the monitoring and warning system are computed. The vehicle models and the DAISY software are computed within a 0.05 sec time frame. All source code is written in C in order to be easily portable to in-vehicle computers.

5.2 The reference driver (RD)
The basic structures of the RD are finite automata and

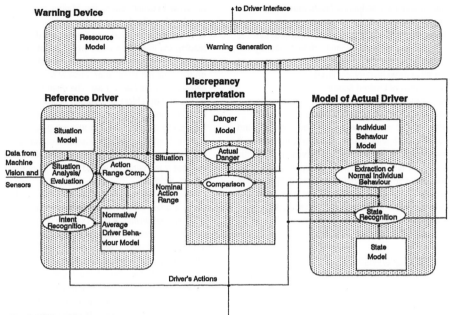

Warning Device

Ressource Model

Warning Generation

to Driver Interface

Reference Driver

Situation Model

Situation Analysis/ Evaluation

Action Range Comp.

Intent Recognition

Normative/ Average Driver Beha- viour Model

Data from Machine Vision and Sensors

Situation

Nominal Action Range

Discrepancy Interpretation

Danger Model

Actual Danger

Comparison

Model of Actual Driver

Individual Behaviour Model

Extraction of Normal Individual Behaviour

State Recognition

State Model

Driver's Actions

Fig. 2. The DAISY architecture

decision trees to represent the generic knowledge about both situation and driver behaviour (Onken, 1991). Decision trees are used to represent rules or static situational aspects and automata are useful for representing sequential events (scripts) or time dependent situational aspects. It is often necessary to integrate both situative and behavioural aspects into a single structure. The concept "Lane Change State", for example, contains the geometric parameters of road and vehicle as well as the driver's actions in terms of steering wheel activity and the resulting dynamic parame- ters. Here, a behavioural model has been integrated to decide as early as possible whether the driver intents to initiate a lane change or is willing to keep the actual lane.

The automat "Lane Change State of Environmental Vehicle" will be described now in some detail. The state transition diagram is given in Fig. 3.

Its principal task is to distinguish between lane keeping and lane change. The transition from lane keeping to lane change has to be detected as early as possible. The under- lying statistical behaviour model has been extracted from driving simulator experiments with 10 subjects (Diepold, 1991).

Knowledge of the probability distributions of lateral position and lateral velocity in either case allows computation of threshold values to decide whether the hypothesis "lane change" or "lane keeping" is true with an error probability of α. An algorithm like this is used to trigger the transitions in the automat.

The lane change state of the own vehicle can be treated analogously. In this case, the parameter space is three dimensional because an additional parameter (the steering wheel angle) is known.

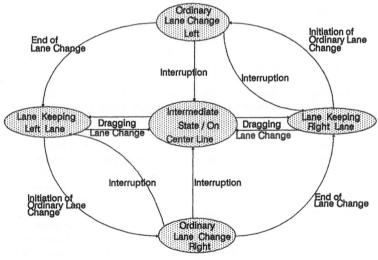

Fig. 3. State transition diagram

So far this has been a description of techniques used in the situation analysis.

Considering the intent recognition it is assumed, that the driver's motivation is predominantly determined by his desired speed. Hence, the intent recognition process can be structured in three hierarchical levels:
- Recognition of Desired Speed
- Recognition of Overtaking Intent
- Recognition of Lane Change Intent.

The desired speed can be determined by observing the actual speed in a situation where the driver is not hindered by environmental issues like vehicles, curves or gradients. This information is supplied by several situation analysis structures. Observing the speed of vehicles overtaken will yield the range of tolerance where the driver is willing to stay behind a slower vehicle. This leads to the recognition of an overtaking intent. From this a lane change intent can be inferred using additional knowledge from the MOAD.

5.3 Discrepancy Interpretation

It is the task of the DI to decide whether, and to which degree, a situation is dangerous, taking into account the situation and the driver's behaviour. It has to assess if there is a deviation from the driver's normal behaviour (as known from the MOAD) and if this deviation is dangerous.

Three principal dangers are considered in DAISY:

I: Collision with preceding vehicle because of not suitable deceleration behaviour

II: Exdeeding of nominal lane because of not suitable steering behaviour

III: Exceeding of road in curves because of not suitable velocity

Computation of the time reserve with respect to a danger avoiding action is the fundamental concept. It is determined by:
- the dynamic state of the system environment/vehicle
- the physical safety margins (friction)
- the parameters of the danger avoiding action.

Depending on the parameters used for computation this time reserve has two aspects: An **a-posteriori time reserve** is defined computed by
- the actual dynamic state variables
- absolute physical safety margins
- maximum possible danger avoiding action.

This measure is the actual distance to the ADB. For danger I and II the a-posteriori time reserve consists of the time-to-collision (TTC) and the time-to-line-crossing (TLC; Godthelp, 1984) respectively minus a particular time depending on the parameters of the danger avoiding action. During normal driving this measure will be far above the human reaction time because the driver normally drives safely.

To describe better the looking-ahead aspect of an actual situation an **a-priori time reserve** is defined, computed by
- actual as well as potentially danger augmenting state variables
- physical safety margins usually claimed by the actual driver
- parameters of the driver specific danger avoiding action.

This concept will be explained in the following for danger I. The time reserve in this car following situation is a function of the actual accelerations and velocities of the two cars, the distance between them, the friction coefficient and the deceleration of a danger avoiding braking action.

To compute the a-priori time reserve, the deceleration of the leading car is set to a value which is not exceeded with a probability of, for instance, 95 %. Hence this value depends on the situation specific probability distribution of the acceleration of the leading car. The distribution is influenced by the kind of vehicle, the relation to other leading vehicles, its driver's intent, the traffic density and the road conditions. Probability distributions like this are given e.g. in Winzer, 1980. In Diepold, 1991 a deceleration behaviour description has been derived from a trial series in the simulator. It turned out that it is mainly governed by the time-to-collision (cf. Färber, 1986).

The deceleration of the danger avoiding braking action is set at a driver and situation dependent value being used while normal driving. A driver can be characterized that he is taking the potential danger into account if he keeps the a-priori time reserve great.

Computing the a-posteriori time reserve the actually measured acceleration of the leading car is used. The deceleration of the braking reaction is assumed to be set at the maximum value depending on the actual friction coefficient. If this time reserve falls under the minimum human reaction time a collision is unavoidable unless the leading car decreases its deceleration or the driver manages to accomplish a danger avoiding steering action. At that point, if a clear picture of the situation can be achieved, autonomous braking is applied during the reaction time period of the driver. If a lane change has been recognized by the automat described above, the time reserve is of course computed only with respect to vehicles in the target lane.

5.4 Warning device and MMI

The warnings from danger I, II and III are subdivided into those for longitudinal and those for lateral guidance. Different concepts of warnings are used for demonstration in the simulator and the in-car realization. If one of the a-priori time reserves of I or III falls under a situation and driver dependent limit the resulting danger will be indicated on the lower part of the simulator outside image screen by a couloured bar of great width. This bar is designed to be percepted by peripheral vision. The coulour ranges from yellow via orange to red in order to give a feeling for the degree of danger. If the driver decelerates, resulting in an increase of the time reserve, the bar will vanish to indicate that the driver has responded to the danger. For in-car realization a haptic display using an active gas pedal is considered.

In case of the a-posteriori time reserve falling below the driver specific minimum reaction time an acoustic alarm will be triggered to indicate the immediate need for a braking action with full possible force. In order to prevent blocked wheels the car should be equipped with ABS.

Concerning danger II the a-posteriori time reserve is computed with respect to the available lateral space. If, for example, lane change is not possible (information supplied from MD) only the width of the own lane is available. Otherwise, the width of the entire road could be used. The warning information is fed directly into the steering wheel by an additional artificial steering force. This haptic indicator is guiding into the nominal direction of steering wheel deflection and the force increases if the a-posteriori time reserve falls below a situation and driver dependent threshold. The force is sufficiently strong to be realized by the driver but not too large to be easily compensated by the driver if he is willing to do so. The increase of the artificial force is sufficiently soft to prevent panic reactions.

5.5 The model of actual driver (MOAD)

The task of the MOAD is two-fold. In the learning phase it has to extract the actual driver's normal individual behaviour (i.e. to determine his NR). Having concluded this phase the MOAD monitors when the driver's behaviour is moving out of his NR in order to determine driver's state.

The concrete procedure will be shown for danger II. Here, the statistic of the a-posteriori time reserve is recorded. The lane keeping/lane changeing behaviour is assumed to be mainly influenced by velocity and road geometry. Fig. 4 shows these dependencies as a result of simulator trials with ten male subjects (Diepold, 1991). It is plotted the mean of the 5%-time reserve. This means that the time reserve remains unter this threshold in 5% of time. Positive or negative time reserves correspond to the danger of exceeding the right or the left lane boundary respectively.

The empirical probability distribution of the time reserve is computed for each combination of situational parameters. Then, the set of α% values of these distributions is extracted forming the empirical description of the individual driving behaviour. Having concluded the learning phase the α% values will be used to adjust the warning threshold for lateral guidance to the individual driver. Moreover, deviation from normal can be detected recording the fraction of time when the time reserve is below the α% value. If this fraction exceeds α% it can be inferred that the driver behaves more dangerously than normal. Additional indicators should be considered to find out the reason for this deviation from normal.

During the entire learning phase the time reserve theshold for lateral guidance warning is set about the human reaction time in order to produce a warning only in an objectively dangerous situation.

6. RESULTS

At the time being the MMI components are in state of final implementation. For spring of 1992 an extensive evaluation phase is planned. Results are expected to be available in summer 1992. Here will be given a demonstration of the essential features of DAISY. A traffic situation with two environmental vehicles has been simulated in the simulator. Fig. 5 shows the time history of essential variables. In order to give an imagination of the relative positions of the vehicles they are depicted symbolically for several stages of the run. As danger III occurs on motorways only at very low friction coefficient or very high speed it has been neglected here. The time reserves have been plotted reciprocally because they are infinite in the often occuring safe case. The following variables are plotted:

a_E = acceleration of the own car [m/s*s]

swa = relative steering wheel angle [rad]

TR1 = reciprocal a-priori time reserve for longitudinal guidance [1/s]

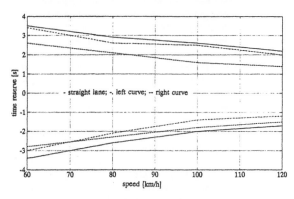

Fig. 4.   5%-time reserve for lateral guidance

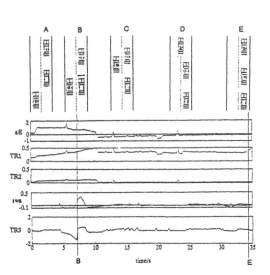
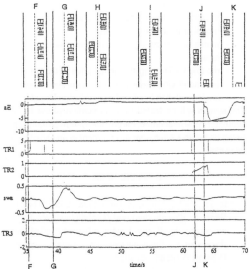

Fig. 5.   A sample run with DAISY in the driving simulator

193

TR2 = reciprocal a-posteriori time reserve for longitudinal guidance [1/s]

TR3 = reciprocal a-posteriori time reserve for lateral guidance [1/s]

There are some remarkable events in course of the run:

B: The driver didn't realize the overtaking car. When he tried to initiate a lane change the a-posteriori time reserve for danger II, which has been computed with respect to the right lane, falls under the alarm threshold resulting in the WD triggering the additional steering wheel force. In consequence the driver accomplished the appropriate steering action.

E/F: Car 1 approaches car 2. Hence, the probability for braking of car 1 increases resulting in an decrease of the a-priori time reserve (- increase of TR1). This will be indicated to the driver via the coloured bar. However, the a-posteriori time reserve does not decrease because the actual acceleration of car 1 remains unchanged.

F/G: Immediatly after DAISY recognized a lane change (observe the steering wheel angle) the time reserves for danger I are computed with respect to the target lane. Since there is no other car these time reserves are infinite (TR1 and TR2 = 0).

J/K: Car 2 initiates a lane change in spite of the own car being close behind. DAISY has recognized this lane change in its early state resulting in the time reserves for danger I being computed with respect to car 2. In this case, even the a-posteriori time reserve falls below the alarm threshold indicated to the driver by the acoustical alarm. After engageing the brake (observe a_E) the time reserves are again infinte (TR1 and TR2 = 0) which means that the danger is no longer existent.

7. CONCLUSION

DAISY, a driver supporting monitoring and warning system has been presented. Its most important features are:

- information processing based upon the driver's cognitive process
- situation analysis module establishing an internal model of the current situation
- mathematical description of danger in terms of time reserve
- module for learning the actual driver's driving style and monitoring of deviations from normal behaviour
- implementation in a fixed-base driving simulator.

The system will be tested in an evaluation phase in the simulator in the near future. Parts of the system are currently being ported to a transputer environment in an experimental vehicle.

8. REFERENCES

Benda e.a., 1983: Benda, H.von, Hoyos, C.G., Schaible-Rapp, A.; Klassifikation und Gefährlichkeit von Straßenverkehrssituationen. Research Report 7320 of Bundesanstalt für Straßenwesen, Dep. Unfallforschung, D-Bergisch Gladbach, 1983

Diepold, 1990: Diepold, S.; Fahrsimulatorversuche zur Ermittlung situationsspezifischer Fahrverhaltensparameter. Thesis at the University of German Armed Forces, Munich, 1991

Dudek, 1990: Dudek, H.-L.; Wissensbasierte Pilotenunterstützung im Ein-Mann-Cockpit bei Instrumentenflug. Doctoral Thesis at the University of German Armed Forces Munich, 1990

Färber, 1986: Färber,B.; Abstandswahrnehmung und Bremsverhalten von Kraftfahrern im fließenden Verkehr, Zeitschrift für Verkehrssicherheit, 32(1), 1986, pp 9-13

Fastenmeier, 1988: Fastenmeier, W.; Ein Katalog von Verkehrssituationen für repräsentative Fahrttypen. PROMETHEUS Report EW-13, Munich, 1988

Godthelp, 1984: Godthelp,H.; Studies on Human Vehicle Control, Institute for Perception TNO, NL-Soesterberg, 1984

Gstalter&Fastenmeier, 1987: Gstalter, H., Fastenmeier, W.; Recognition of Driving Conditions of Drivers. PROMETHEUS Report WP 31 138, Munich, 1987

Kopf&Onken, 1990: Kopf,M. & Onken,R.; A Machine Co-Driver for Assisting Drivers on German Autobahns, 9. European Annual Manual, I-Ispra, Sept. 1990

Kopf&Onken, 1991: Kopf, M., Onken, R. Monitoring and Warning System for Driver Support on the German Autobahn, ISATA 1991, I-Florence, Mai 1991

Näätänen&Summala, 1976: Näätänen, R., Summala, H. Road-user behaviour and traffic accidents, North Holland Publishing Company Amsterdam, Oxford 1976

Onken, 1991: Onken,R.; Monitor- und Warnfunktionen, PROMETHEUS PRO-ART semi-annual report I/91 of the University of the German Armed Forces Munich, 1991

Ross, 1974: Ross, H.E.; Behaviour and Perception in Strange Environments, Allen & Unwin, London 1974

Smiley&Michon, 1989: Smiley, A., Michon, J.A.; Conceptual Framework for Generic Intelligent Driver Support. Report of DRIVE project V1041, Traffic Research Centre, University of Groningen, NL-Groningen, 1989

Thomas e.a., 1989: Thomas, D.B.,Brookhuis, K.A., Muzet, A.G. e.a.; Monitoring Driver Status: The State of the Art. Report of DRIVE project V1004, Cologne, Groningen e.a. 1989

van der Molen & Bötticher, 1988: van der Molen, H.H, Bötticher, A.M.T.; A hierarchical risk model for traffic participants Ergonomics, 31(4), 1988, pp 537-555

Warburton, 1978: Warburton, D.W.; Physiological aspects of information processing and stress. In: Hamilton, V. & Warburton, D.W. (Eds.): Human stress and cognition, New York, pp 33-65

Wickens e.a., 1983: Wickens, C.D., Sandry, D.L., Vidulich, M.; Compatibility and Resource Competition between Modalities of Input, Central Processing, and Output. Human Factors 5(2), 1983, pp 227-248

Winzer, 1980: Winzer, T.; Messung von Beschleunigungsverteilungen; Forschung Straßenbau und Straßenverkehrstechnik, Heft 319, BmV, Bonn-Bad Godesberg 1980

KNOWLEDGE BASED COCKPIT ASSISTANT FOR CONTROLLED AIRSPACE FLIGHT OPERATION

T. Wittig and R. Onken

Universität der Bundeswehr München, Institut für Systemdynamik und Flugmechanik, Werner-Heisenberg-Weg 39, D-8014 Neubiberg, Germany

Abstract. A knowledge based cockpit assistant for flight operation under Instrument Flight Rules (IFR) is presented. The system is aimed at improvement of situation assessment including monitoring of the pilot crew activities and performance increase by computer aids for flight planning and plan execution. The modular system structure is described as well as the individual system modules. A first prototype of this Cockpit Assistant was tested about two years ago in a flight simulator at the University of the German Armed Forces in Munich by professional pilots under realistic IFR-scenarios. The results of these test runs proving enhancements in overall system performance are presented.

Keywords. Human factors; Artificial intelligence; Knowledge engineering; Speech recognition; Simulation.

1. INTRODUCTION

Civil air transportation of today is characterized by flights under Instrument Flight Rules (IFR), since this kind of flight operation guarantees flight execution with almost full independence of the weather conditions. However, among other factors lacking visual references as well as increased automation and complexity of cockpit instrumentation can result in overcharges of the pilot crew. It is a fact that by far the majority of accidents is caused by human errors [1,2]. Statistical data of aircraft accidents and their causes can be correlated with findings on the cognitive behaviour of humans [3]. From this it became evident that electronic pilot assistance has a good chance of becoming effective for:

— situation assessment

— planning and decision making and

— plan execution.

This requires a system design complementing human capabilities and not replacing human control functions generally by automatic ones.

On the basis of this formal knowledge of the user needs a cockpit assistant for IFR operation is being developed at the University of the German Armed Forces in Munich and implemented in a flight simulator. This research, when started in 1988, was aimed at assisting the pilot in SPIFR (Single Pilot IFR) operations and led to a first prototype, called ASPIO (Assistant for Single Pilot IFR Operation) [5,6,8]. Since 1991 a similar advanced Cockpit Assistant System (CASSY) for the two man crew is developed in cooperation with the Dornier company [8].

To achieve the assisting functions CASSY is structured into several modules and integrated into the air traffic system with interfaces to the aircraft, the pilot and the Air Traffic Control (ATC). The major modules and the information flow within this system are described in the following chapter. ASPIO, the first prototype of this cockpit assistant, was tested and evaluated at the beginning of 1990. The experimental setup and the experiment results of this test period are presented.

2. STRUCTURE OF THE COCKPIT ASSISTANT SYSTEM

2.1. Modular structure and information flow

As mentioned before, the requirements for the cockpit assistant made it necessary to structure the system into several task specific modules. Besides the interface to the aircraft the system consists of the following components:

— Dialogue Manager (DM)

— Automatic Flight Planner (AFP)

— Piloting Expert (PE)

— Pilot Intent and Error Recognition (PIER)

— Monitor for Flight Status (MFS), Flight Systems (MS) and Aircraft Environment (ME)

— Execution Aid (EA)

The modules of CASSY together with the information flow are shown in figure 1 and will be described in the following.

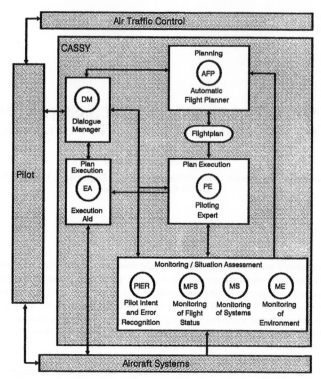

Fig. 1. Structure and information flow of CASSY.

2.2. Interfaces

The aircraft interface establishes the access to all information which can be drawn from all kinds of aircraft systems, including the autopilot. It is represented by a dynamic data pool containing all aircraft relevant data about flight status, autopilot settings, radio navigation and communication settings and status of aircraft subsystems. It supplies the modules of CASSY with the information needed and on the other hand gives the Execution Aid direct access to the aircraft systems in order to assist the pilot in executing the actual flight plan.

The Dialogue Manager (DM) comprises all components for the information transfer between CASSY and the pilot crew, including the management of information flow to the pilot crew.

Extensive use is made of speech communication in either direction. The first generation speech interface (ASPIO) is described in [7]. Speech input is used for acknowledgement and rejection of system recommendations and as a communication medium for instructions to the executional aids of CASSY. For this purpose a speaker dependent speech recognition system is used based on the phraseology of civil aviation. Synthetic speech is used for speech output, with different voices for different categories of assistant messages. More complex information like comprehensive flight plan recommendations is presented visually using one of the multifunctional displays.

Hereby, the Dialogue Manager controls the syntax of the speech input, the priority and category of each speech output message and information to be visually presented to the pilot crew.

The ATC interface is not explicitly realised within the cockpit assistant. All ATC instructions and information have to be transferred to CASSY by the pilot crew through speech input. This is rather easily achieved because the ATC instructions are to be acknowledged verbally by the pilots anyway. There are ongoing developments for a digital datalink between ATC and the aircraft. When this is available a direct link between ATC and CASSY can be established.

2.3. Automatic Flight Planner (AFP)

For every flight, a flight plan must be issued before takeoff. This flight plan can be worked out by the Automatic Flight Planner (AFP) or can be prepared by means of other facilities and then fed into the system as part of the initial conditions.

During the flight the AFP is activated when significant deviations from the flight plan occur because of such events as new ATC instructions not in accordance with the flight plan, adverse weather conditions or system failures. The module can also be activated by the pilot crew who might feed in their intent of changes. Also the PIER module recognizing changes of pilot intentions might initiate variations from the original flight plan. An evaluation of the current situation and its future projection might pinpoint where conflicts with the original flight plan arise. This can result in a problem solving algorithm for the selection of an alternate destination and corresponding generation of a new flight plan. Hereby, route and trajectory planning is performed by the AFP under consideration of aircraft system state and performance limitations. It includes tactical and strategical planning.

The AFP planning results are presented to the pilots as recommendations. If not corrected by the pilots, agreement for the new flight plan is achieved.

2.4. Piloting Expert (PE)

On the basis of the flight plan as generated by the AFP and acknowledged by the pilots, the Piloting Expert (PE) performs the automatic management of flight plan execution. This is carried out by following the instructions of ATC, information from MS or ME about system failures or bad environmental conditions, messages from MFS considering the flight progress and regulations for piloting. Hereby, the Piloting Expert is construed as a model of the pilot crew taking into account the standard pilot activities as well as the individual behaviour. In this way the module determines the expected actions the pilot crew is supposed to carry out during the various flight segments. The modelling is essentially rule-based on the basis of the extensively elaborated and published piloting regulations.

2.5. Pilot Intent and Error Recognition (PIER)

The Pilot Intent and Error Recognition (PIER) compares pilot activities with expected pilot behaviour as generated in the PE.

By use of a continuous situation analysis and evaluation the overall flight situation with regard to flight status, ATC commands and system status is examined and the necessity for changes of the actual flight plan is evaluated. As part of it the module follows up all pilot actions during the flight. In case of discrepancies between pilot activities and the expected behaviour it is verified whether the crew actions tend to reduce those deviations by observing changes in altitude, airspeed, heading or course histories and pilot control actions. If there is no evidence for reduction of deviations, the distinction has to be made whether the crew is acting intentionally or not. For this purpose, the knowledge about the flight situation is utilized and the pilot behaviour is evaluated by use of the fuzzy set theory.

In parallel, the intent recognition component tries to find out the possible crew intentions. Hypotheses for the crew intentions are generated on the basis of the actual flight situation and are assigned with an apriori probability. A great amount of situation dependent criteria are evaluated with respect to the hypotheses concerned by a probabilistic approach.

The distinction whether the crew acts intentionally or not on one side and the possible hypotheses for the crew intent on the other now finally lead to a differential diagnostic. The crew behaviour is only be classified as mistaking in case of no meaningful intention was discovered or if the danger boundary is exceeded. In this case a warning message is transferred to the crew by use of the DM module. If intentional behaviour has become evident but the intention itself

is not completely uncovered a short hint is given to the crew and the module carries on trying to find out the crew intent. Having recognized the crew intent this information finally is transferred to the AFP.

2.6. Monitoring modules

The Monitor of Flight Status (MFS) has got the task to monitor the flight progress in order to be aware of the arrival at any subgoals of the flight plan. Thereby, the flight plan can be actualized and the expected pilot behaviour can be generated in the PE module. There are also standard callouts, usually delivered by the co-pilot in the conventional two-man cockpit, performed via speech messages.

The health status of aircraft systems is monitored by the Monitor of Systems (MS). This module works on the information about defects in aircraft systems in order to derive messages to the pilot crew. Information is also given to the AFP and the PE, since on one side replanning could be necessary and on the other side the tolerances for the pilot behaviour could become different.

Sensor information for the assessment and evaluation of weather conditions and surrounding traffic are gathered in the Monitor of Environment (ME). This monitoring module also identifies deviations from normal with respect to the aircraft environment and reports to the pilot crew and the CASSY modules concerned.

2.7. Execution Aid (EA)

To assist the pilot crew in executing the actual flight plan, the Execution Aid (EA) offers a variety of functions usually performed by the co-pilot. Among these functions are instrument setting, autopilot setting, flap and gear setting or navigational calculations. The EA can be tasked by the pilot crew by use of the speech input.

3. EXPERIMENTAL TESTING

3.1. Experimental setup

The cockpit assistant system implementation in a flight simulator facility at the University of the German Armed Forces in Munich is continuously extended.

A one-seat fixed base cockpit is used. The central computer of the experimental setting as shown in figure 2 is a UNIX IRIS 4D / 140 GTXB Graphics workstation with four central processor units. Aircraft dynamics (6-degrees of freedom model of the HFB 320), autopilot, radio navigation systems and wind characteristics are simulated and a high performance head down instrumentation display is generated. The workstation is also used to run all CASSY modules and to perform the interfacing with speech input and speech output, the stick force simulation unit and a control and display panel. The image outside vision is generated by an additional SkyWriter workstation. Also a radar display for use as a combined ATC controller / instructor workstation is installed.

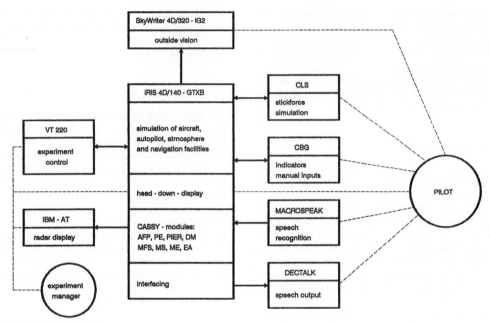

Fig. 2. Experimental setup.

A first prototype implementation of the cockpit assistant was tested at the end of 1989. This Assistant for Single Pilot IFR Operation (ASPIO) differed from CASSY in some respects. So this prototype

— was limited to support the single pilot and not the pilot crew,

— was not able to recognize and consider consciously the pilot intentions,

— was characterized by the use of a simple speech recognition system

— and was realized only for the arrival and approach segment.

This prototype compriced the main functions, though, being well suited for more extensive flight simulator experiments.

The experiments were aimed at proving enhancements in overall system performance, also with regard to safety, pilot workload balancing and pilot acceptance under the most possible realistic conditions. For this purpose, the following criteria were investigated:

— Accuracy of flight

— Pilot errors in situation assessment and system operation

— Duration and quality of planning processes

— Pilot workload

— Pilot acceptance

The pertinent parameters were evaluated for a number of test runs of IFR-approaches. Three different IFR-scenarios were developed which consisted of standard situations as well as of unusual events and emergency situations. A total of nine professional pilots, everyone with a great amount of IFR flight experience, have performed these test runs. Some of the evaluation results are presented in the following [5].

3.2. Evaluation of flight accuracy

As an indicator for flight accuracy, the airspeed deviation was considered to be most appropriate for evaluation. It was supposed to be controlled manually by the pilots.
The evaluation results for the standard deviation of airspeed for each of all nine pilots are depicted in figure 3. It was shown that the improvement in flight accuracy by use of the assistance functions is highly significant.

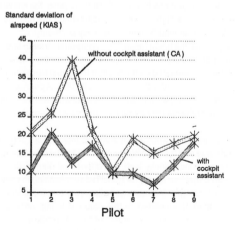

Fig. 3. Standard deviation of airspeed.

198

3.3. Evaluation of pilot errors

Pilot errors could be directly and unambiguously detected. One can say that, by definition, no major deviations from normal flight could be observed for the flights with activated monitoring function of the cockpit assistant. However, without cockpit assistant, pilot errors were detected, also more serious ones.

3.4. Evaluation of planning and decision processes

The time needed for planning and decision-making was determined. The planning activity was triggered by an ATC instruction, which also demanded for a quick reply about the pilot's further intentions. The time between the ATC instruction and the pilot's reply was measured.

In figure 4, the decision time of every pilot without support by the cockpit assistant is shown in comparison to that with cockpit assistant aid. The planning und decision task was to check for an alternative approach procedure under consideration of weather and airport data, after the information was given about the breakdown of the instrument landing system at the destination airport. The differences are obvious. Moreover, the pilots stated that all decisions recommended by the cockpit assistant, made sense.

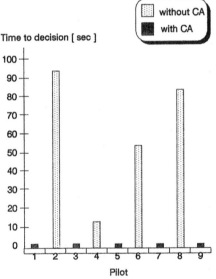

Fig. 4. Time to decision.

3.5. Evaluation of pilot workload

The pilot workload was assessed by means of subjective rating and the introduction of a secondary task. For subjective rating, the SWAT-method (Subjective Workload Assessment Technique) was used. As a secondary task, periodical tapping was specified. The results show slight reduction of pilot workload, but without significance.

3.6. Evaluation of pilot acceptance

Pilot acceptance was evaluated from pilot statements in a questionnaire which was furnished by the pilots after their test flights. In part, the technique of the semantic differential [4] was used, as shown in figure 5. The median values show positive mean reactions, i.e. good acceptance by the pilots. The neutral overall assessment for the component "not distracting/distracting" was explained by the pilots by certain lack of familiarization in system handling, in particular with respect to the specific speech recognition system used for these experiments.

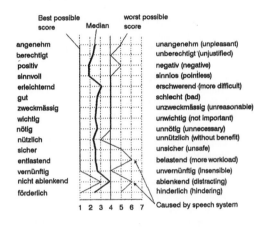

Fig. 5. Semantic differential.

4. CONCLUSION

The human pilot's intrinsic limits of capability and behavioural characteristics of mismatching lead to certain categories of errors and resulting accidents. Therefore, the knowledge about human debits can be exploited for the specification of automatic cockpit aids. Electronic pilot assistants can offer great benefits in monitoring, planning and decision-making for complex situations. Those systems can rapidly offer recommendations to the pilot without getting "tired" or loosing information.

The presented cockpit assistance system for IFR operation is able to understand the situation on the basis of knowledge about facts and procedures of the piloting task environment and the actual data about the flight status and pilot actions. The situation can be evaluated with regard to conflicts concerning the actual flight plan. If necessary, the system derives a revised flight plan as a recommendation to the pilot or can also serve the pilot for plan execution tasks. It also presents advices or warning messages to the pilot crew in case of recognized pilot errors.

The flight simulator experiments for testing a first prototype of this cockpit assistant under realistic conditions led to the conclusion that this kind of system actually can complement human capabilities and is accepted by the pilots. Flight accuracy was significantly improved, adverse consequences of pilot errors were eliminated and also planning and decision-making was improved. It could be shown that pilot workload did not change significantly, although the pilot interface was not optimized at this stage of development.

5. REFERENCES

[1] Rüegger, B. (1990). Human error in the cockpit.
     Swiss Reinsurance Company, Aviation Department.

[2] Chambers, A.B., Nagel, D.C. (1985).
     Pilots of the Future: Human or Computers.
     Communications of the ACM, Vol. 28, No. 11.

[3] Rasmussen, J. (1983). Skills, Rules and Knowledge;
     Signals, Signs and Symbols, and other Distinctions
     in Human Performance Models.
     IEEE-SMC-13, No. 3.

[4] Bergler, R. (Hrsg.) (1975). Das Eindrucksdifferential.
     Verlag Hans Huber, Bern.

[5] Dudek, H.-L. (1990).
     Wissensbasierte Pilotenunterstützung im
     Ein-Mann-Cockpit bei Instrumentenflug.
     Universität der Bundeswehr München, Dissertation.

[6] Onken, R. (1990). Knowledge-Based Cockpit
     Assistant for IFR Operations.
     AGARD GCP Symposium, Madrid.

[7] Dudek, H.-L., Onken, R., Wittig, T. (1989).
     Speech Input/Output for Communication between
     pilot and pilot's assistant system.
     Military and Government Speech Tech.

[8] Wittig, T., Onken, R., Dudek, H.-L. (1991).
     Unterstützung des Piloten im IFR-Flug durch
     wissensbasiertes Assistenzsystem.
     DGLR-Jahrestagung, Berlin.

[9] Lizza, C., Rouse, D., Small, R., Zenyuh, J. (1990).
     Pilots's Associate: An Evolving Philosophy.
     Proceedings of the 2nd Joint GAF/RAF/USAF
     Workshop on Human-Electronic Crew Teamwork,
     Ingolstadt.

[10] Amalberti, R., Champigneux, G. (1990).
     Man Machine Coupling: A Key For Electronic
     Copilot Architecture.
     Proceedings of the 2nd Joint GAF/RAF/USAF
     Workshop on Human-Electronic Crew Teamwork,
     Ingolstadt.

A CONNECTIONIST TRAFFIC SIGN RECOGNITION SYSTEM FOR ONBOARD DRIVER INFORMATION

D. Krumbiegel, K.-F. Kraiss and S. Schreiber

Research Institute for Human Engineering (FGAN-FAT), Neuenahrerstr. 20, 5307 Werthoven, Germany

Abstract.

In this paper, we deal with the development of a support system for vehicle handling based on a traffic sign recognition system. This pattern recognition system consists of three modules dedicated to detection, recognition, and knowledge-based control. Neural networks are used as adaptive classifiers. The stage of development and the applicability of such a system are discussed.

Keywords.

Image processing; knowledge based control; neural nets; parallel processing; pattern recognition; support system; vehicles.

INTRODUCTION

The growth of traffic today demands more and more of the driver's attention. Similarly the consequences of wrong behaviour in traffic have increased. One reason for wrong behaviour can, for example, be the overlooking of a traffic sign. Possible reasons for this are:

- momentary inattentiveness of the driver.

- traffic conditions (e.g. oncoming traffic, black ice), which require full concentration.

- too many traffic signs, which make excessive demands on the driver.

  (see introduction to model test "Fewer Traffic Signs"; BASt 1985)

- high speed, which prevents recognition.

The automatic recognition of traffic signs in real time in unfavourable conditions forms the basis for the preparation of warning and support functions for the driver. This support system for vehicle handling would then be able to recognize discrepancies between the driving behaviour requested by traffic signs and that of the driver and to give corresponding warning signals. It would also be able to assist the driver by displaying the last traffic signs passed. This is a new source of information for the driver. The glut of information reflecting the traffic situation is filtered until the driver has only the facts he needs. Another application is decision aid, for example as component of a navigation system. If the system is able to read direction signs, e.g. "Bonn 45 km", it can lead the driver to his destination.

TRAFFIC SIGN RECOGNITION

If you look at Fig. 1 you can see a simple traffic scene. The task is to recognize and to classify the traffic sign in the right edge of the traffic scene. Even for the human driver it is not possible to recognize the traffic sign at a glance. At first he will search all over the scene for possible traffic signs. Then, he will recognize the object on the right edge. Now, he will bring the object into focus and then, if he knows the sign, he will classify it as traffic sign "Attention". At last he will adapt his driving behaviour.

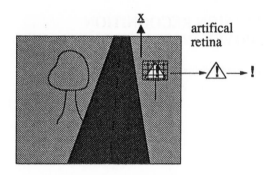

Fig. 1: Traffic sign recognition

For technical systems we need the following description of the recognition task (see Pao 1989).

$$X \to \underline{x} \to f(\underline{x}) \to R(f(\underline{x})) \to c(\underline{x}) = X$$

A class is called X. X discribes the class of the traffic signs "Attention", for example. The pattern $\underline{x}$ is an instantiation of X (Fig. 1). With the preprocessing procedure f() the pattern $\underline{x}$ will be transformed in appropriate features $f(\underline{x})$. This is the input for the recognition $R(f(\underline{x}))$. R() maps the pattern $\underline{x}$ into c_X, the class designation X, if it belongs to this class $(c(\underline{x}) = X)$. For more complex recognition tasks this process can be split up. The partial recognition tasks use different preprocessing procedures $f_i()$ and different classification procedures $R_j()$. They can be executed in parallel.

The technical systems described here aims at emulating the described human visual search behaviour. With the presented concept the first intention is to detect picture areas where traffic signs are likely to occur. These are called "areas of interest" $(\underline{x})$. For this purpose one or more artificial retinas are moved over the picture. This first step is called the "detection phase". If an area of interest is found, this picture segment forms the input for the next step which performs the actual pattern recognition task. This step is called the "recognition phase". At first a pattern recognition module tries to classify the traffic sign form. As soon as a

form has been classified the picture segment can be reduced to the centre part of the traffic sign. This section forms the input for the next module which performs the recognition of the pictograms. This goes on until the traffic sign has been recognized completely.

One aspect of this concept of image processing is that one needs less image data from step to step and this results in a faster processing performance. This is very important to vehicle handling because the recognition of traffic signs only makes sense in real time. When the system observes a sequence of images it can also improve the performance through data pipelining, that is to say, while the pattern recognition modules analyze the area of interest the next traffic scene is being examined by the detection module.

A connectionist approach

The evolved system uses neural networks (NN) to perform the classification task $R_{NN}(f(\underline{x}))$. A pattern recogniton system for the recognition of traffic signs from a rapidly moving vehicle should be able to classify 34 danger signals, 63 instruction signals and 190 direction and additional signs (traffic regulations). However, because of damage, pollution or partial coverage the number of patterns to be differentiated is in fact much higher (Fig. 2). Also, the time of day and weather conditions cause considerable pattern variations. Other variations like unsharpness or wrong aperture setting are

Fig. 2: Poor traffic sign pictograms

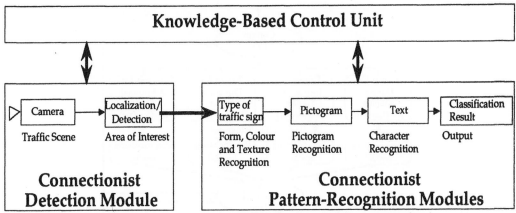

Fig 3: A connectionist traffic sign recognition approach

caused by the camera. Furthermore, the patterns are scaled, since as the pictures are made the distances between car and sign change. It is a feature of neural networks to tolerate such variations of pattern. With their generalization capability neural networks are able to infer unknown patterns from known patterns.

For the classification task $R_j()$ two kinds of neural networks were implemented: on the one hand the backpropagation procedure (BP; see Rumelhart 1986) $R_{BP}()$ and, on the other, the restricted coulomb engery procedure (RCE; see Cooper, Elbaum and Reilly 1982) $R_{RCE}()$. Both procedures are examined and compared with the recogniton of pictograms (see Krumbiegel 1991).

This approach can be divided into three components (Fig. 3). These are:

- Detection module

- Pattern recognition modules

- Knowledge-based control unit

Detection module

The detection module has the task of finding picture areas where traffic signs are likely to occur. The input is the traffic scene $\underline{s}$ and the output is the area of interest $\underline{x}$ ($d(\underline{s}) = \underline{x}$) (Fig. 4). For this different colour extracts of the camera picture furnish indications of the traffic sign's position. As the colours of traffic signs are known and, moreover, of greater intensity than the usual colour nuances in nature, they stand out clearly from the rest of the background, e.g. in rural areas. The

knowledge-based control unit determines certain picture segments. For these segments each colour extract (red, green and blue) is rastered. The centre of colour is determined then the retina is placed on it for each colour extract. The picture segment presented to the retina is the input for a neural network, which is trained with the backpropagation procedure. If the retina is not exactly placed on the possible traffic sign, the net output is the extent of shifting. On the other hand, the extent of shifting is zero if the retina is in the exact position. The knowledge-based control unit evaluates the three net outputs and either induces a shifting of the retinas and a new inspection of the picture segments or passes the determined section $\underline{x}$ on to the recognition modules.

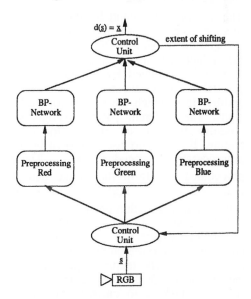

Fig 4: Detection module

203

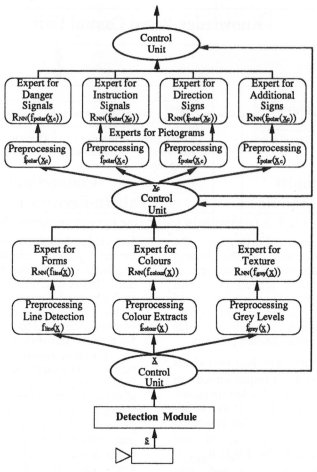

Fig.5: Recognition modules

Recognition modules

The section determined by the detection module forms the input for the pattern recognition modules. Now it will be hierarchically analyzed. In a first stage the section is examined by the modules in parallel if there are forms, colours or textures in it which are typical for traffic signs. The different pattern recognition modules function as experts for each feature. Each expert uses his own image preprocessing procedure $f_i()$. These procedures transform the image data $\underline{x}$ into a feature vector $f_i(\underline{x})$. The form expert can use a line detection procedure, e.g. the Sobel Operation or the Laplace Operation. The colour expert can use the different colour extracts to search for typical colour combinations. The texture expert can use the grey levels of the picture. In this first stage the task is to classify the traffic sign's type. Special guidelines are valid for form, size and colours of traffic signs to permit their recognition. As soon as the form has been classified the picture segment $\underline{x}$ can be reduced to the centre part $\underline{x}_c$ of the traffic sign.

This section forms the input for the next stage, which performs the recognition of the pictograms. For the pictogram recognition preprocessing is needed in order to compensate for translation and rotation variations, since the implemented neural networks cannot compensate for this kind of variation. The implemented image preprocessing procedure is based on a polar coordinates transformation.

If there is no pictogram in the traffic sign and the form indicates a direction sign, a character recogniton will be performed. Since the character font is determined by special regulations it facilitates character classification.

Fig. 5 illustrates schematically one way in which the different pattern recognition modules can act in combination. Further

204

specializations are possible. The experts can be split up to get a higher degree of parallelism.

Knowledge-based control unit

One of the biggest advantages of a knowledge-based control unit is that is uses its knowledge of the problem to perform the task. Thus one can combine the advantages of the connectionst models, namely the adaptation to the problem, the knowledge acquisition by implication and their generalization behaviour, with the knowledge of the problem. With this knowledge many steps can be skipped in the recognition task. For real time applications that is of particular interest.

The knowledge-based control unit can support the detection module by determining probabilities for any picture segments. These probabilities stand for the occurrence of traffic signs. This is possible, since for the installation of traffic signs special guidelines are valid which determine, e.g., location (left or right side of the street or above the lane) and height of installation or dimensions. In connection with information on vehicle and environment the probabilities can be determined for each picture segment. With that the detection module can scan the scene in order of the probabilites. This emulates the visual search behaviour of an experienced human driver.

In the recognition phase knowledge of different types of traffic signs is necessary. Special types of traffic signs use special pictograms. Knowledge of combinations and correlations of several traffic signs is also necessary. The final classification is established by the control unit, which combines a small number of partial results in a rule-based manner. At the same time the knowledge-based control unit is able to manipulate the environment. This means that the unit can control and receive data from additional sensors and can activate warning and support functions for the driver. This forms the basis for the vehicle handling support system.

An experimental pattern recognition system is currently working in a laboratory set-up. For the recognition of form and pictogram the traffic scene is recorded by a RGB-camera and is digitized by means of a frame-grabber. The preprocessing procedures $f_i()$ and the different neural networks $R_{NN}()$ are implemented on transputers. This means that, with regard to further real time processing, partial tasks can be executed in parallel. A user interface was implemented on a Macintosh computer. With this interface the user can combine the different preprocessing procedures with the different classification procedures to a new recognition system $R_j(f_i())$. The detection module and pattern recognition modules are separately developed. Further work is concerned with the integration of the modules by the knowledge-based control unit.

Applicability

Figure 6 illustrates schematically how

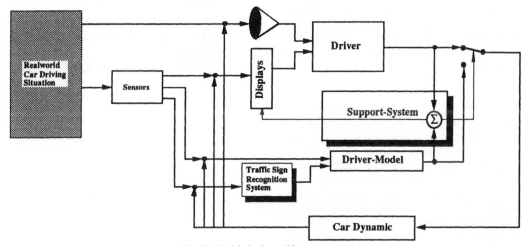

Fig 6: Vehicle handling system

traffic sign recognition can be used during vehicle control. The pattern recognition system will be used as onboard information system for the driver model. In this way the driver model uses the same view as the human driver in order to base its behaviour on the same data as the behaviour of the human driver. The support system is then able to recognize discrepancies between the driving behaviour of the model and that of the driver himself. The driver model may be applied as a trainer for novices, for instance. For this, the driver model needs only the driving behaviour requested by traffic signs. On the other hand, individual driver models can be used to recognize differences between the actual and normal driving behaviour of a special human driver (see Kraiss, Küttelwesch 1992).

Another application is the use of the pattern recognition system for direction signs combined with a navigation system. If a navigation system is able to read the information on a direction sign it can help the driver reach his destination, and if the system recognizes a new sign it is able to update the presented information.

CONCLUSION

There is a wide range of applications for the use of a traffic sign recognition system to support vehicle handling. Two principal problems arise in the development of such a system. First, the task must be performed in real time; second, the recognition system must be robust. The proposed approach takes these problems into consideration by using neural networks in parallel and by combining of connectionist and knowledge-based techniques.

Two kinds of connectionist models were implemented within the framework of the development. In order to compare the perfomance of both models, the different networks, the number of the training cycles, and the generalization behaviour were examined. The generalization behaviour of both connectionist models lead one to assume that it is possible to recognize traffic signs in real-world car-driving situations. After the detection

module and the pattern-recognition modules have been integrated by the knowledge-based control unit, further work is concerned with the evaluation of the device, using real-world video tapes as test material. In addition, the installation in a driving simulator is scheduled, where the installation of an information device on the panel and its acceptance by the driver can be studied.

REFERENCES

Bundesanstalt für Straßenwesen, BASt (1985). Projektgruppe "Modelversuch Weniger Verkehrszeichen"; Weniger Verkehrszeichen, Leitfaden zum Modellversuch des Bundesministers für Verkehr;

Cooper, L. N., Elbaum, C., and. Reilly, D. L (1982). A Neural Model for Category Learning; Biological Cybernetics 45, 35 - 41.

Kraiss, K.-F., and Küttelwelsch, H. (1992). Identification and Application of Neural Operator Models in a Car Driving Situation; (Submitted to MMS'92)

Krumbiegel, D. (1991). Konnektionistische Mustererkennung Anwendungsbeispiel: Erkennung von Verkehrszeichenpiktogrammen; Forschungsbericht FGAN-FAT.

Pao, Y.-H. (1989). Adaptive Pattern Recognition and Neural Networks. Addison-Wesley Publ. Comp., New York.

Rumelhart, D. E., J. L. McClelland & the PDP Research Group (1986). Parallel Distributed Processing, Volume 1: Foundations; MIT Press.

MANUAL CONTROL OF SLOWLY RESPONDING PROCESSES BY USE OF ONLINE SIMULATION MODELS

C. Hessler[1]

Laboratory for Man-Machine Systems, University of Kassel, Mönchebergstr. 7,
D-3500 Kassel, Germany

Abstract. Online simulation models can be used as operator aids for the
control of slowly responding processes. The basic idea is to run a
quick motion simulation in parallel to the real process. The operator
uses this simulation in order to evaluate the effect of future control
inputs. A new graphical interactive man-machine interface is presented
which allows easy manipulation of the model. The basic modes are the
control input mode where the operator evaluates the effect of different
control sequences, and the trajectory input mode where inversion of
the model is used for the computation of appropriate control sequences.
These modes have been applied successfully to a ship manoeuvring task
and to a load distribution task in a combined heat and power station in
laboratory studies.

Keywords. Online simulation; supervisory control; predictive control;
power station control; ship control.

INTRODUCTION

When automating dynamic systems, the process control tasks have to be allocated
optimally to the human operator and to the
automation. Possible reasons for manual
control of dynamic systems are that visual
information has to be used (e. g. airplanes, ships), or that manual control in
normal operating conditions is used as
training for rare disturbance situations
where automatic controllers do not work in
a satisfying way. There exist several
approaches for operator aids in manual
control situations. One way is to linea-
rize and simplify the dynamics by underlying controllers (e. g. dampers for airplanes). Another way is to use predictor
displays (Dey, Johannsen, 1973; Veldt,
Boomgard, 1986) which reduce the effective
order of the system.

In this paper we discuss the application
of online simulation models which the

operator actively manipulates. The pre-
condition for this use of a simulation
model is that the process reacts slowly
enough in order to allow conscious, interactive planning of control inputs to
the operator. The basic idea is to run
the online model in parallel to the real
process in a quick motion simulation.
Thus, the operator may evaluate the
effect of future control inputs in advance. This is to reduce workload and
increase the certainty of control decisions.

There are several examples for this. Sundermeyer (1980) describes an online simulation model as an aid for aircraft pilots in trajectory planning. It computes the effect of the position of way-
points and of the velocity of the plane,
interactively manipulated by the pilot.
An other application is trajectory plan-

The author is now with Siemens AG - KWU R242, Post Box 3220, D-8520 Erlangen

ning in spacecraft control where the operator uses a simulation model to evaluate different activation sequences of the
steering jets (Brody et al., 1988).

In developing such an interactive planning aid, the handling qualities are mainly influenced by an appropriate selection
of the input variables of the model. If
we select the input variables of the real
process as input to the model, planning
using the model requires full understanding of the input-output relationship of
the real process. If we choose the goals
of the planning process as input to the
simulation model, all the reasoning about
the control inputs necessary in order to
reach these goals has to be done by the
simulation itself. Thus, the handling of
this planning aid will be very simple at
the cost of the system's flexibility to
different operating conditions. We avoid
this problem by providing a set of input
modes between which the operator may
choose according to the operating condition. If the operator's goal is covered
by the goal input mode he/she uses this
mode. Otherwise he/she has to find a solution by iteratively modifying some intermediate goals as input until the simulated process fulfills the operational goal.

If one provides appropriate visualisations
for the different modes such a system facilitates the transition between manual
and fully automatic operation. In a manual control situation we distinguish three
different phases of the operator's planning and supervision activities (Bieker,
1987; Sundstroem, 1990):
(1) Selection of the appropriate control
    input sequence. The above mentioned
    goal input mode can be used for this.
(2) Formulation of the expected outcome
    of the control inputs defined in step
    (1). This can be supported by a simu-
    lation model which predicts the pro-
    cess trajectories under the inputs
    defined in step (1).
(3) Supervision if the process follows the
    expectation of step (2), eventually
    switching to the next step of the
    planned control sequence, or planning
    a new one (restarting with step (1)),
    if big deviations between expectation
    and outcome are stated. Operator's

support is provided by repetitively
predicting the process trajectories
and displaying them together with the
originally expected ones.
For every step the operator may decide
whether to perform it alone or use the online simulation, thus adjusting the automation level to his demands. Also a fully
automatic mode is available where the control inputs planned by the simulation
model are given to the process automatically, and where it's effect is supervised automatically, too, by comparing the
online model's prediction with the reference trajectory underlying the planned
control sequence.

In this paper we describe a new man-
machine interface of an online simulation
model which can be used for interactive
planning in all phases of manual control
situations, and which allows flexible
transition between manual and automatic
operation. Two laboratory studies were
performed to evaluate the applicability of
this concept, a ship manoeuvring task and
a load distribution task for a combined
heat and power station. They revealed
some technical problems as well as the
general usefulness of this approach.

GRAPHICAL-INTERACTIVE MANIPULATION

Overview
In practical applications the interaction
modes of the online simulation model have
to be adjusted to the necessities of the
technical process. In order to make the
characteristics of the different modes
comparable, they will be explained for an
idealized control system to which all the
modes can be applied. We stick to
- a MIMO-system (number of output = number
  of input variables);
- with global proportional behaviour, consisting of several time delay systems;
- the control inputs are limited;
- nonlinear, monotonic characteristics.
The operator's tasks are to keep the difference between refernce values and outputs as low as posssible and to adjust the
controlled variable as fast as possible
to the set point in case of step functions
of the reference variable.

For all the modes we use a trend display on a graphics screen showing the trajectories both of input and output variables. The basic idea is that the operator interactively manipulates the curves on the screen. A mathmatical simulation model predicts the effect of future control inputs drawn on the screen, or computes the inputs necessary to follow given trajectories. All the modes use a common set of data describing the different curves on the screen, thus enabling any switching between goal and control input mode. Because of the limited area on the screen and the number of variable parameters, this concept seems to be limited to two or three variables to be controled.

The computations have to be done so fast that the operator manages quasi static curves which show the dependency of input and output variables. He uses two sets of controls, one for the real process, the other one for handling of the input curves. Planning is done by modifying the (quasistatic) picture of the input variables until the output curve show the expected shape. The functions provided for this are
- lowering/raising of curve sections;
- shifting a step laterally;
- dividing a section into two parts;
- deleting a step by sliding it into the neighbour step.
This is accomplished by picking a curve element, binding it to the cursor and then moving it together with the cursor.

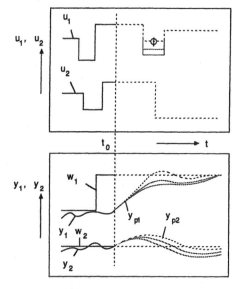

Fig. 1. Control input mode

Control Input Mode
In the control input mode the operator uses a set of controls which allow the drawing of curves on the screen which are interpreted as future (virtual) control inputs. They are fed to a quick motion simulation model which repetitively predicts the effect of these inputs (fig. 1). In correspondence to typical operator inputs of slowly responding processes drawing is limited to horizontal and vertical lines (discrete control inputs). The initial state of the simulation model is taken from the actual measurements of the process variables, thus keeping the simulation close to the real process´ behaviour.

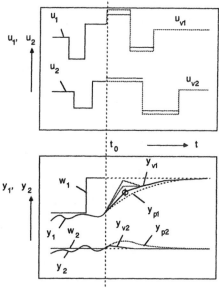

Fig. 2. Trajectory input mode

Trajectory Input Mode
If we use a very precise process model the above described procedure may partially be automated. The operator specifies the duration of the control impulses and draws the trajectories of the output variables he wishes. This corresponds to the "goal input mode" mentioned in the introduction. The control inputs necessary for these curves then are computed automatically by inversion of the process model. Because of the limitations for the manipulated variables (fixed number and duration of impulses, limited controls), not every given trajectory can be realized. Therefore planning is done by modifying the

output curves until the input curves and the corresponding prediction (which gives the nearest realizable trajectories) fulfill the operator's requirements (fig. 2).

For a linear, time-discrete input output model is quite straightforward. The model is given by

$$\underline{y}_{i+1} = \sum_{i=0}^{m} B_i \underline{u}_{j-i} + \sum_{i=0}^{l} A_i \underline{y}_{i-i}.$$ (1)

Inversion is done successively by inserting the samples of the reference trajectory $\underline{y}_{vi}$, k=0, 1, 2, ... The required controls then are given by

$$\underline{u}_{v0} = B_0^{-1} \cdot \left(\underline{y}_{v1} - \sum_{i=0}^{l} A_i \underline{y}_{-i} - \sum_{i=1}^{m} B_i \underline{u}_{-i} \right)$$ (2)

$$\underline{u}_{v1} = B_0^{-1} \cdot \left(\underline{y}_{v2} - A_0 \underline{y}_{v1} - \sum_{i=1}^{l} A_i \underline{y}_{1-i} - B_0 \underline{u}_{v0} - \sum_{i=2}^{m} B_i \underline{u}_{1-i} \right)$$ (3)

etc. In the case one of the computed virtual controls violates its limits the reference trajectory is no longer realizable and has to be adjusted to the limits. The output of the inversion is then coarsened by bundling (averaging) the samples $\underline{u}_{vi}$ according to the arrangement of the impuls durations prescribed by the operator.

Switching Curve Mode

Further simplification of planning is provided by the switching curve mode. The basic idea is to give an overview of the range of possible inputs which are appropriate to reach the operational goal. If we stick to control inputs consisting of sequences of three step functions, the display according fig. 3 is appropriate. The first switching curve (marked as "1" in fig. 1) shows when to switch from the first to the second impuls as a function of the value of the first impuls. Curve 2 indicates when to switch to the stationary value, and curve 3 shows the stationary value itself. Thus, the operator is free to choose the first impuls in the vertical range of curve 1, whereas the horizontal range gives the appropriate switching time.

Using inversion of the process model as described above the switching curves can be computed very easily. For a given reference trajectory (provided by the operator, or generated automatically) the proper control input sequence $\underline{u}_{v0}$, $\underline{u}_{v1}$... is computed. We obtain the function values of the switching curves by averaging sections

of this sequence of different lengths Δt, whereby the section $0 \leq t \leq \Delta t$ is used for the first impuls u^1, the section $\Delta t \leq t \leq 2\Delta t$ is used for the second impuls u^2, and the intervall up to the prediction horizon is used for the stationary control.

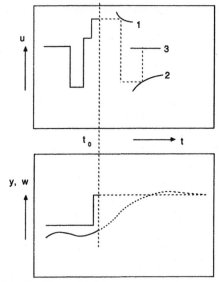

Fig. 3. The switching curve mode

A predictor display can be included easily which takes in account the future control inputs of the operator. It is assumed that the operator will follow the switching curve. Therefore, we get the duration and values of the following impulses by computing the intersection of the initial control input with the switching curve. This infered control sequence is then fed to the quick motion simulation model which predicts the resultant process trajectory. Both assumed controls and prediction are displayed on the screen so that the operator may evaluate several control sequences simply by varying the actual control input in the vertical range of the first or second switching curve.

APPLICATION TO A SHIP MANOEUVRING TASK

In a first study we applied the control input mode to a ship manoeuvring task. The goal of this study was to evaluate the applicability of the above described concept to a more realistic task. Both the control and the trajectory input mode have been implemented in a laboratory ship

simulation running on a VAX 11/750 with a
Megatek graphics system. The interface of
the online simulation model was included
into a simulated radar screen showing the
ship's position and course on a map
display.

The Simulation Model

For both the realtime simulation and the
online simulation model a nonlinear set of
differential equations for the ship's movement in the horizontal plane was used:

$$m\ddot{x} - m\dot{\psi}\dot{y} = \sum X(\delta, \kappa, \kappa_b, \dot{x}, \dot{y}, \dot{\psi})$$
$$m\ddot{y} + m\dot{\psi}\dot{x} = \sum Y(\delta, \kappa, \kappa_b, \dot{x}, \dot{y}, \dot{\psi}) \qquad (4) - (9)$$
$$I_z\dot{\tilde{\psi}} = \sum N(\delta, \kappa, \kappa_b, \dot{x}, \dot{y}, \dot{\psi})$$
$$\psi = \int \dot{\psi} dt$$
$$\dot{x}_f = \dot{x}\sin(\psi) + \dot{y}\cos(\psi)$$
$$\dot{y}_f = \dot{x}\cos(\psi) - \dot{y}\sin(\psi)$$

The state variables used were
- $\dot{\psi}$ angular velocity; ψ track angle;
- $\dot{x}$ longitudinal $\Big\}$ velocity, body axes;
- $\dot{y}$ lateral
- x_f, y_f geodetic coordinates.

The forces X, Y and the rotational momentum N about the z-axis are nonlinear functions of the manipulated variables
- δ rudder deflection;
- κ_b thoughput of thruster;
- κ pitch setting of the propeller.

(Details see Fisher, 1990). The parameters
of the differential equations were adjusted to the behaviour of a 50000 tdw container ship reacting slowly enough to
allow the interaction of the helmsman with
the online simulation model in parallel
to steering the ship itself.

Interaction with the Online Simulation

In the control input mode the helmsman
uses a special set of controls in addition to those controling the real ship.
These controls provide the input to the
online simulation model which runs ten
times faster than realtime. This time factor still provides good handling qualities of the quick motion simulation whereas predictions are obtained fast enough in
order to evaluate future manoeuvres. The
trajectory of the simulated ship is
displayed on the radar screen together
with the real ship's position. The course
of the quick motion simulation is controled by three functions:

- "start" initiates the simulation; the
  state of the simulated ship is set to
  the real one's state (position, velocity etc.).
- "position" is used for backtracking on
  the trajectory of the simulated ship
  with a constant velocity.
- "end" interrupts the simulation. The
  simulated trajectory is preserved, and
  the data describing the input of the
  simulation model are stored. They will
  be used for the proposal of control
  inputs when the helmsman tries to follow the simulated trajectory.

Figure 4 gives an example of the planning
process using this mode.

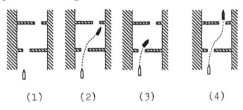

(1)   (2)   (3)   (4)

Fig. 4. Quick motion simulation.

Step (1): The real ship's position.
Step (2): When planning the passage the
simulated ship gets stuck.
Step (3): Backtracking.
Step (4): Trying again.

In the trajectory input mode there is a
bigger difference between handling the
real ship and the simulation model. In this
mode the helmsman enters a few points of
the reference trajectory which then are
connected automatically by cubic splines.
This produces smooth, realizable curves.
For the modification of these curves
appropriate editing functions are provided: Inserting a new point, adding a new
one, moving and deleting an existing one.
In addition, the helmsman has to provide
the points with the attributes desired
course, velocity, reversion of movement,
and assumptions about wind and current.
This is accomplished by a set of menues
where the operator hasto enter the desired values, or where he may accept default values. Planning using this mode is
quite straightforward. It is done by entering a few critical points of the reference trajectory and eventually modifying
them a little bit if the resulting spline
or the prediction of the realizable tra-

jectory doesn't fulfill all requirements,
e.g. if it leads into confined areas. When
following the planned trajectory the in-
verse model computes the appropriate rudder angles and propeller pitch and displays
them as suggestions to the helmsman. Thus
planning of steerable trajectories is un-
burdened of explicitly taking into account
the ship dynamics. The calculations necessary to evaluate a proposed trajectory
are done by the online model.

Inversion of the mathmatical model

In the prototype implementation of the
trajectory input mode the computation of
rudder angle and propeller pitch was accomplished by numerical inversion of the eq.
(4) - (9). The reference trajectory is
given by a parametric curve

$$x_t = x_t(s) \qquad \dot{x} = v = v(s) \qquad (10)$$
$$y_t = y_t(s)$$

as a function of the spline parameter s,
the length of the arc. For the solution of
(4) - (9) with respect to δ, κ we derive
$\dot{x}_t, \dot{y}_t, \psi$ from (10) using the chain rule

$$\dot{x}_t = \frac{dx_t}{ds} \cdot v_t \qquad \psi = \arctan\left(\frac{\frac{dy_t}{ds}}{\frac{dx_t}{ds}}\right) \qquad (11)$$
$$\dot{y}_t = \frac{dy_t}{ds} \cdot v_t$$

etc. The drift velocity $\dot{y}$ cannot be derived from (10). So we use it as the third
unknown. We write the eq. (4) - (6) in the
form

$$A_1(\delta, \kappa, \dot{y}) = \dot{x} - \dot{\psi}\dot{y} - \frac{1}{m}\sum X(\delta, \kappa, \dot{y}) = 0$$
$$A_2(\delta, \kappa, \dot{y}) = \dot{y} + \dot{\psi}\dot{x} - \frac{1}{m}\sum Y(\delta, \kappa, \dot{y}) = 0 \qquad (12)$$
$$A_3(\delta, \kappa, \dot{y}) = \dot{\psi} - \frac{1}{I_z}\sum N(\delta, \kappa, \dot{y}) = 0$$

This 3x3-system of nonlinear equations is
then solved iteratively by linearizing
and solving the linear equations for corrections of initially assumed values of
$\delta, \kappa, \dot{y}$.

Following a planned trajectory

When following a planned trajectory the
appropriate control inputs are indicated
to the helmsman as suggestions. Because of
wind, current and unavoidable errors of
the online model the error between actual
and reference trajectory will increase by
the time. Therefore, an automatic feedback
controler is used which corrects the deviations between reference trajectory and
the actual state of the ship. Thus, the

control suggestions consist of a super-
position of a feedforward control sequence
obtained during the initial planning process, and the feedback control keeping the
ship on the reference trajectory in spite
of disturbances. The feedforward control
is responsible for the approximate course
while the feedback controler adjust small
deviations.

Laboratory Experience

When building up the laboratory test stand
for this manoeuvring aid, we made the
following experience:
- the implementation of the model inversion for a complex nonlinear ship model
  is possible but it reached the limits
  of our computing system;
- the implementation of the quick motion
  simulation used for the control input
  mode was relatively easy; we didn't have
  any performance problems and were able
  to cover all operating situations with a
  single simulation model.

In the trajectory input mode only forward
motion of the ship was treated properly.
In an experiment we made a comparison of
the use of the quick motion simulation and
conventional ship control with engineering
students as operators. In these experiments the performance during different
landing manoeuvres was considerably better
when using the online model than in manual
operation. An important effect was that
the subjects spent much more time on trajectory planning than during manual operation (Hoppe, Posner, 1991). We obtained
similar results in a power plant application which will be discussed in more detail in the next section.

EXPERIMENTAL EVALUATION OF
THE CONTROL INPUT MODE

As a second example the control input mode
has been applied to a load distribution
task for a combined heat and power plant.
The goal was
- to evaluate a nonlinear, adaptive process model (see Hessler, 1992), and
- to evaluate the quick motion simulation
  in a more systematic way than in the
  ship control example.

Man-Machine Interface

The simulated heat and power plant consits of a gas turbine, a waste heat boiler and a steam turbine. A long-distance heat network is supplied by the power plant. The heat supply of the network during shutdown perions of the power plant is maintained by a heat storage. The display used for process control (fig. 5) shows the relevant curves in two windows. The upper window contains trajectories of the manipulated variables (energy production using gas (PGT) and steam (PDT) turbine) and of the heat demand of the long distance heat network (PFW). In the lower window the controled variables (energy production PEL, setpoint PELSOLL) and the content of the heat storage WWS was shown. In the control input mode the operator could draw and manipulate trajectories of future control inputs on the screen. The predictions of the online model were included in the output window. The prediction took into account also the operator's assumptions of the heat consumptions in the heat network drawn on the screen.

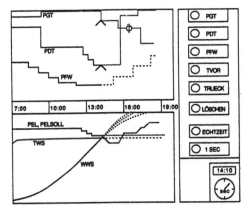

Fig 5. Interface for the power plant

Experimental design

The goal of the experimental evaluation was a comparison of manual process control (M) and use of the quick motion simulation (Z). The operator's task was to balance the generation of electrical energy and heat production in such a way that both the heat storage was filled at a certain time and the demanded electrical energy was supplied. The development of the energy demand was not known in advance to the operator although he was provided with coarse information of approximate times of maxima and minima of the load profiles.

As independent variables we selected the type of process control, M and Z, and the difficulty of the process control task. The dependend variables were the time error t_f by which the planned shutdown time of the plant was missed, and the time t_e spent for the planning of control actions. In order to speed up the experiment (nominal time range: 10 h) the process simulation ran in two different speeds: during the planning of control actions it ran in real time whereas times without planning activities were covered by increased simulation speed. Thus one trial took some 30 minutes. The time spent in realtime mode was classified as planning time. Six engineering students (20 to 30 years old) participated in the experiment.

Results and Interpretation

The learning curves of t_e (fig. 6) show a short learning phase after which nearly constant planning times are reached. The considerably longer decision times when using the online model (Z) are mainly caused by the instrumentalisation of the planning process. The operator draws planned/expected curves on the screen and repeatedly updates them according to the process behaviour. The longer planning times are balanced by much more precise process control. The mean time error t_f is reduced from 121 to 50 min (fig. 7). The interpretation of this improvement is not that the online model produces more predice planning but that explicitly writing down the planned control sequence on the screen provides some kind of external memory to the operator which avoids that he forgets to follow an initially planned sequence during the long time between load changes.

DISCUSSION, CONCLUDING REMARKS

We described a new concept for the graphical-interactive manipulation of an online simulation model. Two examples showed the adaptation of the concept to two application processes. The basic experience is that

(1) the control input mode can implemented easily. In the ship and the power station

example we didn't have any computing
time problems for the quick motion mode.
(2) The trajectory input mode demands
much more implementation effort. For nonlinear processes inversion can be done
only for selected operating situations.
Therefore the flexibility to unforeseen
situations is very limited.

In the experimental evaluation only students, no professional operators participated. Therefore the validity of the results is limited. In spite of this a clear
tendency can be seen:

- The instumentalized planning using the
  online model increases the workload of
  the operator; at least in the power
  station example this seems acceptable.
- Process control becomes more precise and
  consistent because of the precision of
  the online model and its function as an
  external memory to the operator keeping
  conscious the originally planned controls.

There remains a lot of interesting
questions, e. g.

- what is the required precision of the
  online model in order to be a useful
  planning aid ?
- How can we integrate the output of a simulation model with visual information
  (e. g. head-up display, synthetic scene
  presentation in vehicle control) ?
- What is the maximum number of variables
  that can be handled by the operator ?

REFERENCES

Bieker, B. (1987). Wissensrepräsentation
für eine einfache Expertenregelung.
Automatisierungstechn. Praxis 29, 36-43.

Brody, A. R., Ellis, S. R., Grunwald, A.,
Haines, R. F. (1988). An Evaluation of Interactive Displays. AIAA/IEEE 8th Digital Avionics Syst. Conf., San Jose, Cal.

Dey, D., Johannsen, G. (1973). Anthropo-
technische Untersuchung einer Übergrund-
anzeige und eines künstlichen Horizontes
mit Voranzeige zur manuellen Regelung
von VTOL-Flugzeugen. Z. f. Flugwissen-
schaft 21, 140-145.

Fischer, K. (1990). Entwicklung einer
Planungshilfe für Schiffsmanoeuver.
Diplomarbeit I, Lab. f. Man-Machine
Systems, Univ. Kassel.

Heßler, C. (1992). Online-Simulation: ein
Hilfsmittel zum Führen langsamer Pro-
zesse. Diss. Univ. Kassel.

Posner, S., Hoppe, K.-H. (1991). Aufbau
und Bewertung eines Manövrierhilfssy-
stems für die Schiffsführung. Studien-
arbeit, Lab. f. Man-Machine Systems,
Univ. Kassel.

Sundermeyer, P. (1980). Untersuchungen zur
Verlagerung der Pilotentätigkeit auf
eine höhere hierarchische Stufe der
Flugführung. Diss. TU Braunschweig.

Sundstroem, G. (1990). User Modelling for
Graphical Design: Concepts and Prototype Implementation for an User Model
of an Intelligent Graphical Editor.
Lab. f. Man-Machine Syst., Univ. Kassel

van der Veldt, R. J., van der Boomgard, W.
(1986). Predictive Information in the
Control Room. In H.-P. Willumeit (Ed.),
Human Decision Making and Manual Control, Elsevier, North Holland.

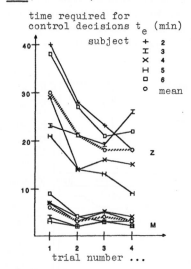

Fig. 6. Learning curves for t_e

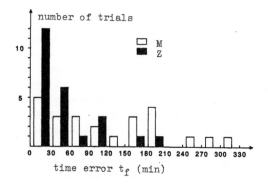

Fig. 7. Distribution of t_f

AIDING THE OPERATOR IN THE MANUAL
CONTROL OF A SPACE MANIPULATOR

J.F.T. Bos*, H.G. Stassen** and A. van Lunteren**

*National Aerospace Laboratory NLR, P.O. Box 90502, 1006 BM Amsterdam, The Netherlands
**Man-Machine Systems Group, Faculty of Mechanical Engineering and Marine Technology,
Delft University of Technology, Mekelweg 2, 2628 CD Delft, The Netherlands

Abstract. When an operator has to control a space manipulator manually he is faced
with several difficulties. Three problems were investigated: Lack of direct vision,
up to 6 degrees-of-freedom (DOF) to be controlled, and slow and complex dynamics.
The results showed that reference lines enhance the 3D perception; further
automation of the insert task led to a more efficient task execution; the display of
the stopping configuration can lead to a faster and safer task execution. The
display of the predicted trajectory led to less energy consumption due to a more
calm control behaviour. The length of the prediction horizon should equal half of
the stopping time.

Keywords. Man-machine systems, telecontrol, robots, space vehicles, human factors,
predictive control.

1. INTRODUCTION

Currently a European spacecraft called Hermes
(Fig. 1), and a space manipulator, the Hermes
Robot Arm (HERA), are being developed. The
manipulator will be used for executing tasks
like grasping and moving objects and as a support for extravehicular activities. The total
length of HERA will be about 11 meter. The
total mass is about 230 kg (Andre, Schoonejans, 1989). Altough normally HERA will be
controlled under supervisory control, a requirement is that under all circumstances a
human operator must be able to control the
manipulator manually.

This paper summarizes the activities and results of the project "Man-Machine Aspects of
Remotely Controlled Space Manipulators" (Bos,
1991). The research dealt with the manual control situation. The design criteria for a man-
-machine interface (MMI) concerned the safe
and economical use of the manipulator. To obtain realistic characteristics of a space manipulator, the specifications of the HERA are
used.

When an operator has to control a space manipulator manually he is faced with several
difficulties: Lack of direct vision, the control of up to 6 degrees-of-freedom (DOF), slow
and complex dynamics, the possible presence of
time-delays, and the presence of flexibility
effects. Therefore, the central question of
the research project was: *In which circumstances and with what aids can the human operator
accomplish a required task in a reliable way?*.

2. TACKLING THE DEFINED PROBLEMS

The problem of time-delays was not investigated, because only in the long term planning the operator will be placed on Earth.
Problems due to flexibility effects were not
investigated, because the expectation was that
the control system could be designed in such a
way that the flexibility modes are excited
only to a small extent. Time-delays and flexibility effects are considered in the follow-up
of this project.

The experimental tools were restricted to the
use of a graphical workstation. Therefore, the
subjects could not have direct vision on the
control scheme. The question arised how graphical displays can be used to provide the
operator with a good three-dimensional (3D)
perception. This was investigated for a *rough
positioning task*, where global views (Fig. 2)
of the scene have to be used. The results are
presented in Section 4.

The problem of controlling up to 6 DOF is most
restricting for a task where a large accuracy
is required, such as an *insert task*, or peg-
-in-hole task. Further automation of the task
is considered in Section 5.

A possible aid for the operator to cope with
slow and complex dynamics is a predictive display. Two kinds of predictive display were
evaluated: a display of the stopping configuration and a display of a predicted tra-

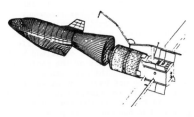

Fig. 1.   Hermes, HERA and the Man-Tended-
Free-Flyer (by courtesy of Fokker
Space and Systems).

jectory. The results are presented in Section 6. An important issue when displaying a predicted trajectory is how to choose a good length for the prediction time horizon. The results of a pilot study towards this issue are given in Section 7.

The above defined research required the incorporation of two different tasks in an experimental facility: a rough positioning task and an insert task. These task are described in more detail in Section 3. In space operations the tasks have to be completed in the first place safely, and in the second place as efficiently as possible. The instruction given to the subjects was according to this desired behaviour.

Because of limited computing and graphic display power the simulated manipulator reflected only the main characteristics: The manipulator consisted of 2 links and 1 End-Effector (EE) and it had 6 DOF. The coriolis and centrifugal forces were neglected due to the small velocities of the manipulator. The DC-motors were modelled as systems which could provide limited torques.

3. EXPERIMENTAL FACILITY

The subjects sat in front of a monitor of a graphical workstation, and they could generate control actions via two displacement joysticks with each three DOF. The subjects controlled the manipulator by providing velocity commands.

In the case of an insert task, visual information was provided by the camera on the EE (see Fig. 3). The picture showed the part of the ORU the camera observes (white surface) and the hole (black surface). Proximity information was displayed at the side of the camera display.

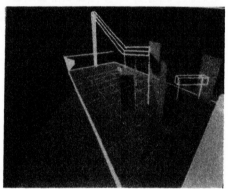

Fig. 2.   Global view for a rough positioning task with obstacles present in the environment.

Fig. 3.   End-Effector camera view.

When the manipulator "hits" the environment, a so called *incident* occurs. The moment the software detects an incident, automatically an emergency stop is initiated. During the time it takes to stop the manipulator the subject's commands are not executed. When the manipulator has come to rest, the so called *recovery phase* is entered. The subject has to control the manipulator back into the safe region, before the task can be continued.

4. VISIBILITY ASPECTS FOR A ROUGH POSITIONING TASK

The operator cannot see the manipulator and its environment directly, i.e. in three dimensions (3D). Apart from TV pictures the use of graphical displays is foreseen. An advantage of graphical displays is that they can provide views from viewpoints where no camera is or can be present. In aircraft (Ellis, McGreevy, 1987) as well in some teleoperation experiments, the use of reference lines, i.e. projections of an object an a base plane or grid, gave good results. The use of reference lines can be motivated by the fact that a light source projects a shadow of an object on the ground. The presence of shadows is one of the depth cues for a human being (Overbeeke, 1988).

In (Kim, 1987) the subjects had to execute a 3D pick-and-place task. They had to control a cylindrical manipulator with 3 DOF. In a first experiment the results obtained by Kim were evaluated for a situation where the operator had to execute a rough positioning task with a space manipulator with 6 DOF. So, not only the position but also the orientation of the EE had to be estimated from the picture. Contrary to (Kim, 1987) the subjects could use two different viewpoints. No obstacles were present

in the environment.

The effectivity of the use of reference lines can depend on the target position. The target position can have a fixed and known position with respect to Hermes (ORU storage location), but the target may also float in space (satellite). The first situation is modelled in the simulation by projecting the target on the base plane too. So, four experimental conditions were defined: EE with and without reference lines, combined with the target with and without reference lines (Bos, van de Klashorst, 1989). The results indicated that projecting the EE on a base plane leads to a safer and faster task execution.

Projecting the EE caused that the subjects switched less from viewpoint. Also the frequency of viewpoint switches decreased when projecting the EE. Statistically, the results were not very significant (5% Sign-test (Sachs, 1982)), contrary to the results in (Kim, 1987). This can have three reasons. Firstly, the task to be executed was not critical enough for misfits in position estimations by the subjects. Secondly, the subjects could compensate the lack of a good 3D perception from one view with no (or less) reference lines by intergrating the two different views more intensively. The higher frequency of viewpoint switches confirms this. In (Kim, 1987) the subjects had only one view available. Thirdly, the lack of a clear difference in completion times can be explained by the small maximum velocities of the manipulator (0.05 m/s). To test the correctness of this explanation for not having obtained very clear results, a small experiment was performed. A subject had to execute the same rough positioning task, but now obstacles were present in the environment (Fig. 2), and the manipulator was unloaded. The maximum EE velocity was 0.1 m/s. The target was projected on the base

plane. It was investigated whether also projecting the EE did lead to a better performance. The results are summarized in Table 1, and they support the stated explanation.
The subjects clearly preferred the presence of reference lines from the EE, both for a known and an unknown target position. The subjects found it hard to estimate the orientation of the EE. Moreover, the subjects experienced a dependency between the pitch and yaw movements. The problem to make a good estimation of the height was enlightened by the use of reference lines.
The viewpoint switches, the number of incidents and the subject opinions all together showed that it is desirable to project the EE on a base plane, especially for a known target position.

TABLE 1. <u>Results for 14 runs with a faster responding manipulator, and with obstacles present in the environment.</u>

ref. lines from EE	number of incidents	compl time [s]	average number of view-point switches	average freq. view-point switch
no	8	109	6.8	0.06
yes	0	73	1.6	0.02

5. ON TASK ALLOCATIONS BETWEEN MAN AND COMPUTER FOR AN INSERT TASK

Generally an insert task has to be excuted in the EE control mode, where the EE frame functions as reference frame, and the tip of the EE or payload as Point-of-Resolution (POR). When using the EE frame as reference frame, the direction of a translation depends on the orientation of the EE. This might cause inefficient control and unsafe situations. A possible solution is the use of the target frame as the reference frame. This can be done when the orientation of the target is known. However, the true limiting factor for the human operator is probably the fact that he has to control up to 6 DOF.
When the relative difference in postion and orientation between the EE and the target frame can be measured, closed loop control schemes can be designed, which reduce the number of DOF to be controlled by the human operator. Consider for instance Fig. 4: when both the current postion (p_1) and the aimed position p_2 are known, the aimed direction is known. Then the operator needs only to provide a velocity setpoint (1 DOF) in the specified direction. A similar scheme can be applied for the control of the orientation. In this way the number of DOF to be controlled by the human operator is reduced to 2.

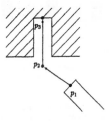

Fig. 4.   2DOF mode.

In the 2DOF mode, the subject provides velocity gains along two known directions: one direction for the position and one for the orientation. The 2DOF mode is split into two phases. In phase I, the approach phase, the ORU has to be positioned in a small neighbourhood of p_2, and the orientation of the ORU has to be very close to the orientation of the target frame. These bounds are such that the ORU can be brought into the hole safely. Phase II consists of bringing the ORU into the hole. A possible factor of influence on the desirability of a task allocation is the difficulty of the insertion, i.e. the relative size of the hole. In the case of HERA a tolerance of 1 to 5 mm is foreseen, when an ORU has to be inserted. Different tolerances used in the experiments led to the same results (Bos, Doorenbos, 1990).
Only very few incidents and jamming situations occurred. The occurances were more or less evenly spread over the experimental conditions. Figure 5 indicates that there were no structural differences in the approach times between the two EE modes. Compared to the EE control modes, the approach times in the 2DOF

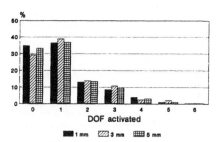

Fig. 5.   Average approach time.

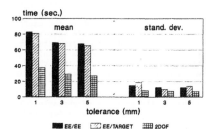

Fig. 6.   Multivariable character of subject 2.

mode were about 50% faster.
The applied control strategy in the EE modes of controlling the orientation first was a result of the experiences the subjects had in the training. In the training sessions they learned that incidents and jamming situations occurred mostly due to orientation misfits. However, the consequence of controlling the orientation first, is that the difference between the two EE modes disappears. The orientation of the two reference frames becomes the same. This explains that the performance was about the same in the two EE modes.
In Fig. 6 the percentages of the approach times subject 2 controlled a specific number of DOF simultaneously are shown. The subjects controlled mostly none or one DOF at the time. The percentage of time they controlled two or three DOF is also substantial, but the subjects rarely did control more than 3 DOF simultaneously.
When the misfits were in the range of the proximity display, the subjects did only control one DOF at the time. In the 2DOF mode the com-

puter takes care of the simultaneous control of the three position and/or the three orientation DOF. This explains the large difference in approach time between the EE modes and the 2DOF mode.

It turned out that, in all control modes, generally the subjects were quite able to complete the insertion in a time optimal manner. An explanation can be that, when entering the hole, the subjects needed to control only one DOF, assuming that there were no or only small orientation misfits left.

6. NON-LINEAR AND SLOW DYNAMICS: AIDING THE OPERATOR WITH A PREDICTIVE DISPLAY

A space manipulator has non-linear and slow dynamics. This can cause problems with the timing of his commands for the operator, which will worsen his performance. A possible aid is a predictive display (Veldhuyzen, 1976; van de Veldt, 1986; Keyser, van Cauwenberghe, 1981). A predictive display shows the operator a future response of the system under control, where a certain assumption is made about the future commands of the operator. Two kinds of predictive display were investigated. Firstly, the display of the stopping configuration, i.e. the configuration in which the manipulator will stop when setting all rate setpoints to zero. Secondly, showing a predicted trajectory of the manipulator over a certain time horizon, where it was assumed that the rate commands remained constant over the prediction horizon.

The subjects had to execute a rough positioning task in the presence of obstacles. The subjects could switch between a *top* view and a *side* view.

The usefulness of the two kinds of predictive display was investigated in relation to the

system dynamics. Two combinations of EE velocities and the mass attached to the EE which are foreseen for HERA, were used in the experiments:
1. A mass of 550 kg.;
   Max. translational velocities of 0.09 m/s;
   Max. rotational velocities of 0.09 rad/s;
   This combination is defined as the *fast situation*.
2. A mass of 3500 kg.;
   Max. translational velocities of 0.05 m/s.;
   Max. rotational velocities of 0.05 rad/s.;
   This combination is defined as the *slow situation*.

In order to cover the whole possible range of dynamics for a space manipulator, the maximum torques were varied. As reference torque vector was chosen the vector (100, 100, 75, 100, 100, 100)^T Nm. Four different vectors of maximum torques have been used: 1, 2, 3 and 4 times the reference vector. In order to be able to compare the results of these experiments with other ones, the times to stop the manipulator, and the corresponding distances were calculated. They varied respectively from 1.5 to 19.3 seconds and 0.1 to 0.68 meter.

Subject 1 served as a pilot-subject, to check the correctness of the experimental design. the subjects trained the task execution starting in a fixed configuration. In the experiments different starting configurations were used to avoid a preprogrammed control behaviour of the subjects. This concept proved to be sound in a pilot experiment (Bos, 1989). However, learning effects were clearly present in the data. Subject 1 needed quite some time to understand the new situations, when faced with the different starting configurations. Therefore, the results of subject 1 were not taken into account in the further analysis. The results indicate that in new situations or

in an unknown environment the combined problems of the 3D perception and the control of 6 DOF are more serious than the problems due to the slow and non-linear dynamics. On the basis of these results the training for the other subjects was adjusted.

Most incidents occurred in the plain situation with a torque factor 1, but only in 2.5% of the runs an incident occurred. About half of the number of incidents were caused by the fact that the subjects took more risk in the task execution when a predictive display was provided: While keeping the same safety level, they tried to optimize their performance with respect to the task completion time and the energy consumption.

Another main cause for the incidents was the difficult 3D perception. The reference lines provided the subjects with good information about how far the EE is above an obstacle and about the horizontal distance to the obstacles located on the base plane. But the subjects had difficulties with estimating the distance of the links to an obstacle, and the distance of the EE to the obstacle which was floating, i.e. the obstacle which was not located on the base plane. These results indicate that a collision detection algorithm is very desirable, if not necessary for a guaranteed safe task execution.

Only in the slow situation for the torque factor 1 both kinds of predictive display did lead to a substantial (ca. 20%) decrease in the completion time.

Figure 7 shows that with the display of the stopping configuration about the same amount of energy was used as in the plain situation. This was also the case in the slow situation. The display of the predicted trajectory did lead to a substantial reduction of about 25% in the slow situation and of about 18% in the fast situation.

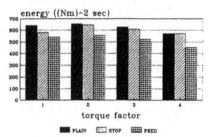

Fig. 7. Average energy used; 3D; fast situation.

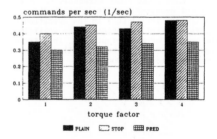

Fig. 8. Average number of issued commands per second; 3D; slow situation.

The display of the predicted trajectory did lead to less energy consumption per second. So, the length of the task execution was not the cause for the differences in energy consumption between the experimental conditions.

The display of the predicted trajectory did also lead to a more calm control behaviour of the subjects (Fig. 8). When the energy is divided by the number of issued commands, the difference between the experimental conditions disappears to a large extent. So, the reduction in the energy consumption is caused by the more calm control behaviour.
The subjects more or less considered the displayed stopping configuration as the one to be controlled. This may explain why, with the display of the stopping configuration, the energy consumption (per second) and the control behaviour (Fig. 8) were about the same as in the plain situation.
In the fast situation the subjects' opinion was that the display of the stopping configuration was not very useful, because the stopping distance was very small. In the slow situation, the experienced advantages of the display of the stopping configuration were the increased safety, the easy end-positioning of the EE in the target region, the ability for a faster task execution, and the increased controllability of the manipulator.
All subjects rated the display of the predicted trajectory as very useful. The main experienced advantages were the increased safety, the increased controllability of the manipulator, and the use of less energy.
Another advantage of the display of the predicted trajectory was that the EE could be better controlled around obstacles. Especially in the 3D situation the predicted trajectory served as a tool to determine whether the EE would safely move over an obstacle.
In the slow situation the subjects controlled more DOF simultaneously when a predictive display was provided, as is illustrated for subject 2 in Fig. 9. This holds especially for the display of the predicted trajectory. The significance of this fact decreased for in-

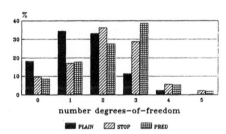

Fig. 9.   Percentages of time that a specific number of DOF is controlled simultaneously; 3D; slow situation; torque factor 1; subject 2.

creasing torque factors. In the fast situation the differences disappeared to a large extent. This may have two reasons. Firstly, in the fast situation the times needed to stop the manipulator were smaller than in the slow situation. Therefore, the subjects were able to control more DOF simultaneously, also when no predictive display was provided. Secondly, in the fast situation the manipulator moved with a larger velocity than in the slow situation. The manipulator arrived much sooner in a critical safety region. When starting a movement of the manipulator the number of DOF which were controlled increased sequentially in time. When a subject had time and confidence to do so, he added more DOF. Therefore, although a predictive display is provided the operator has not enough time or capacity to control more DOF.

7. DESCRIBING THE OPERATOR AS A PREDICTIVE CONTROLLER

An important issue when displaying a predicted trajectory is how to choose the best length of the time horizon. Apart from time consuming experiments with subjects, control theory might be used to investigate this issue. From the existing human operator models the Optimal Control Model (OCM) has been used to study the effects of a predictive display on the performance and behaviour of a human operator (Milgram, Weverinke, 1985; van de Vegte et.al., 1990; Johannsen, Govindaraj, 1980). In all studies the predictive display influences the *quality of the state estimation*, thus influencing the performance of the closed loop system. Another way of thinking can be that a predictive display influences *the determination of the control law*. A predictive dis-

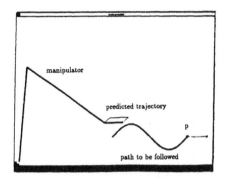

Fig. 10.   The display of the manipulator, predicted trajectory, and the path to be followed.

play influences the commands issued by the operator (Veldhuyzen, 1976; Passenier, 1989, Section 5). Therefore, predictive control theory seems to be more natural to apply in an investigation towards a good choice for the prediction horizon.
Two experiments were performed: one in which the control is performed by a predictive controller, and another in which a human acts as a controller (Bos, van den Bosch, 1991). It was decided to start with a pilot study. Therefore, the subjects had to control a planar manipulator with 2 DOF, and the path to be followed (Fig. 10) was fixed for both the computer and the human.
An inherent problem is that the path to be followed, as presented to the human operator on a display, has no time-scale attached to it, contrary to the reference trajectory as used in predictive control theory. To solve this problem the subjects were instructed to keep the maximum velocity in the horizontal direction.
The results showed that, to cope with the critical point p (Fig. 10), the optimal length of the prediction horizon equalled about half of the stopping time. To follow the path without a deviation, the manipulator has to be stopped at the critical point. In order to spot the critical point in time, the length of the predicted trajectory has to be equal to the stopping distance. Now, Newton's laws of motion explain the factor 0.5.
Although the results did not exactly fit, they showed a large resemblance. So, predictive control theory might serve as a means to describe human control behaviour, when studying the effects of predictive display.

8. CONCLUSIONS

Visibility aspects
In the case of a rough positioning task, the presence of reference lines enhanced the 3D perception, which resulted in a tendency towards a safer task execution.

Task allocation
When executing an insert task, referencing the operators command to the target frame, instead of the EE frame did not improve the human operators control performance. The operator did control the orientation first. In that way the difference between the two reference frames disappeared, before translation commands were issued.

For an insert task, the reduction of the number of DOF to be controlled by the human operator, led to a more efficient task execution.

Aiding the operator with a predictive display
The subjects felt more secure with a predictive display when the time needed to stop the manipulator was larger than about 5 seconds or when the stopping distance was larger than 0.2 meter.

The display of the stopping configuration did lead to a faster task execution when the time needed to stop the manipulator was larger than about 6 seconds or when the stopping distance was larger than 0.3 meter.

The display of a predicted trajectory always did lead to less energy consumption, due to a more calm control behaviour of the subjects. The results indicate that a collision detection system is very desirable, if not necessary for a guaranteed safe task execution.

General
In new situations or in an unknown environment the combined problems of the 3D perception and controlling 5 DOF are more serious than the problems due to the slow and non-linear dynamics.

ACKNOWLEDGEMENTS

The work is supported by the Netherlands Technology Foundation and takes place in cooperation with Fokker Space & Systems BV.
Floris van de Klashorst, Harold Doorenbos en Jaap Jan van den Bosch have contributed to the research. This is greatly acknowledged.

REFERENCES

Andre G , Schoonejans P (1989), *The Hermes Robot Arm : Teleoperation and Control Concept*, Proc. 2nd Eur. In-Orbit Operations Techn. Symp., Toulouse, France, Sept. 12-14.

Bos J F T (1991), *Man-Machine Aspects of Remotely Controlled Space Manipulators*, Ph.D. thesis, fac. of Mech. Eng. and Marine Techn., Delft UT, The Netherlands, ISBN 90-370-0056-8

Bos J F T , Klashorst F van de (1989), *Visual Enhancements for Graphical Displays in the Control of Space Manipulators*, proc. 8th Eur. Annual conf. on human decision making and manual contr., June 12-14, Techn. Univ. of Lyngby, Denmark, pp.119-132

Bos J F T, Doorenbos H (1990), *On Task Allocations Between Man and Computer for an Insert Task with a Space Manipulator*, proc. 9th Eur. Annual conf. on human deciscion making and manual control, Sept., Varese, Italy, pp. 31-40.

Bos J F T, J.J. van den Bosch (1991), *Describing the Operator as a Predictive Controller in a Space Telemanipulation Task*, proc. 10th Eur. Annual conf. on human decis-

cion making and manual control, Liege, Belgium, Nov. 13--16.

Ellis S R, McGreevy (1987), *Perspective Traffic Display Format and Airline Traffic Pilot Avoidance*, Human Factors, vol. 29 (4), pp. 371-382.

Johannsen G, Govindaraj T (1980), *Optimal Control Model Predictions of System Performance and Attention Allocation and their Experimental Validation in a Display Design Study*, IEEE trans. on SMC, vol. 10, no. 5, pp. 249-261.

Kelley C R (1972), *Adaptive Display Using Prediction in Displays and Controls*, Swets and Zeitlinger, Amsterdam, pp. 61-74.

Keyser D, Cauwenberghe A van (1981), *A Self-Tuning Multistep Predictor Application*, Automatica, vol. 17, no. 1, pp. 167-174

Kim , Tendick , Stark (1987), *Visual Enhancements in Pick and Place Tasks: Human operators Controlling a Simulated Cylindrical Manipulator*, IEEE Journal of Robotics and Automation, vol. 3, no. 5, pp 418-425,

Milgram P , Weverinke P (1985), *Model Analysis of Remotely Controlled Rendezvous and Docking with Display Prediction*, proc. 21st annual conf. on manual control, NASA, CP-2428, Moffet Field

Overbeeke C J (1988), *Space Through Movement: Theory*, PhD. thesis, Fac. of Indus. Design, Delft UT, The Netherlands

Passenier P O (1989), *An Adaptive Track Predictor for Ships*, PhD. thesis, Fac. of Elect. Eng., Deflt UT, The Netherlands.

Sachs L (1982), *Applied Statistics*, Springer Verlag, ISBN 0-387-90558-8

Vegte J van de , Milgram P , Kwong R (1990), *Teleoperator Control Models: Effects of Time Delay and Imperfect System Knowledge*, IEEE trans. on SMC, vol. 20, no. 6, pp. 1258-1272

Veldhuyzen (1976), *Ship Manoeuvring under Human Control. Analysis of the Helmsman's Control Behaviour*, Ph.D. Thesis, Fac. of Mechanical Eng., Delft UT, The Netherlands.

Veldt R van der , Boomgaard W van den (1986), *Predictive Information in the Control Room*, Human Decision Making and Manual Control, H.P. Willumeit (editor), Elsevier Science Publ. BV (North-Holland), pp.249-264.

BIOMECHANICS IN AIRCRAFT CONTROL

M.M. van Paassen

Faculty of Aerospace Engineering, Delft University of Technology, Kluyverweg 1, 2629 HS Delft, The Netherlands

Abstract: In this article a linear model for the pilot's neuromuscular system is presented. This model is then combined with a model for the pilot's crossover behaviour, a model of the side stick and a simple model of aircraft roll control. With this model set, three control situations with side sticks are analyzed:
1   The model is compared to data from an experiment with the side stick.
2   The factors contributing to roll ratchet are studied by a root-locus analysis.
3   The control with an active side stick is considered.
The structure of the model corresponds with the structure of the neuromuscular system. In this way the parameters known for parts of the neuromuscular system, such as muscle data, may be used in the model.

Keywords: aerospace control, biocybernetics, man machine systems, modelling, neuromuscular system, side stick controller, roll ratchet.

Introduction

In some recent aircraft designs (Airbus A320, General Dynamics F16), side sticks, flight control computers and fly-by-wire technology have been implemented instead of conventional controls. The main advantage of the application of these new control devices the great freedom that is offered in the choice of aircraft handling qualities. Additional benefits are a lower weight of the control system and the improved view of the instrument panel.

The side sticks as applied in these aircraft may be called *passive side sticks*. The feel characteristics of these sticks are determined by 'passive' elements such as springs, dampers, masses and stops. These sticks do not, as conventional aircraft manipulators do, provide proprioceptive (feel) feedback from the aerodynamic forces on the aircraft's control surfaces. This proprioceptive feedback is considered to be an important cue for the pilot. In those modem aircraft in which side sticks are applied, the loss of proprioceptive feedback is (maybe more than) compensated for by the application of flight control computers, which facilitate the task of stabilizing the aircraft.

An other means of improving the pilot's control of the aircraft could be by re-introducing the proprioceptive feedback on the side stick. This may be done with *active side sticks* (Hosman 1990). An active side stick contains a servo motor which is used to feed back the aircraft's motion to the pilot's hand.

For a better understanding of the interactions between a side stick and the pilot's arm, and the implications of a specific stick design (whether passive or active) an appreciation of the neuromuscular system of the pilot's arm is necessary.

Models of the neuromuscular system have already been used in the description of aircraft control, for example the precision model by McRuer and Krendel, 1974. But these are models that represent the combined effect of the neuromuscular system and the manipulator.

In 1988 at the Delft University of Technology, research was started to develop a model of the neuromuscular system in the pilot's arm. The model should be suited for the description and analysis of aircraft control with side sticks. As a basis for this model, models were adopted from *biomechanics*, a field in which human and animal locomotion and the manipulation of objects is studied. In addition experiments were conducted with a side stick. The data from these experiments were used for model validation and to obtain some parameter values for the model.

Model Description

Reference frame and assumptions.

Control with a side stick in just the one degree of freedom considered here (about the roll axis) is already quite complicated. Multiple bone and joints are involved. The wrist may be flexed and rotated, the humerus (bone in the upper arm) may rotate and the scapula (shoulder blade) may glide over the thorax. In addition to this, many muscles in the shoulder and arm are involved, these are not only necessary to generate a force on the stick, but also to stabilize the shoulder joint and scapula (Van der Helm, 1991). For this study however, a number of assumptions is made to simplify the modelling:

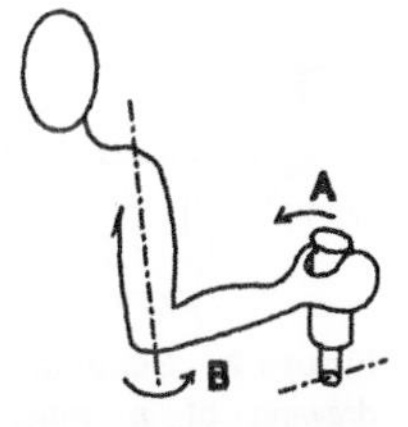

Figure 1 Roll movement of the side stick (A) is accomplished by rotation of the arm about an axis through the upper arm (B).

- The trunk of the pilot may be considered fixed to the airframe.
- Movements in the wrist, nor movements of the shoulder blade are considered.
- Movement of the stick is accomplished by rotation of the upper and lower arm around an axis through the humerus (see fig. 1).

- The action of the (complex) groups of muscles in the shoulder and arm controlling limb and side stick in roll direction may be represented by one, linear, *lumped* model.

In order to avoid problems with the computer implementation, all movements and forces of the model, except those for the aircraft, are expressed as rotations and moments around an axis through the humerus (fig. 1). For example the stick mass is expressed as a moment of inertia around this axis: $I_s = 0.163$ kgm^2. With a lower arm length of $d_l = 0.33\ m$, this means that the apparent mass at the point where the hand grips the stick is: $M_s = 0.163/0.33^2 = 1.5\ kg$.

Side stick and aircraft model.

Side sticks are generally mass-spring and damper devices such as shown schematically in fig. 2. Non-linearities of the stick will not be considered here. For the application in combination with the model of the neuromuscular system, the stick can be considered as a device with a force input and its position as the output. In Laplace transfer function form:

$$H_s(s) = \frac{X_s(s)}{F_s(s)} = \frac{1}{I_s s^2 + B_s s + K_{st}} \qquad (1)$$

I_s is the moment of inertia of the stick, K_{st} is the spring constant and B_s is the damping coefficient of the stick.

For the aircraft a simple model is chosen which represents roll motion with an integrator and a roll time constant. Added to these are an optional delay which represents the processing time in the flight control computer. Also added is a second order system with a bandwidth ω_a and a damping ζ_a. The input is stick position or force, output is the aircraft roll angle:

$$H_a(s) = e^{-\Delta t_a s} \quad \frac{\omega_a^2}{\omega_a^2 + 2\zeta_a \omega_a s + s^2} \quad \frac{K_a}{s(\tau_a s + 1)}$$

$$\begin{array}{ccc} \text{time} & \text{control} & \text{aircraft} \\ \text{delay} & \text{actuation} & \text{model} \end{array} \quad (2)$$

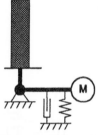

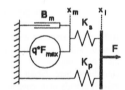

Figure 2 Schematic drawing of a mass-spring and damper side stick.

Figure 3 Mechanical analog of the muscle model. The contractile component consists of a force generator (q^*F_{max}) and a damper (B_m).

Muscle, skin model and limb inertia.

For the muscle a model of the so-called Hill type is chosen. Such a model may best be explained at the hand of a mechanical analog (fig. 3). The three components of the model are:

1 A series elastic component, with spring constant K_s, which represents the elasticity in the muscle's tendons.

2 A parallel elastic component, with spring constant K_p, which represents the elastic properties of non-activated muscles.

3 The contractile component, which consists of a force generator generating a force q^*F_{max} and a damper, with damping coefficient B_m. These represent the contractile properties of the muscles.

The force generator is driven by a neural input signal q, which varies from 0, for inactive muscle, to 1 for fully activated muscle. For the model described here q will also assume negative values, because the model represent muscle activity for both left and right movement of the stick.

For non-linear models (for example described by Winters, 1985 and Hof, 1984) the spring constant of the series elastic component increases with the force in this component, and the damping constant increases with muscle activation. By tensing the muscles against each other (co-contraction) the dynamic properties of the muscles may be changed. This means that the simulation of the neuromuscular system with a linear model for the muscle may only be valid for a constant co-contraction level and for relatively small excursions and force changes. For this study the spring constants and the damping constant of the model in fig. 3 will be taken linear. In Laplace transfer function form the muscle force response to a length change is given as:

$$H_x(s) = \frac{F(s)}{X(s)} = \frac{K_s B_m s}{B_m s + K_s} + K_p \qquad (3)$$

and the response to a change in neural activation is:

$$H_q(s) = \frac{F(s)}{Q(s)} = \frac{F_{max}}{B_m s + K_s} \qquad (4)$$

F_{max} is the maximum force generated by the contractile component. K_s is the series elastic spring constant, K_p is the parallel elastic spring constant and B_m is the damping coefficient of the contractile component.

In the analysis of a previous experiment (Van Paassen 1990) it proved necessary to include a model for the 'skin' between the stick and the bones of the hand. Also for matching the frequency response of this model to the results of (Van Paassen 1991) this skin model proved essential. No literature reference on modelling of the mechanical properties of skin was found; the skin was thus modelled as a relatively stiff spring, with spring constant K_c, and a damper, damping constant B_c, between the stick and the inertia of the limb. Behaviour of the total model is not very sensitive to variation of the parameter values for the skin.

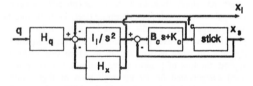

Figure 4 Combination of the side stick, skin, limb inertia and muscle model, called SSLM in fig. 5.

For the discussion of the control and feedback processes in the neuromuscular system, it is easier to form a combination of the stick, skin, limb and muscle model. The structure of this sub-model is shown in fig. 4. The input of this model is the neural activation q supplied to the muscle model, the output is the stick position x_s and the limb position x_l.

Control and feedback.

In human motor behaviour (specifically in the manipulation of objects and in sports) a remarkably wide range of behaviour is observed. On the one hand humans may perform fast ballistic movements. Considering the delays present in the transmission of feedback signals from the limb to the spinal cord and brain, such movements must be executed in an open loop manner. On the other hand precise movements are executed which may only be accomplished in closed loop control. In order to combine these two requirements, the model contains a *conditional feedback loop* (comparable to the one described in Houk & Rymer, 1981) and a *neural compensation block*. An internal model of the stick, muscle, skin and limb inertia is embedded in a feedback loop to form this neural compensation block. The compensation accepts a target position for the limb (x_t) and calculates the control signal q_c for the muscle. In this way an open loop control signal for the muscle is calculated which is *compensated* for the muscle and load dynamics. In addition to this the expected (reference) signals for the limb's sensory organs are calculated.

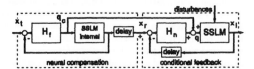

Figure 5 Conditional feedback loop and the neural compensation loop on the basis of an internal model of the side stick, skin, limb and muscle models (SSLM, fig 4).

The conditional feedback loop uses the reference signals (x_r) from the compensation and compares these with the actual signals (x_l). The difference between the two sets of signals is fed back to the muscle activation via H_n.

The sensory organs in the neuromuscular system include the muscle spindles, which measure muscle length, Golgi tendon organs, which measure muscle force, and various sensory organs in the joints and the skin. Only for the muscle spindles suitable models may be found in literature (Daunicht, 1983, Hasan 1983). However these models do not account for possible modifications to the spindle signals in the spinal cord, before these signals are fed back to the muscles. To avoid complicating the model by incorporating a model for the spindle as well as a model for filtering in the brain or spinal cord, only a simple lead-lag term was used which fed back the limb position:

$$H_n(s) = K_n \frac{1 + \tau_{nL} s}{1 + \tau_{nl} s} \qquad (5)$$

A simple lead-lag and gain combination was also used for the filter in the neural compensation loop:

$$H_f(s) = K_f \frac{1 + \tau_{fL} s}{1 + \tau_{fl} s} \qquad (6)$$

Fig. 5 shows the neural compensation loop with the internal model, and the 'external' side stick, skin limb and muscle combination model with the feedback loop. This combination is the *neuromuscular system model*, combined with the stick. Input is a desired or target stick position x_t, output is the actual stick position x_s. This model may be placed in series with the pilot model from equation 7.

Pilot model.
For the description of closed loop and total open loop control, a model of the pilot must be added. The lead-lag model of McRuer, 1973 is used here. As it is used here the output of this model is not the stick position but the target position for the neuromuscular system x_t. The input for the model is the error signal e perceived on a display:

$$H_p(s) = K_{pl} \frac{1 + \tau_{pL} s}{1 + \tau_{pl} s} e^{-\Delta t_p s} \qquad (7)$$

Model implementation.
For the implementation of the model the MATLAB software package was used. The (sub-)models as discussed above are entered in state-space representation. A Pade approximation is used for the time delays in the model. The resulting program is menu-driven. Various combinations may be chosen, such as open loop and closed loop for the filter and internal model in the compensation block. The MATLAB functions APPEND and CONNECT are used to form the final model out of the sub-models.

For analysis of the resulting model, a number of facilities are added to the program, such as bode and time plots, and a root-locus calculation for varying parameters.

Applications

Comparison to measured data.
In an experiment with a side stick (Van Paassen 1991), a set of two transfer functions describing the frequency response of the pilot was measured. The first transfer function H_1 describes the force exerted on the stick in response to an error on the display. The second transfer function H_2 describes the force exerted on the stick in response to movements of the stick. In order to measure these two responses, an experiment was used in which two disturbing functions were present: I_1 is a disturbing force on the side stick and I_2 is a disturbance at the input of the controlled system, see fig. 6.

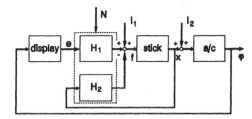

Figure 6 Experiment with two inputs, I_1 and I_2, to determine the pilot's force response to display error H_1 and to stick position H_2.

With the sub-models discussed in the previous chapter, a combination may be made to model H_1, see fig. 7a and another combination may be used to model H_2, see fig. 7b. These combinations were implemented in the program package.

By hand tuning some of the parameters of the model, a reasonable match was found between model and data (see fig. 8). The tuning process was as follows:
- A number of parameters were fixed beforehand, based on Winters 1985 and Van Paassen 1990. Muscle damping coefficient $B_m = 5$ *[Nms/rad]*, parallel elastic spring $K_p = 3$ *Nm/rad*, maximum contractile force $F_{max} = 30$ *Nm*, moment of inertia of the limb $I_l = 0.07$ *kgm²*. The properties of the stick that was used in the experiment were spring constant $K_{st} = 27.2$ *Nm/rad*, damping coefficient $B_s = 2.5$ *Nms/rad* and moment of inertia $I_s = 0.163$ *kgm²*.
- The series elastic spring was tuned to get the dip in H_2 at the right location, $K_s = 27$ *Nm/rad*.
- Then the neural feedback was adjusted. The lead term was adjusted to offset the muscle lag, $\tau_{nL} = 0.4$ *s*, furthermore $\tau_{nl} = 0.17$ *s* and the gain was adjusted to make the model match the low frequency gain of H_2. For the delay a value of $\Delta t_n = 0.05$ *s* was used.
- Then the H_1 response was matched. First the neural compensation was tuned by setting $K_f = 90$, $\tau_{Lf} = 0.4$ *s* and $\tau_{lf} = 10$ *s*. This gave the neural compensation a bandwidth of approximately 20 rad/s.

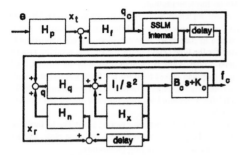

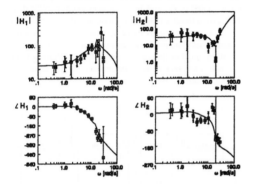

Figure 7a Combination of sub-models to represent the force response to display error H_1.

- Finally the pilot model was adjusted with as results $\tau_{pL} = 0.4$, $K_p = 0.49$ and the pilot delay $\Delta t_p = 0.25$ s and $\tau_{pl} = 0.05$ s.

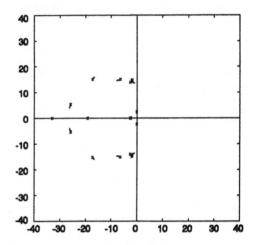

Figure 7b Combination of sub-models to represent the force response to stick position H_2.

the aircraft rotation axis is increased from 0 to 1 m, in steps of 0.05 m. The root-locus results of the analysis are given in fig. 9. For this combination there seems to be no tendency towards roll ratchet. This is in accordance with the findings of Aponso, who found no tendency towards roll ratchet for an aircraft with a passive side stick and position of the stick used as input to the aircraft.

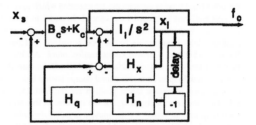

Figure 8 Model response (continuous line) and measured transfer functions for the control task with two inputs.

The model with these parameter values also showed stability for the total closed loop (pilot, neuromuscular system, stick and aircraft). The resulting parameter values were used as the basis for subsequent analyses.

Analysis of factors contributing to roll ratchet.
When side sticks are applied in moving vehicles, such as aircraft, the effect of the vehicle's acceleration on the stick and limb should be taken into consideration. Under certain conditions the acceleration effects can cause high frequency oscillations or even instability, the so-called roll ratchet (Aponso, 1981). The influence of aircraft acceleration on the stick input is modelled here by introducing an equivalent moment on the stick and the limb inertia.

$$M_{eq_I} = -I_I \ddot{\varphi}_a \frac{d_a}{d_I}$$

$$M_{eq_s} = -I_s \ddot{\varphi}_a \frac{d_a}{d_I}$$

(8)

d_a is the height of the stick and lower arm above the aircraft's roll axis, d_I is the length of the lower arm. The acceleration effects are introduced as an extra feedback in the model. The properties of the side stick used in the simulation are $I_s = 0.163$ kgm^2, $K_{st} = 27.2$ Nm/rad and $B_s = 2.5$ Nms/rad. The aircraft properties are chosen as follows: $K_a = 5.5$ and $\tau_a = 0.5$ s. To mimic a fly-by wire situation and the use of a flight control computer a time delay of $\Delta t_a = 0.1$ s is used, and the control actuation dynamics have a bandwidth of $\omega_a = 18$ rad/s and a damping coefficient of $\zeta_a = 0.7$. The behaviour of the total closed loop is considered, and the height of the stick above

Figure 9 Root locus plot for roll ratchet analysis, moving stick, stick and arm height from 0 to 1 m, in steps of 0.05m.

Next a non-moving stick with force input on the aircraft was considered. Since the program as yet does not incorporate such a stick, a similar situation was simulated by increasing the stiffness of the stick from $K_{st} = 27.2$ Nm/rad to $K_{st} = 2720$ Nm/rad. The damping was increased accordingly to $B_s = 25$ Nms/rad. The input gain of the aircraft was adapted, so the force necessary for a certain input remained the same, $K_a = 550$. The pilot gain was then adjusted to obtain good crossover behaviour, $K_p = 0.02$. There still appeared to be no tendency towards high frequency instability.
Then the roll time constant of the aircraft was eliminated, resulting in $1/s$ aircraft dynamics. This is a so-called rate command, attitude hold system. With this change in aircraft dynamics, the pilot's lead term may be omitted. Other parameters were unchanged. As may be seen in fig. 10, instability occurred when the stick and limb were at a height of 0.7 m above the aircraft roll axis.
It was also checked whether instability would occur for the moving stick in combination with the 1/s aircraft dynamics. All parameters were chosen as for the first configuration analyzed for the roll ratchet. Only the aircraft roll time constant and the pilot's lead time constant were omitted. This configuration also produced instability, with the stick and hand at 0.55 m above the aircraft roll axis.

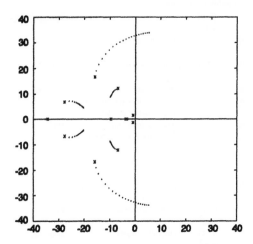

Figure 10 Root locus plot for roll ratchet analysis, very stiff stick and 1/s aircraft dynamics. Stick and arm height from 0 to 1 m, in steps of 0.05m.

It must be noted that the 1/s aircraft dynamics are a limit case of a rate command - attitude hold system. Actual aircraft with such flight control systems will usually have quite complicated higher order dynamics, and possibly a roll pole at a high frequency. The above examples are generalized and simplified cases, which show that roll ratchet may occur both with moving and non-moving sticks.

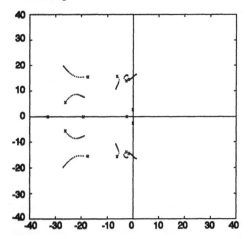

Figure 11 Root locus plot for roll ratchet analysis, moving stick and 1/s aircraft dynamics. Stick and arm height from 0 to 1 m, in steps of 0.05m.

Active side stick possibilities.
With a servo motor in a side stick it is possible to feed back the output of the aircraft to the pilot's proprioceptive system. A possible configuration is displayed in fig. 12.
The input to the aircraft is the force on the side stick and the roll rate of the aircraft is fed back to form the side stick's position. In experiments with such side sticks improvement of pilot tracking performance was demonstrated (Hosman, 1990). Again, as a basis, the parameters found in paragraph 3.1 are used as initial values for the neuromuscular system. The substitution of the passive stick by the active stick implies that a part of the aircraft model (the part that describes roll rate as

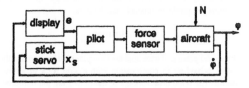

Figure 12 Active stick configuration with feedback of aircraft roll rate on stick position.

result of force on the stick) will be used for the internal model. For this example an active stick with roll rate feedback is chosen. The feedback gain was $K_v = 0.055$ [rad/(rad/s)]. The gain of the aircraft model was adjusted so the input force needed for a specific roll rate remained constant $K_a = 0.055$ rad/Nm. The filtering in the neural compensation loop with the internal model is adjusted to obtain good behaviour of this loop. The results are $K_f = 30$, $\tau_{lf} = 10$, and $\tau_{Lf} = 0.4$. The result is that now the target position of the stick x_t, which is equivalent to the aircraft roll rate, may be controlled by the neuromuscular system. As a consequence, the system left to be controlled by the pilot is now a single integrator. The pilot's lead and lag terms are no longer needed, and his gain may be adjusted to $K_p = 0.1$, so the requirements of the crossover model are met. The open loop transfer function is given in fig. 13. The resulting bandwidth for the closed loop is somewhat higher than for the configuration with a passive stick. This model was also checked for roll-ratchet tendency, with the same procedure as in paragraph 3.2. No tendency towards instability appeared.

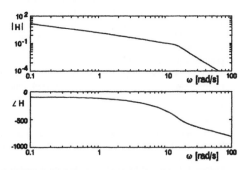

Figure 13 Frequency response of the open loop of pilot and aircraft model, with an active side stick.

When an active side stick with rate feedback is applied, the pilot's neuromuscular system loop may control the stick position, and thereby control the aircraft's roll rate. This makes the control task for the pilot easier, and comparable to the control of a rate command - attitude hold system. Further alleviation of the pilot's task by combining the active side stick with a rate command - attitude hold system seems unnecessary. It may even have negative effects. With active side sticks, as with conventional sticks, the dynamics of the stick must be carefully considered. Controlling the 1/s dynamics of a rate command, attitude hold system, with the force on the side stick, and then feeding back the roll rate on the stick would mean that stick excursion will be proportional to the force on the stick. Thus the dynamics as felt by the pilot would be of a spring-loaded stick, but without any mass nor damping. Not only do these characteristics seem unfavourable, but in experiments with side sticks driven by servo motors at Delft University of

Technology, it was found the a simulation of such a combination produces unstable oscillations.

Conclusions

The model presented here appears to explain the neuromuscular system behaviour in combination with a side stick in aircraft control. The main disadvantage of the model are the numerous parameters that need to be supplied. However because the structure of the model is analogous to the structure of the neuromuscular system, the parameters that may be known from experiments or from literature may be inserted in the model.

For the roll ratchet phenomenon the model shows that this is mainly influenced by the properties of the controlled aircraft. The properties of the neuromuscular system and the design of the stick also may play a role, but roll ratchet could be found in the simulations with both the moving stick and the non-moving stick. Especially the high frequency behaviour of the aircraft is an important factor. In this view a rate command - attitude hold system may be a roll ratchet inducing design choice.

The application of an active side stick with roll rate feedback alleviates the pilot's control task by utilizing control capabilities of his neuromuscular system. The active side stick, in combination with 'conventional' aircraft roll control characteristics, is insensitive to roll ratchet.

References

- Aponso BL, Johnston DE (1988), Effect of manipulator and feel system characteristics on pilot performance in roll tracking, AIAA 88-4326.
- Daunicht WJ (1983), Re-examination of a linear systems approach to the behaviour of mammalian muscle spindles, Biological cybernetics 48, p85-90.
- Hasan Z (1983), A model of spindle afferent response to muscle stretch, Journal of neurophysiology vol. 49 no. 4.
- Helm FCT van der (1991), The shoulder mechanism, PhD Thesis Technical University Delft, Delft.
- Hof AL (1984), EMG and muscle force: an introduction, Human Movement science 3 p119-153, Elsevier science publishers B.V. Amsterdam.
- Hosman RJAW, Benard JB, Fourquet H (1990), Active and passive side stick controllers in manual aircraft control, Proceedings of the IEEE Conference on System Man and Cybernetics, Los Angeles.
- Houk JC, Rymer WZ (1981), Neural control of muscle length and tension. In Brook VB (ed) Motor control. Handbook of physiology section 1 vol. 1, American physiological society.
- McRuer DT, Krendel ES (1974), Mathematical models of human pilot behaviour, AGARDograph no. 188.
- Paassen MM van (1991), Human operator response to visual and proprioceptive inputs is a control task with a side stick, 10th European Annual Conference on Human Decision Making and Manual Control, Liège (Belgium).
- Paassen MM van (1990), Modelling the neuromuscular system for manual control, 9th European Annual Conference on Human Decision Making and Manual Control, Ispra (Italy).
- Winters JM, Stark LW (1985), Analysis of fundamental human movement patterns through the use of in-depth antagonistic muscle models, IEEE Transactions on biomedical engineering, vol.32 no. 10.

Appendix: notations

	(fixed value)		belongs to model:
B_c	6.5 Nms/rad	damping constant	skin
B_m	5 Nms/rad	contractile component damping constant	muscle
B_s		damping constant	side stick
d_a		height of stick and arm above aircraft center of rotation	
d_l	0.33m	length of lower arm	limb
f_c		force output of skin model	
F_{max}	30 Nm	maximum contractile force	muscle
H_1		pilot force on stick response to display error	
H_2		pilot force on stick response to stick position	
I_l	0.07 kgm^2	moment of inertia	limb
I_s	0.163 kgm^2	moment of inertia	side stick
K_a		gain constant	aircraft
K_c	300 Nm/rad	spring constant skin	skin
K_f		gain constant	compensation filter
K_n		gain constant	neural feedback
K_p	3 Nm/rad	parallel elasticity spring constant	muscle
K_{pi}		gain constant	pilot
K_s	27 Nm/s	series elasticity spring constant	muscle
K_{st}		spring constant	side stick
K_v		feedback gain active stick	side stick
q		muscle control signal	
q_c		muscle control signal calculated by compensation loop	
x_s		stick position	
x_l		limb position	
x_t		target position, input neuromuscular system	
Δt_a	0.1s	computational time delay	aircraft
Δt_n	0.05s	time delay	neural feedback
Δt_p	0.25s	time delay	pilot
ζ_a	0.7	damping coefficient control actuation	aircraft
τ_a		roll time constant	aircraft
τ_{fl}		lag time constant	compensation filter
τ_{fL}		lead time constant	compensation filter
τ_{nl}		lag time constant	neural feedback
τ_{nL}		lead time constant	neural feedback
τ_{pl}		lag time constant	pilot
τ_{pL}		lead time constant	pilot
ω_a	18 rad/s	bandwidth control actuation	aircraft

ANALYSIS AND MODELLING OF PILOT AIRPLANE INTERACTION BY AN INTEGRATED SIMULATION APPROACH

P.C. Cacciabue*, G. Cojazzi*, S. Mancini* and E. Hollnagel**

*Commission of the European Communities, Joint Research Centre, Institute for Systems Engineering and Informatics, 21020 Ispra (Va), Italy
**Computer Resources International, Space Division, P.O. Box 173, DK-3460, Birkerod, Denmark

Abstract. This paper presents a preliminary simplified application of the System Response Generator (SRG) concept, i.e. the use of simulation approaches in the form of a software tool for analysing the influence of human decision making and action on accidents evolution. The study is focused on the dynamic aspects of the pilot-airplane interaction and on pilot errors, with reference to the case of a Boeing 747 type aircraft during a well defined transient situation: the Approach phase To Landing (ATL). Three components play a role in the proposed method: the deterministic simulation of the airplane and its control mechanisms; the simulation of the pilot behaviour; and a model for the failures/errors generation. The two simulations of the airplane and of the pilot behaviour are described in detail. The error module has been based on a simple generator of systematic events of omissions of proceduralised actions: hardware failures are not considered. The results obtained by applying the methodology to a well defined case of ATL are then compared with the data obtained performing a "classical" human reliability study for the same case. The potential applications and the outcome of the proposed method are enhanced by the comparison.

Keywords. Man-machine systems; Human factors; Reliability; Safety; Cognitive systems.

INTRODUCTION

The approach to the analysis and design of modern complex systems needs an upgrade of methodological tools, proportional to the connotations which are nowadays associated with the meaning of terms "complexity" and "system". In most technologically advanced environments, one usually considers a "system" as the combination of a machine with all those other elements needed for operating such a machine for a period of time with a certain function. This implies that the meaning of "system" encompass more than the mere hardware components undergoing a physical process; in reality, it includes the actual plant, the humans, the interfaces, the computers, the control and management procedures (Moray et al, 1990). The major "complexity" of a system is identifiable in the many dependencies which are built into its various components. This generates several dependencies, of statistical as well as of causal nature, between the variables governing the system behaviour. The analysis of the propagation "in time and space" of disturbances, such as during the analysis of

transients, incidents or accidents (Mancini, 1986; Sheridan, 1986), thereby becomes particularly difficult.

The result of this discussion is that it is impossible to separate the effects of the human "controller" from the plant dynamics, and thus it is necessary to integrate the human component in the design of the safety and control features at all levels. The use of modelling simulation is a very natural approach for studying human-machine interaction in such contexts. The literature provides several reviews of the various approaches and methods dedicated to the modeling issue: examples of well-structured and complete analyses of the state-of-the-art, made at different times and thus reflecting somewhat different issues and backgrounds, are the works of Pew and colleagues (1977), of Rouse (1980), of Sheridan (1986) and the more recent review of Stassen, Johannsen & Moray (1990).

We are presently following the objective of using simulation approaches in the form of a software tool, which can be used to analyse the influence of human decision making and action on the way in which accidents evolve. We have named the tool under development System Response Generator (SRG). The methodology proposed in SRG is based on a consideration of a thoroughly dynamic evaluation of the man-machine environment, where a major role is played by the simulation of the plant and its control/supervisory mechanisms and by the simulation of the operator behaviour, coupled with a model for the error generation. In the SRG, rather than treating the operator as a system component, both the operator and the process are modeled on equal terms, i.e. using the best available knowledge and techniques. The SRG concept, originally introduced in the nuclear reactors safety domain (Cacciabue & Cojazzi, 1985), is presently being transferred to the analysis of avionic systems (Hollnagel & Cacciabue, 1991) and, more in general, to the study of complex working environments.

This paper presents a simplified application of the SRG concept to the study of pilot errors for safety and reliability analysis. Reference will be made to the case of a Boeing 747 type-like aircraft during a well defined transient situations, such as the Approach phase To Landing (ATL). In particular, we will firstly describe, in more detail, the SRG architecture and objectives; in the second part, a general "default" human model for the SRG will be introduced and the computer simulation of the airplane and of the pilot behaviour adopted for the case under study will be described in detail. Finally, we will compare the results of the SRG approach with a classical Human Reliability Assessment (HRA) technique for the case study of the ATL of a Boeing 747.

THE SYSTEM RESPONSE GENERATOR

The System Response Generator (SRG) is a software environment or tool which uses computer models for specific purposes (Hollnagel, Cacciabue and Rohuet, 1992). The SRG approach is primarily intended to be exploited during system design, with particular regard to the design of safety systems, emergency procedures, information presentation, and task allocation. The interest of the SRG is with the way in which misunderstandings and incorrectly executed human actions can affect the evolution of an accident. As far as the HRA approaches are concerned, the SRG approach offers the characteristics of dynamics, due to the simulation of the man-machine interaction, which are generally lacking in the more classical approaches to human factors analysis. Another important extension of the scope of the SRG is in the post-hoc analyses of accidents that have occurred and in the performance of a batch, or off-line, generation of a very high number of computer simulated sequences including human errors and recoveries. In order to develop the SRG into a generally applicable tool, the architecture of the system is based on four *modules*, which constitute the basic *event cycle* (Fig. 1): the Event Driver, the Process Response Generator, the

Operator Response Generator, and the Response Interpreter.

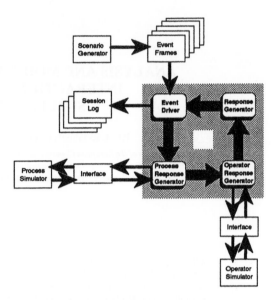

Fig. 1. General architecture of the SRG.

The basic *event cycle* is repeated until the desired output from the SRG is produced or until other conditions for terminating the SRG session are met. In addition the SRG needs other specialized function modules. The most significant of these is the *Scenario Generator*, which provides the necessary details for a given scenario. The input to the Scenario Generator includes: requirement analysis, process probabilistic data, data on human erroneous actions, and process expertise. The output from the Scenario Generator, in the form of Event Frames, contains the information to control the Event Driver as well as the initial parameter values to drive the Process Response Generator and the Operator Response Generator. The results of a case of the SRG are stored in a Session Log for study/analysis purposes. The Process Simulation and the Operator Simulation, with their relative interfaces to the SRG, are not part of the SRG as a tool because they have to be developed "ad-hoc" for the specific systems and working environment under study. However, a "default" operator model is being developed (User Representation Model) for the simulation of the basic processes of perception-interpretation-planning and execution.

User Representation Model. The purpose of the user model in SRG is clearly to regulate or control the performance of the SRG, i.e. to provide part of the control input for the process simulation and is, accordingly, not intended to serve as a true model of human cognition. It is therefore appropriate to talk about a *User Representation* (UR) rather than user model and hereafter we will use the terms *user model* and *user representation* as synonymous (Hollnagel and Cacciabue, 1991). In synthesis, the user model must be able to emulate and control the following categories of *operator functions*: observation/perception, memory (recall of information), interpretation, choice (planning/decision making) and execution of a plan. It is assumed that the performance of each of these functions can be described at three different levels, namely:

- Level-0, where the performance is flawless and correct. This correspond to a user with correct knowledge, infinite capacity and infinite attention.
- Level-1, where the performance can be expressed by means of a continuous function of external conditions such as interface design, structure of procedures, work demands, environmental conditions etc.
- Level-2, where the performance can be expressed in terms of rules or heuristics. This corresponds to a user who is influenced by the perceived conditions or internal states (cognitive states), e.g. assumptions, intentions, biases.

The above classification gives a framework for developing a user model to be suitably included in the SRG. In the study case hereafter described we will restrict ourselves to the formulation of a level-1 model of the pilot. This allows us to define a detailed procedure for all the observations and actions to be performed during the simulation. Possible effects of pilot erroneous actions are considered only in terms of deviations from the level-1 behaviour.

CASE STUDY: THE APPROACH TO LANDING PHASE FOR CIVIL AIRCRAFT

Simplifying Hypotheses and Initial Conditions

The case under study considers the phase of ATL for a Boeing 747, which consists of the descent from 10000 ft to a few hundred feet above ground level. The ATL demands a number of consistent control actions and verifications by the piloting crew in order to prepare the aircraft for the last phase of the flight, i.e. the actual landing on the ground. Given that during such phase the airplane and control system have to assume a well defined configuration, a number of initial conditions and simplifying assumptions have been made. These have an effect on the model of flight dynamics, on the model of the man-machine interface and on the model of the pilot behaviour. Two main *assumptions* have been made which are preliminary to all considerations relative to the ATL phase and, therefore, have to be discussed first. They are:

- All flight systems, such as the mechanical parts and the control and instrumentation systems, are postulated to operate perfectly. with no failures or malfunctions.

- The human model is developed within the boundaries of the UR, level-0 and level-1.

Apart from the two main assumptions, it is necessary to evaluate accurately a number of *initial conditions*, which define the system under study and its initial state. As far as the model of flight is concerned, the *initial conditions* contribute to the simplification of the aircraft equations of motion. Similarly, for the simulation of the interface pilot-aircraft, i.e. the cockpit instrumentation and the on-board commands, the *initial conditions* allow to reduce the vast amount of information available to the one actually needed to manage the ATL phase. The following *initial conditions* have been chosen:

1. The ATL phase develops at low speed.
2. The ATL has a duration of about 10-15 minutes, i.e. negligible with respect to the duration of the entire flight.
3. During the ATL phase the engines are left at minimum working power.
4. The airplane is in "clean" flying condition at the beginning of the ATL.
5. The instrumentation of the cockpit presents the information with no delay.
6. The control actions affects simultaneously all 4 engines of the aircraft.
7. The ATL starts with the airplane in horizontal straight uniform flight conditions, oriented towards and aligned with the runway.

On the basis of the above conditions, a number of *simplifying assumptions* have been made which have strongly affected the models of the airplane flight dynamics and of the pilot behaviour.

The model of the airplane dynamics

The simulation of flight dynamic consists in determining the position and velocity of an airplane at all times (Nelson, 1989). This aim can be achieved by solving the *aircraft equations of motion*, given the initial conditions and the external loads acting on the aircraft. While a detailed description of the mathematical treatment of the problem can be found elsewhere (Mancini, 1991), here we will briefly recall the most significant aspects of the simulation and of the numerical solution algorithms.

The conservation principles of the linear momentum and angular momentum are represented by 12 ordinary non-linear differential equations which have to be solved in order to obtain the position and the flight conditions at all times. For the solution of the system of these equations a predictor-corrector numerical method has been applied: the Adams-Multon of the 4th order. The starter of the numerical procedure has been obtained by the algorithm of Runge-Kutta also of the 4th order. In this way an optimum balance between expected precision and computational speed has been obtained (Conte de Boor, 1965).

The model of the control and instrumentation

For the definition of the commands of the pilot, a number of components for the flight control (*commands*) and for the information transfer (*indicators*) needs to be included in the simulation.

The *commands* allow to control the flight by means of actions aimed at the regulation of the aerodynamic loads, of the propulsion and of the aircraft configuration. The commands govern five major components of the aircraft, namely: 1) Flaps and slats; 2) Landing gear; 3) Engines; 4) Elevator; 5) Stabilizer.

The *indicators* allow the recognition of the state of the airplane, i.e. the flight configuration, the relative speed of the wind, etc. and affect the decision making and action of the pilot. The following *indicators* are simulated: 1. the *altimeter*, for the altitude; 2. the *radar-altimeter*, for a more precise position definition with respect to ground level; 3. the *rate-of-climb-indicator*, for the vertical speed; 4. the *airspeed-indicator*, for the airplane relative speed with respect to the air; 5. the *Instrumented Landing System (ILS)*, for the deviation from the ideal trajectory; 6. the *acoustic signal*, for the radio-signal of a landmark.

The model of the pilot behaviour

The UR model defines the criteria on which to base the simulation of the interaction of the operator with the plant under control. In particular, the interaction amongst the *functions* of the UR, as defined above, and the actual operator behaviour has to be sustained by a *theory of cognition*: the Skill-Rule-Knowledge (SRK) theory (Rasmussen, 1983) has been taken here as a paradigm. Moreover, we have limited our modelling effort to a simple description of automatic (skill) and procedural (rule) knowledge of a pilot for the simulation of the basic functions of the behaviour.

Indeed, on the one hand, the actions on the flaps, slats and landing gear are governed by well defined speed limitations. Consequently, the information read on the airspeed indicator is interpreted as a sign of the state of the airplane, that might satisfy the conditions for the extension of such mobile parts: this corresponds to a procedural behaviour. On the other hand, by acting on the engines and on the elevator, the pilot is able to govern the speed and altitude of the aircraft. In this case, the information read on the indicators allows the pilot to adjust the actions in order to follow a pre-defined plan: this correspond to an automatic response.

In this way the resulting model allows the simulation of adaptive response of the pilot to the dynamic evolution of the behaviour of the airplane.

The study case

We have limited our analysis to the possible omissions of procedural actions, assuming that the automatic actions, carried out by the pilot during the various sub-phases of the ATL, are correctly performed. In other words, the level-1 behaviour is decomposed in two principal kinds of activity according to the distinction between automatic (continuous) and procedural (discrete) events. An example of automatic action is the regulation of the engines throttle during manual flight (this task is usually accomplished by the auto-pilot). For procedural actions we intend those triggered by a specific conditions such as the setting of an instrument or the extraction of the landing gear. In all cases we will consider deviations from the ideal operator affecting only discrete actions. This is done in order to obtain a reasonable number of deviations from the ideal behaviour described by level-0.

The test case has been developed defining precisely the initial conditions, in order to characterise the normative behaviour (i.e. the behaviour without errors) of the pilot and, thus, obtaining the ideal ATL sequence, given the initial conditions. The results obtained studying the consequences of pilots erroneous actions have been compared with the analysis of the same ideal sequence performed with a human reliability methodology, usually applied in probabilistic safety analysis, known as THERP (Technique for Human Error Rate Prediction) (Swain and Guttman, 1983). THERP is an approach which accounts for human errors using probability trees and models of dependence, as it is done in engineering risk assessment methodologies, but also considers Performance Shaping Factors affecting operator actions.

In the remaining part of the paper we will concentrate on the comparison of the results obtained for the case study applying the THERP analysis and the SRG architecture with the models of the airplane and of the pilot described above.

The actual situation to be investigated in the study case is the ATL on a generic airport according to the standard route manual, up to the crossing of the nearby VOR (Very high frequency Omnidirectional Ranging). The study of the ATL phase starts by assuming that the pilot perceives correctly the information from the control tower to start the descent from 10000 ft, and it ends with the crossing of the VOR. Therefore the procedural actions which can occur during the evolution of the ATL, and on which the analysis of the pilot errors will be focused, are the following:

F1   *Extend the flaps to position 1*: as the indicated air speed reduces below 250 Knots, the flaps are extended to position 1 (sub-phase control the descent).

F5   *Extend the flaps to position 5*: as the indicated air speed reduces below 230 Knots, the flaps are extended to position 5 (sub-phase control the descent).

F10 *Extend the flaps to position 10*: as the indicated air speed reduces below 215 Knots, the flaps are extended to position 10 (sub-phase control the descent).

CA *Calibration of the altimeter on QNH*: as the indicated altitude reduces below 7000 ft, the pilot must calibrate the altimeter on the actual air pressure at sea level (*QNH*) (sub-phase control the descent).

LV *Level the flight on 4000 ft*: as the indicated altitude reduces below 5000 ft, the pilot switches from the *control of descent*, instinctive based behaviour, to the *leveling of the flight*.

CL *Execute the approach check list*: as the indicated altitude reduces below 4050 ft, the pilot executes the approach check list and, in particular, he controls whether the altimeter has, or not, been correctly calibrated on the *QNH* (sub-phase flight leveling).

It is important to recall once again that in a dynamic system, such as the one we are considering here, the order in which the actions are actually executed is strictly dependent on the evolution of the system and, thus, on its initial conditions. In particular, different initial conditions will necessarily result in a different order of the actions to be performed.

Two different cases were considered for the comparison of the results obtained by the THERP approach and by the proposed dynamic approach. These two cases refer to two different initial states of the airplane. The first set of initial conditions is associated to a nominal ATL situation. The airplane position and speed are such that there is enough time for the pilot to complete all the prescribed actions and for the airplane to properly adapt to the input coming form the pilot. The second set of initial conditions relates to a much more critical situation in which the airplane is at the same altitude of the previous case, but nearer to the runway. In this case the time is just enough to allow a successful ATL if the entire procedure is properly executed by the pilot, therefore this situation will be hereafter identified as the limit one. Table 1 summarizes the set of initial conditions for the two study cases.

TABLE 1 Initial Conditions in the Nominal and in the Limit Case

Initial data	Nominal case	Limit case
Weight of the airplane	660000 lb.	627000 lb.
Indicated air speed	300 Knots	270 Knots
Indicated altitude	10000 ft.	10000 ft
Distance from runway	45 NM	37.1 NM
Air pressure at sea level	1013 mbar	1018 mbar
Air temperature at sea level	15°C	30°C

In both cases the ATL phase is considered to be successful if the airplane state, at the crossing of the VOR fits the following conditions:

1. Flaps extended to position 10.
2. Altimeter correctly calibrated on *QNH*.
3. Indicated altitude around 4000 ft (± 100 ft).
4. Indicated air speed around 180 Knots (± 10 Knots).

These success conditions can be directly utilized in analysis done with the SRG. The THERP analysis, which is not based on the simulation of the MMI, needs to collapse the conditions three and four in the equivalent condition 3) The Pilot has activated the procedure for leveling the flight. The simulation of the ATL phase, in absence of pilot errors, is obviously successful with both sets of initial conditions, even if, as already mentioned, the resulting order of execution of the six actions is different. In particular, in the ideal case, the order of execution is the following: CA, LV, F1, F5, F10, CL. Due the different dynamics of the system, in the limit case the order of execution becomes: F1, F5, CA, F10, LV, CL.

The THERP Results. In principle, the six different actions and the corresponding errors should generate a tree with 2^6 branches. On the other hand, since the THERP methodology does not necessarily imply the physical simulation of the system, it is assumed that the non-execution of the action "level the flight on 4000 ft" corresponds to the impossibility to reach the correct final state defined by the previous conditions 3 and 4. This, together with the possible error on the action "extend flaps to 10" or on the action "execute the approach check list", will cause the failure for the mission and, thus, enables to stop the development of the corresponding branches in the tree. These *branches* are so labeled with F (Failure) and need not to be further splited. In the *nominal case* the remaining tree is constituted by 26 branches while in the *limit case* it is necessary to consider 32 branches. In both cases we neglected, for simplicity, the possibility of recoveries. In Fig. 2 we report the human error tree for the case of limit conditions. The success branches are 12 and are labeled with S while the remaining 20 branches are labeled with F. The success and failure probabilities of each branch are also reported.

For each action the corresponding probabilities of success have been derived from chapter 20 of the THERP handbook (Swain & Gutmann, 1983), where the basic human error probabilities are defined for prototypical actions. As an example, let us consider the action "Flaps 1" which can be further decomposed in two sub-tasks: the recognition by the Pilot Flying (PF) that the speed has decreased below 250 Knots, and the activation, by the Pilot Not Flying (PNF), of the command performing the extension of the flaps, as the order is received. Therefore, basic error probabilities of these two sub-tasks can be identified respectively in the tables 20-

12 and 20-18 of the THERP handbook. Moreover the following dependencies between actions have been considered:

1. Altimeter calibration on *QNH* => Level the flight.
2. Altimeter calibration on *QNH* => Approach check list.
3. Flaps not 1 => flaps not 5.
4. Flaps not 1 and flaps 5 => flaps 10.
5. Flaps not 1 and flaps not 5 => flaps not 10.
6. Flaps 1 => flaps 5.
7. Flaps 1 and flaps 5 => flaps 10.
8. Flaps 1 and flaps not 5 => flaps not 10.

Where by the symbol => we intend that the probability of success/failure of the first action affects the probability of success/failure of the second one. Moreover, *low, medium* and *high* dependencies are properly combined in order to treat correctly the situations 1 to 8. The final numbers reported in Fig. 2 reflect the human basic error/success probabilities properly modified by the appropriate dependencies. Given that the success of ATL is obtained by adding the sequences leading to success, the resulting success probability **S** is about 99.5% (0.99487). Correspondingly, the failure probability F (i.e. 1-S) is about 5.13 10E-3. The same success probability is obtained in the nominal case, where 12 branches are associated to the success condition and the remaining 14 are failures.

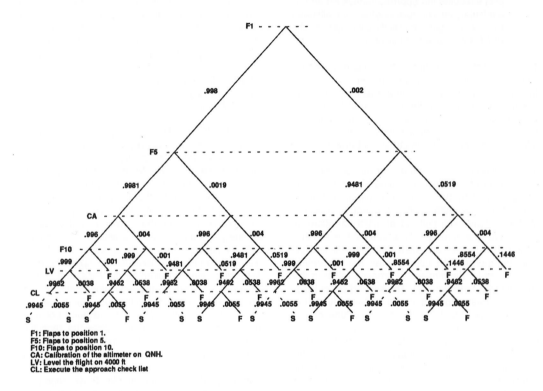

F1: Flaps to position 1.
F5: Flaps to position 5.
F10: Flaps to position 10.
CA: Calibration of the altimeter on QNH.
LV: Level the flight on 4000 ft
CL: Execute the approach check list

Fig. 2. Results of THERP analysis

The SRG Approach. The same two situations, with nominal and limit initial conditions were analyzed by considering the coupled dynamics of the pilot-airplane and the effect of the erroneous procedural actions on the airplane state. The maximum simulation time was of 500 s. but the sequences could obviously reach the success or the failure condition before this time. The time step of integration of the flight equations is a variable one of 0.25 s. or less and the time step adopted for the pilot model is of 1 s. The nominal case resulted in a tree of 34 sequences with 12 successes. The different number of sequences with respect to the THERP analysis is just due to the fact that the eight

sequences in which the flaps were not extended to position 10 were not pruned form the tree, as was done in the THERP, in order to evaluate the state of the airplane at the crossing of the VOR. The success probability is obtained by adding the probability of the success sequences and is 0.99487 this results identical to the one of the THERP analysis, as could have been foreseen, since in this case, there are no temporal limits on the execution of the pilot actions.

The limit case is much more interesting and needs a more complete description. In Fig. 3 we report the corresponding event tree obtained by means of the

SRG approach. It is important to notice that the tree is not the result of the work of the analyst but can be easily obtained from the results of a single run of the simulator. The different actions are now represented with symbols since each node is associated to the relative time, expressed in seconds from the beginning of the sequence, at which the action take place. The elementary failure probabilities are the same adopted for the THERP analysis and are thus not reported on the respective branches. It can be noticed that the number of sequences taken into account from the dynamic analysis is now 38

instead of the 32 that result from the THERP tree. Again this discrepancy is due to the fact that, in the dynamic analysis the sequences with flaps not to 10 are of failure, but are not pruned from the tree. Actually not all these sequences are considered since the ones resulting from conditions which are never met in the dynamic analysis are lost. The main point is that there are now only 7 success sequences, while the THERP tree presented 12 successes so that five success sequences of the THERP have now been lost.

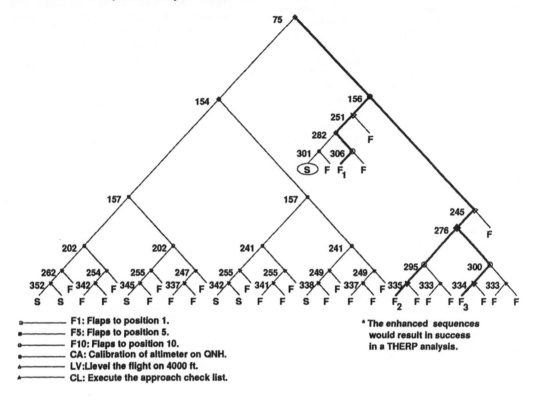

F1: Flaps to position 1.
F5: Flaps to position 5.
F10: Flaps to position 10.
CA: Calibration of altimeter on QNH.
LV:Llevel the flight on 4000 ft.
CL: Execute the approach check list.

* The enhanced sequences
would result in success
in a THERP analysis.

Fig. 3. Results of SRG analysis

One of these success sequences, the one in which S is surrounded by a circle, is lost since in its evolution the condition *execute the approach check list* is never met, but the sequence remains a success. This means that the final altitude of the airplane is between 4050 and 4100 ft so that the success condition is met but the altitude at which the approach check list is executed (4050 ft) is not reached. The remaining four success sequences have become failure sequences and are labeled as F1, F2 and F3, notice that F1 takes the place of two success since the splitting due to approach check list is not executed. As we explained before, the success conditions were adapted to a logic condition in the THERP analysis, in particular, the conditions 3. and 4. were considered equivalent to the activation of the procedure of leveling the flight. This was done since THERP analysis cannot treat in

detail the physics of the system. With the integrated simulation of the pilot-airplane, it is possible to follow the real evolution of the system, by so doing it results that the sequence F1 does not match the speed limit (condition 4) while F2 and F3 end with an indicated altitude incorrect (Condition 3).

In sequence F1, the omission of flaps at position five, causes an insufficient decrease of the speed of the airplane, so that when "flaps to ten" is correctly executed by the pilot the airplane is already too near the airport and there is not enough time to match condition 4. Sequences F2 and F3 result in a leveling of the flight at a wrong altitude, since the calibration of the altimeter at 7000 ft was omitted, and the air was not in standard conditions. At the execution of the approach check list the pilot recognizes the error and calibrates correctly the altimeter, but again there is not enough time to correct the altitude since the airplane is already in proximity of the airport.

CONCLUSIONS

This paper has presented a simplified application of the SRG concept to the study of the Man-Machine Interaction. The proposed methodology is based on a consideration of a truly dynamic simulation of the interaction between the system and the human operator. A preliminary application in the avionics domain has been presented with reference to a Boeing 747 type airplane to study in detail a well defined transient situation, such as the Approach phase To Landing. The case study has been strongly simplified, as far as the actions of the pilot are concerned, with the objective of analysing the phase with the proposed dynamic methodology and a with more classical human reliability technique (THERP). In order to obtain comparable results, the success event has been defined very accurately and the same basic error probabilities, derived from the THERP handbook have been applied in the THERP analysis and in the proposed approach.

Two different cases were considered, according to nominal and limit initial conditions for the airplane state. In the nominal case a THERP analysis has easily been developed and the analysis by the proposed approach yields the same overall success probability. In the case of limit initial conditions the success probability is different. This happens because some sequences, which appear as successes in the THERP analysis, are actual failures when the time variable is taken into account. In other words, the time dimension, which is currently neglected in plant as well in human reliability studies can be easily taken into account if a simulation of the MMI is performed.

ACKNOWLEDGMENTS

The work described in this paper is sponsored by the CEC as part of the programme on Science and Technology for Environmental Protection (STEP), DG XII, Directorate General for Science, Research and Development.

REFERENCES

Cacciabue, P. C. and G. Cojazzi (1985). Analysis and design of a nuclear safety system vs the operator time constraints, *Proceedings of 2nd IFAC Conf. on Analysis, Design and Evaluation of Man-Machine Systems*, Varese, Italy, 10-12 September 1985, Pergamon Press, Oxford.

Conte de Boor, S.D. (1965). *Elementary numerical analysis: an algorithmic approach*, McGraw-Hill, New York.

Hollnagel , E. (1991). The Phenotype of Erroneous Actions: Implications for HCI Design. In G.R.S. Weir & J.L. Alty (Eds.), *Human Computer Interaction and the Complex System.* 73-121. Academic Press, London.

Hollnagel, E., P.C. Cacciabue (1991). Cognitive Modelling in System Simulation. Proceedings of Third European Conference on Cognitive Science Approaches to Process Control, Cardiff, September 2-6, 1991.

Hollnagel, E., P.C. Cacciabue and J-C. Rouhet (1992). The use of an integrated system simulation for risk analysis and reliability assessment. 7th International Symposium on Loss Prevention, 4-8 May, 1992, Taormina, Italy.

Mancini, G. (1986). Modelling humans and machines. In *Intelligent Decision Support in Process Environments*, E. Hollnagel, G. Mancini and D. D. Woods (Eds.), NATO ASI Series, Springer-Verlag, Berlin, FRG.

Mancini, S. (1991). Applicazione di modelli per la valutazione dell'affidabilità umana al comportamento di un pilota di un velivolo civile. Tesi di Laurea in Ingegneria Auronautica, Politecnico di Milano.

Moray, N., W.R. Ferrell and W.B. Rouse (Eds.) (1990). *Robotics, Control and Society*, Taylor & Francis, London.

Nelson, R.C. (1989). *Flight stability and automatic control*, McGraw-Hill, New York.

Pew, R.W., S. Baron , C.E. Feehrer and D.C. Miller (1977). *Critical Review and Analysis of Performance Models Applicable to Man-Machine-Systems Evaluation.*, BBN Report No. 3446, Cambridge, MA.

Rasmussen, J. (1983). Skills, Rules and Knowledge: signals, signs and symbols; and other distinctions in human performance model. *IEEE Transactions on Systems, Man, and Cybernetics*, SMC-13, 3, 257-267.

Rouse W. B., (1980). *Systems Engineering Models of Human-Machine Interaction*, North Holland, Oxford.

Sheridan, T. B. (1986). Forty-five years in man-machine systems: history and trends, *Proceedings of 2nd IFAC Conf. on Analysis, Design and Evaluation of Man-Machine System.* Varese, Italy, 10-12 September 1985, Pergamon Press, Oxford.

Stassen H.G., Johannsen G., Moray N., (1990). Internal Representation, Internal Model, Human Performance Model and Mental Workload, *Automatica*, 26, No 4, 811-820.

Swain, A. D. and H.E. Guttman (1983). *Handbook on Human Reliability Analysis with Emphasis on Nuclear Power Plant Application.* Draft Report. NUREG/CR-1278. SAND 80-0200 RX, AN. Final Report. Sandia National Laboratories, Albuquerque, New Mexico.

METHOD FOR THE PROBABILISTIC FAILURE ANALYSIS
OF A MAN-MACHINE SYSTEM

G. Heslinga

Risk and Reliability Analysis Group, N.V. KEMA, P.O. Box 9035, 6800 ET Arnhem, The Netherlands

Summary Probabilistic failure analyses of complex installations are usually made for a plant in a steady state. The current techniques are however too limited for non-steady state situations (e.g. in case of start-up of a plant). This paper will present a possible technique for evaluating the safety of a man-machine system that is brought to another state for which procedural actions have to be performed. It consists of combining a method for the identification of the systems states that may occur due to human errors in following the procedure incorrectly and a systems analysis of these system states. An example related to a man-machine system for locking out electrical power for maintenance is presented.

keywords: automation, Boolean algebra, ergonomics, error analysis, event trees, fault trees, probability, safety.

INTRODUCTION

Probabilistic failure analyses are more and more performed to evaluate risks involved in different types of industry. Extensive studies started in the nuclear industry (USNRC, 1975) to assess the risk of a plant in a steady state (delivery of a product under normal stationary conditions). They are well known under the name of Probabilistic Risk Assessment or Probabilistic Safety Assessment. These studies are now more and more started to be used in other types of industry, such as the process industry.

The technical design usually forms the basis for performing a Probabilistic Safety Assessment. Starting from the design undesired consequences and their causes are considered. The analysis of human errors as part of a Probabilistic Safety Assessment is usually done by considering the non-performance of an activity. This may be the result of simply an error of omission, i.e. the action is not performed at all; but also an error of commission, such as a selection error implying that another switch is selected, can lead to the non-performance of the desired activity. The assessment of human performance in this way as part of a Probabilistic Safety Assessment is usually gathered under the name Human Reliability Analysis: the probability is calculated that an activity or procedure is performed correctly, yes or no (Swain & Guttmann, 1983). Single human errors are usually considered in a Human Reliability Analysis.

The effects or consequences of a human error are usually not explicitly considered in a Human Reliability Analysis. It may occur that the effect of a diagnostic error is analyzed or that outcomes of certain errors of commission are taken into account. This is however usually performed ad hoc and not in a structured way by means of a Failure Mode and Effect Analysis (CISHC, 1977).

The structural analysis of all human errors that may have effects on the plant status and may lead to initiating events become much more important in a situation in which procedures are being performed extensively. This occurs when a plant has to be brought to another steady state, such as during starting up and shutting down. Instead of starting with a systems analysis of a technical design leading to a consideration of the technical failures and the human errors involved, the procedure of human activities becomes the starting point (Heslinga, 1988).

In addition to the probability of the human error, the evaluation of the outcomes of human errors becomes essential. Since the risk (combination of error probability and effect) of human performance becomes relevant instead of the reliability, it is better to use a term Human-Performance Risk Analysis or Human-Performance Safety Analysis instead of Human Reliability Analysis. The techniques for a Human-Performance Safety Analysis are not as wide spread as for a Human Reliability Analysis (Heslinga, 1988). The techniques are particularly very limited for an analysis of a sequence of human errors. Most undesired consequences occur however as a combination of human errors (e.g. the accident at Three Miles Island and the accident with the Herald of Free Enterprise).

This paper will present a method for making a probabilistic failure analysis of a system that is brought from one state to another state in which procedural human activities form an essential part (e.g. start-up). Since analysis of both man and machine is performed the term probabilistic failure analysis of a man-machine system (MMS) will be used. The method implies the application of two techniques, viz. a technique for making a Human-Performance Safety Analysis and a technique for making a systems analysis of the technical system. The next paragraph will present the method followed by an application.

235

METHOD

The method consist of three parts (see Fig. 1). The first part uses the human activities as a basis for which an error analysis, i.e. a Human-Performance Safety Analysis in this case, is made. Once this analysis is performed the undesired system states a human can cause by erroneously performing the desired activities are known; this information is used in the second part which consists of a systems analysis. The final part is a combination of the results of the first two parts.

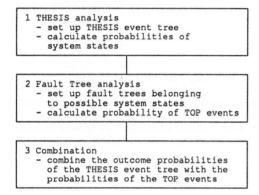

Fig. 1   A method for making a probabilistic failure analysis of a man-machine system.

THESIS analysis

The method that was developed for making a Human-Performance Safety Analysis of procedural activities is termed Technique for Human-Error-Sequence Identification and Signification, THESIS for short (Heslinga, 1988). The technique will be clarified in the example in the next chapter and its principle is presented now briefly. The basis of the THESIS method is formed by the creation of modules of the human actions that are performed according to procedures. A module is an event tree of one action showing the errors that can be made in performing that action, the outcomes of the errors and the possibilities that are available to the human controller to rectify his errors. This step is to a great degree similar to a Failure Mode and Effect Analysis (CISHC, 1977) as performed for technical systems.

Next, the modules are combined leading to the event tree of the complete procedure. This usually implies a reconsideration of outcomes, since sequences of errors may lead to entirely different outcomes than single human errors. Moreover, this step presupposes a detailed process knowledge so as to be able to predict those outcomes.

Although the preceding qualitative approach may already give sufficient insight into the weak spots in procedures and man-machine interfaces used in following the procedures, it may be necessary also to quantify the probabilities of the outcomes. This is done by assigning probabilities to the human errors and the recovery possibilities to rectify errors made. Dependencies between human activities are taken into account

and the probabilities that the outcomes occur are obtained by combining the event probabilities.

Fault tree analysis

The systems analysis which is performed next is a Fault Tree analysis (USNRC, 1981). A fault tree shows how a predefined consequence, the TOP event, is caused by a number of basic events; this is contrary to an event tree which shows how a disturbance, an initiating event, may develop into certain consequences.

A fault tree is set up for each possible state of the technical system as caused by the human errors. These possible states are derived from the THESIS analysis. Application of Boolean algebra then gives the probability of the TOP events.

Combination of results

The method ends with a combination of the probability to reach certain state (i.e. to obtain a particular fault tree) as obtained from the THESIS analysis and the probability of the TOP event from the Fault Tree analysis.

APPLICATION OF THE METHOD TO A MAN-MACHINE SYSTEM

An analysis was performed of three different MMSs (three different system designs and three related procedures) for an organization in the Netherlands. Each MMS can be used to lock out electrical power to prevent that power is supplied to a motor while a serviceman is doing maintenance on the motor. The analysis of only one design and related procedure will be presented here. The analysis according to the method described above assesses the probability that power is supplied to the motor while being maintained possibly leading to the service-man being injured.

The electrical circuit belonging to the lockout system is shown in Fig. 2. The main power can be shut off by putting the main power switch to OFF or by having the relay M opened. The relay is controlled by a circuit that contains a field switch, a process interlock and a panel switch. The field switch will be called FIELD, the panel switch PANEL and the main switch MAIN. To close the relay, the process interlock must be closed, and the PANEL and the FIELD must be in RUN-position. Either one of the switches is put to the START-position which causes the relay to close. The switch will automatically return to the RUN-position after it is released and the relay will remain in the CLOSED-position. This design contains a BY-PASS within reach of the FIELD for checking purposes as will be clarified in the procedure below.

The procedure to bring the electrical circuit to a safe state is as follows:
1   Go to the motor in the field and adjust the FIELD to the STOP-position if the motor is running.
2   Note the motor identification number.
3   Go to the Main Control Centre and select the right MAIN using the motor identification number.
4   Switch off the MAIN.
5   Attach a label and a padlock.

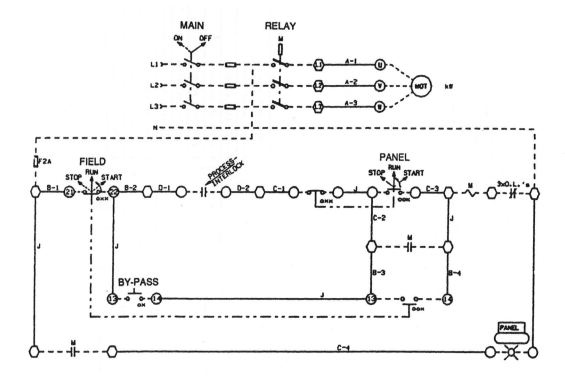

Fig. 2   Electrical circuit of the lockout system.

6   Go to the motor in the field and make a start attempt using the FIELD and pushing the BY-PASS simultaneously.
7   Check that the motor does not start.
8   If OK, start work.
This procedure will be analyzed further in this paper.

THESIS event tree

The method starts with making small event trees, THE-SIS modules. These modules are then combined to draw the complete THESIS event tree. A module can be made of each step or of a short combination of steps. In this case four modules can be made (see Fig. 3):
module 1: step 1
module 2: steps 2, 3 and 4
module 3: step 5
module 4: steps 6, 7 and 8
A module may consist of five human errors (Heslinga, 1988): a sequence error, an error of omission, a selection error a handling error and an extraneous activity. Here, the modules only contain an error of omission and a selection error as analysis of the procedure showed that the other error types are not relevant. In addition, a module may contain events expressing the state of the system if they influence the following errors to be made (e.g. module 1, first event) or if they can lead to the possibility to rectify errors made (e.g. module 4, latter event). Due to the check, the repair-man may discover that he has done something wrong after which he may perform (a part of) the procedure again.

Combining the modules leads to the event tree of the

procedure, see Fig. 4. Certain branches may be combined if they lead to the same outcome. Eight different states are possible on following the procedure correctly or incorrectly:
Z1: MAIN in OFF-position with padlock and label and FIELD in STOP-position.
Z2: MAIN in OFF-position with padlock and label and FIELD in RUN-position.
Z3: MAIN in OFF-position with padlock and label and FIELD in RUN- or STOP-position.
Z4: MAIN in OFF-position without padlock or label and FIELD in STOP-position.
Z5: MAIN in ON- or OFF-position without padlock or label and FIELD in STOP-position.
Z6: MAIN in OFF-position without padlock or label and FIELD in RUN-position.
Z7: MAIN in OFF-position without padlock or label and FIELD in RUN- or STOP-position.
Z8: MAIN in ON- or OFF-position without padlock or label and FIELD in RUN- or STOP-position.

These states are placed as outcomes of the THESIS analysis at the offshoots of the tree. The outcome Z1 represent the desired successful outcome, the other Zi's represent a failure outcome. Besides these outcomes, the possibility to rectify an error, expressed as R, is also present. This is however not treated as a separate outcome as this will finally lead to one of the eight states just-mentioned.

It is noted that a wrong motor may be selected at the start of the procedure. The tree remains the same since the selection error of switch implies the selection of a FIELD not corresponding with the motor selected.

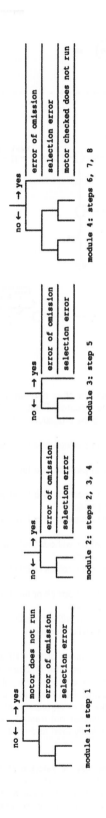

Fig. 3 THESIS modules of the procedure.

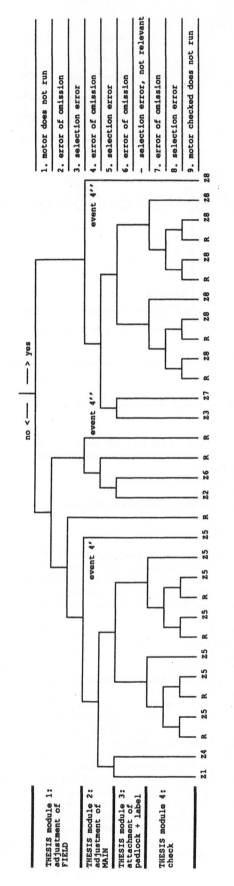

Fig. 4 THESIS event tree of the procedure.

Selecting an incorrect motor only implies that the wrong motor is being maintained, not in a less safe state (apart from somewhat higher probability that the PANEL is adjusted by someone else but this effect is from a lower order).

Human-error probabilities to quantify the tree (Table 1) were adapted from several tables from Swain and Guttmann (1983). Event 5 is a combination of a reading error (information at the FIELD is incorrectly read or forgotten when the repair-man enters the Main Control Centre) and a selection error in the Main Control Centre (additionally to remembering a correct FIELD number). The probability of the event that indicates the state of the motor (event 1 and event 9) was obtained from the organization for which this study was done.

TABLE 1 The probabilities used to quantify the THE-SIS event tree (the probability to follow the NO-direction)

EVENT	DESCRIP-TION	PROB-ABILITY	SOURCE
event 1	motor does not run	0.5	Plant-specific
event 2	error of omission	0.01	Table 20-7 item 3 upper bound
event 3	selection error	0.001	Table 20-12 item 2 lower bound
event 4'	error of omission	0.01	Table 20-7 item 3 upper bound due to dependence effects
event 4"	error of omission	0.003	Table 20-7 item 3
event 5	selection error	0.004	0.001 (Table 20-10 item 2) + 0.003 (Table 20-12 item 2)
event 6	error of omission	0.001	Table 20-7 item 3 lower bound due to dependence effects
event 7	error of omission	0.001	Table 20-7 item 3 lower bound due to dependence effects
event 8	selection error	0.001	Table 20-10 item 2 lower bound
event 9	motor checked does not run	0.5	Plant-specific

Multiplication of the branch probabilities yields the probabilities of the eight outcomes. Dependence is accounted for by adapting the branch probabilities on the basis of failure or success at a preceding event (this

was done for event 4). The effect of the R probability is ignored since this value appears to be too low after calculation). The outcome probabilities are presented in Table 2, first column.

Fault trees

Eight fault trees can be drawn, based on the eight different states just-mentioned. However to save space only one is drawn here from which the others can be deducted (see Fig. 5).

Depending on the state that occurs due to human errors certain parts of the fault tree are not applicable, e.g.:
- if the state Z4 occurs, the FIELD is in the STOP position making basic event G not relevant (technical failure to come accidentally in the RUN-position is neglected because of its relatively low probability),
- if state Z2 occurs the FIELD is in the RUN-position, hence basic event G has probability of 1,
- event X can either be E, H or I depending on the state of the system (neglecting combinations because of their low contribution).

The equations for the TOP events become:

$$Z1 = H(A + BDF)$$
$$Z2 = H(A + B(DF+C))$$
$$Z3 = H(A + B(DF+CG))$$
$$Z4 = I(A + BDF)$$
$$Z5 = E(A + BDF)$$
$$Z6 = I(A + B(DF+C))$$
$$Z7 = I(A + B(DF+CG))$$
$$Z8 = E(A + B(DF+CG))$$

Table 3 contains the data used for quantification of the fault trees. Failure rates were obtained from USNRC (1975). The time intervals and the probabilities were obtained on the basis of plant-specific information. The probability that relevant switches (Q_c, Q_d and Q_i) are adjusted by someone else accidentally was calculated by combining information on how often switches on the relevant panel have to be operated, the number of similar switches and the generic value for the human-error probability of a selection error.

The probabilities of the TOP events are presented in Table 2, second column.

TABLE 2 The probabilities of the outcomes and the probabilities of the TOP events

Outcome probabilities	TOP event probabilities
P(Z1) ≈ 0.49	T(Z1) = 6.5•E-12
P(Z2) ≈ 5.0•E-3	T(Z2) = 9.2•E-11
P(Z3) ≈ 0.50	T(Z3) = 7.1•E-11
P(Z4) ≈ 4.9•E-4	T(Z4) = 1.8•E-8
P(Z5) ≈ 1.1•E-3	T(Z5) = 2.7•E-5
P(Z6) ≈ 5.0•E-6	T(Z6) = 2.6•E-7
P(Z7) ≈ 5.0•E-4	T(Z7) = 2.0•E-7
P(Z8) ≈ 2.5•E-3	T(Z8) = 2.9•E-4

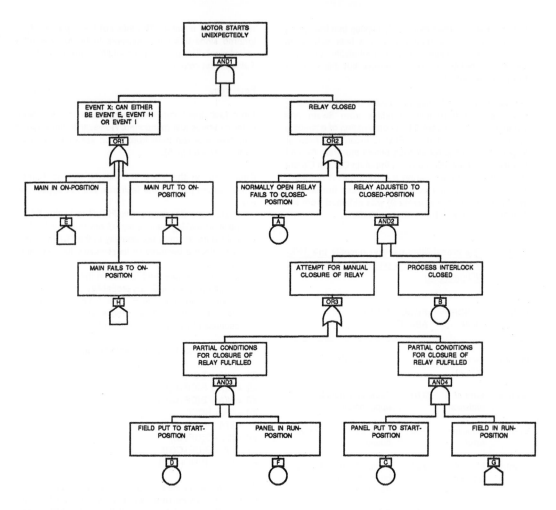

Fig. 5 Fault tree of the electrical lockout system.

It is noted that a number of basic events were neglected when it was directly clear that their probability was relatively low. E.g. the combination that the FIELD fails to the RUN-position and a premature closing of the contact between B-3 and B-4 (Fig. 2) was neglected. This event is similar to the FIELD being put to the START-position. However, this latter probability will be much higher. A similar reasoning accounts for the bypass being failed to the CLOSED-position. The interlock and the PANEL in the RUN-position have the same effect but have a much higher probability.

Combination of the results of both analyses

The probability that the motor starts running unexpectedly is calculated by combining the results of the preceding sections. The probability, indicated by Prob, is:

$$\text{Prob} = \sum_{i=1}^{8} P(Zi) \cdot T(Zi) = 7.5 \cdot E\text{-}7$$

TABLE 3 The probabilities, failure rates and time intervals used to quantify the fault trees

BASIC EVENT	RATE, PROBABILITY	TIME INTERVAL
A normal open relay closes	$\lambda a = E\text{-}8/hr$	$Ta = 36$ hr
B process interlock closed, i.e. no interlock	$Qb = 0.95$	
C PANEL put to START-position	$Qc = 5 \cdot E\text{-}4$	
D FIELD put to START-position	$Qd = 5 \cdot E\text{-}5$	
E MAIN in ON-position	$Qe = 0.75$	
F PANEL in RUN-position	$Qf = 0.75$	
G FIELD in RUN-position	$Qg = 0.75$	
H MAIN fails to ON-position	$\lambda h = E\text{-}8/hr$	$Th = 36$ hr
I MAIN put to ON-position	$Qi = 5 \cdot E\text{-}4$	

DISCUSSION

The paper has presented a possible technique for making a probabilistic failure analysis of a situation in which a system is brought to another state and in which procedural human activities form an essential part. The method can be regarded as an extension of the reliability methods used so far since the outcomes of human errors are explicitly considered instead of only human errors and its reliability aspects. Moreover, the procedure of human activities forms the start of the analysis instead of the technical design as is normally the case in a Probabilistic Safety Assessment.

It is noted that the method applied so far has only considered one event of interest, i.e. the unwanted delivery of electricity to the motor during maintenance. If a number of different TOP events were of interest, the probability of the TOP events have to be combined with their weights to come to a real Probabilistic Safety Assessment of non-steady state situations.

The method was developed for making an evaluation of comparable MMSs. The electrical lockout system evaluated in this paper and the MMS that was presented was one of three possible designs. The two other MMSs were treated similar to the one in this paper. The analyses and results of these two MMSs are not presented here to keep this paper compact. The most safe system, the system with the lowest probability that the service-man may become injured during maintenance was selected on the basis of the results.

A number of problems occurred in applying the technique, in particular in the THESIS part. One of the most important problems is the extensiveness of the event tree if the procedure to be analyzed becomes very long. The experience is that the method in which all the error sequences are being considered can only be applied to relatively short procedures. The complete analysis of the three different MMSs took about two man-weeks; the procedures related to the other two MMSs not presented here were comparable in length to the one treated in this paper. Much longer procedures may be analyzed but the time involved is expected to be much longer. Perhaps single human errors and their outcomes should then be considered. A case-study has shown that this may also lead to satisfactory results under certain conditions (Heslinga and Stassen, 1992).

Besides the problem that is related to the method presented here, a number of general problems occur in applying reliability techniques for the assessment of human activities. The problems can either be of a qualitative type (were all possible events included?) or of a quantitative type (can a person be assigned an error probability like for a technical component?). Doubts can be expressed for both types: an analysis as with THESIS can only predict the foreseeable events; it is also questionable to which extent a person can be assigned a probability as for a technical component.

The idea is however that insight is gained into the MMS problems in a structured way. Not the absolute number for the probability is of importance but more the safety value of a particular MMS as compared to another nearly similar MMS. This relative use of probabilities makes the uncertainty in human-error probabilities somewhat less relevant. Instead of only applying ergonomic guidelines the method helps in understanding the weaknesses of MMSs and helps in selecting the best man-machine design.

This is important in a time in which systems are made more and more safe because of the consequences involved in case of malfunctioning. Due to this we are less able to learn from experience with similar systems in the past. The method should be regarded as a contribution to the application of ergonomics for coming up with the best man-machine system and to try to determine to what extent systems should be automated.

CONCLUSIONS AND STATEMENTS ABOUT FUTURE APPLICATIONS

The investigations made during the development of the method presented here have resulted in a wide variety of conclusions. The most important within the framework presented here are:
(1) Probabilistic failure analyses are a helpful tool in obtaining a better understanding of the man-machine system; their application can highlight the weak spots in procedures and in man-machine interfaces and may help in deciding to what degree a man-machine system should be automated.
(2) The quality of human performance can currently less easily be improved on the basis of feedback of experience from the past, so that the use of predictive probabilistic failure analyses becomes more and more a prerequisite in future.

REFERENCES

CISHC (1977). A guide to hazard and operability studies. Chemical Industry Safety and Health Council, London.

Heslinga, G. (1988). Technique for human-error-sequence identification and signification (Dissertation). Delft University of Technology, Delft.

Heslinga, G., and H.G. Stassen (1992). The prediction of human performance safety with event trees. To appear in IEEE Transactions on Systems, Man and Cybernetics, July 1992

Swain, A.D., and H.E. Guttmann (1983). Handbook of human reliability analysis with emphasis on nuclear power plant applications (Report NUREG/CR-1278). United States Nuclear Regulatory Commission, Washington.

USNRC (1975). Reactor safety study. An assessment of accident risks in U.S. commercial nuclear power plants (Report WASH-1400). United States Nuclear Regulatory Commission, Washington.

USNRC (1981). Fault tree handbook (Report NUREG-0492). United States Nuclear Regulatory Commission, Washington.

MODELLING THE HUMAN FACTOR

J.A. Stoop

Safety Science Group, Faculty of Philosophy and Technical Social Sciences,
Delft University of Technology, Delft, The Netherlands

Abstract.
Over the last 20 years major changes have occured in the Dutch sea fishing industry.
Accident registration shows an increasing involvement of fishing vessels in collisions at
sea, due to a number of factors. There is a need for modelling human behaviour especially
with respect to the decision making and control strategies which are applied by the man
on the bridge.
By the application of human error theory to normative task analysis for the man on the
bridge, insight is gained factors which may contribute to accidents.
Requirements with respect to safety, mental load and human behaviour are formulated on
the conceptual level. It gives a prospective tool to foresee use of a system before it
becomes operational. Demands are put on workstation layout, technical equipment,
procedures and training.
Pilot studies concentrated on the structure of the information presentation and
processing, the workstation layout and the vision lines inside and outside the bridge.
Finally, the contribution discusses the strategies of a social and organizational nature,
required for the approach of the residual risks and side effects involved with the chosen
design solutions.

Paper submitted to the 5th IFAC Conference, The Hague, The Netherlands, 9-11 June 1992.

Keywords.
Automation, cognitive systems error analysis, human factors, prediction, safety, ship
control.

SUMMARY

The objective of the project, presented in this
contribution is twofold. First, the project aims
at the improvement of working conditions on board
of a fishing vessel by reduction of mental under-
and overload of the man on the bridge. Second the
study aims at reduction of the number of
accidents on and with fishing vessels in which
bridge activities are involved. The project
emphasizes on the man-machine interaction as a
dominant accident causation factor. The project
is developed in a step by step procedure, from
analysis of verdicts of the Dutch Admiralty
Court, to several technical solutions for
redesign of the bridge. The foreseeable effect on
safety by influencing the human factor is placed
in a wider context of socio-technical
circumstances in which the solutions are to be
implemented.

1. INTRODUCTION

The North Sea is one of the busiest sea areas of
the world for commodity transport. It also offers
proper conditions for a large variety of fish.
The North Sea proved to be also one of the
prospective areas for the production of oil.
Conflicts of interest are therefore bound to
arise between concentrated fishing activities in
the proximity of shipping lanes and in the
vicinity of offshore production platforms.
Over the last 20 years, major changes have
occurred in the Dutch sea fishing industry by the
changes in ship size, installed engine power,
fishing techniques and the application of microelectronics in the support of the bridge tasks.
After a period of growth and scaling-up the
fishing industry in the Netherlands faces a
period of cost reduction and quality improvement.
Due to the European Community policy fishermen
are limited in their fishing days and have
individual catch quota. The Dutch Arbo law
(working environment law) has acquired the force
of law for the fishing industry in november 1991

and this will have serious consequences on board.
Environmental problems are also likely to
generate guidelines and legislation in the near
future. A favorable climate therefore is growing
for improvements in bridge lay-out and equipment.

2. PROBLEM ANALYSIS

Data supplied by the Shipping Inspectorate show
an increase in the proportion of fishing vessels
involved in collisions on the North Sea during
the last decade. Beside these collisions there
has been an increase in the number of 'near
misses' (Van der Sloot, 1988).

year	1978	1979	1980	1981	1982	1983	1984	1985	1986
tot	22	26	28	20	15	30	32	39	29
fv	9	9	10	10	9	12	22	23	20

Accident statistics show that fishing vessels are
involved in collisions 3 times as often as might
be expected from their numbers when compared with
other types of vessels. Therefore the fishing
industry was selected as the target group for
further research.
Analysis of the 59 Verdicts of the Dutch
Admiralty Court between 1983 and 1988 concerning
fishing vessels, revealed 6 specific accident
types which were performed under 4 different use
conditions (Stoop, 1990). In almost every
accident bridge activities were involved.
Therefore, the bridge was selected as the most
important workstation on board to be examined.
Although related to bridge activities,
occupational accidents were excluded from the
project and dealt with in a separate project
(Hoefnagels et al, 1988). Since beam trawlers
proved to be involved almost exclusively, the
attention was restricted to this type of vessel
and fishing technique.

The work load on the bridge has been changed
drasticly by the concentration of tasks on a one-

243

manned bridge as there are navigation, communication, position finding, engine monitoring, energy supply and control of the fishing process itself. The tasks are performed continuous in a two hours repeated work cycle during four a five days a week. A number of accident factors were typical for these fishing vessels, such as poor procedures for watch keeping, unofficial collision avoiding manoeuvres and poor VHF radio discipline. Frequently mentioned accident causation factors were overfatigue, wrong interpretation of the radar and poor bridge layout.

Further analysis indicated common background factors in carrying out bridge tasks in three respects.

1. dense traffic, poor visibility conditions and preoccupation with the fishing task are aggravating circumstances in normal conditions and may cause work overload and human error in task performance
2. routine navigation tasks under normal conditions may cause a work underload in which human error may cause serious problems
3. the involvement of inexperienced and unqualified personnel in navigation tasks generates a specific problem

For these reasons mental work load and human error were selected as dominant accident causation factors and analyzed further in-depth.

3. RETROSPECT AND PROSPECT

Accidents can be defined as 'undesired deviations' from the 'normal' process. Although deviations occur in practice, they may have their origin in design assumptions concerning the 'normal' use, in technical limitations in the design of equipment and workstation layout, in work procedures and regulations, in training or education. Shortcomings in design, mismatch between human capabilities and required skills by inadequate training, ambiguous legislative regulations may put impossible and contradictive demands on the 'proper' or 'normal' performance of a task. Therefore, not only the 'undesired deviation' should be examined, but the 'normal' task itself as well.

In such a reflection on tasks and deviations lessons can be learned from the past. Traditionally, in aviation, railroad and marine shipping, in-depth investigation of accidents has proven itself as a valuable source of insight in accident causation mechanisms. However, this retrospective approach is not enough if the future use has to be foreseen of systems which are not operational yet. Prospective approaches have to be developed to facilitate a reliable prediction of future use.

3.1. A retrospective approach

In the existing situation two approaches were possible.

First, study can be performed into the actual use of equipment on the bridge and the mental work load which occurs in the performance of the tasks during the several phases of the fishing process. During several trips the mental load was estimated by field observation of the use-frequency of the equipment by the man on the bridge (Van der Sluis, 1991).

These studies revealed a clear difference in attention for the various instruments between the several subtasks on the bridge discriminating between steaming, fishing and hauling of the nets as well as a difference in mental load during these phases. During steaming free at open sea under favorable conditions a mental underload was possible, while during fishing and hauling an overload could occur under aggravating circumstances as poor vision, heavy seas and dense traffic (see Fig. 1).

Second, study can be performed into the possibility of human error during the performance of a task. To picture the activities on the bridge a normative task description was established, which shows the main and subtasks in a time sequence and distinguishes the different phases of the journey. This description enables us to evaluate the tasks and activities and their possible interference and to establish the eventual workload. There are two principles to relate tasks to errors.

First we can learn by experience by applying human error theories to the analysis of accidents that have already occurred. Second we can apply human error to try to predict 'foreseeable' use and error, before accidents have occurred. Such

theories should therefore have both an explanatory and a predictive potential.

Models from cognitive psychology as developed by Reason and Rasmussen, better than models of risk taking behaviour, proved to be able to link normal behaviour, human error and risk inducing factors which may lead to accidents (Rasmussen, 1980; Reason, 1987). These theories were successfully applied on accidents in road traffic and gave a satisfactory explanation (Hale et al, 1988). To verify the retrospective potential, the generic Error Modelling System (GEMS), as developed by Reason, was applied to the normative bridge task as a predictor to human error. Verification of the results was achieved by a comparison with accidents registered by the Dutch Admiralty Court. With the GEMS model 26 (72%) of 36 collisions were explainable, which is high, since not all accidents were caused by human error as the dominant accident causation (Heinrich, 1988). Sometimes the skipper had fallen asleep, the watchman had left the bridge or was occupied with side activities as preparing meals. These causes are not included in the notion 'error'.

3.2. A prospective approach

Verification of the prospective potential of human error models however proved to be difficult. These models could be valuable if they enable the designer to asses his design critically before it is implemented. Problems of methodological nature occurred, as well as problems of a more practical nature (Hale and Hommels, 1989). First experiences with models like GEMS shows that, even with a course level of task analysis, they can be used to generate a large number of potential errors (Heinrich, 1988). While analysis of these potential errors can lead to useful design recommendations, the danger is that there are relatively so many that the designer gets swamped with information and does not know where to begin with modification of his design. What is currently missing from the theories is systematic information about human recovery; which types of error are most (and least) likely to be spotted and recovered by the man on the bridge or compensated for, so that they do not lead to accidents. This would allow designers to do the necessary sensitivity analysis to see which features most need modification. It would also help them in making choices where their proposed design change is likely to reduce the probability of one potential type of error, but increase that of another. Psychologists should supply designers with rules which give an adequate description of the behaviour of users and the specific conditions under which these rules are applied (Hale et al, 1990).

4. SOLUTIONS

In this project only solutions of a technical nature are elaborated. Solutions of a social, legal or organizational nature are considered additional to technical solutions and have to be developed when technical improvements are expired.

The technical solutions are developed in a modular form; several packages of technical solutions are present, ranked to their increasing complexity, effect and required resources. Each package restricts itself to one sub-problem area, thus facilitating the skipper-owner to select the solution which fits the most to his needs, possibilities and cost-benefit considerations. In such a step by step approach eventually conceptual changes are possible in the vessel layout and equipment.

In the redesign of the bridge, three pilot studies were done on respectively the level of an isolated instrument; the low cost ARPA, the level of instrument integration; the fishermans' user interface and the level of layout of the bridge; the beamer bridge 2000.

4.1. The low cost ARPA

It is possible to reduce the probability of collisions by the installation of a device, linked to the radar, which sets off an alarm when another ship is coming too close. The ARPA being such an instrument is already compulsory on ships of more than 10.000 tons, but is expensive. The fishing industry is an outstanding target group for extension of the use of this sort of instrument, but it will need to be cheaper if it is to be installed on the fleet without the compulsory intervention of the Shipping Inspectorate. It has proved possible to develop such a device, although the dimensioning of the filter parameters is of utmost importance for the prevention of false tracks by clutter at a short distance from the user ship (Van der Sloot, 1988).

If such an instrument were to be introduced we would need to reckon with a number of residual risks and side effects (Van der Sloot 1988). The residual risk is determined to a great extent by the reliability of the collision alarm. To be credible the radar and collision detector must function highly reliably, otherwise vessels will be missed or false alarms will erode the confidence in the alarm.
It is an open question whether man or machine should perform the correct steering manoeuvre to avoid collision once the alarm has been given. This dilemma is comparable with the use of 'stick pushers' in aviation.
The device also offers no comfort if the bridge is unmanned. It could also lead to overreliance in the equipment. Among younger generations of mates 'blind' navigation by radar is more the rule than the exception. Consideration should be given to trying to compel the mate to make a periodic check of the information presented by the equipment, by looking at the world outside. The device does not necessarily prevent collisions. In merchant vessel traffic the 'ARPA assisted collision' is already known as a phenomenon.
Finally a new accident scenario is introduced for fishing vessels. It would be possible to install a navigation alarm on the deck, triggered by the collision alarm. The bridge might then remain unmanned until a warning was given. If, in addition the steering manoeuvres were taken over by an autopilot, dangerous situations could definitely be conceived of. Such a situation is plausible during intense deck activities or a period of rest after a heavy trawl. Blind faith in the instrument, supplemented by fatigue, could cause a low level of attention. An emergency situation might then cause panic and incorrect intervention, if it was not clear which event had led to an alarm.

4.2. A fishermans' user interface

Modern electronics may add a wide range of extra facilities to equipment. This results in an increased information stream and makes operation of the equipment more complex. Interfacing also creates an overlap of information. The outcome of research in the merchant marine was that the presentation and operation of equipment could be improved. Screen presentation joining several data shows only relevant information to the watchman. More detailed information can be called up or will be generated by an alarm system. A user interface designed for the fisherman integrates radar, plotter, engine and sounder data on a screen and offers methods to manipulate this data (De Vries, 1990).

Because of the conventional bridge layout is not altered, two units are needed: a main unit at the work station for fishing and steaming and a slave unit at the workstation for gear handling. The 19 inch main screen contains a combined videoplotter and radar presentation field, and two smaller fields on the right-hand side. An alarm field is at the bottom of the screen. The plotter-radar field is continuously present and offers information on position, course, speed and identification of objects. The smaller pages may present more detailed information from the radar, engine, fishing gear or echosounder. According to scenarios of stress and boredom of the watchman, a hierarchic information and alarm system will operate. The equipment can be tuned to the actual vessel conditions with respect to:
- good or poor visibility
- quiet or busy traffic situations
- skilled or unskilled watchman
- fishing or steaming mode.
(see Fig. 2)
The slave unit uses the same data, but displays only the most important part. The slave unit is used only during gear handling for which engine and winch data are given extra emphasis. Navigation tasks can be performed simultaneously and in case of malfunctioning, the slave unit can take over. Eventually, it was found out by expert consultancy among fishermen, ergonomists and electronic manufacturing experts that

presentation of data divided over two screens instead of one was preferable. The presentation will be clearer, especially for the plotter-radar field. As a conclusion for this study a setup for two screens at a single workstation was made: one for navigation data and one for engine and fishhold data, etcetera. The realization of the system requires only a software update to modernize the entire system and may guarantee an up-to-date presentation in the future.

4.3. Bridge layout

In the development of a new bridge some constraints were formulated for the new design. The redesign of the bridge should anticipate on the trend to the bridge as the operational centre of the vessel and therefore fit in with the project 'Beamer 2000', an innovative fishing vessel design for the year 2000.
The design should be modular with respect to the integration of the information and the bridge layout so, eventually in a step by step approach, it may lead to a new bridge concept. Design requirements which were to be fulfilled were with respect to hardware, software and to workload.

The first step concentrates on the redesign of the navigation workstation of the skipper (Buijs and Van der Sluis, 1991). This design proposal contains state-of-the-art equipment and is characterized by the lack of duplicate instruments. The console is redesigned, accordingly to experiences in the process industry and merchant marine. It replaces a straight port and starboard console alongside the skipper in the middle of the bridge by a curved console with standardized panels for fishing, navigation, videoplotter and communication. Infrequently used instruments are placed in a ceiling panel. The console is placed slightly to the starboard side, in view of anti-collision regulations. The arrangement of the equipment over the panels is based on the analysis of the watchmans' task (according to the Woodson method). A separate winch control panel at the front still exists.
A full scale model was constructed and supplied with navigation equipment and demonstrated on an inland shipping and fishing trade exhibition. Simulation tests with fishermen are planned for the near future.

The second step contains the implementation of the integrated users' display in the redesigned console. The introduction depends on the previously mentioned development of the integrated screen.

The third step is the redesign of the layout of the bridge and contains a complete review of the wheelhouse construction. The redesign of the bridge is a part of the BEAMER 2000 project which envisages a complete redesign of the fishing vessel on arguments of safety, working conditions and environmental aspects. In the Beamer 2000 project not only the bridge as such is considered, but also the interaction with other working stations on board is taken into account (Stoop and Veenstra, 1992).
The results of a pilot study show considerable differences between a BRIDGE 2000 and a conventional bridge (Van der Sluis, 1991):
- the bridge is raised one meter for improved vision
- glass windows set back at front for vision on decks
- the navigation equipment is mounted close to the windows
- on port and starboard set back of bridge front for vision on side decks
- split console on front with two workstations containing 19-inch screens, presenting integrated navigation data
- new administration workstation next to the chart table
- navigation data presented in the skippers cabin
- reduction of winch control handles by combined functions. (see Fig. 3)

This bridge is only feasible in the case of a new build vessel. The electronic equipment is not yet dedicated for such a design. The bridge is a pilot study and should be tested at full scale simulations. Momentarily only a 1 to 10 model is available for evaluation.

5. CONCLUSIONS AND CLOSING REMARKS

5.1   Conclusions

Redesign of work stations can make use of accidents as an entrance to the problem. However, further problem development is required, especially with respect to the analysis of human behaviour and human error. Retrospective methods enable us to draw up design requirements which lead to technical solutions. These technical solutions must be accompanied by additional safety measures of non-technical nature since otherwise the management of residual risks and side effects encountered is not possible. Prospective methods with respect to human error seem to have good perspectives but are in their first stages of development. Redesign should not be restricted to the bridge deck alone. There are relations to the working deck, to the processing and storage of the catch and to future trends in automation and registration. An integral approach of safety is only possible when the vessel as a whole is considered, as done in the BEAMER 2000 approach.

5.2.   Closing remarks

This contribution started with an overview of the data of the Shipping Inspectorate over the years 1978 until 1986.
If we follow the figures of the Verdicts of the Dutch Admiralty Court, based on these data over the years 1984 until 1990 there is a remarkable change.

year	verdicts	fishing vessels	of which collisions
1984	39	14	10
1985	34	19	16
1986	39	16	13
1987	18	3	2
1988	31	7	2
1989	24	4	2
1990	22	5	3

This decrease in collisions with fishing vessels with a magnitude of about 5 is not the result of the redesign of the bridge, nor the equipment. These redesigns still have to be implemented in the near future. The decrease is the result of the introduction in 1987 of the Sea Days Directive by the Dutch Government; a limitation of the days fishermen are allowed to be at sea. By this directive it is no longer profitable to go out during aggravating conditions and as a result, peak workloads are diminished and risk taking behaviour is no longer profitable. The number of accidents is consequently lower. Although efforts on redesign may be promising, there is always a higher order in the system which has opportunities to influence the safety level far more drastically than any lower order.

REFERENCES

A.M. Buijs and A.M. van der Sluis, 1991
Bridge layout beamer 2000. International Council for the exploration of the sea. C.M. 1991/B:20.

A.R. Hale, B.W. Quist and J. Stoop, 1988
Errors in routine driving tasks: a model and proposed analysis techniques. Ergonomics, 1988, vol.31, no. 4, 631 - 641.

A.R. Hale and J. Hommels, 1989
Modellen voor het voorspellen en beheersen van verkeersgedrag. Rapport, Technische Universiteit Delft.

A.R. Hale, J. Stoop and J. Hommels, 1990
Human error as predictors of accident scenarios for designers of road transport systems. Ergonomics, 1990, vol.33, nos 10/11, 1377 - 1387.

J.P. Heinrich 1988
Ergonomisch brugontwerp: De methodologie om te komen tot het Pakket van Eisen. Rijksinstituut voor Visserijonderzoek. Rapport TO-88-11.

W. Hoefnagels, J. Stoop, K. Bouwman and F. Veenstra, 1989
Veiligheid in de zeevisserij. Rapport, Technische Universiteit Delft, fase 1: Informatie en analyse, september 1989, Fase 2: Synthese en evaluatie, juni 1990.

J. Rasmussen, 1980
What can be learned from human error reports. In: K. Duncan, M. Gruneberg and D. Wallis (Eds), Changes in working life. London: Wiley.

J.T. Reason, 1987
Generic Error Modelling System (GEMS): a Cognitive Framework for Locating Common Human Error Forms. New Technology and Human Error. Edited by J. Rasmussen, K. Duncan and J. Leplat. John Wiley & Sons Ltd.

B.J.R. van der Sloot, 1988
Verkenningen naar een Low-cost aanvaringsalarm. Technische Universiteit Delft, afstudeerverslag faculteit Elektrotechniek.

A.M. van der Sluis, 1991
Bruginrichting Kotter 2000. Rapport, Rijksinstituut voor Visserijonderzoek/ Technische Universiteit Delft.

J. Stoop, 1990
Safety and the Design Process. Doctoral Thesis, Delft University of Technology, April 1990.

J. Stoop and F.A. Veenstra, 1992
Kotter 2000, veiligheids geïntegreerd (her)ontwerpen: methode KINDUNOS. Rapport, Technische Universiteit Delft/Rijksinstituut voor visserijonderzoek RIVO-DLO, in press.

B. de Vries, 1990
Een unit voor de visser. Rapport, Rijksinstituut voor visserijonderzoek/ Technische Universiteit Delft.

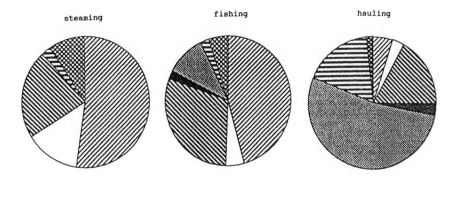

steaming fishing hauling

☑ navigation
☐ communication
◪ manoeuvring
▥ propulsion
▨ gear handling
⊟ monitoring
▨ support systems

Fig. 1 Bridge activity.

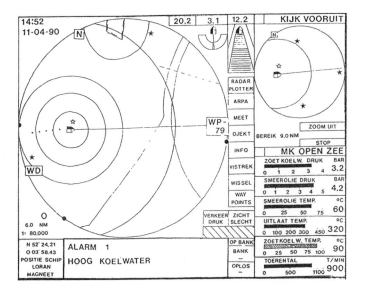

Fig. 2 Navigation Unit.

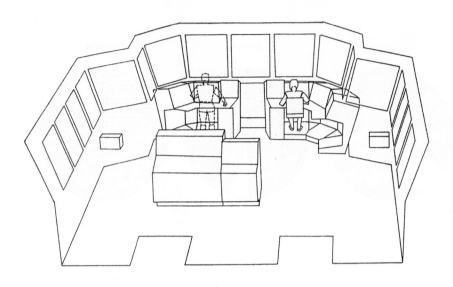

Fig. 3 Beamer Bridge 2000 in perspective.

THE DESIGN OF MULTIMEDIA INTERFACES FOR PROCESS CONTROL

J.L. Alty and M. Bergan

Computer Studies Department, Loughborough University of Technology, Loughborough, Leicestershire, LE11 3TU, UK

Abstract. The new problems which are likely to arise when designers begin to use multimedia interfaces in process control interface design are discussed. It is pointed out that there is little current understanding of how to choose which media for which interface task. It is suggested that a characterisation of operator knowledge requirements (derived from task analyses) could be compared with media characteristics to generate media possibilities. These could then be modified using psychological principles. Some examples of the approach are given.

Keywords. Process control: Multimedia; Interface design; Task analysis; Human factors.

THE NEED FOR A NEW APPROACH TO INTERFACE DESIGN IN PROCESS CONTROL

At the present time the role of the operator in process control is undergoing a major shift in emphasis. Advances in control and information technology are shifting the operator from being the key element in the control loop to a new role as a plant supervisor and trouble shooter. Many routine control functions, for example, have recently been taken over by direct digital control systems and the operator's key goal is now maximising Meantime Between Upsets rather than maximising the Meantime Between Component Failure. In their earlier role, the operators required a step by step view of the process. Now, in addition, they need a higher level, overall view. The interface designer thus needs to provide more intelligent advisory systems (for example GRADIENT, Alty and Johannsen 1987, Alty 1989) and higher levels of interface abstraction which are easier to use (Alty et al 1985). The former requirement is expected to be met through the implementation of knowledge-based technology and the latter will need to utilise new presentation technologies of which the multimedia approach must be a serious candidate for consideration.

Single-media interfaces have already begun to show serious shortcomings in this new environment both with respect to attention capture and with effective information display. It is hoped that multimedia user interfaces will enable operators to reduce their cognitive overload by spreading information processing across different modal channels and by using information renderings most appropriate to their information communication needs. At present this statement is effectively of an act of faith, since we have very little information on what the real value of a multimedia approach might be. However, it is probably a reasonable assumption that an interface using multiple media, in the right circumstances, will be better than a single media.

Human beings when given the option always adopt a multimedia approach presumably because such an approach offers some clear advantages.

The enabling technologies for multimedia operation such as Bit-mapped Displays, Device Independent Graphics, Audio facilities, Data Compression, Control and Synchronisation, Networking, and Authoring Tools are now, effectively in place. A number of Multimedia toolkits, therefore, are now entering the marketplace. For example, Apple have introduced QuickTime on the MacIntosh and there have been recent multimedia extensions to Windows. SUN has also announced multimedia facilities on SPARC workstations.

Given that we can now deliver a variety of media simultaneously such as graphics, pictures, (whether moving or still), text, voice output, sounds, diagrams, etc., the important issue is when to use which medium (singly or in combination) in order to achieve some operator interface goal and to achieve this goal in an "improved" way. This is the real multimedia challenge for designers in process control and it is the contention of this paper that we do not yet know quite how to approach this challenge. Without real understanding, we run the risk of providing answers to questions such as: "what can we build using a multimedia approach ?" rather than "what sort of multimedia approach will really benefit the operator function in process control ?".

SO WHAT'S DIFFERENT ABOUT MULTIMEDIA ?

Some may point out that our approach to interface design has always been somewhat arbitrary. We have often tried out the technology without understanding its communicative power, and have somehow learned from that experience. So is the multimedia situation any different ? Why do we not just try out many interfaces and learn from experience ?

Unfortunately, the multimedia situation is different and here are the reasons why.

In the early days of computing, we only had scrolled text as an output medium. Although we did not fully understand how to use the medium effectively we had, at least, a thorough understanding of text and had used it for communication for a few hundred years. When graphics became a possibility we still had a solid experience behind us of graphical communication. Colour was the first medium extension which caused problems. Many human beings do not understand the use of colour for communication. Most of us have difficulty in matching clothes or wallpaper. Only a few human beings are skilled enough to use colour effectively (we value them as artists). Therefore, the early colour interfaces were disasters with colour being used because it was there not because it was needed.

Yet, even at that time, there existed a considerable body of knowledge on how to use colour effectively. The problem was that very few computer scientists or engineers bothered to consult it. With an arrogance which seems to pervade the subject of interface design, the designers thought they knew best. It is only recently that we have rediscovered the fact that the secret of using colour is to use it sparingly. Psychologists and Artists knew this truth when colour was initially used in interfaces, but hardly anyone asked their opinion.

Now we are moving into the multimedia age. If we knew little about colour we know nothing about the use of multimedia in communication. The production of films and video sequences is a highly skilled activity. Not only does the average human being have very high standards by which to judge such media, in addition there are highly developed mechanisms for achieving communication. The quality of most home videos illustrates the point very well.

It may be appropriate for multimedia interface designers to seek the help of the professionals in these areas. This is a point made by Oren (1990) who stated:

"Building a new medium is a talent-limited process because we are early in the evolution of genre, searching for seminal visions and rallying points. As with Edison and film, it is unlikely that the inventors of the technology will be the creators of such visions. This is more likely to fall to the authors, educators, and artists who have something to say and who find that the new medium offers them a way to say it for the first time"

Some equivalence of literacy is required in every medium and most computer people are not well-skilled in film or video presentation language. Take the practice of visual editing and montage. There is a well-known language, for example, a cut between shots indicates a new perspective, a fade indicates passing of time and a split screen highlights an act of comparison. Understanding such facts is like understanding the alphabetic code in reading. Developing material in a multimedia situation without such knowledge is like trying to read before the alphabet is fully understood (Alty 1991).

Although Oren's remarks were primarily aimed at the educational area, they almost certainly apply (perhaps in a different sense) to the process control area.

A MULTIMEDIA INTERFACE - WHAT DOES THIS MEAN ?

The word "Multimedia" has now become commonplace in the language of interface design. Everyone knows what it means - or do they ? The more the subject is examined the more the terminology appears ambiguous using words such as channel, medium, mode, multimodal, hypertext, hypermedia and so on. But what exactly is a "medium" ? We know that a medium carries a message, we talk of the "film medium", the "television medium" and the "medium of the press". In physics, a medium supports the transmission of energy without being physically changed itself, in other words a medium is a carrier. Although the medium is not changed during message transmission, the nature of the carrier does affect what can be transmitted and at what speed. This is also true of interface media.

To transmit a message to someone else you have to create a disturbance of some sort which can be detected by the other person. With electrical transmission, the disturbance can be transformed during the transmission process. Thus a physical disturbance which moves a key on a keyboard can result in a rearrangement of pixels on a display screen many miles away (i.e. a movement becomes a light intensity rearrangement). What really defines a medium of communication is the combination of syntax, semantics and pragmatics of the language used to communicate. These, in turn, are constrained by the creation process, the transmission process and the sensing process although in human-computer communication the creation process is often driven by a computer program which puts relatively few constraints on the syntax (however available computer power can seriously constrain the language syntax possibilities e.g. animation).

At present we do not really understand the capabilities and drawbacks associated with the different languages of communication offered by multimedia workstations. Questions such as "in this instance should we be communicating by sound or vision ?" or "would a graph or a table be better at this juncture ?" cannot be answered satisfactorily. One problem with such questions is that they cannot be answered in abstraction. There is no well defined way of deciding which media should be used for communicating particular types of information.

Indeed, it has already been established that the effectiveness of communication depends upon the nature of the information being sought be the reader rather than the efficiency of different media for transmitting information (DeSanctis, 1984). It therefore becomes important to know why the operator is seeking particular information. For example, if it is important for an operator to detect a maximum or minimum in a set of data, then a graphical display must be better than a table. If browsing of data is important for the operator then an audio output will not be very helpful. The operator's goals thus dictate requirements which the media designer must respect.

The problem thus has to be approached from two aspects (Fig. 1.) - what do the operators want the information for, and how do the constraints of a particular medium enhance or diminish the capability of the operators to succeed in their goal ?

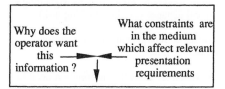

Fig. 1. Competing Requirements

Although quite a number of papers have been produced which discuss multimedia implementations, the reasons for choosing such representations are almost never discussed nor evaluated. For example, Tanaka el al (1988) discuss a learning environment for maintenance with pedagogical interfaces for enhancing the performance of operators carrying out the maintenance of switchgear. The system uses a videodisc and colour graphics as well as more conventional media. They say "Reality in learning is very important because impressions of operators are much influenced by how the results are shown". Presumably this is why they chose to provide real pictures of maintenance parts, however, there is no real justification of why such a medium was chosen, or any evaluation of the success of the system. This comment is not meant to be particularly critical of this paper which presents an interesting approach for the development of a learning environment. It could

be applied to most papers which involve the implementation of multimedia interfaces.

WHAT SHOULD WE BE EXPLOITING IN THE MULTIMEDIA APPROACH ?

A key question is, therefore, when to use which media and in what combination to achieve the maximum effect. Although there are no definite answers to the above question, there are some pointers from research. It is well known that some media transmit certain kinds of information better than others. Deathersage, (1972), has given guidelines on when to use audio or visual display techniques.

Deathersage (1972)	
Use Auditory	Use Visual
simple message	complex message
short message	long message
later reference	no later reference
events in time	locations in space
immediate action	later action
visual overloaded	audio overloaded
location too bright	location too noisy
operator movement	no movement

Edwards (1988) has developed auditory interfaces for disabled users. His word processor contains the audio equivalents of menus, windows, scroll bars etc.. and can be operated by a mouse utilising auditory feedback. He points out that short term memory is much more important in auditory interaction than in visual interaction.

It is usual to think that the visual always dominates over the auditory channel, but this is not always the case. Prompted by work which indicated that weather forecasts were recalled better from the radio than from television, Wagenaar et al (1984) combined pictures and words in a set of bi-sensory presentations. Although recall was better than with individual channels, it was not as good as might have been expected from a combination of the channels. We really need information on when visual and auditory channels cohere or interfere.

Marmollin has summarised when different renderings might be appropriate (shown in Fig. 2).
(see end of paper)
Fig. 2. Possible Use of Different Media
(Marmollin 1991)

When a problem is new to a user, a representation which allows exploration seems to be best. On the other hand when the user has a lot of experience the textual form seems to be better. Between the two, visualisation is preferred. Such visualisation appears to be an important tool in the design process.

SPECIAL REQUIREMENTS IN PROCESS CONTROL

Most of the current literature about multimedia approaches originated in the educational sector. This educational multimedia approach involves the use of videocomputers, interactive television, and electronic books to provide a richer educational environment. Such educational applications tend to be activated by the learner alone where the user of the multimedia system is in complete control. Thus media channel conflicts (such as two voice outputs required simultaneously) do not normally arise.

This situation is different in Process control where one might suddenly have simultaneous requests for the use of the same media resources by different tasks of equal importance. In this situation there are serious multimedia resource constraints and resource scheduling is of key importance. The application of multimedia techniques to process control is therefore new and unproven.

The PROMISE project (Alty and McCartney 1991) is attempting to address these issues. The project (PRocess Operator's Multimedia Intelligent Support Environment - ESPRIT project no. 2397) is concerned with harnessing the two key technologies of IKBS and multimedia to improving process control interfaces (the multimedia interfaces delivering advisory information from a set of knowledge-based systems). However the emphasis in PROMISE is on multimedia development rather than the IKBS development.

The PROMISE system addresses the resource contention issues in the following manner. The actual presentation of an object depends upon a number of factors :

- designer preferences. Each object is actually an AND/OR tree containing alternative representations (with weightings)
- user preferences. The user (or operator) may have preferences, e.g. permanent (such as colour blindness) or temporary (such as noise in the control room)
- media resource availability. At different times different media may be overloaded.
- HCI rules. Cognitive rules which may affect what is desirable or sensible in the situation.

A resource manager resolves these conflicts and presents an appropriate object. In cases of high overload such resource management is essential. When there are excess resources available, the best choice of presentation can be made.

The PROMISE multimedia system is now installed in two industrial sites and experiments are in progress to determine guidelines for multimedia design. Some of the comments below are derived from our early experience.

HOW CAN WE CHARACTERISE DIFFERENT INFORMATION COMMUNICATION REQUIREMENTS AND MEDIA ?

Fig. 1. outlined a possible approach to developing a multimedia design methodology. We now try to develop this approach further.

In process control operators may require a number of important information communication features to achieve their goals, for example:

- **Information spread over a number of dimensions:** A graph, for example, requires two dimensions for its presentation. Usually these are achieved using X and Y coordinates visually, but we might use a sound pitch as one variable and time as another. Alternatively we might use one visual dimension and sound pitch as the other. Tables and Pictures also require two dimensions.

- **Ordering or sequence:** A table can differ from a graph in that order might not be important. Order can be on all dimensions or a subset. A Picture or Diagram usually has strict ordering on both dimensions.

- **Gross features:** there may be crucial gross features such as maxima, minima, spikes, step increases or decreases etc..

- **Relationships:** There may be a strict relationship between dimensions (as in a reproduction of a painting). However a graph or diagram might not lose its information content if it were stretched in one dimension. This might be thought of as a real-world connection

- **Dynamic:** The rendering of information may need to change in some defined manner such as displaying a fluid flowing along a pipe. Two dimensional vision has advantages here, though sound (often when used with vision) can be used quite effectively for signalling a critical distance (say in a docking task).

- **Static:** no relationships change during rendering time

- **Persistence:** Does the information need to exist for some time so that it can be examined at some future time. Graphs and Diagrams are persistent. Voice output is not. A continuous klaxon is persistent though it might not be desirable.

- **Instance:** We may only need to know the current instant value of something (this is the opposite of persistence). Sound is satisfactory for this property.

- **Information Content:** this could be measured in bits. How much information needs to be communicated ?

- **Urgency:** How long has the operator to understand ?

- **Context Change:** Does a context change need to be signalled ? Traditionally sound has had a powerful role as a context changer.

There are, of course, many more. Hovy and Arens in a draft paper (1990) have examined the need for such a vocabulary and have suggested measures such as Carrier Dimension (dimensions required to exhibit the medium), Internal Semantic Dimension (dimensionality of information presented), Temporal endurance (persistence), Granularity (continuous, discrete) and Urgency.

We need to map these information requirements onto appropriate media. Thus we need another list of media characteristics related to their capability of supporting such requirements, for example

- ICON - is 0-D, has no ordering, limited relationships, static, is persistent and has limited information content (an ICON could also be dynamic).

- BEEP - is 1-D, no ordering, could be related to real sound, static, could be persistent and has very limited information content.

- PICTURE - is 2-D, strict ordering, probably strongly related to real world, static, persistent, high information content

- VOICE OUTPUT - is 1-D, strict ordering, strongly related to real world, dynamic, non-persistent, medium information content.

People will argue with some of these but the list illustrates the general idea.

THE APPROACH TO MULTIMEDIA INTERFACE DESIGN

The approach, we suggest, should follow that in Figure 3 *(see end of paper)*

Fig. 3. A Suggested Approach towards a Methodology

A Task Analysis (Diaper 1989, Payne and Green 1989) will be carried out on the operator tasks to establish the information needs. These needs will then be characterised in terms of a defined information processing need using a characterisation scheme based upon the list outlined earlier. The capabilities which the various media provide are then matched to these needs to obtain a set of multimedia possibilities. Finally, an envelope of defined interfaces can be derived by checking the possible interfaces against our knowledge of the human cognition system.

Use of a particular medium will depend, of course, upon operator goals. The goals of the operator could be Interpretation, Problem Comprehension, Task Performance, Decision Quality, Speed of Comprehension, Decision Speed, Recognition and Recall, or Viewer Preference (DeSanctis 1984).

TWO EXAMPLES - COMPARISON AND PATTERN MATCHING

In process control, tasks are usually quite complex and it is difficult in a short paper such as this one to give a realistic but simple example of using the approach. However, here is an example concerned with comparing two nearly equal written maintenance procedures to find out the differences. Further, assume that the differences are small but significant. This is a difficult job for a human being. It is slow, tedious and very error prone. The task is simple enough from a knowledge point of view. The text is compared sentence by sentence, until a difference is observed. Then it is noted and the task proceeds. A complication might be that paragraphs are in a different order or missing. The task could be done by comparing each word in sequence. However this would be exceedingly slow, so most operators try to do the task one sentence (or part sentence) at a time.

The information is 1-D with strict ordering unless a global view of paragraphs is required in which case it is 2-D. It is static and we only need a medium sized instance to work with. A simple visual comparison approach would seem appropriate. Why do human operators find it so difficult ? The problem arises

from a human cognitive characteristic - that of short term memory. When the operator tries to examine the first document he or she has to commit this to memory and switch gaze to the second document. At this point the second sentence has to be read in and compared. Usually short-term memory is saturated and major errors occur.

We can get round the problem by using a multimedia approach. The 1-D ordering suggests a combined visual and auditory approach. Visual and auditory channels can complement each other in certain tasks so we avoid the movement of gaze from one document to the other by supplying the content of the second document in the auditory channel. In other words, the system reads over one document whilst the operator follows the other on the screen. At a stroke we speed up the task by a high factor and simultaneously reduce the error rate.

To deal with paragraph sequencing problems (a 2-D problem) we need a visual approach for the second document only (to be used when there appears to be a paragraph sequence problem. The instant nature of the sound channel does not handle the persistent nature of this task.

One of the problems at the present time is not having enough psychological information to go on. Our approach must therefore be part analytical and part prescriptive. What we need are more general principles for multimedia design. It may be that the example above can be generalised to a detailed comparison of two structured objects (i.e. the comparison of two identical plant subsystems).

The second example of an important multimedia requirement which we derived from the operators in PROMISE relates to the importance of pattern matching. In one of our exemplars the operators have to take regular tests of the salt content of a solution. This is done by taking a sample, adding a chemical which deposits the salt. This is then centrifuged in a narrow tube and the length of the column gives a salt reading. The operators were most unhappy that only the reading would be carried over to future shifts.

It turned out that many other factors influenced the operators when interpreting the test, e.g. the colour, the texture etc.. In other words there was a 2-D, strict ordered, relationship with the real world which can only be satisfied with a full colour picture of the test output.

This strict real-world relationship is quite common, and many computer implementations have ignored it. Indeed the advance of automation in control rooms has progressively detached the operator from the physical world of the plant. For example, when an operator shuts a valve, it would be very helpful to have the actual sound of the closing of the valve relayed to the control console as well. In the PROMISE system we intend to have a microphone icon which can be dragged over the process control diagram so that operators can listen to particular sets of equipment when control actions are initiated.

CONCLUSIONS

The approach discussed in this paper describes the first steps in setting down a methodology for multimedia interface design. What is needed is a much better characterisation of the interface information requirements, a better characterisation of what different media can deliver, and an increased understanding of the psychological principles underlying the human cognition system. We feel that our approach offers a way forward and this will progressively be refined as more experimentation is carried out.

ACKNOWLEDGEMENT

This work was carried out with support under a grant from the European Commission ESPRIT programme in project P2397 PROMISE which we are pleased to acknowledge. I would also like to acknowledge the work of Terry Mayes, (Mayes 1989) who produced an interesting internal PROMISE report on Critical M4I Issues from which some of the information about the human psychology of different media is derived.

REFERENCES

Alty, J.L. (1989). The GRADIENT Dialogue System: Providing better interfaces for process control. In P. Salenieks (Ed.) Computing Technologies , Ellis Horwood, Chichester, pp 75 - 101.

Alty, J.L., (1991). Multimedia: What is it and how do we exploit it ?. In D. Diaper and R. Winder (Eds.), Proc. HCI'91, Cambridge University Press. pp 31-44.

Alty J.L., and G. Johannsen, (1987). Knowledge based dialogue for dynamic systems. In R. Isserman R. (Ed.) Proc. 10th IFAC World Congress on Automotive Control, (Munich), pp 358-367.

Alty J.L., P. Elzer, O. Holst, G. Johannsen, and S. Savory, (1985). Literature and User Survey of Issues related to Man-machine Interfaces for Supervision and Control Systems. ESPRIT P600 Pilot Phase Report (available from CRI Copenhagen).

Alty J.L., and C.D. McCartney, (1991). Design of a Multi-Media Presentation System for a Process Control Environment. In E. Kjelldahl (Ed.), Proc. of the Eurographics Workshop (Stockholm 1991), Chapter 22, Springer Verlag.

DeSanctis,G., (1984), Computer Graphics as Decision Aids. Decision Sciences. Vol 15, pp 463 - 487.

Diaper, D., (1989). Task Analysis for Knowledge Descriptions (TAKD): The Method and an Example. In D.A. Norman and S.W. Draper (Eds.), Task Analysis for Human-Computer Interaction, Lawrence Erlbaum Associates.

Edwards, A. D. N. and I.J. Pitt, (1990). Adapting the MacIntosh and Other Graphical User Interfaces for Blind Users. Special Edition of InfoVisie Magazine Vol 4, No 3.

Hovy E., and Y. Arens, (1990). When is a Picture worth a Thousand Words? Allocation of Modalities in Multimedia Communication. AAAI Symposium on Human Computer Interfaces, Stanford 1990.

Marmollin, H., (1991). Multimedia from the Perspectives of Psychology. In E. Kjelldahl (Ed.), Proc. of the Eurographics Workshop on Multimedia, to be published (Springer Verlag).

Mayes, A.T., C. Dolphin and J.L. Berzal, (1989). Critical M4I Issues. PROMISE technical Report No. PRO/HCI/11/1

Oren, T., (1990). Designing a New Medium. In B. Laurel (Ed.), The Art of Human Computer Interface Design, Addison Wesley, N.J., U.S.A., pp 467 - 479.

Payne, S., and T. Green, (1989). Task-Action Grammars: The Model and its Developments. In D.A. Norman and S.W. Draper, (Eds.), Task Analysis for Human Computer Interaction, Lawrence Erlbaum Associates.

Tanaka, H., H. Muto, J. Yoshizawa, S. Nishida, T. Ueda, and T. Sakaguchi, (1988), ADVISOR: A Learning Environment for Maintenance with Pedagogical Interfaces to Enhance Student's Understanding. In H. J. Bullinger, E.N. Protonatorios, D. Bouwhuis and F. Reim, Proc. of EuroInfo'88, Athens, North Holland, Amsterdam, pp 886 - 891.

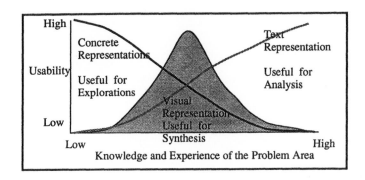

Fig. 2. Possible Use of Different Media (Marmollin 1991)

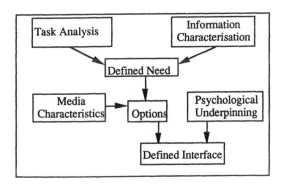

Fig. 3. A Suggested Approach towards a Methodology

Fig. 2. Desirability of Different Media (Marcus, 1992)

Fig. 3. A Principled Approach towards a Methodology

A SOFTWARE SYSTEM FOR DESIGNING AND EVALUATING IN-CAR INFORMATION SYSTEM INTERFACES

G. Labiale*, K. Ouadou** and B.T. David**

*INRETS, 109 Avenue S. Allende, 69675 Bron Cedex, France
**Ecole Centrale de Lyon, Departement Mathématiques, Informatique et Systèmes, 69131 Ecully, France

Abstract. To integrate the requirements of ergonomists, car designers, and computer scientist for psycho-ergonomic studies and design of in-car information systems, we have developed a set of software tools called "IN-CAR-DISPLAYS". The purpose of this article is to describe the functionalities and the applications of this set of tools. More precisely, this software system provides the appropriate tools which include a wide range of functions which enable : easy and rapid prototyping of displays for different information sets and different types of dialogues for in-car information systems, testing of possible physical designation devices, real-time processing for dynamic management of different modalities used to present information and for dialogue purposes, recording of driver reaction rates and manipulatory responses during dialogues to assist with evaluations of responses. The original interest of this work is that its software development is based on a systemic philosophy which integrates the points of view of in-car human factor specialists and experts in user interfaces.

Key words. Software tools; display systems; man-machine systems; human factors; system theory; vehicles; In-car information system; user interface prototyping.

INTRODUCTION

Currently many attempts are being made to introduce information and communication resources to cars. These include, for example, navigational and road guidance system, traffic information system, obstacle detection system, cruise control and malfunction warning system, all of which are additional to conventional in-car equipment such as telephones and the ubiquitous radio-cassette player.

The integrative structure, already adopted by some car manufacturers, implies ergonomic studies into how information is presented and the dialogue resources that drivers can use. It is essential to define the operating principles for the interface to ensure visibility, legibility, comprehension and comfort and to contribute in this way to improving driving safety for all drivers when they are collecting information and dialoguing with the system .

Some ergonomic guidelines exist to define the design of instruments fitted to the dashboard (Galer & Simmonds, 1984; Green, 1983) and information systems (Labiale, 1990). It would appear, nevertheless, that these are either too general in their scope or too specific and even

inadequate. Particularly, the M.M.I. guidelines used to define man/machine interfaces (McCormick & Sanders, 1982) cannot be applied to a real driving situation. So as to define ergonomic guidelines for informational displays on the in-car screen - structures and contents of the messages, icons, maps, texts,... - and the modalities of dialogues in driving cabs, it became essential to develop a software tool enabling rapid prototyping and evaluation of designs and dialogues for interfaces. These need to be able to integrate all different existing and future information systems. Research in this field is still relatively limited and, to our knowledge, only two experiments concerning of design software have been undertaken, both using the Macintosh "HyperCard" (Green et al. 1990; Stadelmeir 1989). Although extremely user-friendly, their functionalities remain relatively limited, particularly when programming functions which "HyperCard" does not normally handle, i.e measuring driver reaction times in milli-seconds and the fact that the images are in black and white. To meet the requirements of automobile engineers for human factor studies and designs, we have developed a set of software tools called "IN-CAR-DISPLAYS".

The purpose of this article is to describe the functionalities and the applications of this set of tools (including the generator used for IN-CAR-DISPLAYS and interface dialogues). This system was developed on and for a compatible PC microcomputer, which is interesting as it is highly efficient, economic and can be installed in a car. The additional interest of this work is that its approach is based on a systemic philosophy which integrates the points of view of in-car human factor specialists and experts in computer system interfaces. This approach has enabled the construction and validation of several interactive application architecture models.

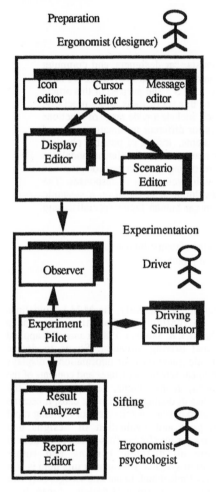

Fig. 1. The functional architecture of "In-car-displays"

FUNCTIONS OF THE SOFTWARE SYSTEM

"IN-CAR-DISPLAYS" is a system designed for used by all categories of people interested in such interfaces - ergonomists, psychologists, designers, engineers and drivers. For each category the system provides the appropriate tools which are included in a wide range of functions which enable:
- Easy and rapid prototyping of screen designs for different information sets and different types of dialogues for different in-car information systems,
- Testing of possible physical designation devices (touch screen, joystick, trackball, keyboard,...),
- Real-Time processing for dynamic management of different modalities used to present information and for dialogue purposes either based on data defined previously, or on data collected in real time,
- Recording of driver reaction rates and manipulatory responses during dialogues to assist with evaluations of responses and dialogues in various real car driving situations or in laboratory driving simulators.

For practical purposes, the system includes a set of coordinated and complementary tools such as the display editor, scenario editor, the scenario execution function, the result logging function and the result analysis tool (Fig. 1).

The Display Editor

The display editor is used by the operator to prepare images composed of varied graphic items. The aim is to determine the best presentation and organization of information systems designed to assist with driving. Using the range of editing possibilities, the operator can define all the presentation characteristics, the way the page behaves and the items included in each page.
Items are graphic objects which, depending on their nature, are dedicated to the acquisition of information (knobs, keys, lists,...) to restitution (instruments and gauges,...) or to presentation (texts,...). More specifically, the display editor enables the definition of :
- the characteristics of the display page - i.e. the size, the background colour, contrast, brilliance, etc...,
- the characteristics of the items, - i.e. the shape, the size, the colour, the content, the quantity...; the operator uses a graphic tools to construct geometric shapes such as circles, squares, triangles, etc., analog counters (circular or straight-line shapes - the instrument pointer characteristics) and digital counters. Furthermore, to facilitate the preparation of these pages, the display editor enables the operator to choose the spatial organization of the items.
The behaviour of the displays and the items can be described using scripts and constraints. Items can be attributed predefined static behaviour or dynamic behaviour patterns. More accurately, this means, in fact, defining passive, reactive or active items:
- the passive status concerns static information which cannot be modified by a driver command (for example, a function title or a list of instructions),

- the reactive status represents information which reacts to a choice made by the driver such as a system command (for example the choice of an element in a menu, ...),
- the active status concerns information which is changing continuously, periodically or asynchronously. Examples include an item which changes colour if an instant occurs in the car, a pointer which moves in the speedometer, the position of the vehicle on a map, etc...

The Scenario Editor

The scenario editor is used to elaborate an experimental scenario. This script describes the overall behaviour of the application depending on the driving context and the driver's actions. A scenario is broken down into a set of elementary sequences which correspond to the pages displayed and - even more precisely - to the items on each page. The scenario enables the operator to select the page displayed, to define the experimental settings, the items to be used on the page (visible or not), a set of messages that the page must trigger in response to specific events (either predefined or defined by the operator, such as traffic information on the simulator, external operator intervention,...), page priority, etc... The script of the scenario basically defines the type of scenario (functional : prerecorded or experimental : manual intervention by the operator) and the set of events used to trigger each elementary sequence (simulator flags or intervention by the operator during execution).

The main functions of the editor enable the operator to prepare elementary sequences and to set up a range of scenarios easily. It is, in fact, possible to create or adapt sequences (the availability of conventional editing capabilities, such as cut, copy and paste facilitates this task) and to set up and manage the libraries for sequences, events and messages. Moreover, the editor enables the operator to test scenarios immediately: in this way it is possible to judge the visual aspects of the sequence, navigation within the scenario, response to events, detection facilities, etc...

The Experiment Pilot

The experiment pilot is the resource used to pilot the experiment set up by the operator and to observe scenario execution interactively with the driver. On a functional level, it is basically a automaton which handles the sequencing of the elementary scenario sequences depending on their type, i.e. a sound signal to warn the driver, to present a page, to collect driver commands and, if necessary, to display a warning message and activate the next sequence.

There are three possible types of scenario :
1 - the experimenter guided scenario : the experimenter is responsible for triggering the sequences in the scenario by deciding at which moment to input the instruction, to activate an alarm message or to display a new sequence. The experimenter specifies the time for which messages and pages are displayed.
2 - the driver guided scenario : drivers are free to trigger the elementary sequences by their successive choices of items in the various pages.
3 - the driving situation guided scenario : the events which can be interpreted by the scenario are related to flags defined in the simulator circuit or in the real road.

The function of this monitoring unit is to observe the behaviour of the scenario and of the driver so as to collect a maximum amount of information about the sequences and the stress and reactions of drivers in relation to these sequences. The monitoring unit records all the following information in a file that can be used during the analysis phase :

.the experimental context (driver, situation type, ...),
.the execution context (current sequence, validation errors, ...),
.the time taken by drivers to manipulate items or messages which includes the perception time (i.e. the time required to read and understand a page and the related instruction) and the movement time (corresponding to the manipulation of the designator).

CONCEPTION AND ARCHITECTURE

The Systemic Approach

We used the systemic approach to design the system (Morin, 1986). This means that the system is considered as an entity communicating with its environment, able to receive inputs and to produce results which are usable by this environment. These enables different specialists i.e system analysts, ergonomists, drivers, psychologists, and designers to intervene.

Based on this view, our aim was to develop a systemic model ensuring a higher level of inclusion as the system evolves. This generally occurs during the following phases :
1 - to define the system and observe objectives likely to evolve as the system capabilities become better understood,
2 - the definition of objectives led to the definition of system boundaries to identify the system within its universe and isolate it for study,
3 - identification of the environment,
4 - identification of the data flows at the system boundaries,
5 - identification of the internal structure, the components and the pilot of the system,
6 - determination of diagnosis criteria for the system pilot.

By using this approach that we have been able to present the objectives for the system, its limits and its modules in the preceding paragraphs. If this approach does not guarantee security, it does nevertheless provide a good procedural context enabling high quality system design and easy communications between the different specialists concerned (as we shall see in the following paragraphs). Let us now review the models used to design the internal structure and the components of the system together with the constraints which led to the selection of the hardware and software configurations.

Software Architecture Design

The design of this system and its software architecture reflect state of the art work in interactive applications. During the design phase we gave priority to interactivity and upgradeability and we applied the fundamental principles conventionally used for User Interfaces (UI) :
1 - logic separation between the interface and the application,
2 - physical dialogue devices are virtualized,
3 - the principle is to reuse the components of one interactive application in another ,
4 - extents to functionalities,
5 - use of similar interfaces for all tools,
6 - massive usage of resource files.

The Object Oriented Approach

The IN-CAR-DISPLAYS system is considered as a set of tools for prototyping and ergonomic evaluation. This has led us to define a User Interface Management System (UIMS) (Pfaff, 1985) and to consider IN-CAR-DISPLAYS as one specialization of this UIMS. One of the objectives is to avoid redundancy by the definition of stable structures in the system (data and behaviour used by all components) and thus the requirement that all system activities should be fully integrated. We have developed a central core which is a object-oriented approach (Cox, 1986; Meyer, 1990). This core is the result of two complementary studies on the architecture of interactive applications, one concerning specialized editors (Ouadou,1990) and the other on interaction models (David & al.,1991).

In fact, we consider that all the editors basically behave in the same way and that they can be developed by the specialization of a generic contextual editor. This approach enables the creation of the generic tools required for system activities and to dedicate them to the specific needs of each existing tool, as well as ensuring their extension to future tools. This is made possible by the fact that we consider the architecture of each editor as the cooperation of homogeneous objects, i.e. possessing a homogeneous model such as

PAC (Coutaz, 1987) and in our case EPAC (David, 1990). The EPAC (Extended PAC) model is based on the extension of the PAC model. The PAC model is based on three-facetted homogeneous objects (Presentation, Abstraction and Control) which appeared to us to be insufficient to make accurate descriptions of behaviour, as it combines accurate components as constraints, assistance and intelligent behaviour within the control facette. In the EPAC model these are considered as separate discrete facettes and are considered naturally, automatically and with predefined data and behaviour patterns. The number of facettes in EPAC is dictated by current application requirements.

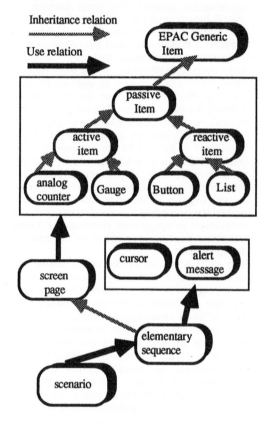

Fig. 2 : objects structure

Core Structure

These are the reasons which led us to adopt an object-oriented approach to build the system core. This makes it possible to define generic objects, which can be reused, and to dedicate them progressively using heritage, overload and polymorphism mechanisms. This approach, applied to the different elements presented above, gives the following results : the scenario for each experiment is composed of elementary sequences which are in fact specialized pages, each of which is made up of different items. Each element is characterized by specific attributes and behaviour;

the Figure 2 shows the object structure of the core. Each of the objects is a descendant of the EPAC interactive class and in this way we obtain an homogeneous architecture.

In this structure, the various editors enabling the manipulation of these objects are, in fact, more completely finished objects using different, more general classes. Our objective was to elaborate visual programming tools (Myers 1986, 1989a, 1989b). By considering that these are, in fact, specialized editors based on the same representational model (Nierstrasz & al. 1990a) i.e. a graph and commands such as create, describe or delete an object are included in each editor and most of the other commands such as copy, paste or cut are derived from them. The execution model is specialized by the semantics and by the constraints specific to vehicles. The specialization of the presentation is also related to the field of applications concerned.

Using this approach, we have been able to organize the various tools into a class structure, the root of which is a generic editor which combines all the common methods and data.

Hardware and Software Environment

The choice of the hardware and software were one of the major constraints for the IN-CAR-DISPLAYS system because of the high inter-relationship with the environment (cab restrictions, technological availability,...). The system was designed to run on low-cost, robust, hardware with a small screen giving good colour graphic definition and to support several designator devices (trackballl, joystick, mouse, touch screen,...). At the start of the project, we selected a compatible PC (AT 80386, 25 MHz) with good (VGA) colour graphics definition and enough slots for any additional connections required. In actual fact, our software can run on a portable PC with a minimum configuration of 2MB RAM and an Intel 80286 processor.

From a software point of view, we opted for the Microsoft Windows environment as it is a good development context for interactive applications, facilitating the use of different I/O devices as they are virtualized. We started with MS-Windows 2 and programmed in C. Recent developments (MS-Windows 3.0 and the availability of new programming languages in its environment, particularly turbo-Pascal and C++) led us to reprogram the system in C++. In this way our reuse and extension objectives are now easier to implement than with the first version. Current and future upgrades of MS-Windows particularly the multi-media support capabilities will give rise to interesting extensions to this work.

APPLICATIONS

Ergonomists wish to discover the most appropriate presentations of interfaces and dialogue resources for in-car information systems in different driving situations (real or simulated). To do this they can use the IN-CAR DISPLAYS to elaborate different types of page designs and dialogues for the interfaces and to record reaction times and motor responses by drivers. Furthermore, with video recordings and the eye recorder, they can use additional tools to analyse the different visual exploration strategies used by drivers and therefore improve interface definition. IN-CAR-DISPLAYS also enable psychologists to study problems relating to visual perception, mental load in double task situations and motor regulation, taking account of the various designation devices.

In this document we present one example of generic application fields concerning page design wich illustrates some of the graphic possibilities, such as an icons menu (Fig. 3).

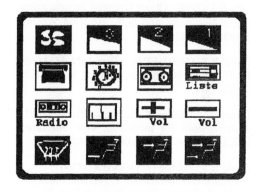

Fig. 3: Icons menu designed with Display Editor.

CONCLUSION

IN-CAR-DISPLAYS constitutes a precious methodological tool for studying and evaluating interactions between drivers and in-car display systems; it will enable human factor studies and thus contribute to the preparation of recommendations concerning the presentation and dialogue for interfaces.

However, some improvements must be integrated in the future, particularly the possibility of generating and manipulating sound messages, the availability of a help function and a user training programme (ergonomists,...), as well as the performance levels of the algorithms.

The development of this flexible and upgradeable set of in-car interactive application software tools has required communication and pooling of pluridisciplinary concepts and targets for ergonomists, system analysts and engineers. In the spirit of our systemic approach, commmunications between the different participants were essential so as to define all the problems, to identify all the requirements and to ensure that the approach could be adapted over time. From this point of view, the design phase of this job was also a contribution to cognitive engineering (Norman, 1986) which combines psychology with data processing.

Finally, this approach, and the software architecture which it implies, opens possibilities for integrating the IN-CAR-DISPLAYS software package into future "intelligent" systems, particularly the "intelligent" interface for assistance in driving (Labiale, David, Ouadou, 1991) through the use of the EPAC interaction model and the "intelligent" design of in-car interfaces which will determine the best ergonomic designs for presenting data to car drivers.

REFERENCES

Coutaz J. (1987). PAC, an Implementation Model for Dialog Design. Interact'87, Stuttgart, pp. 431-436, sept. 1987

Cox B.J. (1986). Object Oriented Programming-An evolutionary approch. Addison-Wesley, Reading, Mass., 1986.

David B.T. (1991). EPAC, Extended PAC Model for Dialog Design. Research Report MIS-ECL n° 91-03, 1991.

David B.T., Ouadou K., Sadou S., Vial C. (1991). A framework for intelligent user interfaces. OZCHI 91, nov. 1991, Sydney, Australia

David B.T., Labiale G., Ouadou K. (1991). Integrated Environment for Ergonomic Studies of on-board display. Proceedings of 24th ISATA International Symposium on Automotive Technology and Automation, on Intelligent vehicle-highway systems, Florence, may, 1991, pp 623-630.

French R.L. (1986). Historical overview of automobile navigation technology. 36th IEEE Vehicular Technology Conference, Dallas, 1986.

Galer M., Simmonds G. (1984). Ergonomics aspects of electronic instrumentation. A guide for designers SP-576. Warrendale , PA: Society of Automotive Engineers, Inc.

Green, P. (1988). Ergonomics of automotive displays. Tutorial (T26) presented at the International symposium on optical engineering and industrial sensing for advance manufacturing technologies .Bellingham, WA.

Green P., Boreczky J., Seung-Yun Kim (1990). Application of rapid prototyping to control and display design. USA, SAE paper n° 900470, pp. 23-44.

Labiale G. (1990). Psycho-ergonomie de l'interface conducteur -automobile. Revue générale de l'Electricité, n° 9, 35-39.

Labiale G., David B.T., Ouadou K. (1991). L'interface "intelligente" d'informations et de communications dans le poste de conduite automobile. Communication at the colloque "Sciences humaines et Intelligence artificielle", Université de Lyon II.

McCormick E.J., Sanders M.S. (1982). Human Factors in Engineering and design. Aukland : Mc Graw Hill.

Meyer B.(1990). Conception et programmation par objets pour du logiciel de qualité. Inter-Editions (Ed.), 1990 Paris.

Morin E. (1986). La méthode: la connaissance de la connaissance. Seuil.(Ed.)

Myers B.A.(1986). Visual programming, programming by example, and program visualization: A taxonomy. In proceedings of the ACM CHI'86 conference on human factors in computing system (Boston, Mass., Apr.) ACM, New york, pp 59-66.

Myers B.A.(1989a). Encapsulating interactive behaviors. Proceedings of the conference on human factors in computing system (SIGCHI 89), Apr. 1989, pp 319-324

Myers B.A.(1989b). The state of the art in visual programming. In Kilgour A. & Earnsnaw R. (Ed.) Graphics Tools for Software Engineers, Cambridge University Press 1989.

Nierstrasz O., Dami L., De Mey V., Stadelmann M., Tsichritzis D.C., Vitek J.(1990a). Visual Scripting, towards interactive construction of object-oriented applications. In Tsichritzis D.C., Object Management, pp 315-331, centre universitaire d'informatique, University of Geneva, July 1990.

Norman D.A. (1986). Cognitive Engineering. In Norman, D.A & Draper, S.W. (Ed.) User Centered System Design: New Perspectives on Human Computer Interaction. Erlbaum, Hillsdale, New Jersey.

Ouadou K. (1990). Génération d'éditeurs contextuels. Research Report MIS-ECL n° 90-01, 1990.

Pfaff G.E. (1985). User Interface Management Systems. Springer-Verlag, 1985.

Parviainen J.,Case R.E., Sabounghi L.R.N. (1989). Future Mobile Information Systems : Potential Applications and Systems Features Vehicle Navigation and Information Systems Conference, Toronto, Sept. 1989.

Stadelmeier S. (1989). Using Hypercard as interface simulator. Proceedings of Interface 89. pp. 277-282.

HUMAN-COMPUTER INTERACTION IN CONTEXT: PHYSICIAN INTERACTION WITH AUTOMATED INTRAVENOUS CONTROLLERS IN THE HEART ROOM

E. Moll van Charante *,**, R.I. Cook *,**, D.D. Woods*,**, L. Yue* and M.B. Howie**

*Cognitive Systems Engineering Laboratory, Department of Industrial and Systems Engineering,
The Ohio State University, Columbus, Ohio, USA
**Department of Anesthesiology, The Ohio State University, Columbus, Ohio, USA

Abstract. One result of recent research on human error and disaster is that the design of the human-machine system, defined broadly, modulates the potential for erroneous action. Clumsy space use of technological powers can create additional mental burdens or other constraints on human performance that can increase the chances of erroneous actions by people especially in high workload, high tempo operations. This paper describes studies of a computer based automated device that combine critical incident analysis, bench test evaluations of the device, and field observation in order to understand physician-device interaction in the context of heart surgery. The results link, for the same device, user group and context, three findings. First, the device exhibits classic human-computer interaction flaws such as lack of feedback on device state and behavior. Second, these HCI flaws actually do increase the potential for erroneous actions and increase the potential for erroneous assessments of device state and behavior. The potential for erroneous state assessment is especially troublesome because it impairs the user's ability to detect and recover from misassemblies, misoperations and device failures. Third, these data plus critical incident studies directly implicate the increased potential for erroneous setup and the decreased ability to detect errors as one kind of important contributor to actual incidents. The increased potential for error that emanates from poor human-computer interaction is one type of latent failure that can be activated and progress towards disaster given the presence of other potentiating factors in Reason's model of the anatomy of disasters.

Keywords. Human Error; Computer Interfaces; Medical information processing; Automation; Man-machine Systems; Human Factors.

INTRODUCTION

Improvements in our ability to monitor, understand, and influence complex processes have encouraged the development and introduction of more automated systems. For example, advances in the understanding of blood pressure mechanics and heart rate regulation, improved ability to monitor these variables and newly developed drugs with specific influences on these variables, have permitted the design of sophisticated intravenous infusion controllers capable of more precise intravenous drug delivery. Similar developments are underway in commercial aviation (new automation in the flight management systems of glass cockpits; cf., Wiener, 1989; Sarter and Woods, 1991), air traffic control (plans for closer, more "efficient" aircraft spacing), manned spaceflight (Woods et al., 1991), and electrical power generation and distribution. The objective of developing

these new more autonomous systems is to augment operators by shifting activity from the operator to the automated device. It is an article of faith among device developers that their introduction will reduce demands and workload on the operator by taking over some part of the operator's task.

We have been studying the impact of technology on practitioner performance in four domains – commercial aviation, spaceflight, electrical power generation and anesthetic management during surgery. These studies have shown empirically that the assumption that new automation automatically reduces practitioner workload and improves overall performance is not justified. New automated devices create new cognitive and ergonomic "costs" at the both the individual operator and team level. The costs may involve new or changed tasks such as device setup and initialization, configuration control (e.g., software version control), or operating sequences. Cognitive demands change as well, including tracking automated device state and performance, new data management tasks, new attentional demands, new communication tasks and new knowledge requirements. These costs represent new levels and types of operator workload. Frequently, new automated devices produce a condition called clumsy automation by Earl Wiener (1989). Clumsy automation or the clumsy use of technology is a form of poor coordination between the human and machine. Here the automation benefits accrue during workload troughs while automation costs occur during high criticality or high tempo operations. Overall, the clumsy use of technology creates new operational complexities that appear to produce increments in workload and decrements in practitioners' ability to track changes in their environment, especially during high tempo, high criticality periods. Significantly, these deficits can create new opportunities for human error and produce new classes of system failures.

In this paper we report on our studies of a new automated intravenous infusion controller used to achieve more precise administration of vasoactive intravenous drugs in order to influence patient hemodynamics during cardiac surgery. Our investigations combined critical incident analysis, bench test evaluations of the device, and field observation in order to understand physician-device interaction in context. The results show that (a) the device exhibits a variety of classic human-computer interaction flaws (Norman, 1988), (b) these flaws increase the potential for various types of erroneous actions and impair the physician's ability to detect and recover from errors, and (c) these "latent failures" (Reason, 1990), when combined with other factors, contribute to critical incidents.

This paper focuses on the cognitive analysis of device use in context based on field observation of device use coupled with bench test evaluations. The results of a critical incidents analysis is reported for the anesthesiologist community in Cook, Woods and Howie (1992). A re-design of the device that illustrates how to correct error increasing HCI deficiencies has been completed as well (Yue, Woods and Cook, 1992).

COGNITIVE ANALYSIS OF DEVICE USE IN CONTEXT

Critical incidents reports obtained in a large teaching hospital (Cook, Woods and Howie, 1992) identified a group cases associated with an infusion controller (Rateminder, Critikon Corporation) used to control the flow of blood pressure and heart rate medications to patients during heart surgery. Each incident involved delivery of drug to the patient when the device was supposed to be off or halted. Detailed debriefing of participants suggested that, under certain circumstances, a device would deliver drug (sometimes at a very high rate) with little or no evidence to the user that the infusion was occurring.

Bench tests and field observation were used to identify (1) characteristics of the device which make its operation

error prone and difficult to observe and (2) characteristics of the context of cardiac anesthesiology which interact with the device characteristics to provide opportunities for unplanned delivery of drug. The bench tests of device behavior and operating room observations of device use were mutually supportive. Observation suggested additional investigations about how the device behaved in particular circumstances. The human-computer interface related deficiencies identified in bench testing helped to direct operating room observations to test whether these problematic aspects did in fact lead to greater potential for erroneous actions and erroneous assessments of device state.

Infusion Control and Cardiac Surgery. Generally, in cardiac surgery, the anesthesiologist monitors the patient's physiological status (e.g., blood pressure, heart rate) and administers vasoactive drugs to control these parameters to desired levels based on patient baselines, disease type and stage of cardiac surgery. The vasoactive drugs are administered as continuous infusion drips mixed with intravenous (IV) fluids.

The device in question here is one type of automatic infusion controller which regulates the flow of fluid by adjusting a mechanical occluding clamp around standard intravenous (IV) tubing. The rate of flow (drops per minute) is determined by counting the drops that form in a drip chamber, comparing the measured rate to a user entered value, and adjusting the occluding clamp to achieve the target rate (Figure 1). If the device is unable to regulate flow or detects one of several different device conditions, it is programmed to cease operation (close the occluding clamp to stop flow) and emit an audible alarm and warning message. Three front panel controls are available: the target rate setting control, an ON/OFF button and a START button; however, the latter two perform multiple functions. A front panel display indicates the user entered target rate, a brief alarm message, and a "running" indicator (a large icon which flashes when the device is running but does not indicate actual flow or flow rate). The front panel also includes an unlabeled green light emitting diode (LED) that blinks when a falling drop of fluid is detected in the drip chamber. In clinical use up to six devices may be used simultaneously to control six different drugs (Figure 2); the individual drug flows are mixed together in a manifold and delivered to the patient in a single IV line (Figure 1).

Methods. Bench tests were used to identify characteristics of the device which make its operation error prone and difficult to observe. These tests required instrumenting the device to provide recording of its operation and then performing experiments to demonstrate device characteristics over a wide range of normal and abnormal conditions. Abnormal conditions included possible misoperations and device failures. In particular, we determined the device's ability to regulate flow with different rate settings, its ability to function in the presence of back pressure (a clinically relevant condition), and the speed with which it adjusted flow rates and detected deliberately induced faults. The device possesses multiple alarm states, many of which relate to detectable faults in setup. These alarm states and normal operation were mapped using simplified finite state diagrams (Figure 3 shows the finite state diagram for just three of the possible alarm messages; note that this is only one of many finite state diagrams that were needed to capture the complete space of device states/modes and operating sequences). The performance of the device was also determined when it was in motion; the devices are used during transport of patients from the cardiac surgical suite to the intensive care unit. The testing covered a period of three months and included over one hundred different experimental combinations.

To inform and complement the bench tests of the device, several physicians were observed setting up and using the device under actual conditions. Two authors observed cardiac surgical cases, at first together and later independently, and recorded user interactions with the devices. Users gave informed consent regarding the data collection and the study was approved by the institutional committee regulating research with human subjects. Each discrete user interaction with the device was recorded, along with details regarding the part of the surgical procedure and other user activities. Particular attention was given to the occurrence of alarms and changes in drip rates. The early combined records suggested high concordance between independent observers, and prior research on the conduct of cardiac surgical cases (Cook, Woods, Howie, 1990), supported the validity of the data gathering method.

Over 25 device setups and 10 complete surgical cases were recorded. User setup and operating sequences were transcribed and user and case identifying characteristics removed. The cases were heterogeneous; in some the infusion devices were not used at all while in others several infusions were started and stopped throughout the procedure. Data from this study were analyzed using process tracing techniques (Woods, 1992). In particular, physician-device interaction was mapped using diagrams which compare actual device operation, the image the device presents to the user, and user actions and intentions in context (Suchman, 1987).

Bench Test Results on Physician-Device Interaction.
External indicators of the device operation make it difficult for users to assess or track device state and operation: (1) there is no indication of the actual rate of fluid delivery (the large digits on the front of the display refer to target and not actual rate), (2) a large moving element on the display suggests the actual rate of delivery but is misleading (it's flash rate seems to indicate the target rather than actual rate). Actual delivery of drug is indicated by the small unlabeled. The field study results confirmed that users confused indications of demanded drop rate with measurements of actual rate.

During operation, the controller initially overshoots the desired rate of flow, providing a small bolus of drug during startup (Figure 4). While this amount is small (in most cases amounting to 12 to 20 drops of fluid), the effect may be clinically significant when the target rate is low (e.g., 5 drops per minute) and the drug potent. Moreover, as the field observations established, in nearly

all cases physicians will start infusions at low levels and titrate to a desired effect. In addition, the device requires variable lengths of time to and reach steady state at the target value. For example, at slow rates of flow the adjustment may take up to sixty seconds to reach steady state. Because the period during which the device searches for control is relatively long and not indicated to the user, users sometimes made changes in drip rates faster than the controller can accommodate them. The observation data showed that the indication to the user of target rather than actual rate strongly suggests to users that the device reaches its target rate very quickly.

Any motion of the device during operation can cause drops to be missed and leads the controller to lose its setpoint and begin searching to regain control. This inevitably results in either restarting with an initial overdelivery of drug (as a safety feature the device will close the occluding clamp, reducing flow to zero, and then begin to search for the target that has been set) or shutdown and an alarm. As a result, during transport from the operating room to the intensive care unit drug delivery will be quite erratic with frequent stoppages and overshoots of target rate. The field data showed that, given the lack of feedback, physicians were unaware of the seek mode of device behavior and the erratic control during patient transport.

Overall, the lack of visible feedback and the presence of hidden modes combined to hide device state and behaviors from the physician, i.e., they were unaware of various controller behavioral characteristics such as overshoot at slow target rates, seek behavior, erratic control while moving.

The bench testing revealed that there is a problem with ambiguous alarm messages in this device. Some messages are nuisance alarms. Several different alarm messages can be displayed for the same underlying problem; the different messages depend on operating modes of the device which are not indicated to the user (see Figure 3 for one example).

It was relatively easy to fool the device droplet sensor with a continuous stream of fluid. The sensor depends on the lens effect of a droplet of fluid but at high flow rates the droplets coalesce into a continuous column of fluid which does not register as a drop. This condition of free flow is detected as the absence of drops and indicated with either NO FLOW or OCCLUDED alarm messages. Because the droplet sensor surrounds the droplet chamber which contains the only indication of fluid flowing in the system, the use of the controller obstructs the users view. Moreover, the sensor is located some distance from the display panel and controls, making visual checks of the device operation even more difficult. The combination of the opaque interface and the inability of the sensor mechanism to discriminate between no/low flow and free flow proved to be a major factor in one of the critical incidents.

The front panel controls are complex and perform multiple functions. For example, the ON/OFF button performs different functions depending entirely on device states which are only partially indicated to the user. The START button provides both an initiation function and the ability to smoothly adjust the target rate while an infusion in progress (this unlabeled feature requires pushing two buttons simultaneously).

The bench tests also confirmed a specific engineering design flaw (as opposed to design flaws from a human-computer interaction point of view). This flaw, which appeared to be the result of a software error, prevents the device from completely occluding the IV tubing under specific circumstances and can permit flow of drug even when the device is apparently powered off. Details of this fault, which were discovered following a clinical incident, were communicated to the manufacturer and were corrected in an upgrade supplied to this institution. The manufacturer insisted that the fault would only be significant if users failed to follow certain setup procedures contained in the user manual.

The bench tests also revealed a variety of mapping problems. The mapping between fluid bags, tubing sets, the different parts of the infusion controller, and the connectors to the manifold are inconsistent and ambiguous (Figure 2). This type of deficiency can be expected to lead to misassemblies where part of one assembly is connected to a portion of a different infusion controller set up.

Overall, the bench test results identified several classic human-computer interaction deficiencies in the infusion controller (Norman, 1988; Cook et al., 1991):

~ Lack of feedback on device state and behavior.

~ Complex and arbitrary sequences of operation.

~ Ambiguous alarms.

~ Poor mappings between state and actions and between the parts of a single assembly.

Field Study Results on Physician-Device Interaction. Users encountered virtually every difficulty identified during the bench tests while using the devices in the field. Many of these problems were related to the initial setup of the device. The data from the field study also was examined to understand how users tailored their behavior to compensate for device deficiencies or weaknesses (Woods and Cook, 1991).

Setup of the devices is complex, involving up to 20 separate steps for each controller (up to six controllers may be used for a single surgical case). Because setup is complicated and because one or more of the controllers may need to be used unexpectedly and quickly to control patient hemodynamics, the infusion controllers are setup well in advance of the surgical procedure. This means that errors in device assembly and configuration may not become apparent until the devices are used, in some cases hours after setup.

Interestingly, users seemed quite aware of the potential for error and difficulties associated with device setup.

Two main strategies for setup were observed in which similar components of each device were setup for all of the controllers before proceeding to the next component, and space, in which a complete device was setup before moving to the next. The setup of two different device sets by different users is shown in Figure 5. A critical step (close occluding clip wheel) which produces vulnerability to the software flaw described above was frequently missed by both users (an omission of an isolated action).

A variety of misassemblies were observed including loading the wrong occluding clip assembly, i.e., the assembly for one infusion controller setup is loaded into another infusion controller, and mounting the wrong drop detector, i.e., the detector of one infusion controller is mounted on the drop chamber for a different infusion controller assembly. Both misassemblies produce a dissociation between regulation and sensing of flow rate across two drug/IV sets and the potential for inadvertent drug delivery. A similar misassembly did lead to unintended drug delivery in one of the critical incidents.

Alarms were remarkably common during device operation. In one sequence of about five minutes duration there were at least a dozen alarms from a single device. It is significant that these alarms were not simply repeats of the same message but a variety of different messages; this sequence provoked a variety of user responses. It is important to note that, given the lack of feedback, when alarms recur, it is very difficult for the physician to determine whether or not the device has delivered any drug in the intervening period.

Users were sensitive to the fact that the devices could, under what appeared to them as ill-defined circumstances, deliver drug unexpectedly and sought to protect the system from these states by keeping various stopcocks and clamps in the IV lines closed until the devices were actually needed. This strategy itself was seen to provoke alarms since users sometimes started the devices without first clearing the IV line of obstructions; this invariably resulted in an alarm from the device indicating that flow was obstructed.

It was also observed that users were seldom able to devote undivided attention to the operation of the infusion controllers. The operating room is a busy place and the circumstances under which the drugs were required were also ones in which many other physician actions were required and where physician attention was focused elsewhere. Because the pace of this activity was faster than the time required for the controller to detect and announce faults, alarms always constituted an interruption of physician attention to other tasks. Users would start (or adjust) an infusion and then proceed to another task, and later they would be alerted by an alarm that the previous action had been unsuccessful (alarms generally occurred about 2 minutes after the triggering condition began). This alarm necessitated attentional scheduling, switching attention to the tree of infusion devices, locating the appropriate device, recall of the prior actions, and troubleshooting of the device. The most common troubleshooting approach was simply recycling the device (effectively restarting the controller's search for the target rate). This approach seemed to be efficacious, probably because it incidentally silenced alarms for some time and because intermittent occlusion (temporary occlusion or backpressure) was common. The complex cycle of adjusting the device, turning to other tasks, device alarm, and refocusing attention to the device group led in at least one instance to a user misidentifying the faulty device.

Particular factors in the environment constrain operators to use the devices in certain ways (e.g. set up precedes use by as much as two hours). The device design does not support these constraints. For example, the START ME alarm appears about two minutes after the device is setup and is a strong prompt to turn the device off since setup and use are always separated by much much more than two minutes. In this environment, START ME is in fact an indication that the device needs to be turned off, not that it needs to be started. Moreover, since it is delayed by two minutes (a long time in this world) the operator will be doing other things by the time it appears. The message was never observed to indicate correctly what the operator should actually do.

The device design, when considered in the context of use, encourages the user to turn the controller off following set up until that device is actually needed. However, misassemblies can occur that exist as latent failures in the system waiting for a trigger to become an active failure, i.e., an inadvertent drug delivery. Since the device is off, no alarms will be generated. This dynamic played out in one of the critical incidents investigated in Cook et al. (1992).

The most intense periods of device use also were those time periods of highest cognitive load and task criticality for the physicians, i.e. the time period of coming off cardio-pulmonary bypass. It is precisely during these periods of high workload that the automated devices are supposed to provide assistance (less user workload through more precise flows, smoother switching between drip rates, etc.). However, this is also the period where the largest number of alarms occur and where device troubleshooting is most onerous. As we have found in other operating room devices (Cook et al., 1990) and in other complex domains (Wiener, 1989; Sarter and Woods, in press; Woods et al., 1991), close study of device use in context revealed that this automated device exhibits the classic properties of the clumsy use of technological possibilities.

The field observations showed that the HCI deficiencies identified in the bench tests actually do increase the potential for erroneous setup and operation of the device. The field observations also showed that device HCI deficiencies increase the potential for erroneous assessments of device state and behavior. The potential for erroneous state assessment is especially troublesome because it impairs the physician's ability to detect and recover from erroneous actions and setups.

DISCUSSION

One assumption of recent research on human error and disaster (e.g., Reason, 1990) is that the design of the human-machine system, defined broadly, modulates the potential for erroneous action. The results of this series of studies directly link, for the same device and context, HCI design deficiencies to increased potential for erroneous actions and impaired ability to detect and recover from errors. The results, when combined with a study of critical incidents for the same device and context, directly link the increased potential for erroneous setup and the decreased ability to detect errors as important contributors to critical incidents. The increased potential for error arising from poor human-computer interaction can be seen as one type of latent failure that can reside within a complex human-machine system. Activating this type of latent failure in the presence of other potentiating factors leads incidents nearer to disaster. The data in this particular case provide direct support for Reason's (1990) pathogen model of the anatomy of disasters (cf. also, Woods, 1990).

As new automated devices are developed and deployed the apparent simplicity of the computer interface tends to hide significant underlying complexity (Cook et al., 1991). Furthermore, one can see in this, as in other cases that we have investigated (Woods and Cook, 1991), how new technology that glosses over substantive issues in human interaction can increase the operational complexity of the overall system. This technology induced complexity increases the demands on practitioners despite superficial putative benefits claimed for the new device or technology.

The field study portion of this research also reveals that practitioners are not passive recipients of new technology; rather, they actively tailor the device itself and their work to accommodate the new device. Because practitioners are responsible agents in the domain, they work to insulate the larger system from device deficiencies and peculiarities of the technology. This occurs, in part, because practitioners inevitably are held accountable for failure to correctly operate equipment, diagnose faults or respond to anomalies even if the device setup, operation, and performance are ill-suited to the demands of the environment. In this study users tailored their behavior to avoid problems and to defend against device idiosyncrasies. However, the results also show how these adaptations may be only partly successful. The adaptations, while useful in narrow contexts, were often brittle and unable to cover the wide range of circumstances actually found in the domain.

One may ask how the present study was able to reveal so many design deficiencies and problems associated with device use. The answer seems to us to lie in the methodology of study which focuses on understanding device use in context. This type of context bound approach depends on a hybrid of techniques involving critical incident investigations, evaluating device function and behavior under controlled conditions, and field studies of device use. Understanding the demands placed on practitioners by the environment is key both to the design and the evaluation of infusion controllers and, by extension, of other computer-based devices as well.

There is a tendency to view new automated devices as operator aids which make fewer demand on operator cognition than the manual tasks they replace. While this may be true in a narrow sense, the new device, when considered in the full context of the user's environment, also creates new cognitive and physical demands on practitioners. The device described here is particularly difficult for operators to setup, monitor and supervise effectively. Many of its design flaws are readily apparent in retrospect. Yet devices such as this one are commonplace and widely used. Interestingly, flaws in the device are not so apparent that it is rejected by users at first glance. Moreover, because the environment is complex and fast paced, when problems do occur their source is frequently difficult to identify:

~   faults may be missed – it was the presence of medically knowledgeable cognitive engineers working in close cooperation with anesthesiologists on another project that enabled recognition and follow up of the critical incidents that occurred,

~   faults may be ascribed to other agents (often the users!) – it was only after a series of incidents had occurred that there was general recognition that the problem lay in device design given the context of the heart room and not with individual users,

~   faults may be rationalized as the "learning curve" for introducing any new device, or

~   faults may be seen simply as "unusual" events with no larger significance.

Especially in the operating room, where poor quality human-computer interface is the rule rather than the exception, the poor performance of a single device may seem to be routine, i.e., just another obstacle that physicians are supposed to work around to provide safe and effective patient care.

Discovering the characteristics of the device presented here required substantial time (the present research represents over two person-years of technical effort) which is far beyond the resources of most users who are, after all, asked to work with many different devices. The apparent simplicity of the device given the interface characteristics was a formidable barrier to understanding how the device actually works both in general and in a particular context. For example, extensive bench tests were required to penetrate the opaque barrier in order to map the state transition space for the device.

More importantly, the interactions between device characteristics and user environments are necessarily heterogeneous. As the complexity of the domain increases, device roles must necessarily become more specialized as is required by Ashby's Law of Requisite Variety. The use of general purpose devices therefore forces the practitioner to undertake the task of shaping the device to conform to the requirements of the task.

The heterogeneity of user environments must be matched by the heterogeneity of devices; as the tasks become more diverse, the user aids must diversify in step. Paradoxically, this is a time in which many devices are becoming generic: the controller described here is used in many different parts of the hospital, for many different purposes and it may function well in some of them.

The purpose of this paper is not to indict a device or even a class of devices, nor is it to call into question the value of automated controllers. Rather it seeks to focus attention on the difficult problem of providing useful assistance to operators of complex, high risk processes (Norman, 1990). It is clear, from this study, that it is possible to produce controllers which have an apparent quality of engineering "reasonableness" but which actually combine with the operating environment to produce the potential for new kinds of faults and failures. The challenge to the engineering and human factors communities is to find ways to characterize the practitioner's context with sufficient precision and detail to permit the design of user aids which assist rather obstruct high workload, high tempo task performance.

ACKNOWLEDGEMENTS

The opportunity to investigate and follow up the critical incidents with the research reported here arose because of another research project to build a Corpus of Cases on Human Performance in Anesthesia sponsored by a grant from the Anesthesia Patient Safety Foundation to the third author. Additional support for this study was provided by The Ohio State University Department of Anesthesia. We thank the Universiteit van Utrecht, Geneeskunde Faculteit, Utrecht, The Netherlands and Professor Smalhout for making it possible for Eric Molle Van Charante to do his research practicum in the Cognitive Systems Engineering Laboratory and the Department of Anesthesia at The Ohio State University. We are very grateful for the assistance of the many physicians who gave their time and shared their experiences with us in the course of this research. Address correspondence to D. D. Woods, Department of Industrial and Systems Engineering, The Ohio State University, 1971 Neil Avenue, Columbus, OH 43210, U.S.A. (email: woods@csel.eng.ohio-state.edu).

REFERENCES

R.I. Cook, S. Potter, D.D. Woods, and J.S. McDonald. Evaluating the human engineering of microprocessor controlled operating room devices. Journal of Clinical Monitoring, 7:217-226, 1991.

R.I. Cook, D.D. Woods, and M.B. Howie. The natural history of introducing new information technology into a dynamic high-risk environment. In Proceedings of the Human Factors Society, 34th Annual Meeting, 1990.

R.I. Cook, D.D. Woods and M.B. Howie. Unintentional delivery of vasoactive drugs with an electromechanical infusion device. _Journal of Cardiothoracic and Vascular Anesthesia_, 6:1-7, 1992.

D.A. Norman. _The Psychology of Everyday Things_. Basic Books, New York, 1988.

D.A. Norman. _The 'Problem' of Automation: Inappropriate Feedback and Interaction, Not 'Over-Automation'_, Philosophical Transactions of the Royal Society of London, B 327, 1990.

J. Reason. _Human Error_. Cambridge University Press, England, 1990.

N. Sarter and D.D. Woods. Pilot Interaction with Cockpit Automation: Operational Experiences with the Flight Management System. _International Journal of Aviation Psychology_, in press.

L.A. Suchman. _Plans and Situated Actions: The Problem of Human-Machine Communication_. Cambridge University Press, Cambridge, England, 1987.

E.L. Wiener. _Human Factors of Advanced Technology Glass Cockpit Transport Aircraft_. Technical Report~117528, NASA, 1989.

D.D. Woods. Risk and Human Performance: Measuring the Potential for Disaster. _Reliability Engineering and System Safety_, 29:387–405, 1990.

D.D. Woods. Process Tracing Methods for the Study of Cognition Outside of the Experimental Psychology Laboratory. In G. A. Klein, J. Orasanu and R. Calderwood, editors, _Decision Making in Action: Models and Methods_, Ablex, New Jersey, 1992.

D.D. Woods and R.I. Cook. Technology-induced complexity and human performance. _Proceedings of IEEE International Conference on Systems, Man, and Cybernetics_, IEEE, 1991.

D.D. Woods, S.S. Potter, L. Johannesen and M. Holloway. _Human Interaction with Intelligent Systems: Trends, Problems, New Directions._ Cognitive Systems Engineering Laboratory Report 91-TR-02, February 1991.

L. Yue, D.D. Woods and R.I. Cook. _Reducing the Potential for Error Through Device Design: Infusion Controllers in Cardiac Surgery_. Cognitive Systems Engineering Laboratory Report 92-TR-01, The Ohio State University, Columbus OH, January 1992.

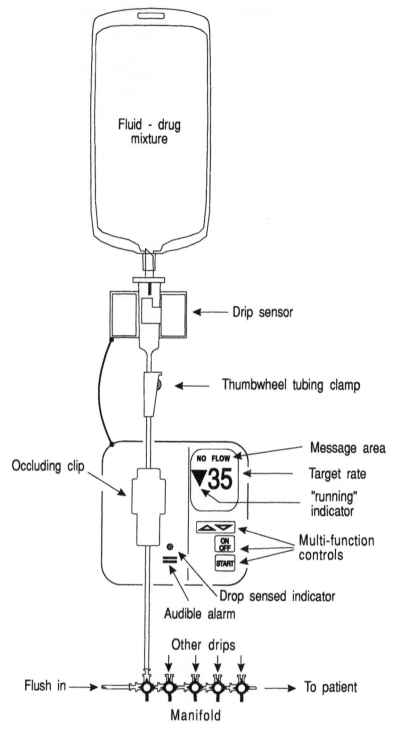

Fluid - drug
mixture

Drip sensor

Thumbwheel tubing clamp

Message area

NO FLOW

▼35

Target rate

"running"
indicator

Occluding clip

Multi-function
controls

ON
OFF

START

Drop sensed indicator

Audible alarm

Other drips

Flush in ⟶

To patient

Manifold

Figure 1. The assembled drug administration set up including the infusion controller. The latter includes three components: the drip sensor, the occluding clip and the controller unit with multi-function controls and LCD display area.

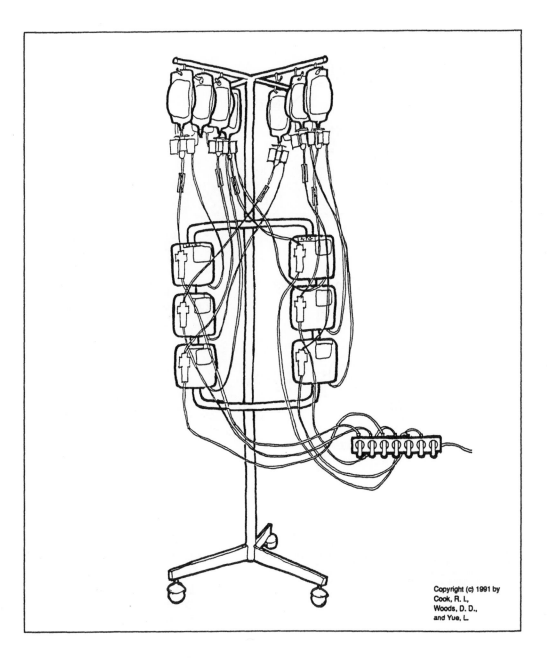

Figure 2. A drawing of the tree of six infusion controllers as would look in preparation for use in cardiac surgery.

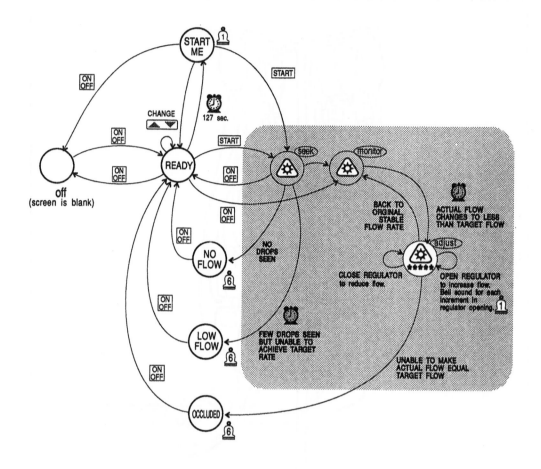

Obstruction to flow alarm states

Only messages in circles are visible to the user. Note that the triangular icon is used to indicate that the device is operating and gives a rough indication of the target but not actual drip rate. Note the multiple functions of the ON/OFF button. For simplicity, numerous states and state transitions not shown.

LEGEND

- (message) Distinct state in which "message" appears on display
- Display icon indicating running state
- An interval of elapsed time
- Alarm bell rings six times
- (state) Internal state (not visible to user)
- [name] Press of the named control button

Copyright 1991 by R.I. Cook, M.D.

Figure 3. State transition diagram for alarms related to obstructions to flow.

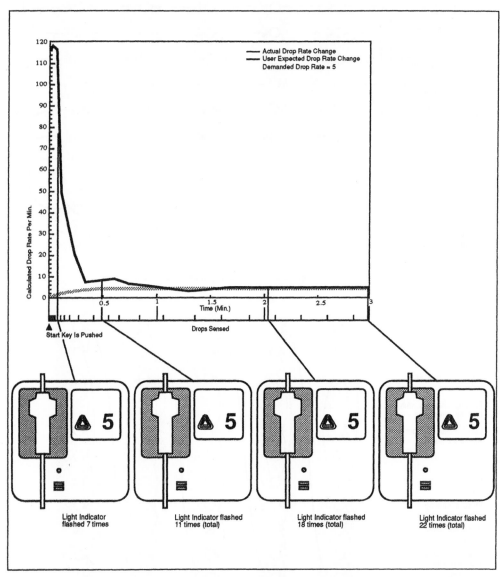

Figure 4. Compares data about the initial overshoot behavior (blackcurve) of the infusion controller at low rates and the anesthesiologist's conceptual model of what the device does (graycurve). The triangle indicates when the device was started; the tickmarks along the time axis indicate when drops were actually sensed.The data plotted is from one typical case run in the bench tests. The drawings of the device represent four snapshots during the overshoot behavior. One can see how the interface hides controller behavior.

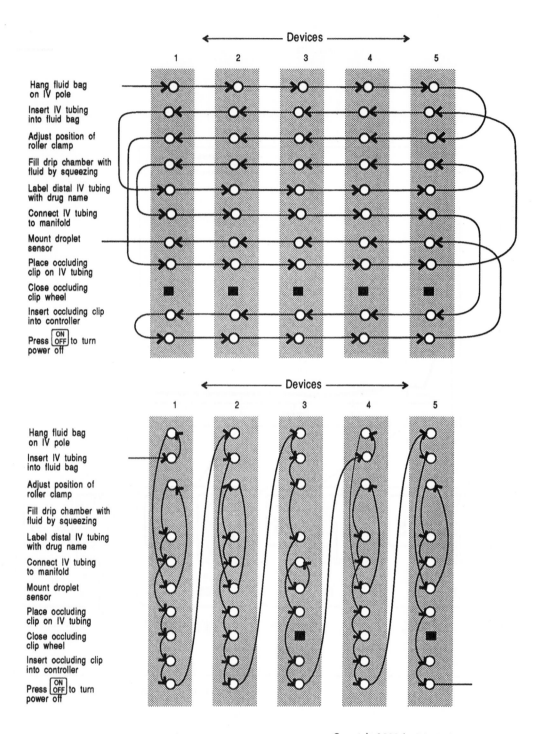

Figure 5. Charts two different device setup strategies used by physicians. The data is from field observations of actual device set up. Note that one step in the "procedure" is frequently omitted in both strategies.

FIELD EVALUATION OF KNOWLEDGE BASED SYSTEMS: THE MEDICAL SYSTEM PLEXUS

C. van Daalen

Delft University of Technology, Faculty of Mechanical Engineering and Marine Technology,
Laboratory for Measurement and Control, Mekelweg 2, 2628 CD Delft, The Netherlands

Abstract. Very few medical knowledge based systems have undergone evaluation in the
operational environment. An important reason for the limited number of reported field
evaluations is the fact that extensive laboratory evaluation will have to precede the evaluation of
systems in field. The aspects of field evaluation which will be considered in this paper are the
performance of the man-machine system and the interaction between the user and the system. This
will include a discussion of the field evaluation of the knowledge based system PLEXUS. PLEXUS
is a medical knowledge based system, designed to assist physicians with the diagnosis and
treatment planning of nerve injuries in the neck.

Keywords. expert systems; medical systems; man-machine systems; computer evaluation; field
evaluation

INTRODUCTION

The evaluation of a knowledge based system should be a
continuous process which has to be carried out in
parallel with the development of the system. During the
beginning of the development, evaluation will be
directed towards obtaining information for
improvement of the system. In the final stages of
development, evaluation will be aimed at investigating
whether the system statisfies the design objectives.
This also implies that the evaluation will progress from
informal studies to a formal investigation. This paper
considers the formal evaluation of knowledge based
systems in the field. A large amount of informal and
formal evaluation work will have to precede the
evaluation of systems in the target environment. This
will involve validation studies of the knowledge based
system, where it should be shown to a satisfactory
degree that the behaviour of the system is correct with
respect to the specifications made of the system (Shwe,
1989). Furthermore, the user interaction should be
investigated, and the performance of a system in a
laboratory environment shoud be evaluated, where the
safety, accuracy, reliability and transferability are of
importance. Few systems have reached the level of
maturity which is required to justify a field evaluation.
Descriptions of actual field evaluation studies of
medical knowledge based systems may, for example, be
found in Adams and others (1986), Murray (1990) and
Sutton (1989a).

Field evaluation should involve investigation of the
system in the target environment by measuring the
impact of system use on the quality of the user's
decisions and the impact on the results of the user's task
performance. Furthermore, the man-machine interaction
is of the utmost importance and should also be

evaluated. This will involve assessment of the
functionality, acceptability and usability of the system
and includes the quality of the human-computer
interface. In addition, cost-benefit analysis should be
carried out, and the impact on the organisation and the
environment, and legal and ethical aspects should be
assessed.

In this paper, the performance of knowledge based
systems in the operational environment will be
discussed, and the evaluation of user interaction will be
considered. This will be illustrated using the knowledge
based system PLEXUS. The discussion of cost-benefit
and legal and ethical aspects is beyond the scope of this
paper.

PLEXUS is a medical knowledge based system which is
being developed at Delft University of Technology
(Jaspers, 1990). The objective of the system is to assist
neurologists and neurosurgeons with the diagnosis and
treatment planning of patients with a brachial plexus
injury. The brachial plexus is an intricate network of
nerves situated in the neck, which innervates the
muscles of the shoulder, arm and hand. Most brachial
plexus injuries are sustained by young men during
motorcycle accidents. Since the network of nerves has a
very difficult structure and many different sources of
information have to be combined to be able to
diagnose these patients correctly, PLEXUS has been
designed to assist physicians with this task.
Furthermore, as the neurosurgical reconstruction
possibilities are relatively unknown, the system also
advises on a possible treatment plan for the patient.

PLEXUS has been constructed using the rule based
expert system shell Delfi2+, which has been developed
at the Faculty of Applied Mathematics and Informatics

at Delft University of Technology (Swaan Arons, 1991). PLEXUS consists of a diagnostic and a treatment planning module. The diagnostic module contains production rules for a rough localisation of the injury. The exact localisation is then carried out using a tree-search algorithm incorporating heuristic search. The treatment planning module is completely rule based. Preliminary evaluation of the problem solving performance has been carried out, the results of which have been described by Jaspers (1990). Presently, an international peer review of the performance is being conducted (part of laboratory evaluation), consisting of a variant of the "Turing test".

Another major focus of attention in the development of PLEXUS has been the design of the user interface. PLEXUS has a graphical user interface which requires no computing or typing experience. In hospitals a special paper form is often used for entering brachial plexus data onto. This form is simulated on the computer screen and used as a basis for data entry. Data entry is carried out using the mouse of the computer and choosing the appropriate answer possibility. Examples of screens of the graphical interface are shown in Figs. 1 and 2 (at the end of the paper). After the physician has entered all relevant data into the computer using the interface, a consultation with the knowledge based system is carried out. The advice which is produced by the knowledge based system is then shown to the physician in the form of text and is also shown in a graphical representation of the anatomy of the brachial plexus.

Many medical knowledge based systems have been designed using the cognitive tool as "prosthesis" paradigm (Woods, Roth and Bennet, 1990), where the advice consists of one or more answers, which the user can either accept or reject. The version of PLEXUS which is being evaluated clinically is no exception to this. In order to adopt a more "instrumental" approach, a critiquing system is presently being developed. The design which is used will influence the evaluation process.

Presently, the knowledge based system PLEXUS is being evaluated clinically in five different hospitals in The Netherlands. The investigation of the clinical efficacy and the evaluation of the man-machine interaction will be described below.

CLINICAL EFFICACY

For an investigation of the clinical efficacy of the man-machine system, an evaluation design will have to be formulated. The design of such a study involves a sequence of choices which have to be made. A framework containing items of relevance in an evaluation design for medical knowledge based systems is shown below. Most of the items in the framework have been mentioned by O'Moore and others (1990), but for this purpose, the items have been ordered and modified.

-Selection of a goal for evaluation
-Experimental setup, involving
- choice of experimental unit
- selection of test input
- selection of way to enter test data
- specification of control group
- specification of standard of performance
- specification of variables to be measured
- analysis of the results
-Bias and confounding variables

O'Moore and others (1990) also mention that the role of those involved must be defined, and legal and ethical issues must be considered. The above framework will be used for the discussion of the evaluation of clinical efficacy. The discussion of each item in the framework will consist of a general discussion of the item, after which, the choices made for the evaluation of PLEXUS will be considered.

Selection of a Goal for Evaluation

Before carrying out an evaluation procedure, the goal of the evaluation must be specified. The evaluation goal will depend on the objective of the system and the stage of the evaluation. The objective of a medical knowledge based system will often be to assist users with a certain task. Therefore, one of the goals of a formal evaluation will consist of investigating whether the system indeed satisfies this objective. However, the desired influence of the system on aspects of the user's task needs to be defined exactly. For instance the fact that user and system should significantly improve on the diagnostic performance of the user without the system, and that user and system should never reach unsafe conclusions.

The final objective of the knowledge based system PLEXUS is that it should assist physicians with the diagnosis and treatment planning of brachial plexus injuries. The object of the clinical efficacy evaluation is to investigate whether physicians' diagnoses and treatment plans improve significantly through use of the system.

Experimental setup

The default design for an interventional trial is the randomised controlled trial with double-blinding (Wyatt and Spiegelhalter, 1990). However, there is a number of limitations which inevitably influences the design which can be used for the evaluation of knowledge based systems. In this kind of evaluation the physician's decisions are evaluated, whereas in drug trials a therapy is evaluated. Furthermore, the user is usally free to accept or to reject the recommendation given by the system (Spiegelhalter, 1983). Spiegelhalter discusses experimental studies, where a controlled trial takes place with balanced allocation to the control and experimental groups, and quasi-experimental studies where performance is measured before and after introduction of the system (Spiegelhalter, 1983). Examples of studies for which historical controls have been used are described Adams and others (1986) and by

Murray (1990) who also incorporates a period when the system is withdrawn.

The general setup used in the clinical evaluation of PLEXUS is the following. A number of physicians has been asked to use the knowledge based system for about a year, i.e. until October 1992. They have agreed to enter the data of all traumatic brachial plexus injuries, which they see during this year, into the computer. After having entered the data, they are asked for their own opinion about the diagnosis and treatment plan for the patient. Following this, they carry out a consultation to obtain advice from PLEXUS. The physicians then are asked for their opinion about the advice given by the knowledge based system (see Fig. 3). At the end of the evaluation period all diagnoses and treatment plans will be judged blindly by a number of experts in the domain of brachial plexus injuries.

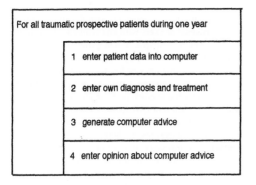

Fig. 3. PLEXUS field evaluation of clinical efficacy.

The patient data are entered into the computer by the physician by means of the graphical interface which runs on an Apple Macintosh at the hospital. The knowledge based part of PLEXUS runs on a central workstation at Delft University. In this way it is possible to enter the data and to receive advice locally, and to perform the reasoning and to keep track of the evaluation process centrally. The details and rationale of the choice of evaluation design are discussed in the sections below.

Choice of experimental unit. According to Spiegelhalter (1983), the choice of the experimental unit is a difficulty in designing trials of decision aids. Spiegelhalter gives a rough indication of how to choose the experimental unit. The patient is chosen as the experimental unit if the system provides immediate information and recommendation useful for a particular patient. If the system educates the physician about careful data collection, clinical judgment and awareness of performance then the physician is chosen as the experimental unit, and if it generates an awareness among a small group of physicians, the group is chosen. In technical domains, the choice of experimental unit may be more obvious. Adelman (1991) states that, in field experiments, organisational units would if possible, be randomly assigned to situations with and without the decision support, and

that their performance is measured when it is stable. The performance of the organisational unit is the unit of analysis.

For the evaluation of PLEXUS , there are presently two neurosurgeons, one neurologist and two neuro-physiologists from two academic and three large hospitals in The Netherlands cooperating in this evaluation study.

Selection of test input. Task difficulty should be as representative of the operational environment as possible (Adelman, 1991). In medical domains prospective patient data will be used in a field evaluation. It will be difficult to determine, however, to which extent the representativeness has been satisfied.

The test input for the evaluation of PLEXUS will consist of all prospective traumatic brachial plexus injuries the five physicians see over the period of one year. Since the yearly rate of brachial plexus injuries in The Netherlands amounts to about 150, the total number of test cases will at most be about 25. The reason for inviting the larger hospitals to take part in the evaluation is the very low number of brachial plexus lesion occurrences. This low number will limit the generalisation possibilities of the evaluation outcomes.

Selection of way to enter test data. The object of the evaluation is to determine the efficacy of the man-machine system when used in the target environment, therefore the data should be entered and the system used by those who would also perform these tasks if the system were in routine use.

In the case of PLEXUS, the system has been designed for use by the physicians themselves, therefore the data are entered and the system used by the physicians during the field evaluation.

Specifying the control group. To form the control and experimental groups, balanced allocation to the situation with and without a knowledge based system may be used (experimental study). However, quite a large number of experimental units may be needed. Therefore, some researchers use historical controls (Adams and others, 1986; Murray, 1990). This means that the control data are collected during a period prior to the introduction of the knowledge based system. However, in these designs it is more difficult to control external factors (Adelman, 1991).

For PLEXUS the total number of physicians involved in the trial is very small, and it will not be possible to find a balanced control group. Therefore, the physician will first assess the patient without the system and then will use the knowledge based system to obtain an advice. This means that the physicians will carry out consultations with and without the system. This may cause a learning effect where the physician may learn from using the knowledge based system, if the man-machine system indeed shows a superior performance.

Specifying a standard of performance. In some domains

an objective standard of performance may be available. Sometimes, the actual results become known after a certain period of time, whereas the knowledge based system should be able to assist at a time when the actual results are not known yet. This will probably apply more often to technical domains than to medical domains. In medical domains the real result often remains unknown. However, in certain areas it is possible to approximate the final outcome, for instance using the discharge diagnosis of the patient as the standard.

In the domain of brachial plexus injuries it is not possible to obtain the actual diagnosis and optimal treatment plan for the patient. Therefore a panel of experts will be asked to assess the performances with and without use of the knowledge based system, blinded from the origin of the output. This will introduce errors of unknown validity and intra- and interexpert variability. Furthermore, to make blind evaluation possible, the output provided by the system and the physicians will have to be coded so that it will not be possible to distinguish the source of the output, this will also introduce errors into the measurement.

Specifying variables to be measured. The variables which are to be measured depend on the goal of the field evaluation study. In most evaluation studies, the outputs provided by computer-assisted users are compared to outputs provided by unassisted users. As mentioned above, the output will often consist of a decision or a number of decisions to be taken. However, it is not satisfactory to solely measure performance, because in medical domains, the diagnosis may be correct, but it may for instance delay the treatment of the patients, so that final patient outcome may be worse (Wyatt, 1987). Therefore, it will also be necessary to measure, for example, final patient outcome and speed. The measurement of the quality of the decisions can often take place at an earlier stage than the measurement of the patient outcome, and as is seen above, improvement in diagnosis (decision) does not necessarily imply an improved patient outcome. The additional object of a system may be to reduce the number of special investigations or to reduce the use of resources. These variables should then also be measured. Cost-benefit measurements and other important measurements, such as determining impact on users, organisation and environment will not be considered here.

The object of PLEXUS is to assist physicians with the localisation of the brachial plexus injury and to indicate whether conservative or surgical treatment should be carried out. At present, PLEXUS does not offer extensive explanations to the physician. The system is not meant to reduce the number of special investigations or to reduce resources, the goal is to increase the quality of diagnoses and treatment plans. It is not possible to measure patient outcome, because it is not known what the alternative would be if another treatment had been carried out and there are not sufficient patients to base statistic analysis on. Furthermore, it may take more than two years of recovery until the final result has been reached.

Analysis of the results. Depending on the variables which have been measured, the appropriate statistical analyses should be carried out. The choice of the experimental unit has to be taken into account during the statistical analysis. There is a number of researchers who indicate methods which may be used for laboratory performance evaluations, where the performance of the knowledge based system is compared to the standard. Indurkhya and Weiss (1989) describe a number of models for measuring output performance of medical knowledge based systems. One approach which is often used in medical literature is the positive-negative correctness model, in which accuracy, sensitivity and specificity are calculated. Bernelot Moens and van der Korst (1991) describe and compare a number of methods which has been used in a performance study of a Bayesian decision support system for rheumatic disorders.

Field evaluations are concerned with the situation where the performance of users with and without a knowledge based system is compared. When a true experiment is used, for which there is only one factor with two levels, i.e. whether a system has or has not been used, and there is one dependent variable, a paired t-test can be used to assess whether the average performance difference is significantly different (Adelman, 1991). Adelman also indicates methods of analysis which may be used for other evaluation designs. If there is no objective standard of performance, a panel of experts may be involved in judging the results. In these cases it is necessary to calculate the inter-expert and intra-expert variability. The Kappa coefficient (Cohen, 1968) is often used as a measure of agreement in these cases.

A number of aspects of performance will be analysed for PLEXUS. Firstly, it will be investigated whether physicians using the knowledge based system produce significantly better diagnoses and treatment plans than when the system is not used. The cases for which the physician alone rates higher than when assisted will be further investigated qualitatively. The rate of compliance with the system will also be determined. Since experts will be involved in the analysis of the data, the inter- and intra-expert variability will have to be taken into account. The limited amount of data and the relatively small number of test sites will make generalisation of the results difficult.

Bias and confounding variables

There is a number of effects which may influence the validity of field evaluation results. Some of these effects will be analysed below.

-Carry-over effect (Wyatt and Spiegelhalter, 1990): This is the positive effect on performance due to the education of users by the knowledge based system. According to Wyatt and Spiegelhalter (1990) this effect may be compensated by raising the size of the experimental unit, or by quantifying the effect by studying alternating knowledge-based system and control periods. The latter depending on the trial either being multi-centre with randomized or asynchronous periods, or there being no significant changes during the trial period.

For PLEXUS, the carry-over effect will inevitably bias the results of this investigation, since the physicians will be diagnosing the patients both with and without the system. However, since the physician's performance without the system may improve through this effect, if a significant difference, between assisted phyisician and the unassisted physician, is present, the conclusion will still be justified. Using the present evaluation design it is not possible to quantify the influence of this effect.

- Hawthorne effect (Wyatt and Spiegelhalter, 1990): This is the effect of increased performance when people know that they are being studied. The effect may be quantified by performing a low-profile baseline study. According to Spiegelhalter (1983), the Hawthorne effect will tend to decrease any relative benefit of the system.

Since, during the evaluation of PLEXUS the users are studied in both test and control situations, the Hawthorne effect will probably also tend to decrease the relative benefit of the system. It is not possible to quantify the impact which this effect will have on the final result.

- Secular trends (Wyatt and Spiegelhalter, 1990): These are changes in the measures of interest which occur during the evaluation period, and which may influence the outcome of the study. This effect is especially large when using historical controls, when studies run for a long period of time, or when there are changes in the way a particular task is carried out.

The PLEXUS experiment will run for the period of a year, and historical controls are not used. Therefore, it seems unlikely that any secular changes will occur.

- Feedback effect (Spiegelhalter, 1983): This is the influence created by giving feedback to the user. If the user is given feedback about his performance during an evaluation, the user's performance may improve over the evaluation period.

Since only preliminary expert evaluation has been carried out for PLEXUS and extensive formal laboratory evaluation is presently ongoing, some feedback to the physicians is necessary if unforeseen situations occur. When a patient's data are entered into the computer and a consultation with the system is carried out, all data are also given to one of the collaborating experts. Should the need arise, the experts will contact the physicians about the case. This will also influence the results in the same direction as the carry-over effect which is described above.

- Checklist effect (Wyatt and Spiegelhalter, 1990): The knowledge based system may encourage a more complete and structured data collection. According to Sutton (1989b), this may account for most benefits claimed for computer aided diagnosis. Kent and co-workers (1985) have assessed the influence of a

computer-based chemotherapy treatment consultant on the completeness of trial data. Adams and others (1986) have used a design in which there are four different groups, one group which used structured data collection forms, one group which used forms and the diagnostic aid, one using forms and receiving feedback, and one using forms and the diagnostic aid and receiving feedback.

In the evaluation of PLEXUS, the physician is asked for his own diagnosis after all the patient data have been entered into the computer, this means that in the control situation, the knowledge based system is used for data collection but not for consultation. Therefore in this evaluation, the data collection is not part of the measurement. It does mean, however, that the influence of structured data collection is not assessed.

- Interface: If the physician enters data using the interface, some errors in system output may arise due to errors or misunderstanding during data entry. This will cause the knowledge based system to perform at a lower level than it would if optimal data entry was carried out. An estimate of the influence may be obtained by having the knowledge engineer also enter the patient data and observing the differences in the data files and in the system's output.

The PLEXUS evaluation involves the physicians entering their own data into the computer. Therefore, errors due to misinterpretation may occur. At the end of the evaluation period all patient data will be entered into the system by the knowledge engineer to observe the influence of this effect on the output.

- Coding (Chandrasekaran, 1983): In evaluations which involve blind judgement of the output, the outputs may have to be coded in order for the source of the output to be unrecognizable to the judges. This coding of the answers will introduce subjectivity into the measurement.

For the evaluation of PLEXUS, it is necessary to involve experts in the judgment of the outputs. These outputs will have to be coded to be able to carry out the blind evaluation. The size and direction of the errors created by this procedure are unknown. It may be possible to have the judges indicate which source they think the answers are from, and to then estimate whether the blinding of the outputs has succeeded.

- Expert judgment: If experts are involved in the evaluation, this will introduce errors of unknown validity of the judgments, and intra- and interexpert variability. The variabilities may be approximated, however the validity of the judgments will not become known. Furthermore, to eliminate errors due to circularity (Wyatt and Spiegelhalter, 1990), the system must not be evaluated by those who have developed it.

For the evaluation of PLEXUS, this will also influence the results. It is necessary to use independent experts, and to calculate the variabilities.

- Parochial bias/transferability issues : The tasks used

to test the system may not be representative. Furthermore, the users testing the system may also not be representative of the complete population. This will introduce errors into the measurement.

The PLEXUS system is being tested in five different hospitals. However, these are located mostly in the western part of The Netherlands. Since these hospitals are among the larger hospitals and may receive the more serious cases, neither the physicians nor the test cases may be representative of the complete population.

- Trial size (Wyatt and Spiegelhalter, 1990): The trial size will influence the statistical conclusion validity of the evaluation study.

The PLEXUS evaluation study involves five hospitals and will probably only include about 25 cases. Therefore, it is questionable whether sound quantitative results will arise from the experiment. Due to the nature of the domain and the circumstances, these problems are unavoidable. Furthermore, due to the limited amount of test sites and data, the results will be difficult to generalise.

ACCEPTABILITY

Many knowledge based systems have been developed, however, very few knowledge based systems are used in actual daily practice. Especially knowledge based systems in medical domains suffer from a lack of acceptance. Very little empirical research has been carried out concerning user demands regarding knowledge based systems. An exception is the study carried out by Teach and Shortliffe (1981), in which, amongst other things, they assess the demands of 146 physicians and 129 nonphysicians regarding medical knowledge based systems. The main demand emerging from their study is the presence of an explanation capability. Other important items were found to be that systems should be portable and flexible, and that they should display an understanding of their own medical knowledge. In the evaluation of any knowledge based system, it is necessary to investigate to which extent the system satisfies the user demands. A questionnaire may be used to help assess user demands and to investigate whether the system satisfies these demands.

For the PLEXUS project, 3 questionnaires have been designed (using Oppenheim, 1986). The object is to assess the acceptability of the knowledge based system. The first one is a short questionnaire in which the user is asked whether he would use knowledge based systems in daily practice, which domains he would use these for, and also whether the physician would use PLEXUS if it was available.

The other two questionnaires are more extensive and should be viewed together. The first one concerns physician's demands (requirements) regarding knowledge based systems in general. There are 32 items which have to be judged on a Likert-type scale. The other questionnaire concerns PLEXUS. The same 32 items are stated, however, in this case the question is to what extent the physician finds that PLEXUS satisfies the 32 requirement items. In this way it may be established whether the physician's requirements are actually met by the knowledge based system. Seventeen of the demand statements are based on a study of the literature concerning acceptance of knowledge based systems (Daalen, 1988), and the other 15 items form a replicate of part of a study carried out by Teach and Shortliffe (1981). In order for a comparison with Teach and Shortliffe's study to be possible, all the demand items which have been used by Teach and Shortliffe have also been used for this investigation. The items used in the present study range from 'knowledge based systems should be connected to hospital information systems' and 'physicians should be able to inspect the knowledge in a knowledge based system in advance' to 'knowledge based systems should not increase the time spent per patient'.

Results

Since there are only five physicians involved in the field trial, the requirements questionnaire will be distributed among a larger number of physicians. A preliminary investigation has been carried out among 10 neurologists and 25 rehabilitation workers. In two separate sessions, one for the neurologists and one for the rehabilitation workers, taking approximately half an hour, a lecture was given on the topic of knowledge based systems in general and PLEXUS in particular. At the end of which they were asked to complete the acceptability questionnaire.

There is a number of factors which makes it difficult to carry out a comparison with the American study. For instance, the questions have been translated into Dutch which may have implications for the results of the comparison. Furthermore, the American study was carried out more than ten years ago, and in this preliminary study, the lecture about the specific PLEXUS application may also have influenced the results.

In the study performed by Teach and Shortliffe (1981) a scale from -2 to +2 was used. In order to be able to carry out the comparison, an interval scale was also used in the present study, although the use of an interval scale here is questionable. The preliminary study shows that the 10 neurologists' most important demand is that 'knowledge based systems should demand little effort to learn or use', with average score 1.91 and standard deviation 0.30 The rehabilitation workers also regard this demand as being most important, with an average score of 1.92 and standard deviation 0.28. In the study carried out by Teach and Shortliffe (1981), for the physicians this demand had an average score of 0.35 and standard deviation 1.20, and it is the item with the ninth highest average score. When comparing Teach and Shortliffe's results, for the physicians, to that of both previous groups, a significant difference is present $(p < 0.01)$ between the average scores on this item. This part of the study will be carried out for a larger number of physicians and the results and implications will be investigated more extensively.

According to Woods, Roth and Bennet (1990), problems with user acceptance are often seen to have to be solved by measures outside the characteristics of the knowledge based system, whereas considering the human and machine as parts of a joint cognitive system suggests these problems are often symptoms of an underlying deficiency in the cognitive coupling between the human and the machine. This aspect also deserves further investigation, since the present study will not allow solid conclusions to be drawn in this area.

HUMAN COMPUTER INTERFACE

PLEXUS, as most other medical systems, does not incorporate a negotiation model, but gives output in the form of a number of answers. This means that in fact the human is seen as the interface between the environment and the machine (Woods, Roth and Bennet 1990). Although this may not be the most desirable situation, it will be necessary to investigate whether the man-machine interface does allow the physician to carry out these tasks correctly and to study the physician's opinion regarding the interface. For PLEXUS this means that the input and output possibilities of the system have to be evaluated, rather than a more extensive evaluation of man-machine communication.

The user interface of the knowledge based system PLEXUS has been assessed at the following three levels: the presentation level, the interaction level and the knowledge level.

The first level, the presentation level has been investigated by means of a questionnaire including, for instance, questions about the size of the screen and presentation on the screens.

The second level, the interaction level, includes a questionnaire to study the physicians' opinions about the way to start the program, the way to move around the patient file and the methods used to enter the information into the system. Furthermore, video observations have been used during the training sessions. A training session lasted for about one and a half hours, during which the physician was asked to enter an imaginary (or retrospective) patient into the computer. The video tape will give information about the interaction between the physician and the computer.

The third level is the knowledge level, for which the semantics of the interface has been studied. During the video observations mentioned above, the physicians could freely indicate their opinion about the content and order of the questions and about the answer possibilities. As was mentioned in the section on clinical efficacy, the knowledge engineer will also enter the patient information into the computer interface, using the patient files. The differences in the database and in the computer advice will also provide information about the interaction level and knowledge level. Possibly interaction errors (errors in the second level) and errors of meaning (errors in the third level) may be distinguished.

CONCLUSIONS

There is very limited experience in the field evaluation of medical knowledge based systems. Research into field evaluation of knowledge based systems is still in its infancy. By illustrating the methodology used in the clinical evaluation of the problem solving performance and user interaction of PLEXUS, we expect to contribute to the discussion on the field evaluation of knowledge based systems. The evaluation of PLEXUS will finish during autumn 1992, by which time the full data on the problem solving performance and user interaction should be available.

ACKNOWLEDGEMENT

The field evaluation of PLEXUS has been made possible by Apple Nederland B.V. who have provided five Apple Macintosh computers and Apple StyleWriters.

REFERENCES

Adams, I.D., M. Chan, P.C. Clifford, W.M. Cooke, V. Dallos, F.T. de Dombal, M.H. Edwards, D.M. Hancock, D.J. Hewett, N. McIntyre, P.G. Somerville, D.J. Spiegelhalter, J. Wellwood, and D.H. Wilson (1986). Computer aided diagnosis of acute abdominal pain: a multicentre study. British Medical Journal, 293, 800-804.

Adelman, L. (1991). Experiments, quasi-experiments, and case studies: A review of empirical methods for evaluating decision support systems. IEEE Transactions on Systems, Man, and Cybernetics, 21, 2, 293-301.

Bernelot Moens, H.J., and J.K. van der Korst (1991). Measuring performance of a Bayesian decision support system for the diagnosis of rheumatic disorders. In M. Stefanelli, A. Hasman, M. Fieschi, and J. Talmon. (Eds.), Proceedings of the 3rd European Conference on Artificial Intelligence in Medicine, Maastricht. pp. 150-159.

Chandrasekaran, B. (1983). On evaluating AI systems for medical diagnosis. The AI Magazine, summer, 34-37.

Cohen, J. (1968). Weighted Kappa: nominal scale agreement with provision for scaled disagreement or partial credit. Psychological Bulletin, 70, 4, 213-221.

Daalen, C. van (1988). Factors influencing medical expert system acceptance. Report N-284, ISBN 90-370-0016-9, Delft DUT.

Indurkhya, N., and S.M. Weiss (1989). Models for measuring performance of medical expert systems. Artificial Intelligence in Medicine, 1, 61-70.

Jaspers, R.B.M. (1990). Medical decision support: an approach in the domain of brachial plexus injuries. PhD. Thesis, Delft University of Technology, ISBN 90-370-0028-2.

Kent, D.L., E.H. Shortliffe, R.W. Carlson, M.B. Bischoff, and C.D. Jacobs (1985). Improvements in data collection through physician use of a computer-based chemotherapy treatment consultant. Journal of Clinical Oncology, 3, 10, 1409-1417.

Murray, G.D. (1990). Assessing the clinical impact of a predictive system in severe head injury. Medical Informatics, 15, 3, 269-273.

O'Moore, R., J. Brender, P. Nykanen, J. Talmon, B. Barber, K. Clarke, P. McNair, R. Smeets, J. Grimson (1990). Methodology for evaluation of knowledge based systems. KAVAS(A1021). Report E.M. 1.2. EEC AIM Office, 62 Rue de Treves, Brussels.

Oppenheim, A.N. (1986). Questionnaire design and attitude measurement. Heinemann, London.

Shwe, M.A., S.W. Tu, and L.M. Fagan (1989). Validating the knowledge base of a therapy planning system. Methods of Information in Medicine, 28, 36-50.

Spiegelhalter, D.J. (1983). Evaluation of clinical decision-aids, with an application to a system for dyspepsia. Statistics in Medicine, 2, 207-216.

Sutton, G.C. (1989a). How accurate is computer-aided diagnosis? The Lancet, october, 905-908.

Sutton, G.C. (1989b). Computer aided diagnosis: a review. British Journal of Surgery, 76, 82-85.

Swaan Arons, H. de (1991). Delfi: Design, development and applicability of expert system shells. Ph.D. Thesis. Delft University of Technology. ISBN 90-6275-734-0.

Teach, R.L., E.H. Shortliffe (1981). An analysis of physicians attitudes regarding computer-based clinical consultation systems. Computers in Biomedical Research, 14, 542-558.

Woods, D.D., E.M. Roth, K.B. Bennet (1990). Explorations in joint human-machine cognitive systems. In S.P. Robertson, W. Zachary, J.B. Black (Eds.). Cognition, computing, and cooperation. Ablex Publishing Corporation, New Jersey. pp. 123-158.

Wyatt, J. (1987). The evaluation of clinical decision support systems: A discussion of the methodology used in the ACORN project. In J. Fox, M. Fieschi, and R. Engelbrecht (Eds.), Proceedings of the 1st European Conference on Artificial Intelligence in Medicine. Marseilles. pp. 15-24

Wyatt, J., D. Spiegelhalter (1990). Evaluating medical decision aids: what to test and how? In J. Fox and J. Talmon (Eds.), System engineering in medicine. Springer Verlag. 13 pp.

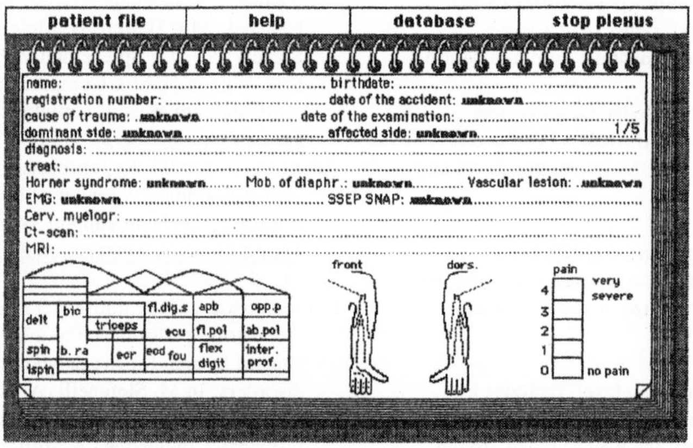

Fig. 1. Representation of the paper form on the computer screen.

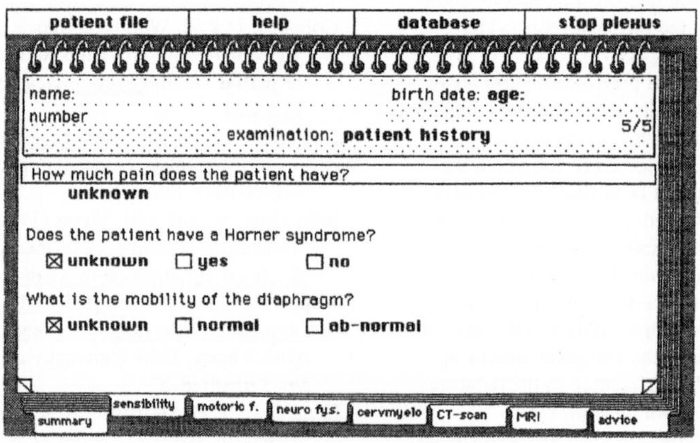

Fig. 2. Screen on which patient data may be entered using mouse of the computer.

DO WE NEED 'MAN' IN THE MAN-MACHINE SYSTEMS?
IF SO, WHY AND FOR WHAT?

Chairman: J.E. Rijnsdorp

The chairman (John Rijnsdorp) put forward a number of issues, among others a subdivision of the 'machine', according to Fig. 1.

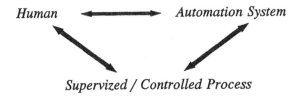

Figure 1. Generalized structure of
Man-Machine System

The panel members followed with short introductions. Neville Moray made a plea for symbiotic intelligence, in a cooperation between man and ('intelligent') machine. Carlo Cacciabue stressed the importance of the real-life situation in comparison to simulation studies of reliability and safety. Lena Mårtensson summarized a Delphy study about the differences between humanization of technology and human engineering, conducted by Tom Sheridan. She also advocated to start with training in an early phase of system design.

The latter topic raised several comments from the audience. Dr. Monta mentioned the important contributions of operators during early training sessions. Gerben Heslinga and others noted the difference between the electric power sector, where early training is common practice, and the process industries, where often training is very much neglected. Alex Levis pointed out the influence of legal procedures and certification processes. Can expert systems be put into prison? According to the chairman, the expert system designer certainly runs this risk.

Jan Wirstad asked for a forecast: how will future implementations of advanced automation influence the issue? Göran Olsson mentioned the difficulties of simulation in the paper industry, where some subprocesses are simply not understood at all.
Neville Moray stressed the importance of research on teams or groups. He also gave some experiences of work with 'microworlds'. He and others indicated that we neglect Man-Machine Systems problems and solutions in Third World countries.

All in all it was a nice discussion about not too many topics. There was a rather large audience, which might have put a barrier to some younger people to participate.

USER INTERFACE DESIGN:
TECHNOLOGY LED OR USER DRIVEN

Chairman: J.L. Alty

The key issue in this round table discussion was how far the user could be at the "centre" of User Interface Design. On the one hand a user should dictate the interface requirements but on the other hand how can a naive user demand appropriate facilities when not familiar with the technology.

A model of technological advance was suggested where user requirements led to technological advance which then suggested further user requirements leading to further technological advance etc.. However, it was felt that there were some advances which were solely technology driven - often it is the other way around: technology is looking for applications.

It was also true that technologists, playing with new tools, sometimes come up with innovative approaches to interface design (e.g. the spreadsheet).

The solution seemed to be better education of users about technological opportunities, but it was an open question as to how much technological knowledge was needed.

Users also need time to assimilate new ideas. The GRADIENT project was a good example. It took three years for the users to be convinced that the technology offered new worthwhile approaches. However, once the users were convinced, the introduction of the technology was more effective.

The similarity with planning changes in ones own house was highlighted to illustrate the differences between a user-driven or technology-led situation. A house holder is assisted by an array of experts - plumbers, architects, builders, joiners. In the technology led situation these people form themselves into a committee and tell the householder what he can have. In the user-led case the user decides upon advice from the experts. It is common to act in the latter way when supervising house alterations. The former situation, however, is still very common in computing. The difference is probably because we are better educated in what we want or can do with house technology than with computer technology.

The computer situation is further complicated by a vast set of manufacturers trying to make money through the introduction of new technology. Such people are not driven by user needs and spend a great deal of their time trying to generate needs for technology. Although this happens in other fields the rate of change is much more rapid in computing.

The issue is therefore not so much one of educating users, but at what level of abstraction should they be educated. The discussion concluded with a heated discussion on this topic without resolution.

AUTHOR INDEX

Alty, J.L. 249, 285
Appeddu, T. 139

Baosheng Hu 9
Bergan, M. 249
Blaauboer, W.A. 89
Bos, J.F.T. 215
Brinkman, P.L. 89

Cacciabue, P.C. 227
Chi-Cheng Cheng 115
Cojazzi, G. 227
Cook, R.I. 263

David, B.T. 257
Debernard, S. 95
Decortis, F. 61

Elzer, P.F. 183

Findler, N.V. 145

Happee, R. 133
Hayakawa, H. 31
Heslinga, G. 235
Hessler, C. 207
Hollnagel, E. 227
Howie, M.B. 263

Inoue, K. 71
Iwai, S. 47

Johannsen, G. 19

Kasiński, A.J. 77
Katai, O. 47
Katona, F. 83
Kjaer-Hansen, J. 61
Kleinman, D.L. 127
Kohda, T. 71
Kopf, M. 189
Kraiss, K.-F. 121, 201
Krumbiegel, D. 201
Küttelwesch, H. 121

Labiale, G. 257
Lee, J. 43
Levis, A.H. 9, 19
Lockett, J.A. 61

Mancini, S. 227
Massimino, M.J. 109
Millot, P. 95
Minguy, J.L. 157
Moll van Charante, E. 263
Monta, K. 31
Moray, N. 9, 43
Mossink, J.C.M. 101

Naito, N. 31

Onken, R. 189, 195
Ouadou, K. 257

Papenhuijzen, R. 53
Pattipati, K.R. 127
Peeters, M.H.H. 101
Pete, A. 127

Rabardel, P. 157
Reising, D.V. 139
Rijnsdorp, J.E. 283

Sanderson, P.M. 139
Sawaragi, T. 47
Schreiber, S. 201
Sheridan, T.B. 1, 109, 115
Smets, G.J.F. 169
Stassen, H.G. 19, 53, 215
Stoop, J.A. 243

Vámos, T. 83
van Daalen, C. 275
van der Mast, C. 163
van Hal, G. 177
van Lunteren, A. 215
van Paassen, M.M. 221
Vanderhaegen, F. 95
Versendaal, J. 163

Weisang, C. 183
Wendel, I.E.M. 177
Wewerinke, P.H. 151
Wheatley, K.H. 61
Wieringa, P.A. 89
Wittig, T. 195
Woods, D.D. 263

Yue, L. 263

Zinser, K. 183

287

KEYWORD INDEX

Printed and bound by CPI Group (UK) Ltd, Croydon, CR0 4YY

03/10/2024

01040320-0013